W9-AEB-480

Turtles of

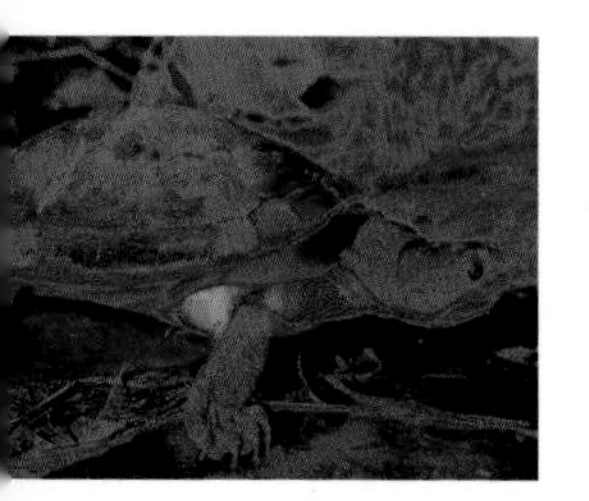

the World

Franck Bonin
Bernard Devaux
and Alain Dupré

Translated by Peter C. H. Pritchard

The Johns Hopkins University Press
Baltimore

Published with the support of the French Ministry of Culture—Centre national du livre.

Ouvrage publié avec le concours du Ministère français chargé de la culture—Centre national du livre.

Editions Delachaux & Niestlé
2 rue Christine
75006 Paris (France)

 Published 2006
Printed in Slovenia on acid-free paper
9 8 7 6 5 4 3 2 1

The Johns Hopkins University Press
2715 North Charles Street
Baltimore, Maryland 21218-4363
www.press.jhu.edu

Library of Congress Cataloging-in-Publication Data

Bonin, Franck.
Turtles of the world / Franck Bonin, Bernard Devaux, Alain Dupré ;
translated from the French by Peter C. H. Pritchard.
p. cm.
Includes bibliographical references and index.
ISBN 0-8018-8496-9 (hardcover : alk. paper)
1. Turtles. I. Devaux, Bernard. II. Dupré, Alain. III. Title.
QL666.C5B57 2006
597.92—dc22
2006003119

A catalog record for this book is available from the British Library.

Translator's Note

I am very pleased to have played the role of translator for this book, written by three good friends of mine. As a turtle scientist myself, I felt I had an advantage in that I could actually visualize the thousands of details of turtle morphology and decor that the authors describe, and this helped greatly in rendering the accounts in English. On the other hand, my scientific involvement in the subject matter also presents a challenge in that, here and there, one could be tempted to edit as well as merely to translate. To do this comprehensively, checking all details from the original literature, would have been a recipe for paralysis, and certainly it would be unethical to substitute my opinion for that of the authors when both opinions could be justified. But to leave minor typos or small errors unchecked would not serve the reader usefully, and I have made some such changes and corrections. The main challenges were those cases in which the authors were not fully up-to-date with conservation measures that had recently been taken for some endangered turtle taxa; in such cases, I have tried to give a brief summary of such recent initiatives, as far as they were known to me. But in the last analysis, the book is theirs, and I am merely the translator.

Monographs on the chelonians of the world have a long history. Examples that spring to mind include Walbaum's *Chelonographia* (1782); Schoepff's *Historia Testudinum* (1792); Schweigger's *Prodromus Monographiae Cheloniorum* (1812); Bell's *Monograph of the Testudinata* (1831; published only in the form of certain fascicles, but the plates were included, with additions, in Sowerby and Lear's *Tortoises, Terrapins and Turtles,* 1872); Duméril and Bibron's *Erpétologie Générale* (vol. 1, 1834); Boulenger's *Catalogue of the Chelonians* (1889); Siebenrock's *Synopsis der rezenten Schildkröten* (1909); the three versions of Mertens and Wermuth's annotated checklist

Schildkröten, Krokodile, Brückenechsen (1955, 1961, 1977); Ernst and Barbour's *Turtles of the World* (1989); King and Burke's *Crocodilian, Tuatara, and Turtle Species of the World* (1989); Iverson's book of detailed turtle and tortoise range maps, *Checklist with Distribution Maps of the Turtles of the World* (1992); Rogner's two-volume *Schildkröten* (1996); Marchan Formelino's *El maravilloso mundo de las tortugas* (1998); and the recent three volumes of Vetter's *Turtles of the World* published in English and German by Chimaira (2002, 2004, 2005). Other excellent surveys of the world's chelonians—Obst's *Turtles, Tortoises, and Terrapins* (1986) and Orenstein's *Survivors in Armor* (2001)—have taken a biological rather than a systematic approach to the subject. My own contributions were *Living Turtles of the World* (1967) and *Encyclopedia of Turtles* (1979).

So why do we need another? There are two reasons. The first is that the books I have mentioned span more than two centuries. Obviously, new discoveries each decade, perhaps even each year, add sufficiently to available information—and to the known species list—to make even the most worthy productions at least partially obsolete quite quickly. And the second point is that there are many different viewpoints, approaches, and emphases in the works I have listed. Some offer nomenclature and related taxonomic details; others take a zoogeographic approach; some look at turtles in captivity; some cover anatomy; and some rely upon extensive photographs and brief, succinct text. It is also true that some of these books are a pleasure to read, whereas others serve mainly for reference. The oldest ones may not be useful for current biological insight, but collectors value them greatly not just for their insight into the history of science but also for the extraordinary quality of the engraved illustrations.

Where does the present volume fit into this formidable sequence of monographs? First, being the most recent of the entire list, it is also the most up-to-date. This is most important in matters of taxonomy and nomenclature, and the authors have gone to great pains to consult with ranking experts on taxonomic status and have done an excellent job in including newly described species from many parts of the world as well as incorporating taxonomic changes of all kinds. Where these are controversial and the dust has not yet settled, this too is noted. Other special merits include the discussions of survival status and conservation priorities for almost all species, and the painstaking details of external appearance, morphology, and markings are indeed encyclopedic in their scope.

Two important players in my own background and professional life passed on while this translation was in progress, and I dedicate my part in this contribution to their memory. One is Raoul Larmour, my French teacher at Campbell College, Belfast, during my school days in the 1950s, who died at Christmastime 2005 in his ninetieth year. The other is John Behler, curator of reptiles at the Bronx Zoo (he always refused to call it the Wildlife Conservation Society), a global leader in turtle conservation, my successor as chairman of the IUCN Turtle and Tortoise Specialist Group, and like myself a 1943 baby. He died in early 2006. It sometimes seems that he personally helped just about every young herpetologist in the country, and we all miss him greatly.

Finally, my warm thanks to Vince Burke of the Johns Hopkins University Press for recruiting me to undertake the major task of translating this book and to Herb von Kluge for donating the best French dictionary I have ever seen—Heath's *Standard French and English Dictionary,* edited by J. E. Mansion and published in two volumes by D. C. Heath in 1934. No modern dictionary that I have found is equal to the task of translating some of the elegant, rare words that the authors not only know—but use.

Peter C. H. Pritchard

Contents

General Biology

According to the Linnaean system, which classifies animals by phylum (plural: phyla), each with branches corresponding to classes of species that show fundamental similarities, the reptiles are characterized as "crawling" species, breathing with lungs and showing poikilothermy (i.e., variable body temperature). They are divided into four orders: the Squamata (lizards and snakes); the Crocodilia; the Rhynchocephalia (a single living genus); and the Chelonia (turtles).

According to the cladistic systematics proposed by Willi Hennig,[1] one can determine the monophyletic branches, or clades, on the basis of single shared derived features. The reptiles are thus revealed to be derived from the sauropsidans by the key feature of a ventral keel, the hypophysis, on the cervical vertebrae.[2] The sauropsidans evolved into two clades, the turtles on the one hand—amniote reptiles recognized by the carapace enclosing the trunk and formed from two sections, the ventral plastron and the dorsal carapace—and the diapsids on the other, defined by the presence of two temporal fossae behind the orbit on each side. The latter became further split into the lepidosaurians, which gave rise to the squamates and the rhynchocephalians, and the archosaurs, which became the crocodilians and the birds. The result is that the crocodiles are closer to the birds than to any other so-called reptiles. Furthermore, the turtles are on a very different track from that of the crocodiles, snakes, and lizards. One might also note that the turtles are missing a key characteristic feature of reptiles, according to Linnaeus's definition, in that they do not typically crawl on their bellies. For this reason, the turtles must be considered to be sauropsidans. Further combinations of characteristics help define this group: the lizards have two lungs and paired hemipenes; the crocodilians have two lungs and a single penis; the snakes have a single functional lung and paired hemipenes; the turtles have two lungs and a single penis; and *Sphenodon* has paired lungs and no penis.

The ancestors of the sauropsidans were the cotylosaurs, which appeared 345 million years ago. They were characterized by a short body, a heavy skeleton, and a skull composed of several massive bones and lacking temporal fossae. They walked like the living salamanders, with the body and tail thrown into sinusoidal curves and the belly dragging on the ground. It seems that the body was covered with scales, or at least forerunners of scales. The eggs would have been hard-shelled, freeing the animal from dependency on aquatic systems. Why did the carapace evolve, and why did the turtles form a separate lineage? It appears that the ancestor of the lineage was a small amniote from South Africa, *Eunotosaurus africanus*, which lived about 260 million years ago. According to D. M. S. Watson, this animal developed enlarged and thickened ribs, which anticipated the formation of the carapace. Nevertheless, *Eunotosaurus* was equipped with teeth, and its shoulder bones were external to the ribs. According to Michael Lee,[3] *Captorhinus* is the most plausible "inventor of the carapace." This reptile, about 600 mm in length, offers an especially telling feature: its shoulder girdles were located inside the rib cage. Its cervical vertebrae were correspondingly reduced in number to just 5. This form disappeared by the end of the Triassic. The next significant player was *Bradysaurus,* a much bigger animal, about 3 m in length. Its vertebral column was covered with bony nodules. It also had 5 cervical vertebrae but just 14 dorsal vertebrae,

(1) In 1950, Willi Hennig proposed a new system of classification called cladistics.

(2) Lecointre G. and Le Guyader H., *Classification phylogénétique du vivant,* éditions Belin, Paris, 2001.

(3) Edward Cope named *Proganochelys quenstendti* as the ancestor of all turtles. For an update, see Michael Lee's 1994 article "The Turtle's Long-Lost Relatives" in *Natural History* magazine.

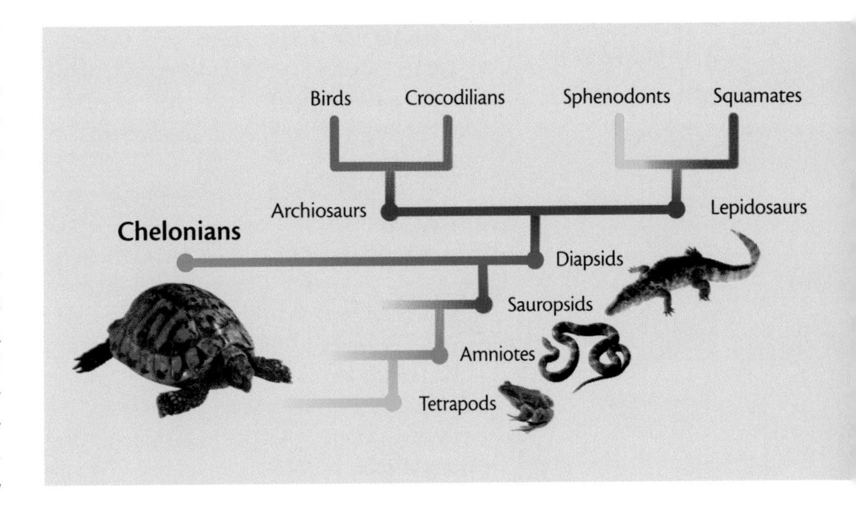

The cladogram shows the lineages of the various Sauropsids.

(1) See Guillermo Rougier, Marcelo de la Fuente, and Andrea Arcucci (1998) regarding *Palaeocheris talampayensis.*

and the shoulder girdle was expanded into a sort of belt that encompassed 3 vertebrae. Another genus, *Anthodon,* which came somewhat later (248 million years ago) was just 900 mm in length, but it had extensive bony armor. The body was almost entirely covered, on the back and the sides, by a mosaic of bony plates, in anticipation of a carapace of the future. This arrangement of bony plates certainly proved to be effective, in that, 30 million years later, *Anthodon* had apparently evolved into *Proganochelys,* considered to be the ancestor of all turtles and tortoises.

Proganochelys

The sauropsidan *Proganochelys,* about 900 mm in length, had all the key features of a present-day turtle. It had 8 cervical vertebrae and 10 dorsal vertebrae. The girdles had slid completely within the rib case, the latter having become an all-encompassing bony box, covering the back, the sides, and the belly. The ossification even extended to the tail and part of the neck. We are now at the beginning of the Triassic, 220 to 230 million years ago. Turtles of those days still had small palatal teeth, like the cotylosaurs, but they lost them before much more time passed.

The ancestor of all turtles, *Proganochelys,* is shown here. The original life size was 800 mm long. With spines on the neck, head, and tail, it was well protected against attacks by Cretaceous predators.

Two groups of turtles developed: those that retracted the neck in a horizontal plane, which gave rise about 210 million years ago to the living pleurodires, of which the ancestral form is *Proganochelys quenstedtii;* and those in which the neck is retracted vertically, in an up-and-down fashion, which gave rise, also about 210 million years ago, to the majority of living turtles, the cryptodires, whose ancestor is thought to be *Proterochersis robusta.* Recent studies have demonstrated that the fusion of the pelvis with the carapace was a primitive trait common to all turtles, and that this is a feature that has been lost in the cryptodires.[1] The vertical flexion of the neck, characteristic of the cryptodires, came a little later, but it was already developed 185 million years ago, in *Kayentachelys,* and somewhat later, about 156 million years ago, *Platychelys* appeared, destined to become the ancestor of the pleurodires. Nowadays the most ancient of the living turtles are the African pelomedusids, appearing about 120 million years ago.

The marine turtles have a somewhat more complex history. We know that several sea turtle families that are now extinct were able to occupy the oceans of the world before they disappeared. The living sea turtles are derived from an inconspicuous branch of the cryptodires, differentiating about 135 million years ago, at the beginning of the Cretaceous. This group comprised five families, of which just two survive today: the Cheloniidae, appearing about 55 million years ago, and the Dermochelyidae, perhaps 5 million years younger and represented by the single living species *Dermochelys coriacea*. Among the turtles that we know about today, the newest forms are the Emydidae (which include the European pond turtle, *Emys orbicularis*), and the oldest are the Podocneminae. The true, land-living tortoises first appeared about 65 million years ago, at the start of the Tertiary. The early family Chelydridae (80 million years ago) resembles the ancestral form *Proganochelys* in the thickened, ridged shell and the heavily armored tail. It comprises the famous American snapping turtles, *Chelydra* and *Macrochelys*. The family Trionychidae, 85 million years old, has undergone a curious reverse evolution, losing parts of the bony carapace, including the peripheral bones and all of the scutes, the shell being covered with a leathery skin in which are embedded remains of the bony plastral elements. These are the soft-shelled turtles, *Apalone* and so on.

As of late 2004, about 330 species of living chelonians were recognized, all characterized by a bony enclosing corselet consisting of three parts: the carapace, the plastron, and the two lateral bridges, which may be either more or less rigid or in some cases may connect the carapace to the plastron by strips of ligamentous tissue. Furthermore, as already mentioned, certain species (*Apalone,* etc.) evolved a continuous covering of soft or leathery skin on the "shell," whereas the leatherback sea turtle, *Dermochelys coriacea,* developed a leathery mantle, laid upon a continuous sheet of small mosaic bones, itself underlain by vestiges of most of the usual shell bones. The weight of adult chelonians covers an extraordinary range, that of the 100 mm *Homopus signatus* from South Africa being only about 80 g, whereas the giant leatherback may reach 500 kg and a total length of 2 m (carapace to about 180 cm). Longevity is less variable; all species are potentially long-lived (half a century or more), and certain species may exceed 100 years. Fortunate individuals of certain terrestrial species have been known to exceed 150 years.

Reproductive strategy continues the heritage of the cotylosaurs from which the turtles evolved; copulation may occur in the water or on land, but all turtles (except *Chelodina rugosa,* page 14) nest on land. This may be the only time when the marine turtles (and also certain freshwater species) actually venture onto terra firma. Some species may deposit just a single egg (*Chersina angulata, Homopus boulengeri,* and *Malacochersus tornieri,* for example), and others, including some of the marine species, may lay up to 150. The eggs range from 20 mm in length for some of the softshells to 90 mm in *Rhinoclemmys,* but the largest species do not necessarily lay the largest eggs. Incubation normally takes from 60 to 100 days, at a temperature between 26°C and 32°C, but extreme values include the 28-day incubation of *Pelodiscus sinensis* or the 664 days recorded for *Chelodina expansa*.

Among the sauropsidans, the turtles are the group that inhabits the greatest range of habitats, some of them, like the sub-Saharan *Centrochelys sulcata,* living in arid deserts, while others, including the leatherback, *Dermochelys coriacea,* may be found in northern waters close to Greenland. Their poikilothermy does not prevent turtles from tolerating very low temperatures, thanks to morphological and behavioral adaptations (muscular activity, fat bodies, thermal inertia, burrowing, hibernation), or very high temperatures, to which they respond by burrowing, slowing of metabolism, mud or sand baths, or discharge of urine. Their adaptive plasticity is surprising and allows them to tolerate quite varied ways of life. Turtles occupy a very large portion of the planet, excluded only from the coldest regions. They are found in all the continents, with the exception of Antarctica. One of the tortoises, *Dipsochelys elephantina,* or the Seychelles tortoise, may even claim the distinction of having the highest density of vertebrate biomass per hectare in the world, with more than 21 tons of animals per square kilometer (on Aldabra), higher than the 17 tons/km^2 calculated for elephants, buffalo, giraffes, and lions together in the Serengeti of Tanzania.

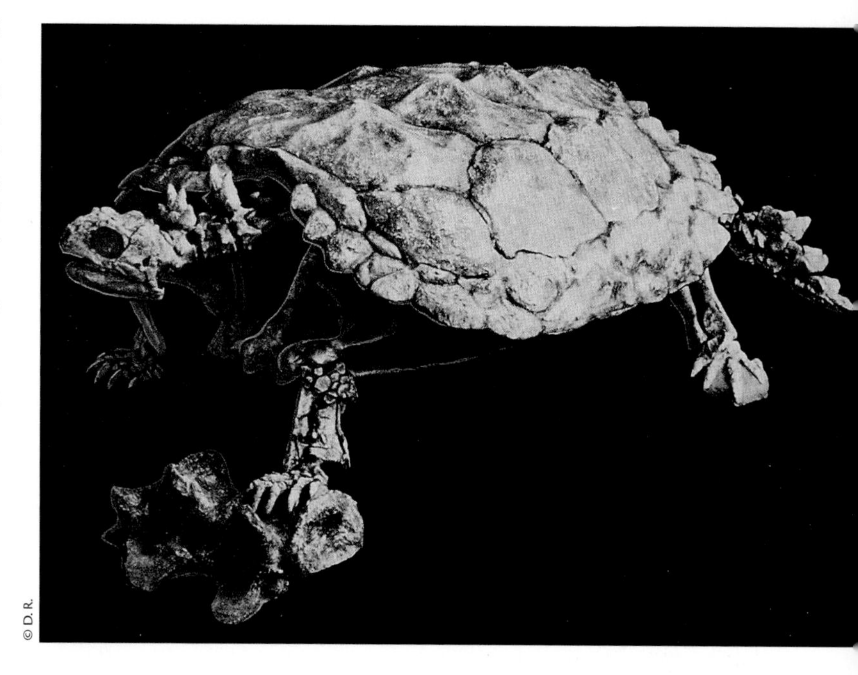

The skeleton of *Proganochelys* is heavily ossified, with vertebral keels and a tail equipped with sharp spines.

Skeleton

The carapace, of dermal origin, is evolutionarily a development of the rib cage. The turtle is the only extant vertebrate in which the shoulder girdles and the pelvis are located within the rib cage, and this evolutionary innovation is evident even in *Captorhinus,* dating back 260 million years. The carapace is typically very rigid and constitutes a protective enclosing box, but it is sometimes flexible and hinged to facilitate the passage of eggs or to allow the animal to close up the shell and retreat further from harm. The evolutionary process has undertaken many experiments to extend the fundamental protective function of the encasement of the turtle body. Some turtles have abandoned the massive domed shell in favor of developing a leathery skin. Others, by contrast, are able to close themselves up completely by means of hinged anterior and posterior lobes of the plastron (*Terrapene, Cuora,* etc.). The hinges are formed from bands of connective tissue that replace certain bony sutures. Most of these lines of flexion form in the plastron *(Terrapene, Pyxis),* but one genus *(Kinixys)* actually features a hinge across the rear of the carapace.

The neck is always formed from eight vertebrae and may be short but is extremely lengthened in such forms as the Australian snake-necked turtles *(Chelodina).* The neck may be withdrawn in a vertical plane (in the cryptodires) or in a horizontal manner (pleurodires). In certain species, the thickening of the base of the neck prevents the animal from retracting or concealing the head, as in *Dermochelys coriacea* or *Caretto-*

chelys insculpta, these being marine or at least aquatic species that presumably have no further need to hide their head and neck within the bony shell itself. In the land tortoises, the carapace is, by contrast, very strong and may represent up to half the weight of the entire animal, as in *Centrochelys sulcata* or *Chelonoidis nigra*. Some parts of the shell may reach a thickness of 50 to 100 mm in some of the largest tortoises. This may occur in *Chelonoidis nigra,* for example, at certain points in the overall structure of the shell, such as the gular area of the plastron.

The bony shell is a calcified structure, covered with a thin layer of keratinous scales, similar to our own fingernails. The layer of keratinous material serves as a physical defense against external threats (fire, injury, microorganisms, fungal infections) and also serves in species identification and in crypsis, or camouflage. The patterns of these scales or scutes may often be similar to the substrate upon which the animal lives. Moreover, the color of the scutes often reflects the general color of the skin of the animal and of the environment that it inhabits. Many different designs may be observed on turtle carapaces, following the selective pressures to which the animal has been subjected. Thus the African spurred tortoise *(Centrochelys sulcata)* is of a uniform orange-yellow, identical to the color of the sandy substrate where it lives. The European pond turtle, *Emys orbicularis,* has numerous little yellow dots that resemble the reflections of sunlight from the water of the marshes that it frequents. The yellow radiations against a black background in *Centrochelys elegans,* from India, correspond to the dry leaves of a local plant. Even the name given by Linnaeus to *Testudo graeca* refers to the angular markings of the shell, which resemble a Greek mosaic.

The cornified outer shell can be compromised by fungal disease or injuries or can even be destroyed completely, with only the bony layer left for protection. In the event of injuries to the carapace, if the animal survives the event, the two layers can each grow back to some degree. The underlying tissues will calcify, taking on a cardboard-like appearance, and finally will re-form into tissues just as strong as the original shell. The keratin also reforms and sometimes may be confused with the bony reconstruction, resulting in a sort of crazy design or mosaic of scar tissue, keratin, and exposed bone demonstrative of the vitality of the carapace and its capacity to repair itself. The replacement of a section of the bony shell may take at least a couple of years, depending on the species.

Organs

The organs of a turtle are not radically different from those of any other vertebrate, including ourselves. Nevertheless, turtles do not have teeth or external ears, and they have a single ventral orifice, or cloaca. But the beak of a turtle is a good replacement for actual teeth, and the animals certainly have internal ears, hidden beneath a scaly membrane behind the eye. Turtles have excellent hearing, and this is especially true of certain freshwater emydid species. The cloaca serves for discharge of waste material, for the extension of the penis in males and of the clitoris in females, and as the intromission canal for the female during copulation. The male penis is often of quite large size, and indeed in some of the small species it may reach half the length of the animal. It is frequently equipped with a horny tip, which facilitates access to the interior of the cloaca of the female. The spermatozoa may remain viable

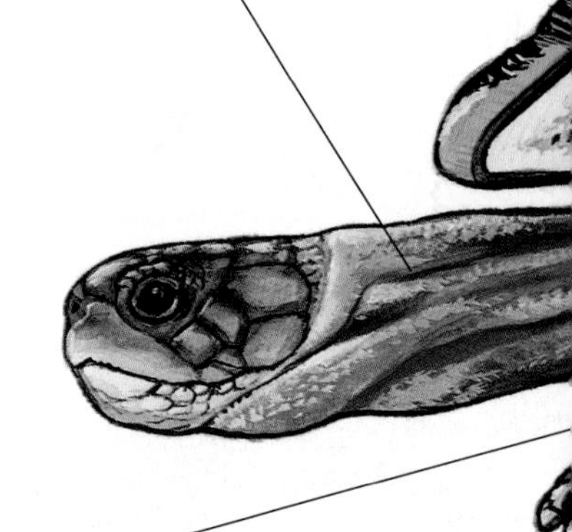

in the folds of the uterine cavity for a maximum of six months (in *Testudo graeca*), making possible the production of fertile eggs long after copulation has taken place.

The heart is formed differently from our own. It is wide and flat, with a rounded tip, and has only three chambers. Its shape has apparently been an inspiration to paleontologists, who have given a certain fossil shell the name "turtle heart." The heart beats about 20 times per minute when the turtle is under optimal temperature, and so it is significantly slower than in human beings. The lungs are large and situated near the uppermost part of the dorsal cavity. The respiratory function occurs through the action of the muscles of the limbs and viscera, in that the rib cage is rigid and is incapable of changes in shape. The liver is located below the heart, in the lower part of the body cavity. This organ was keenly sought after by sailors and hunters in former times, its position near the plastron facilitating extraction of a small piece from which to judge its overall size and taste. After this procedure, the turtle might be cut open completely to extract the entire organ or otherwise was released alive with this incision in the plastron, doubtless to creep away to a slow death. The oviducts are large and contain eggs of a size and degree of development expressive of whether they have been fertilized or not; if they have been fertilized, the body cavity may seem to be almost entirely occupied by the huge oviducts and their shelled, mature eggs.

The bladder has an important function in providing moisture to soften the soil during nest excavation. If the turtle lives under arid, desiccating conditions, it may excrete a white, thick, yogurt-like liquid composed of almost pure uric acid. Tortoises are able to drink large quantities of water at a time, which they keep in reserve in their stomachs. When a tortoise is severely dehydrated, it may drink as much as 50% of its body weight. But

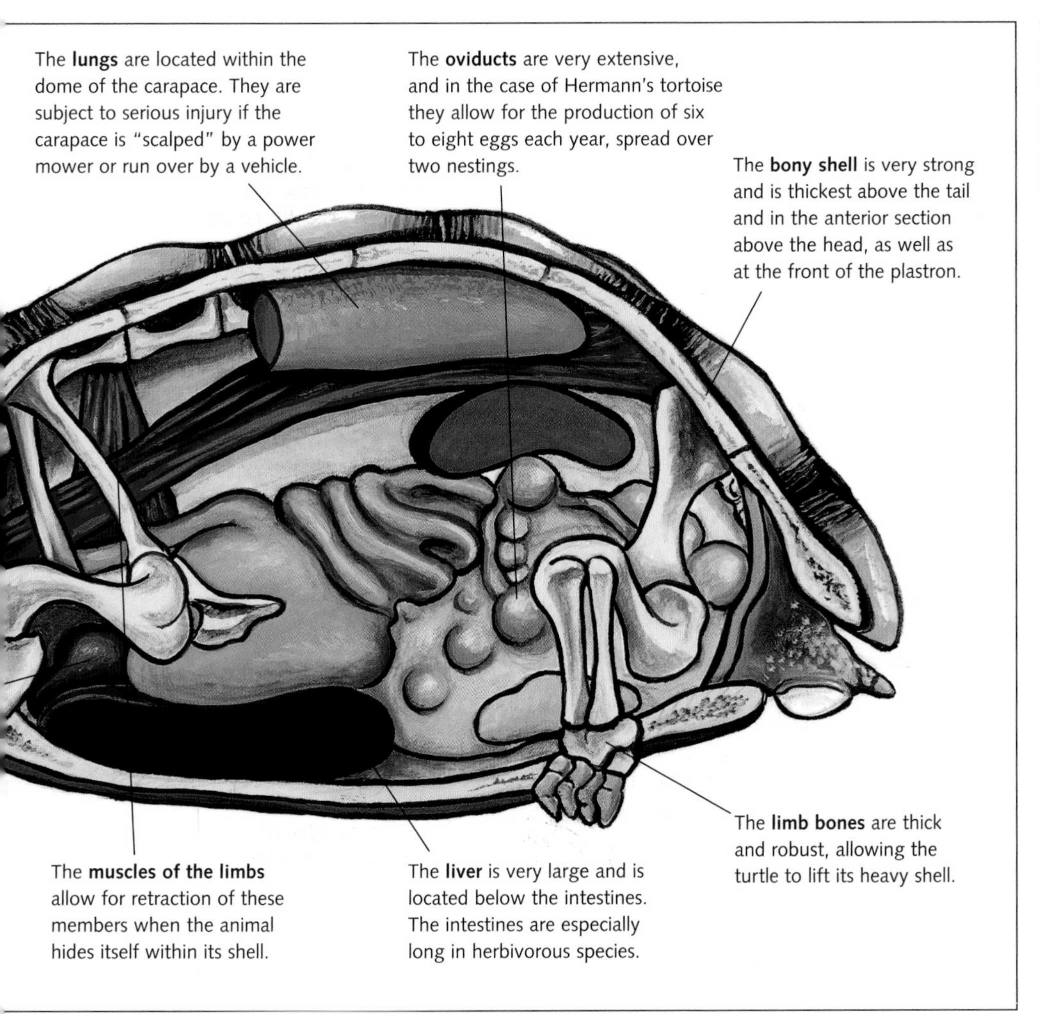

This diagram shows the internal organs of the turtle, with the two lungs in the upper part and the large liver below. Also shown are the powerful muscles that withdraw the neck and limbs.

tortoises can easily survive without eating or drinking for several weeks, living off their reserves of water and fats.

In many turtles, communication between individuals occurs through olfaction. The sensing and the production of odors play an important role in the recognition of potential sexual partners in the great majority of species, especially the aquatic ones. All turtles, apart from the terrestrial testudinids, have two pairs of glands located between the anterior and posterior borders of the plastron (Rathke's glands). In certain species, these glands release a liquid with a strong musky odor *(Kinosternon),* which may serve as a deterrent to predators or perhaps also may function as a means of intraspecific communication in the course of the reproductive cycle. In two tortoise species of the genus *Gopherus (G. agassizii, G. berlandieri),* there is a pair of mental (chin) glands, more developed in the males than in the females, which may discharge a proteinaceous exudate during the breeding season.

Senses

Outside of these specific olfactory communication functions, the sense of smell in turtles appears to be poor, and the vision is not especially sharp. The freshwater species are more farsighted than the terrestrial species, at least among the most modern turtles of the family Emydidae. As with all the sauropsidans, turtles react mainly to movement rather than to shapes. Vision is most acute in those wavelengths of light corresponding to the orange and red end of the spectrum, which may explain why tortoises are attracted by such foods as tomatoes, cherries, and watermelon. Hearing appears to be rather good, and the freshwater species may have good ability to detect minimal sounds at rather long distances. The ear of the green turtle, *Chelonia mydas,* is receptive to low frequencies, in the region of 60 to 1,000 Hz. In the leatherback turtle, the auditory range is much more reduced, approximately to the range of 300 to 500 Hz. It all depends upon the environment in which the turtle lives, but turtles in general perceive low frequencies, as well as ground vibrations.

In some cases, turtles may detect thermal changes through their plastron. Nerve endings inside the plastron permit female turtles to evaluate the warmth of the substrate to guide them to an advantageous nesting site. As with the snakes, ground vibrations are detected with great acuity, which may explain the rapid departure of turtles that one approaches without making a sound, as well as their panicky responses to earthquakes. Turtles have a keen sense of orientation, and this is especially the case with the marine species. This ability to navigate very long distances is due to the presence of magnetite crystals in the brains of certain species. But sea turtles may also orient by taking note of currents; the salinity of the sea; the magnetic polarity of the earth, the moon, and the sun; and even the topography of the shore near the beach selected for nesting. Certain tortoises are equally capable of relocating themselves after being moved for several kilometers. The phenomenon of homing poses questions for the release of turtles if the animals are close to their original home range. Probably there is much still to learn in the field of the senses of turtles, especially since each species, living as it does in its own unique environment, has developed a precise adaptation to its environment and will have developed certain senses more than others.

Hermann's tortoise, chopped by a brush-clearing machine, shows calcification and new ossification built upon the existing shell. This new ossification allows for good closure of the shell. It took about two years for the new bone to reach this stage.

Sexual Dimorphism

Generally speaking, male turtles are larger than females in the species that reach great size, while the reverse is true of small species. For example, in *Chelonoidis nigra,* the Galápagos giant tortoise, the male is twice the size of the female, whereas in *Testudo hermanni* the females are one-sixth to one-eighth larger than the males.

Males usually have a larger tail, thicker and wider than that of the female. In the latter sex, the plastron is flat, whereas in many species the males have, to some degree, a plastral concavity. This concavity is very obvious in the large terrestrial species *(Chelonoidis nigra, Centrochelys sulcata).* In certain freshwater turtles (*Trachemys,* etc.), some of the claws on the forelimbs are much longer in the males than in the females; these are used to stroke the face of the female during courtship. In some marine species the anterior flippers of the male each bear a strongly hooked claw that allows the animal to hold on to the shell of the female. In the sea turtles this claw is single, but there are two on each flipper in *Carettochelys,* the Australian pig-nosed turtle.

Male tortoises often vocalize during copulation, the nature of the sound being very different from one species to another and often being species-specific *(Testudo kleinmanni, Manouria emys).* In the land tortoises the supracaudal scute is often recurved anteriorly at its lower edge in the male but is flatter in females, allowing more space for the eggs to pass. Coloration may differ between the sexes, especially on the neck and head, as in *Callagur borneoensis*. But the most reliable feature is the position of the cloaca, which is closer to the abdomen in females and closer to the tip of the tail in males.

There is also some behavioral dimorphism, the males in general being active and seeking copulatory opportunity, whereas the females are more passive. In many species, the male proceeds through a complex sequence of behaviors, which includes numerous episodes of approach, contact, nibbling or biting, and various preliminaries such as head bobbing, sniffing, and so on, whereas the female is often unreceptive, seeks to hide, or remains retracted into her shell. The details of courtship are in general described separately for each species in this work, because there are major variations. The common theme is that the male introduces his penis into the cloaca of the female, after which his sperm penetrates her ovaries, permitting the development of fertilized eggs within the oviducts. Copulation may take place on the nesting grounds themselves, these being the only places where the males may be sure of finding females, as is the case with the marine species.

Metabolism

The turtles and tortoises, being poikilothermic, do not have the advantage of homoiothermy, whereby mammals and birds regulate their internal temperature in a fashion independent of external environmental conditions. To keep their temperature within a nonlethal range (between 5°C and 45°C), the turtles bring to bear a series of physiological and behavioral adaptations. Their poikilothermy results in a tremendous savings of energy needs for the maintenance of life—generally a reptile needs only one-thirtieth of the energy intake of a mammal of the same body mass—and this allows them to adapt to extreme environments. Their metabolism can slow down, allowing for further energy savings. In addition, turtles have various adaptations that allow them to modify their internal temperature and stabilize it in a way comparable to that of the homoiotherms. These features include the protection and isolation offered by the carapace, the thick layer of insulating fat, and the heat generated by muscular activity, not to mention their ability to hide themselves or to cover themselves with mud, water, sand, or urine as means of thermoregulating. A leatherback turtle has been found with a deep body temperature of 20°C in Greenland, in waters of 7°C., Hermann's tortoise *(Testudo hermanni)* has a minimum internal temperature of 1.7°C and a maximum of 37°C. It appears that 45°C is lethal for all chelonians. Freshwater turtles in the north of the United States and in Canada produce in their cells a natural antifreeze, related to glycerol, which allows them to tolerate temperatures down to −7°C for several weeks. A part of the body of the cell becomes frozen, but without significant crystallization, and when warmer times return, the functions of the animal return to normal.

The optimal temperature for a terrestrial chelonian is about 28°C. At such temperatures, the heart rate is about 30 beats per minute, and the respiratory rate is about 20 breaths per minute. As the temperature drops, the heart rate drops rapidly and may be reduced to only one beat per minute. Pulmonary action may cease completely. Oxygenation, in practically all aquatic species, can be effected through the skin or through oral, pharyngeal, or cloacal membranes. When this is occurring, the turtle remains for long hours at the bottom of the water, cloaca or mouth agape, without coming to the surface. The poikilotherms thus possess a metabolism slower than that of the homoiotherms, and some turtles can remain without food for several weeks or even months. Land tortoises are herbivores in general, the freshwater species carnivores, and the marine species feeders upon fish or eaters of coral, jellyfish, or other invertebrates, but numerous exceptions exist, as discussed in the body of this work. Digestion in a poikilotherm is as slow as its metabolism, and passage of food through the digestive tract of a tortoise may take about eight days on average.

Water needs are reduced in sauropsidans by their scaly integument, which reduces evaporation and prevents perspiration, and also by their slow metabolism. Water loss is also limited, thanks to the ability to discharge wastes in a very dehydrated form, which may have the consistency of yogurt and which contain virtually pure urea. The land species can subsist upon the water content of the food they eat. In the marine turtles, excess salt is secreted by the lachrymal glands, which is the real reason why sea turtles seem to be weeping when they come on land to nest. To escape extremes of temperature, the turtle may hibernate or estivate. Turtles in temperate climates hibernate for several months each winter, and those in very hot climates may estivate for at least several weeks. In both cases, we are witnessing a physiological adaptation, not a biological necessity. When the temperature reaches as high as 35°C, the turtle seeks a cool retreat, digs a burrow, or covers itself with soil or sand. Loss of water brings about a drop in metabolic rate, and the animal goes into a state of estivation. Motionless in its cool retreat, the turtle ceases feeding and movement for several weeks, perhaps even several months in certain African species that bury themselves in the mud *(Pelomedusa)*. On the other hand, when the temperature drops below 15°C, the metabolism of the turtle slows down, and the animal ceases to feed. It seeks a place of concealment. When temperature drops further to 10°C, real hibernation starts. Completely immobile in a cavity in the ground or under a covering of dead leaves, the turtle enters a stage seemingly close to death. It has no need to eat or to drink, and the hibernation may last for several months. Energy needs are essentially zero, and its weight is virtually the same at the onset and at the end of hibernation.

Ethology

Turtle behavior has not been adequately studied, but numerous specific observations of behavioral details will be found in the species accounts in this book, as a result of studies done upon one species or another. We know, for example, that *Manouria emys*, a large Asiatic tortoise, makes a pile of vegetative debris in which it will burrow and make its nest, and then it will protect its nest from predators by attacking intruders. It also emits sounds at the peak moment of copulation, and there is an exchange of calls between the male and the female. An Australian freshwater turtle *(Chelodina rugosa)* actually lays its eggs underwater, but embryonic development is suspended until such time as the waters dry up at the end of the rainy season. Another Australian species, *Pseudemydura umbrina*, digs its nest with its forelimbs, not the hind ones as in other species. Some have evolved an appearance and behavior that allow them to resemble vegetable debris, like the matamata *(Chelus fimbriatus)*, and another *(Macrochelys temminckii)* offers a natural bait, or lure, on the floor of the mouth, a small, mobile blood-filled excrescence that attracts unwary prey. Certain freshwater turtles *(Trionyx triunguis)* are able to enter estuarine or marine conditions to nest on oceanic beaches like the sea turtles. Others dig deep burrows to find cool conditions and adequate humidity *(Gopherus, Centrochelys sulcata, Agrionemys (Testudo) horsfieldii)*. Still others may flatten the body to the point of developing a flat, flexible shell, allowing them to creep under boulders or into cracks to hide themselves *(Malacochersus tornieri)*.

There is not much family life per se among turtles, and not much true territorial or dominance behavior, but certain aquatic turtles may live in small, stable groups. One also notices gregarious tendencies among the Aldabra tortoises *(Dipsochelys elephantina)* and the Galápagos giants *(Chelonoidis nigra)*. The Aldabra tortoises show decidedly communal survival strategies and display unique ritualized behavior patterns. And some sea turtles will form immense assemblages, or *arribadas*, as a manifestation of group nesting; this occurs notably with *Lepidochelys olivacea*, in Costa Rica and in India.

Threats and Protection

Turtles are among the most exploited and abused animals in the world. This exploitation and abuse have occurred at the hands of almost all civilizations and since ancient times. Slow, docile, easy to capture, and providing such products as abundant meat, oil, fat, bony shells, and valuable scutes in some cases, turtles have been and still are being killed in many parts of the world, often in a very cruel fashion, because their lack of facial expression can lead to a false belief that they "do not suffer." The oldest civilizations (e.g., in China) have exploited turtles for divination purposes, but turtles are equally widely exploited in India, throughout the Pacific, and in the Caribbean, sometimes for food, sometimes for medicines or as aphrodisiacs. Several species, mainly in the Indian Ocean, have been completely exterminated by humans in recent centuries (*Cylindraspis borbonica, C. peltastes, C. vosmaeri, Dipsochelys arnoldi, D. hololissa,* etc.). Today the main threats are exploitation for food and commerce; urban development; desertification; and collection for resale. Globalization accelerates the process and intensifies the efficiency of the trade. Huge shipments of turtles originate in Asian countries, to the profit primarily of China, the greatest consumer of turtles in the world. Another, more insidious threat comes from the status of turtles as garden animals in our Western imagination and culture. The proliferation of zoos and menageries, the fashion for keeping reptiles in captivity, and the thirst for new kinds of pets have brought about an exponential increase in the exploitation of reptiles and particularly of turtles, which are considered to be lovable, gentle things. We personally believe that they should remain and be recognized as wild creatures in the ecosystems to which they are adapted, and it is wrong to think of them as pets, like rabbits or guinea pigs. Thinking of them only in this manner increases commerce and traffic in turtles in ways that can surely be harmful. There are many poor areas of the world, like Madagascar, India, and Africa, where people systematically devastate their own turtle fauna for profit. It is sobering to recognize that two-thirds of turtle species are threatened or endangered today, and the hardest hit of all are the freshwater turtles of Southeast Asia.

Just 50 years ago, there were some hesitant first steps in the development of turtle protection programs, led by a handful of distinguished naturalists and herpetologists who valued and emphasized the conservation aspects more than laboratory study or public display of turtles and tortoises. Twenty years later, pangs of conscience were also detected in a good number of developed countries as well as in the world of science. Indeed, conservation biology developed as an entirely new branch of science. Professional associations were started, centers for protection of turtles opened up, and international conferences were held, all seeking to find ways to reverse the progressive loss of turtles during the preceding decades. Attention ranged from the need for habitat protection to the need to release head-started or displaced animals in suitable areas for their survival. The most important message to bring to the public was the need to appreciate these delightful animals, yet to have them remain in the wild rather than in private gardens. But the fundamental problems continue: the expansion of trafficking; the widespread desire of certain people to obtain rare species for their own private pleasure; and the image of the turtle as a domestic animal, an image widespread in certain clubs and captive centers, to the detriment of the biology and the natural ecology of the species. Protecting turtles thus requires that their habitats be protected, that commercialization be opposed, and that attitudes of certain people be encouraged to change with the times. This book takes a novel and unambiguous stand: it will not be a manual written to facilitate the keeping of turtles in captivity, but on the contrary it is written to be a description of turtles in the wild. The accounts and photos in this work are those of wild animals, observed in their own ecosystems, and the section on protection of each species has the goal of enhancing the awareness of the reader to the needs and conservation of this or that species *in its original habitat.*

The classification of species in this book incorporates many new nomenclatural concepts proposed right up to our date of publication and is based upon the recent nomenclature proposed by Dr. Roger Bour of the National Museum of Natural History in Paris. From one country to another, there may be minor disagreements about nomenclatural details, but we have sought the advice of the most competent herpetologists and systematics specialists in the world. Nevertheless, these differences of opinion will continue to exist among both specialists and naturalists.

Pleurodira

The infraorder Pleurodira (Cope, 1864) includes two families, the Chelidae (Gray, 1825) and the Pelomedusidae (Cope, 1868), which represent the most ancient of all turtles, appearing 120 million years ago. Their unifying feature is the lateral or horizontal plane of the retraction of the neck, which bends to one side between the anterior parts of the carapace and the plastron.

Chelidae Gray, 1825

This family is a little more specialized than the Pelomedusidae, the more primitive sideneck family. The Chelidae are characterized by a more specialized skull, cervical vertebrae different from those of the Pelomedusidae, and above all a neck that is often extremely long, especially in certain Australian species. They live today in the Southern Hemisphere, in South America east of the Andes, in Australia, and in New Guinea.

Acanthochelys macrocephala

(Rhodin, Mittermeier, and McMorris, 1984)

Distribution. This species is restricted to a small range within the upper Rio Mamoré Basin in central Bolivia, reaching as far as the upper drainage of the Rio Paraguay southwest of Mato Grosso in Brazil, in the Pantanal area. Buskirk (1988) found that it also occurred in Paraguay.

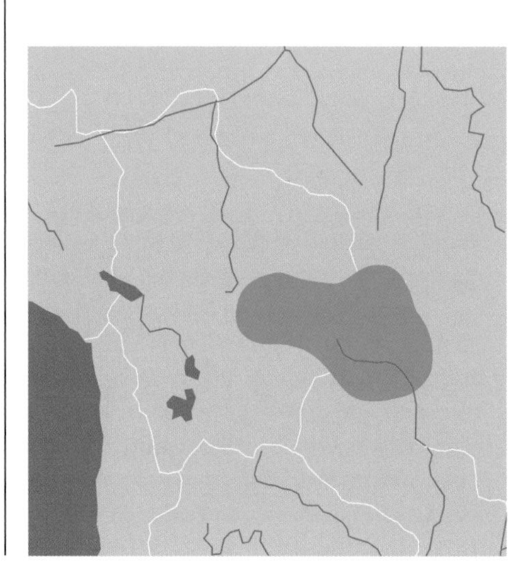

Common name

Pantanal swamp turtle

Description. As the vernacular name implies, *A. macrocephala* (especially old females) has a bigger head than the other members of the genus. The species is of medium size (females to 295 mm), with an oval carapace flattened in the area between the second to the fourth vertebral scutes. The marginals are not serrated, but the individual marginal scutes above the limbs are enlarged. The midmarginal scutes may form a shallow gutter on each side. The overall color is dark brown to almost black, relieved by brown markings in some individuals. The juveniles have a light carapace, with light brown radiating lines on each scute. The plastron and bridges are dark yellow, becoming darker along the sutures, especially in young specimens. The anterior plastral lobe is wider than the posterior one, the latter being strongly notched in the anal region. The wide head is brown to grayish on top and cream to yellow below, but without a clear demarcation between these two areas. The tympanum and the lower jaw are yellow, with

© T. Vinke

This turtle is of medium size and is elongated and flat. The head is rather large, and the neck bears numerous tubercles.

This species is covered with large scales, forming a sort of helmet. The snout is rounded and short, and the eyes are located well forward.

gray or orange spots. The iris is the color of burned bread, without a dark horizontal bar. There are enlarged scales on the top of the head. The neck is gray-brown above and yellow below, sprinkled with pointed conical tubercles on the upper surface. The limbs are strong, gray above and yellow below, with orange spots. They are covered with wide scales. Conical tubercles are present on the back of the thighs. Females are larger than males.

Natural history. This species, like *A. radiolata,* frequents marshes, wetland areas (including the Pantanal), and slow streams. It subsists largely upon snails, which it crushes easily with its wide jaws.

Protection. This species has been placed on the IUCN priority list, and its status, distribution, and conservation are in need of further investigation.

© T. Vinke

Acanthochelys pallidipectoris

(Freiberg, 1945)

Distribution. This species is found in the Chaco region in Argentina, and it may also be found in southern Paraguay and in Bolivia.

Description. The Latin name refers to the pale plastron, which is largely cream to yellowish in color, but this species is most remarkable for the massive spurs that adorn the thigh region, equally well developed in males and in females. This is a small species, about 175 mm in length, with an oval carapace, rather flat, and dorsally depressed between the first and fifth vertebrals. In the adults the fourth vertebral is narrower than long, al-

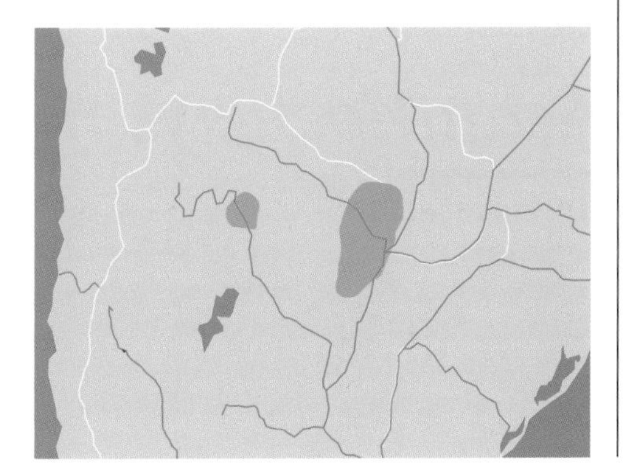

Common name

Chaco side-necked turtle

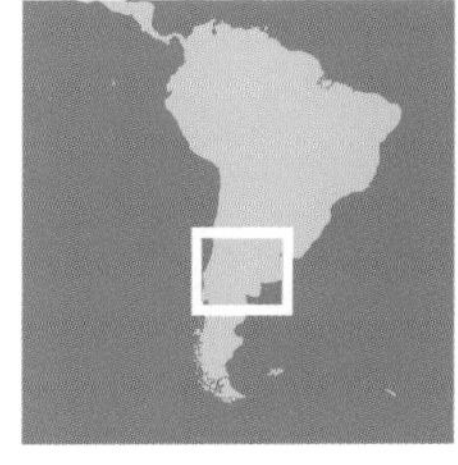

Notice the long neck, folded to one side, and the widely distributed conical tubercles. There are two barbels on the chin.

though the reverse is the case in the juveniles. The marginal area is smooth and unserrated, and the marginal scutes are enlarged and form gutters along the sides. The overall color of the shell is dark yellow, light brown, or gray or even olive, with darker seam lines. The marginals are underlined with a yellow border, especially toward the rear. The plastron and the bridges are yellow, with dark bands along the sutures, which can become wider with age, leaving only a small yellow blotch restricted to the center of each plastral scute. The anterior lobe is larger than the posterior, and there is a wide anal notch. The head is wide, with a yellow central line and a brown line on each side. Toward the tympanum, the color is yellow, and the top of the head is covered with wide scales. The iris of the eye is usually white, but certain observers have reported the iris to be yellow or even greenish. The top of the head is gray brown, with large conical tubercles on the sides, and the ventral aspect is yellowish. The chin has two minuscule papillae. The limbs are yellow, covered with large scales. On the rump are two groups of large spurs, one in each group being larger than the others, especially in males. The limbs are strongly webbed, with five claws on the forelimbs and four on the rear. Sexual dimorphism is little developed in this species.

Natural history. This turtle occurs in slow streams, lakes, lagoons, and oxbows. It is most often spotted when rains cause the pools and lakes to overflow, carrying the turtles into the adjoining forests. This species may be more terrestrial than the other *Acanthochelys.* Its reproductive details are unknown. The diet appears to be primarily carnivorous, including insects, worms, and tadpoles. When threatened, the animal retracts its head quickly, and the posterior tubercles protect the hindquarters from predators. Courtship is straightforward: the male approaches the female, neck stretched out, and grips the carapace of the latter with all four limbs; then he lays his head upon that of the female to bring her to a state of acceptance. Copulation has been observed in September, October, and November. The eggs are ovoid and about 23 × 26 mm in size.

Protection. The numbers of this species seem to be somewhat limited in the natural environment, and it is sought after for the pet trade. This has caused it to be placed in the IUCN Red List of Threatened Species. It is protected in Argentina.

Acanthochelys radiolata (Mikan, 1820)

Distribution. This species is restricted to Brazil, in the Atlantic drainages of the Rio São Francisco, in the provinces of Bahia, Minas Gerais, and Mato Grosso, as far as the suburbs of the city of São Paulo. One record from near the source of the Rio Xingu in Mato Grosso needs to be confirmed.

Description. This is a small turtle (maximum carapace 200 mm), decorated with radiating streaks on the carapace (hence the name *radiolata*), especially in young specimens. The shell is elliptical, very flat, and dorsally depressed between the second and fourth vertebral scutes. The marginals are wide, both anteriorly and posteriorly,

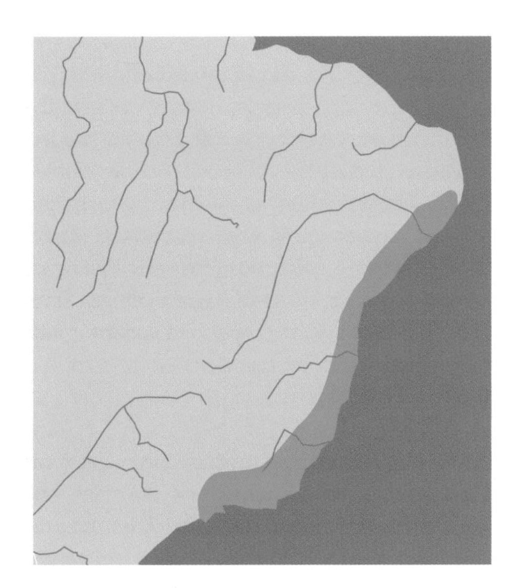

while the midmarginal scutes are narrowed and, in old specimens, form shallow gutters. The posterior marginals are lightly denticulated and notched. There is a modest keel on the supracaudal area. The radiating markings are found on each vertebral and costal scute, extending from the front to the back edges of each one. The general coloration is often dark gray, almost black, or sometimes a very dark green. The plastron and the bridges are yellow, but each scute has a dark spot or sometimes even black streaks. The sutures may have dark borders. The anterior lobe of the plastron is often raised in front and larger than the posterior lobe. The anal notch is very strong.

The head of this turtle is greenish to gray-brown above and yellow on the sides and below. The jaws, which are neither hooked not serrated, are dark yellow, with darker spots. On the top of the head are numerous small, irregular scales. The nose is short, rounded, and slightly pointed, and the irides are white. There is a pair of barbels on the chin. The neck is green or dark brown on the top and is adorned with small tubercles, which become less numerous and smaller posteriorly and ultimately disappear at the base of the neck. The limbs are covered with large scales, and the thighs are equipped with small, pointed tubercles. The upper edge of the limbs is dark green to dark brown, and the underside is yellow. The hatchlings, 40 mm in length, have a brown carapace with marginals spotted with yellow. The plastron is yellow, and each scute has a dark marking that extends to the sutures. The bridges are yellow, with two dark points. Perhaps because of the rather large size of the head, these turtles are sometimes confused with juvenile *Phrynops,* but the presence of large scales on the limbs of *A. radiolata* will always differentiate them.

Natural history. This shy turtle prefers slow-moving waters, oxbows, and impoundments with a soft bottom where it can bury itself. The species has been found in a *tepui* lake at 850 m altitude. It is very nervous and hides at the least disturbance. It is sometimes seen taking the sun or wandering in search of prey. The diet appears to be exclusively carnivorous, and a wide variety of prey is consumed. There are no data on reproduction.

Protection. This species is not in particularly high demand by collectors. There are no data on the status of the different populations. Like other freshwater turtles, *A. radiolata* suffers from water pollution, fires, deforestation, and urban sprawl. Studies on its status are urgently required.

Common name

Brazilian radiolated swamp turtle

This species has a rather short neck, white irises, greenish to bronze colors on the dorsal aspect, and cream to yellow below.

© J. Maran

Acanthochelys spixii (Duméril and Bibron, 1835)

Distribution. This species ranges from the Rio São Francisco as far as southern Brazil and into Uruguay, in the provinces of Rocha and Tacuarembó, and in Chaco and Formosa in Argentina. It has been introduced by humans into western Argentina, at Mendoza and close to the frontier with Chile, at the foot of the Andes Mountains.

Description. The English name refers to the tubercles and spurs of the thighs, which resemble a ball of spines. This is the smallest species of the genus, about 170 mm in length. Elliptical in form, the carapace has a shallow groove running between the first and fifth vertebral scutes. The marginals are wide in front and behind and are narrowed in the midsection. Posteriorly, they are entirely smooth-edged. Each scute of the carapace is sculptured with concentric annuli. The supracaudal is somewhat recurved. The overall coloration is very dark gray to black, and it is possible to distinguish yellow flares along the base of the pleurals. The plastron and the bridges are equally dark gray or black, and some individuals have a light yellow streak along the midline. The anterior lobe of the plastron is somewhat raised in front and is smaller than the posterior lobe, which has an anal notch. The head is gray and covered with numerous scales organized into three or four rows, especially above the tympana. The snout is short and inconspicuous. The jaws are neither hooked nor serrated and are some shade of yellow. The eyes are rather well separated, with white irises. The top of the neck is covered with long, pointed tubercles, which become shorter and less distinct on the sides and below. Two small gray barbels are present on the chin. The limbs are strong and covered with large scales of a dark gray color. The thighs are armed with several rows of pointed tubercles. The feet are well webbed. Sexual dimorphism is poorly developed, but the males have a slightly concave plastron and a thicker tail than the females.

Natural history. This turtle lives in swamps, in lagoons bordered by dunes, and in ponds and lakes, often with heavy vegetation. The details of reproduction have not been described. Even during sunny periods, it does not leave the water but merely ventures into the shallows, where it remains for long periods, its carapace partly exposed above the water. The diet includes snails and tadpoles, but some vegetable matter is also consumed. This turtle often has numerous skin parasites. Copulation has not been described, but the females may nest a good distance from the water, laying about seven rough-shelled, almost spherical eggs, about 25 to 27 mm in diameter. The incubation period may be about 150 days but varies according to latitude. The vertebral depression is not evident in the young; indeed, certain individual juveniles, by contrast, may show a slight median keel. There are radiating striae both on the carapace and on the plastron.

Protection. It is unfortunate that many of the areas where this turtle lives are being heavily developed, with conversion of habitat both to urbanization and to deforestation, as well as the development of industrial fishing ports and large sporting complexes. IUCN is worried about the status of this rare species, which is currently being captive-raised as a conservation measure at the Taim Ecological Research Station in Rio Grande do Sul, Brazil, and in the Santa Teresa National Park in Uruguay. These projects envisage the creation of refuges and reserves in Argentina and in Brazil, similar to the Lagos dos Patos (Duck Lake) wetland in Brazil, that will assure better protection for this species.

Common name

Spiny-neck turtle, black spiny-necked swamp turtle

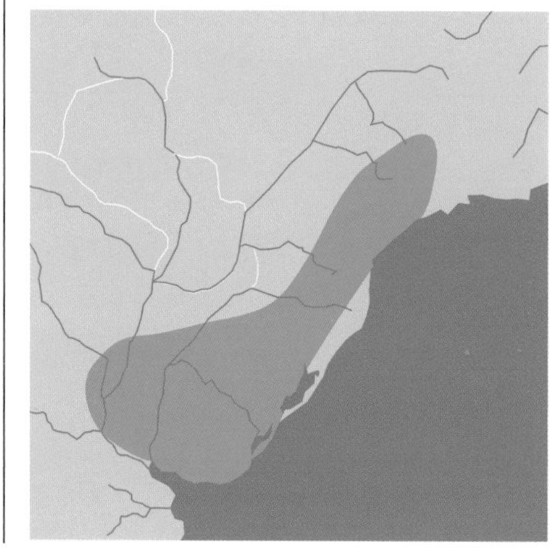

© F. Bonin

In this photo the pointed spurs on the neck can be seen, but the forelimbs and hind limbs are equally equipped with "spines," doubtless for protection from predators.

Batrachemys (Phrynops) dahli

(Zangerl and Medem, 1958)

Distribution. This species is endemic to Colombia and is found only in the valley of the Río Sinu, in the region between Montería and Sincelejo, in the state of Córdoba.

Description. The English name is descriptive; *P. dahli* has a large, toad-like head. This turtle is rare and poorly known. Of medium size (210 mm), the carapace is oval, widened behind, and of a dark olive-brown color. Some adult specimens show a slight median keel. The plastron is large, with the front lobe wider than the hind lobe, and the latter includes a distinct anal notch. The bridges are wide, cream to yellow in color, and lightly pigmented with gray along the sutures (as are the underside of the marginals and the plastron itself). The head is large, wide, and powerful, with a prominent snout. The upper jaw has a slight hook, and there are two light-colored barbels on the chin. The top of the head is dark gray-brown, and the underside cream to yellow. The eye has a dark horizontal bar that continues along the side of the head but disappears at the neck. Males are smaller than females; their plastron is narrower, and their head less enlarged.

Natural history. The species occurs in shallow forest ponds and slow-moving creeks and is not found in rivers and lakes. It may cross long distances of dry land, and it estivates in the forest un-

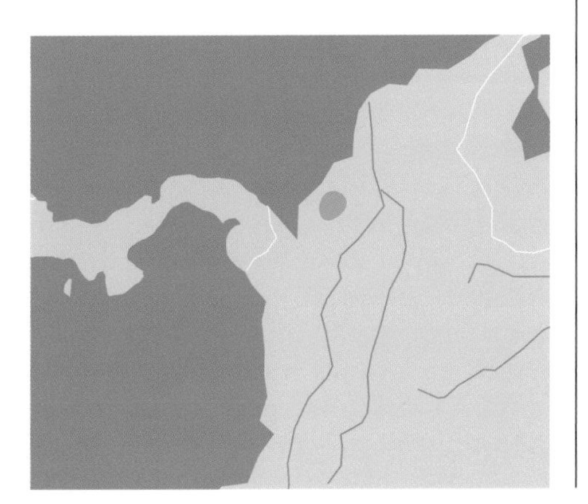

Common name

Dahl's toad-headed turtle

der dry leaves. It is a bottom feeder, chasing after snails, worms, other invertebrates, and small fish with its neck stretched out and snout pointing down. The diet seems to be entirely carnivorous and includes carrion. The courtship is poorly studied, but copulation occurs in June to July and nesting in September to October, with a maximum of six white, elliptical eggs, averaging 32 × 26 mm in diameter.

Protection. The people of Colombia apparently do not consume this species, but it is threatened by deforestation and fires created by the endless hunger for new pastureland. IUCN is drawing up plans to study and protect this species. The Instituto Roberto-Franco in Villavicencio, not far from Bogotá, has attempted captive breeding and head-starting programs, as yet without definitive outcome. The lack of political stability in Colombia makes it difficult to develop long-term research plans. This species is included on the list of 24 "most endangered" turtle species in the world, drawn up in 2002 by the Turtle Conservation Fund.

Batrachemys heliostemma

(McCord, Joseph-Ouni, and Lamar, 2001)

This is a recently described species, listed by Anglophone writers as a "toad-head." In early 1990, William W. Lamar, working near Iquitos in the Amazon drainage of Peru, noticed that some of the *Phrynops* he collected were quite different from others from the same area. At the same time, Roger Bour found a young specimen in a forgotten corner of the storerooms of the Natural History Museum in London. The scientific name derives from the Greek *helios* (sun) and *stemma* (crown), in reference to the luminous bands that mark the head and neck of the species.

Distribution. This species occurs in Ecuador, Peru, the southwest of Venezuela, western Brazil, and southern Colombia and in the upper tributaries of the Amazon Basin (including the Rio Barra, the Rio Tapiche, and the Rio Blanco).

Common name

Western Amazon toad-headed turtle

Description. This is a medium-sized turtle, without marked sexual dimorphism. Males measure 200 to 310 mm, while females measure 250 to 310 mm and are differentiated solely by their shorter, thinner tail. The oval carapace is very dark brown to black and has a slight vertebral depression. The posterior marginals are slightly upcurved, forming shallow gutters. The coloration of the head, although visible only in young specimens, is the feature that distinguishes this species from *B. raniceps,* with which it is sympatric. In the young *B. heliostemma,* a wide, light yellow band originates at the upper lip, runs along the nasal region, and divides into a V behind the eyes, the branches extending to the base of the neck. A small yellow spot also occurs at the tip of the upper jaw, conspicuous against the dark brown of the skin. The chin bears two light-colored barbels. The iris is dark, with no dark crossbar. The upper jaw has a small cavity into which the point of the lower jaw fits perfectly. The plastron, somewhat narrowed, is dark and bordered with yellow, like the underside of the marginal scutes and the outline of the carapace. There is a large anal notch.

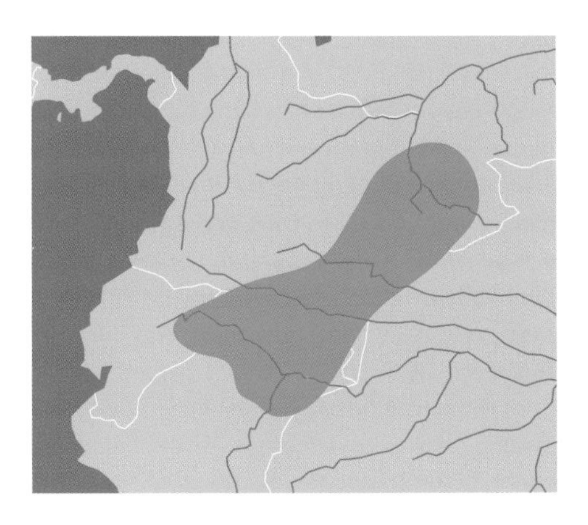

Two color morphs are identifiable among the juveniles: some are completely dark, whereas others exhibit the color pattern described above. The upper and lower surfaces of the forelimbs are uniformly dark, almost black, as is the upper surface of the hind limbs. The latter, however, lighten up to a yellow color in the femoral region. The neck is dark and bears fine tubercles. These bright colors are progressively lost as the animal grows and ages. The five digits of the forelimbs are armed with black claws, although only four of the digits of each hind limb are so marked. All four limbs are strongly webbed.

Natural history. We know very little about the behavior of this newly described species. It lives in sympatry with many other chelids within its overall range and doubtless finds a slightly different ecological niche in different areas. It has been observed in flood-free Amazonian forest, but only close to creeks or forest ponds. It seems to prefer shallow, clear water rather than deep and turbid ponds. The diet appears to be carnivorous (dead fish, etc.). It actively hunts at night. When captured, it sometimes secretes a powerful musk by way of defense.

Batrachemys (Phrynops) nasutus

(Schweigger, 1812)

Distribution. The range extends from Suriname into French Guiana and with a narrow extension into Brazil along the Oyapock River.

Description. Along with *Phrynops geoffroanus,* this species was described a long time ago by Schweigger (1812). It has long been considered as a subspecies *(P. nasutus nasutus)* to distinguish it from *P. nasutus wermuthi* (Mertens, 1969), a form currently known as *Batrachemys (Phrynops) raniceps* (Gray, 1855). This is a medium-sized species (250 mm), but Peter Pritchard measured a very large specimen from Suriname at 317 mm. The carapace is oval and somewhat widened toward the rear. In some old animals, one may see a vertebral depression along the three central scutes. The coloration is blackish brown, without markings. The plastron is large and of a light yellow color, which extends to the lower side of the marginal scutes. It can become almost black, but the sutures and the rim remain light. The anterior lobe is slightly larger than the posterior one, and there is a marked anal notch. Sexual dimorphism is slight, but the males do have slightly concave plastra. The head is very wide, with large, dark eyes circled with white. The head scales are distinct. The upper surface of the limbs and of the head is gray-brown, and the underside yellowish. There is a dark band on the neck and chin. There are two light mental barbels. The limbs are somewhat small and weak; each forelimb has five fine, pointed claws, and each hind limb has four. Tubercles are present on the posterior of the thighs.

Natural history. This species has been spotted by few observers, its distribution being restricted and its habitat difficult to prospect. As with other turtles of the region, it apparently does not estivate, and it is at home both in swiftly flowing waters and in waterholes in the forest. It is never seen basking or even walking away from the water. In

© F. Bonin

This small specimen shows the cream-colored plastron, with its extensive, asymmetrical black markings.

Common name

Common toad-headed turtle

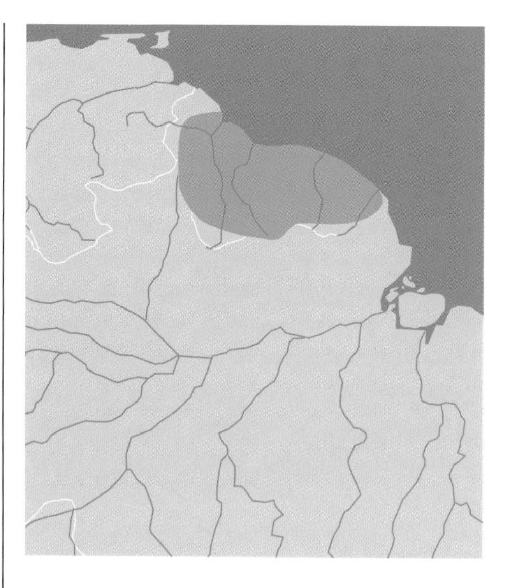

contrast to what some writers have stated, it is not in the slightest bit aggressive when captured. It has been recorded in French Guiana at Saint-Laurent-du-Maroni, Attachi-Baca, and the place called La Crique Voltaire. It is frequently found with leeches attached. The diet has often been characterized as carnivorous (crabs, beetles, etc.), but certain individuals have been seen to regurgitate philodendron seeds. This would tend to place this turtle as an opportunistic omnivore, perhaps because the habitats that it frequents are rather short of prey. The female lay six to eight eggs, about 25 mm in diameter. The juveniles, which emerge from nestings primarily in July and August, present a brown carapace, lighter than that of the adults, with a moderate median keel.

Protection. The status of this species is not well defined, but it is not protected at the international level, nor at the national level in French Guiana. The only regulation is a ban on export.

The recent discovery (2005) by Roger Bour and Hussam Zaher of ***Mesoclemmys perplexa*** in the Parque Nacional da Serra das Confusões in northeastern Brazil adds to the known diversity of this genus.

Batrachemys (Phrynops) raniceps

(Gray, 1855)

Distribution. The range of this species is very wide, extending from the mouth of the Amazon in Brazil to southern Venezuela, the eastern part of Colombia (Guainía, Vaupés), eastern Peru, northeastern Ecuador, and the north of Bolivia (Pando). Nevertheless, Roger Bour advises us that this distribution is not homogeneous and could well encompass isolated colonies that may ultimately be recognized as distinct subspecies.

Description. Formerly known as a subspecies of *Phrynops (Batrachemys) nasutus,* this taxon was promoted to species level in 1987 by R. Bour and I. Pauler. It is very similar to the species mentioned above. This is a turtle with a flat carapace, dark gray to blackish in color, and 250 to 320 mm in length. There is no vertebral keel; by contrast, adult animals have a shallow depression in the three midvertebral scutes. The plastron is rather large and is variable in color, but the ground is generally light yellow, decorated with dark areas clustered mostly toward the rear. This coloration intensifies as the turtle become older. The head is

© R. Bour

The species of *Batrachemys* (formerly *Phrynops*) are sometimes difficult to distinguish from each other. This turtle has no median keel. The head is wide and flat, with a well-developed temporal region. The eyes are located high and up front.

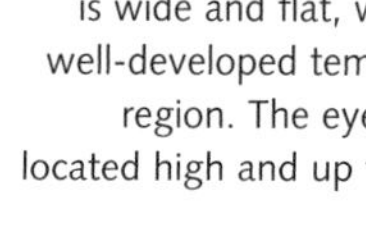

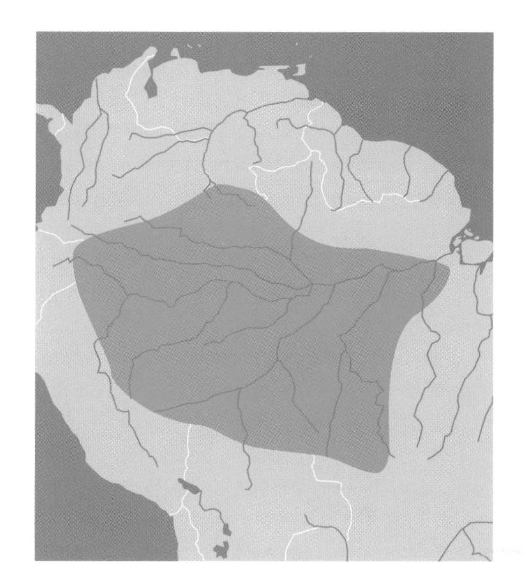

wide, nearly rounded, and the temporal muscles are very developed and enlarged toward the back of the head. The top of the head, gray to black, has an irregular scale pattern. Two black lines extend from the tip of the snout to the tympanic region, separating the dark colors of the upper part of the head from the lighter hues of the chin and neck. The snout is slightly pointed, the eyes dark, large, and placed very high. The upper jaw has a slight notch and is yellow, like the lower one. The chin, always marked with dark pigment toward the center, has two small light-colored barbels. The neck is often covered with small blunt tubercles.

Natural history. This species prefers forest ponds, creeks, slow streams, and even muddy swamps. It is certainly carnivorous, but its biology needs much more investigation. It lays six to nine eggs in a shallow nest. The young have a slight keel on the carapace and measure 58 to 60 mm in length. They have a dark hourglass-shaped mark on the top of the head.

Protection. Certain writers believe that this species is not often eaten by people, but its distribution is sufficiently wide that the species surely encounters human populations that would not hesitate to eat it. Peter Pritchard explained that it was subject of a directed fishery in Peru, where even young individuals were consumed, shell and all.

Common name

Amazon toad-headed turtle

The carapace is wide and very flattened, with a depression marking the first three vertebral scutes. The dark color and subtle appearance of the animal make for excellent camouflage in the swamps and creeks where it lives.

Batrachemys (Phrynops) tuberculatus (Luederwaldt, 1926)

Common name

Tuberculate toad-headed turtle

Distribution. This turtle is endemic to Brazil and confined to the northeast, in the provinces of Ceará, Pernambuco, Bahia, and Minas Gerais. It is widely distributed throughout the Rio São Francisco system and several coastal rivers farther south.

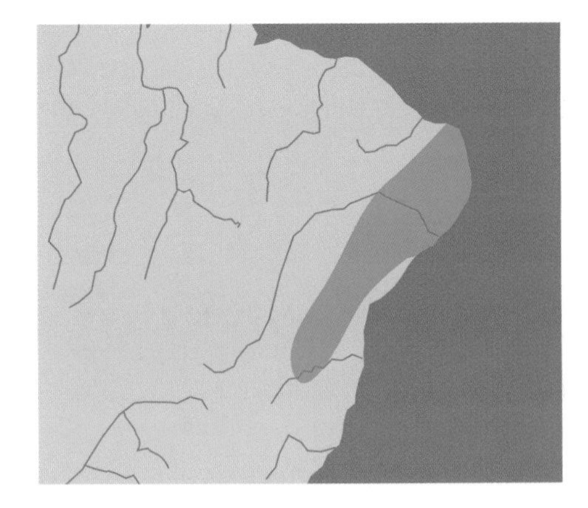

Description. The carapace of this species is slightly less flat than that of the other *Batrachemys* and may have a very sight median keel, bordered on each side by a shallow longitudinal depression. The marginals are slightly upcurved above the tail. The shell is grayish to dark brown, sometimes becoming almost black. The surface of the shell often has a rugose appearance, by reason of the concentric markings on each scute. The plastron is well developed. The two lobes, of which the anterior one is very rounded in front, are about equal in size, and the entire plastron is larger than the outline of the carapace. The plastron is yellowish in color, like the ventral surface of the marginal scutes, and has a dark gray median suture. This coloration darkens with age. The anal notch is very well developed. The head is gray above, and the scales have dark spots outlined with yellow and pink. The region between the eyes is covered with smooth skin. The jaws are light yellow, with a lighter band on the upper one. The underside of the neck is lighter than the upper side. The neck is covered with the conical tubercles that give the species its name, but in certain individuals they may be lacking. The head is very wide, with bumps in the tympanic region, which are covered with irregular scales.

Natural history. This turtle seems to prefer slow-moving waters. In diet it is strictly carnivo-

This specimen hardly shows the conical tubercles that are present in certain individuals and give this turtle its name. On the other hand, the temporal region is expanded and has numerous irregular scales as well as light markings.

rous, and in captivity it prefers earthworms, insects, and fish. Females lay between 5 and 10 eggs, about 31 × 25 mm in size.

Protection. Although present in a densely inhabited area of Brazil, namely the eastern coast, this turtle has not been studied. Serious studies of its biology and conservation status are necessary. In its natural environments, one can imagine that it is subject to numerous threats, especially habitat degradation.

Batrachemys (Phrynops) zuliae

(Pritchard and Trebbau, 1984)

Distribution. This endemic species has only been found in a small area in northwestern Venezuela, where P. Pritchard considers it to be a relictual form. This tiny territory is bordered to the south by the Andean foothills, to the west by the Sierra de Perijá, and to the east by Lake Maracaibo. To the north is a lowland region that was formerly well watered but today is completely arid.

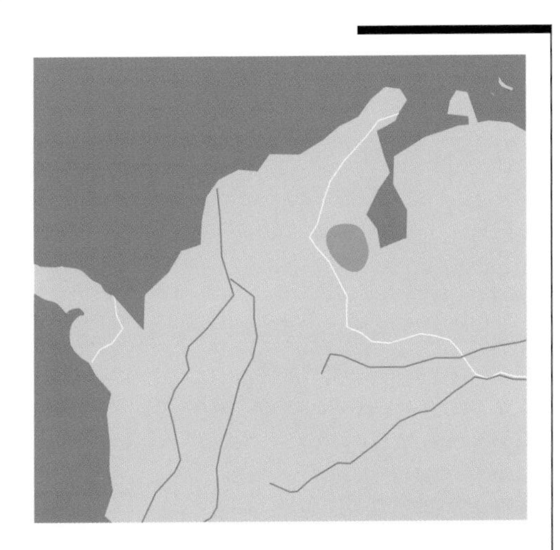

Common name

Zulia toad-headed turtle

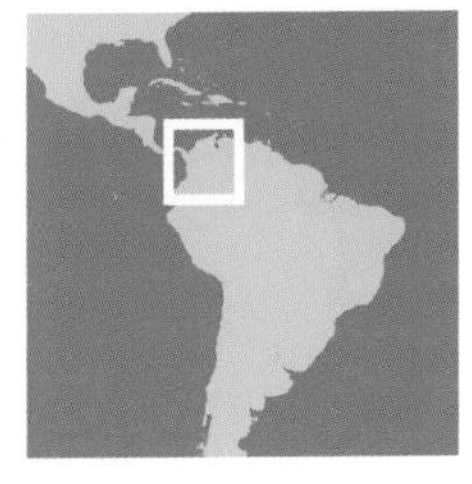

Description. This is a turtle of medium size; the males do not exceed 208 mm nor the females 279 mm. The carapace is oblong and unkeeled, and there is a slightly flattened area between the

© F. Bonin

There is no midline keel in this species. The carapace is wide, expanded, and flat, and the neck bears a large number of small round tubercles. The overall effect is one of neutral gray, allowing the animal to disappear into the background of its habitat.

center of the second vertebral scute and the fourth vertebral. The carapace of males is widened toward the rear, but this is not the case in females. The coloration is uniformly dark gray. The anterior lobe of the plastron is wider than the posterior one, but the whole structure is somewhat reduced in comparison with those of other *Batrachemys*. The bridges are narrow, and the anal notch is very distinct and sharply angular, forming two points at the rear of the animal. The plastron is yellowish, ornamented with several brown stains that also occur under the marginal scutes. The head is very large and may reach a width equal to 30% of the length of the shell. In color, the head is dark gray on top and has dark lines extending from the point of the snout to the tympanic region and passing right through the eye. The throat is lighter and has a slightly granular texture, although the top of the head is covered with small, irregular scales. The chin is provided with two small barbels. There are five claws on each forelimb and four on the rear limbs, and the digits are short and webbed. The males have a smaller head than the females, as well as a narrower intergular scute and a thicker tail. Some individuals have a pinkish orange tinge to the scales of the limbs.

© R. Bour

Natural history. This turtle lives in humid areas and quiet watercourses. One is equally likely to find it on land, as it seems to take frequent walks, particularly in forested areas. It has the habit of burying itself very deeply in moist soil or in mud, and agricultural workers are likely to find it in the course of their work. The population size of this relictual species is difficult to estimate. It is not sought after for food or commerce, in that its homely appearance and foul smell discourage capture. In diet, we assume it to be a complete carnivore. Seven to eight eggs per clutch are laid, measuring about 31 × 37 mm.

Protection. It is difficult to say whether this species is threatened. It may well be attacked, on occasion, by one of the two local crocodilian species, because scratch marks made by teeth have been observed on the shell. It has been listed by IUCN mainly because studies are needed so that we can have a better understanding of its status and overall range.

Bufocephala (Phrynops) vanderhaegi (Bour, 1973)

The photo shows the extraordinary "toad head," flat and wide—and powerful. The eyes are on the upper surface, and the wide mouth reinforces the toadlike impression.

This is the sole species in a new genus, *Bufocephala,* from the Latin *bufo* (toad) and the Greek *kephalê* (head). The species is named in honor of Maurice Vanderhaege, a turtle hobbyist and friend of Roger Bour's.

Distribution. This turtle occupies a range delimited by the Rio Paraguay Basin (Mato Grosso) of Brazil, up to the northeast of Argentina and to Paraguay (Rio Paraná).

Description. The males reach 220 mm, and the females 262 mm. The oval carapace is slightly domed, with the highest point at the level of the third vertebral scute. A slight flattening may be detectable along the midline, and there may be a

faint median keel. The coloration is dark, almost black, sometimes with some reddish tints. The plastron is rather large, the anterior lobe being larger than the posterior and curved upward in front. The head is rather small and has a pointed snout. The tympana are less conspicuous than in *Batrachemys*. The top of the head is black, although the chin and the jaws are light gray, with perhaps a degree of orange or pink vermiculation. The throat is pale in some individuals but is black in others. The chin bears a pair of small, light-colored barbels. The iris is dark, encircled by silvery gray. The neck is covered with small, slightly projecting tubercles. The limbs are gray or brown and are small. The forelimbs have five claws; the hind limbs have just four, but they are larger. Apart from the slight difference in size, sexual dimorphism is very limited.

Natural history. This species inhabits shallow lakes and ponds with dense vegetation but may also be found in forested areas and in areas cleared for agriculture. In diet, this turtle is partially carnivorous. It moves around by day and may be seen moving rapidly on dry land. In contrast to *Mesoclemmys (Phrynops) gibba,* which it closely resembles, this species is of an irascible disposition, and both the adults and the juveniles try to bite when picked up. One nesting has been observed in Paraguay; the elliptical eggs, three in number, measured 35 × 28.5 mm.

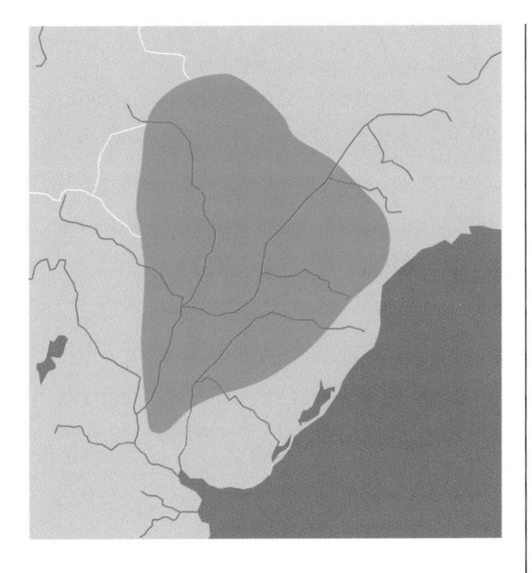

Protection. This turtle is doubtless consumed by the local people in the three nations that it inhabits. IUCN classifies it as a priority species.

Common name

Vanderhaege's toad-headed turtle

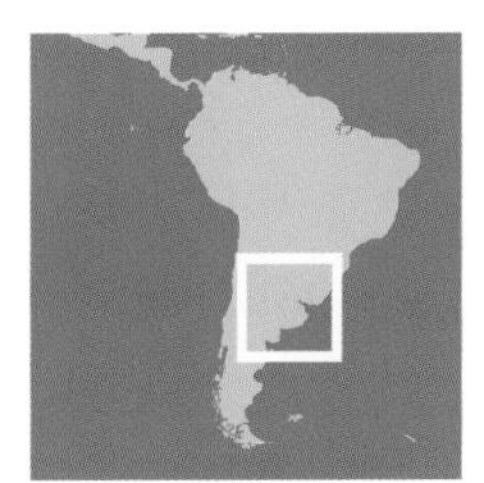

© R. Bour

Here the two chin barbels in this species can be seen, as well as its large eyes, placed very high on the head. The head and neck are powerful and are well protected by the thick scales and the numerous tubercles.

The Australian Chelid Species

John Cann, after diving in a small creek near Sydney.

© B. Devaux

John Cann is without question the top specialist in the turtles of Australia. The Australian herpetologist, in his mid-sixties, is a former Olympic medalist, diving champion, and snake handler, and he has traveled all over the immense continent of Australia to make observations on poorly known reptiles. He has discovered and described numerous turtles in his home country, including Rheodytes leukops *and* Elusor macrurus, *both of which were so distinctive that they had to be placed in new genera. His photographic skills are extraordinary, and the appearance of his book* Australian Freshwater Turtles *in 1998 was a major event for everyone interested in turtles. But the systematics of the chelonian fauna of Australia are not yet finalized, and in the years to come more species and subspecies, many already known to and observed by John Cann, will be added to the known fauna of this happy continent blessed with a myriad of freshwater turtle species.*

The Australian turtles form a distinct and very ancient group, and all except one are pleurodires. Only *Carettochelys insculpta,* the pig-nosed turtle, is a cryptodire. Today there are no land turtles (or tortoises) in this great continent; the last known terrestrial chelonians from Australia, of the genus *Meiolania,* became extinct about 2,000 years ago *(Meiolania mackayi* of New Caledonia, from the Koné period). These turtles were very large and had skulls uniquely equipped with a pair of massive lateral horns.

At the present time the Australian chelonian fauna includes 17 species and several genera, most of which were quite recently discovered. The work of Harold Cogger in the 1970s and the subsequent publication of *Australian Freshwater Turtles,* by John Cann, give us an encyclopedic knowledge of the turtles of this continent. The taxonomy of Australian turtles will continue to evolve. Species and subspecies have already been recognized, although they have not been described or named, and it is not unlikely that completely new ones will yet be found, given that this continent is vast and not all of its rivers have yet been inventoried. In this work, we discuss only those species accepted by the top specialists at the time of preparation of this publication in 2004.

In Australia, one finds turtles totally different from those elsewhere on the planet, a phenomenon resulting from the isolation of this territory due to continental drift. Many of the Australian Chelidae have extremely long necks, which may equal or even surpass the length of the carapace *(Chelodina oblonga).* Relatively recently, a turtle bearing a huge, thick, ossified tail that attains half the length of the entire animal was discovered (*Elusor macrurus,* Cann and Legler, 1994). In the northwest of the country lives the strange pig-nosed turtle *(Carettochelys insculpta),* which has characteristics that seem to place it somewhere between the Trionychidae and the marine turtles. We also note *Pseudemydura umbrina,* which is in the subfamily Pseudemydurinae and is the smallest turtle in Australia (150 mm). *Pseudemydura umbrina* presents a textbook conservation story, having been reduced to a mere 50 individuals by 1953, making this the rarest turtle in the world. The ethology of Australian turtles is similarly unique; one turtle *(Chelodina rugosa)* actually lays its eggs underwater, and another digs its nest cavity with its forelimbs instead of the back ones *(Pseudemydura umbrina).*

Symbolism. The Australian turtles form an intimate part of the imagination of the original people of the country, the aboriginals. Numerous artifacts reflect turtle themes—gravestones, paintings, sculpture. The turtles are often rendered in a style or method known as anatomical, because the image of animals may show the interior of the body, with the muscles, the internal organs, and so on, demonstrating a precise knowledge of their physiology. Furthermore, the turtles have never been considered as taboo or sacred animals; they have always been—and still are—widely consumed by the aboriginals, and their representation in art allows the people to appropriate the special qualities of the turtles before they hunt and consume them.

Protection. All turtle species are protected by Australian law, and they may not be caught, sold, or consumed, but the aborigines benefit from a legal exception that allows them to continue with traditional exploitation of turtles. Nevertheless, this capture pressure is minimal and does not endanger the turtle populations. Certain of the species found in Australia also occur in Papua New Guinea and in West Papua, where they are more heavily exploited. Most of the turtles from these regions that find their way to Western countries are collected in New Guinea. The main threats within Australia itself come from introduced predators, including foxes and buffalo. The foxes are one of the scourges of Australia and are among the worst killers of turtles. John Cann also tells us that the overgrazing of pastures by the buffalo and the destruction by these animals of riverbanks and wetlands, especially in the far north of the country, are also causes of decline of certain species, through the process of nest destruction. Fires are devastating too, as is increasing urbanization, both around Perth and in the eastern part of the country, and development of agricultural lands and proliferation of mining activities contribute to habitat destruction and to turtles becoming ever more rare and depleted. And the extreme rarity of some species is the very factor that attracts Western collectors, with the result that these populations become increasingly fragmented and endangered.

Chelodina expansa Gray, 1857

Distribution. This species occupies a wide, boomerang-shaped range in the southeast of Australia, including the Murray River up to mid-Queensland (Rockhampton). It occurs far from the coasts almost throughout its range, apart from the extreme east—there it can be found on Fraser Island, where possibly it was brought by human agency.

Description of the genus *Chelodina.* These turtles are all pleurodires and have very long necks, the most extreme being *Chelodina oblonga*. They are of medium size, but one species *(C. expansa)* reaches a length of 480 mm. The genus is noteworthy for having only four claws on the forelimbs—not five, as in other Australian turtles—and for having the two gular scutes meet each other in front of the large, hexagonal intergular. When they are disturbed, these species can secrete a milky, musky liquid, which is expressed from four plastral glands, two in front and two at the back. These are mostly aquatic turtles that do not leave the water except for egg-laying purposes, in contrast to *Emydura* and *Elseya*. They are diurnal and carnivorous (fish, mollusks, crustaceans). They live in shallow waters and wetlands, and in slow-flowing creeks. The clutch consists of a dozen eggs, on average, of elongate shape. As with all long-necked turtles, they rest on the bottom of the water with the shell immobile and hidden by leaves or mud, and only the neck moves when the turtle strikes at its prey.

Description of *C. expansa.* This is the largest species of *Chelodina,* reaching a maximum length of about 480 mm. The carapace is flat and is expanded toward the rear, as the name *expansa* implies. In old animals the carapace may be flattened or depressed toward the fourth vertebral scute. John Cann reports a specimen weighing as much as 10 kg. The plastron is narrow and corre-

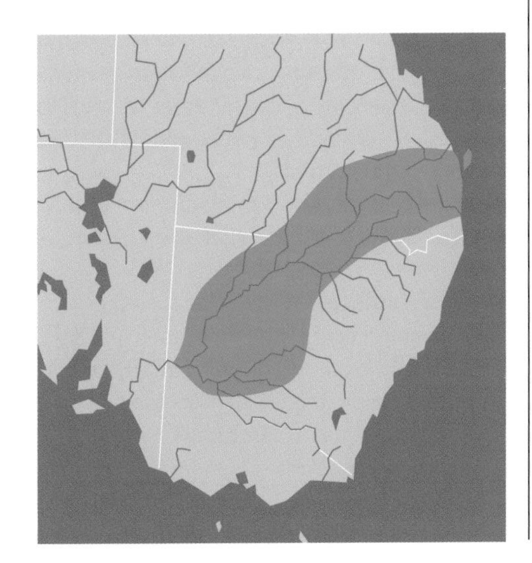

Common name

Broad-shelled river turtle, giant snake-necked turtle

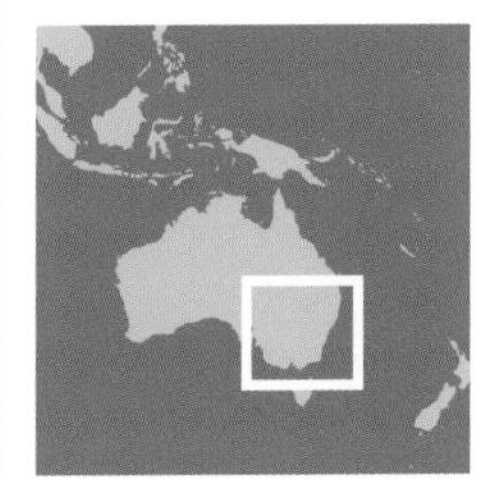

© G. Kuchling

This monstrous animal shows its long, snakelike neck and its very flattened, wide shell, with strongly expanded marginal scutes.

sponds only to about one-third of the surface of the carapace. The head is both wide and flat and perhaps even strongly depressed between the eyes and the rear of the skull. The eyes are small and placed very far forward on the head. The neck is very long, wider and more muscular than that of *C. longicollis*. There are no neck tubercles, but the skin of the neck is somewhat granulated. The overall coloration is dark brown to greenish, with fine dark reticulations visible on the lightest-colored specimens, that is, the young ones. Individuals from Fraser Island are black, with a light line along the marginals. In nature, the animals are often covered with mosses and algae, which conceals them completely in their substrate. The plastron is whitish to brownish, with no markings or dark pigment. Under the chin there are two rudimentary barbels, rarely four. Juveniles are a lot like *C. longicollis,* being black, with light spots on the marginals and on the plastron, but these spots are whitish, whereas they are more orange in *C. longicollis,* and the head is flatter and longer.

Natural history. This species has an extraordinarily long incubation period of up to 664 days, although this single maximum observation may be atypical and extended by the liberating rains coming late and by delays caused by difficulties with emergence from the nest. The usual incubation period is more like 192 days. The young grow rapidly in the first year, as do many Australian turtle species. It is difficult to observe this species in the wild, because it lives in the mud and under layers of dead leaves and other components of the substrate, not far below the surface. Concealed in this way for long periods of time, it awaits the passage of prey swimming past its door. Underwater movement is unique: it walks along the bottom, with long strides, the neck stretched out in a straight line, and it does not swim. When it is disturbed, it buries itself in the bottom substrate, kicking up a great cloud of detritus that effectively conceals it. If it is picked up, it thrusts its heavy neck and head rapidly from side to side, as if to upset the grip and balance of whoever is holding it. It lays between 5 and 30 oblong eggs, about 35 × 25 mm in size, sometimes as much as a kilometer from its normal aquatic habitat. It feeds upon snails, small fish, and shrimp. On Fraser Island it seems to prefer small crayfish.

Protection. Its great size and easy adaptation to captivity make this species a favorite among hobbyists. In nature, it is hard to census populations because they hide very effectively, making precise counts difficult or impossible.

Chelodina kuchlingi Cann, 1997

Distribution. This species seems to occupy a small area in the King Edward and Carson Rivers at Kalumburu, within the Kimberly region of Western Australia.

Description. This turtle carries the name of a renowned Perth herpetologist, Gerald Kuchling. The species is known from only a single specimen. The oval carapace is 230 mm in length. It resembles that of *C. rugosa* but differs in the following points: the shell scutes exhibit radiating sculptured markings, light in color against a dark brown background. The growth rings are very easily visible and are light orange, contrasting with the edge of the scute and somewhat similar to those of *C. novaeguineae*. The oval shell is slightly pinched in front and widened toward the rear. The plastron is orange-yellow. The head, wide and grayish, seems similar to that of *C. rugosa*.

Natural history and protection. No data are available.

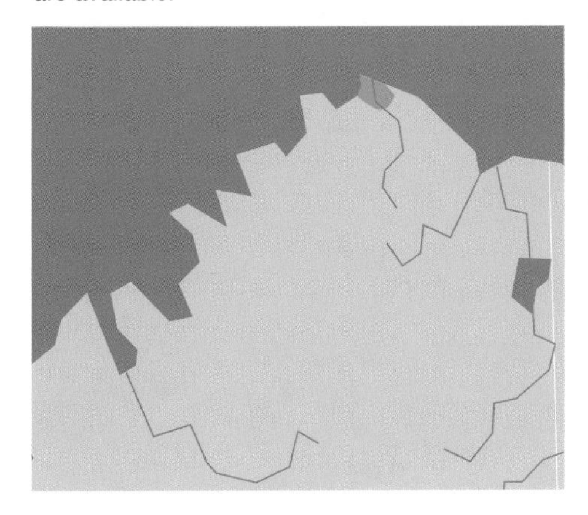

Common name

Kuchling's long-necked turtle

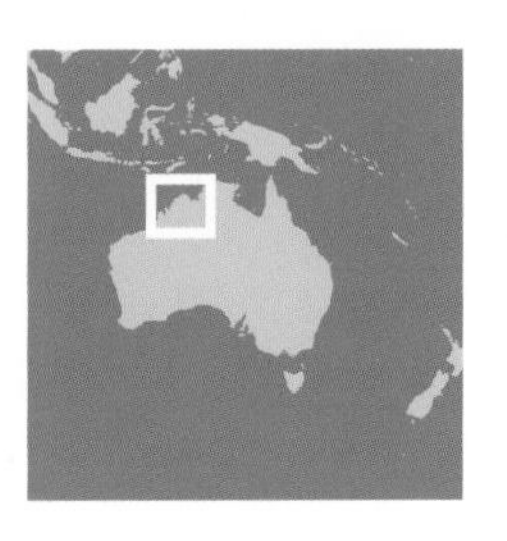

Chelodina longicollis (Shaw, 1794)

Distribution. This species has a wide range in the east and southeast of Australia, from central Queensland (Charters Towers) to as far as Adelaide in the south.

Description. This is a rather small species, with an oval carapace, wider behind than in front and light brown to orangish, sometimes even greenish. One often sees a groove along the middle three vertebral scutes, especially in old animals. Some individuals accumulate a heavy growth of aquatic plants and algae. The marginals may be upcurved into gutters, especially the lateral ones, which become correspondingly very narrowed. The plastron is cream to yellowish, with very broad black bars along the seams. The plastron is narrow—hardly half the length of the carapace—and incorporates a strong anal notch. In spite of the scientific name *longicollis,* the neck is not the longest seen in the genus. The neck is also very slender and covered with medium-sized tubercles, which are rounded, flattened, and regular in their arrangement. The head is long and hardly wider than the neck; is a yellowish or greenish gray, with the underside strikingly yellow; and has a triangular ridge, very smooth, extending from the rear of the skull to the point of the neck. The eyes are located close to the nostrils, and the iris forms a distinctive white ring around the eye. The mouth is short, and the lower jaw whitish to yellow-

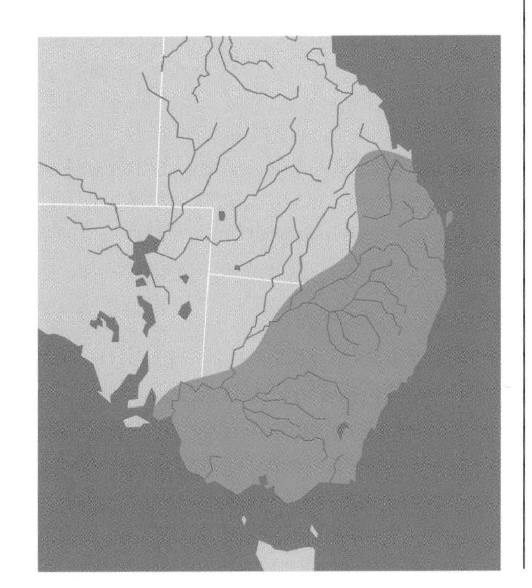

Common name

Eastern snake-necked turtle, long-necked turtle

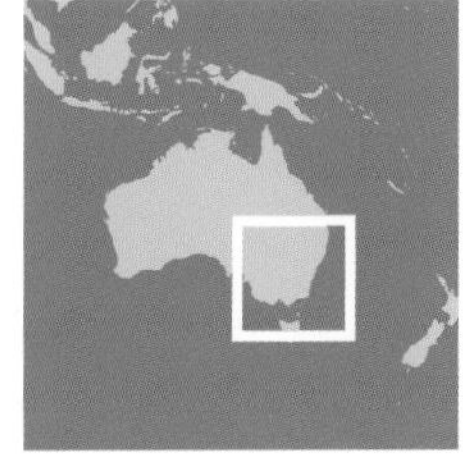

The head of *C. longicollis* is small and is carried at the extremity of the long, threadlike neck that gives the species its scientific name. The remainder of the shell is flattened and wide, but not really elongate.

ish. Sexual dimorphism is minor in this species, with a slight plastral concavity in males, and a minor difference in the location of the cloaca relative to the tip and the base of the tail; coloration and size are the same in the two sexes. The hatchlings are almost entirely black, but they have orange or reddish markings on the plastron and the underside of the marginals and an orange line along the neck.

Natural history. This is a shy turtle that lives among mats of vegetation and mud at the bottom of ponds and slow-moving streams. But it is also somewhat curious and may be seen extending its head above the surface of the water as well as swimming at the surface. It nests from the end of September to December, laying between 8 and 24 small eggs, 30 × 20 mm in diameter, at a well-drained location near the edge of the water. The female will urinate copiously to soften the ground around the nest cavity. The young first see the light of day in January, February, or March and weigh about 4.5 g. Incubation lasts from 120 days to 6 months. As with all Australian turtles, the onset of rain is the key factor to trigger emergence of hatchlings. Specimens in hand sometimes secrete a very smelly milky fluid, but other individuals will never do this.

Protection. This well-known and curious turtle is considered easy to raise by hobbyists, and it is resistant to maladies and the stress of captivity, so it is quite sought after as a pet. The long neck is also a feature that makes it attractive to turtle lovers. The species is widely exhibited in Australian zoos and is often seen in Western animal collections, suggesting that there is a continuing drain upon wild populations, in that there is no breeding or raising center in existence.

Chelodina novaeguineae Boulenger, 1888

Distribution. This species occurs primarily in the south of Papua New Guinea and West Papua, but it also occurs in northeastern Australia, from the Cape York Peninsula to the Bowen area, and on the Gulf drainages of western Queensland, around Mackay. A closely related form, *C. mccordi,* occurs on the island of Roti, near Timor, and another member of the complex, *C. pritchardi,* occurs in extreme eastern New Guinea.

Description. This species is very close to *C. longicollis* but is a little bigger (300 mm) and more circular in shape, with a carapace wider behind and a larger head, an obvious rounding of the prefrontal region, and a wide maxillary surface. The coloration is often brown, orange, grayish, or sometimes blackish, although many of the variations reflect the different habitats from central New Guinea to Queensland in Australia. Some animals have strongly defined patterns of radiat ing streaks extending from the areola of each scute. The tubercles on the neck are more rounded and flattened than those of *C. longicollis*. In old individuals, the marginal scutes at the sides are sometimes recurved upward, although they are, by contrast, curved downward, toward the ground, in young animals. The plastron is uniformly yellow, with fine black sutures and hints of black on the edges of the scutes. The head is brown to orangish or blackish and is rather wide in the temporal area, with the usual long neck typical of the genus, and there is sometimes a depression behind the eyes. The temporal ridge is composed of a regular series of wide, flat scales.

© G. Kuchling

This shy species has a small head and a rather short neck. Note the strongly rounded anterior aspect of the carapace, the vertebral depression, the deep hollow between the eyes, and the pronounced snout.

Certain animals from New Guinea have a vestige of a horny beak on the snout. The young of this species are truly multicolored on the plastron, the neck, and the marginal scutes; these areas may have wide irregular yellow to orange or sometimes reddish markings, the rest of the animal being black. There are also radiating rugosities on the carapace, which disappear with age.

Natural history. This turtle eats snails, which may explain the broadening of the lower jaw, and also consumes insects, shrimp, and aquatic vegetation. In the lagoons of Australia between the Daly and Newcastle Rivers, which are completely dry for part of the year, it estivates for many months, in the same manner as *Pelomedusa* in Africa. It reappears after the onset of the rains. Nesting has been observed in September to October (10–20 eggs, 32 × 22 mm), and incubation lasts about nine weeks. The young weigh about 5.5 g.

Protection. This species is seen more and more in Western animal collections and is heavily collected in New Guinea. The status of populations in both West Papua and Papua New Guinea needs to be examined.

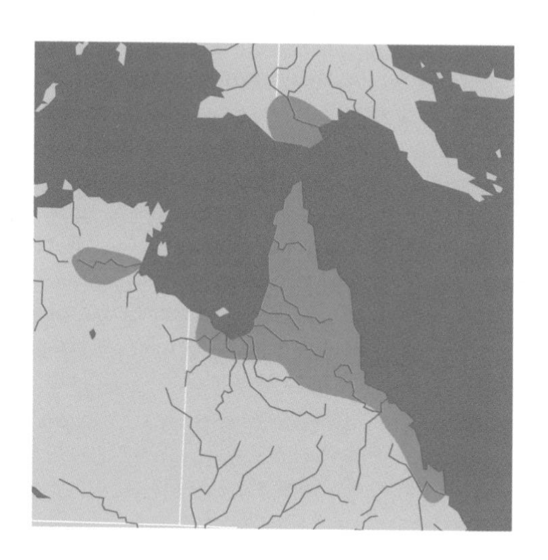

Common name

New Guinea snake-necked turtle

Chelodina colliei Gray, 1856

Distribution. The range of this species is confined to the southwestern corner of Australia, around Perth, from the Hill River to the Fitzgerald. Nevertheless, it is abundant within this area and is found in practically all kinds of bodies of freshwater: swamps, slow-flowing rivers, small lakes, and particularly in the Swan River.

Description. The correct name for this species is still debated. J. Cann prefers the long-used name *Chelodina oblonga,* which is certainly appropriate, but it has been suggested that the type specimen is not in fact the Western Australian species, and this presents some problems. The name *C. colliei* (Gray, 1856) definitely refers to a Perth-area (Swan River) turtle, so we will use this name. This species has an extremely elongated carapace, somewhat wider behind than in front; an extraordinarily long neck, which may exceed the length of the carapace (400 mm for the neck, 350 mm for the carapace); and a wide head. The neck is muscular and thick at the base, and the head alone measures one-third the length of the neck. Even the young have an immense neck, almost as wide and thick as the carapace itself. So it is understandable that this neck cannot be completely retracted between the carapace and the plastron. The texture of the shell is rather smooth, sometimes finely vermiculated, and in nature the shell is often covered with algae and aquatic moss, which helps the animal to pass unnoticed. The coloration ranges from greenish yellow to gray-bronze or dark maroon, with small dark spots in examples with a relatively light carapace. The long head is wide at the rear, and the eyes are placed very near the front. One often sees two slight knobs toward the rear of the skull and a shallow depression behind the eyes in older animals. The color is greenish to brown. The neck as a whole is granular in texture, with small rounded excrescences. The upper side is greenish, the lower yellow. The chin barbels are rudimentary. This species does not secrete foul-smelling fluid, even though it has the usual four musk glands typical of the family Chelidae. The males have a tail that is slightly longer and thicker than that of the female and perhaps have a slightly concave plastron, but in general there is little sexual differentiation. The feet are wide and well webbed, equipped with strong claws. This turtle often travels from one body of water to another in search of food but does not indulge in bottom-walking, as does *C. rugosa.* The young are very attractive; the carapace is almost black, steel gray, or dark green, and the underside is yellow from the throat to the limbs. There are often dark spots on the lighter areas. A bright yellow line runs along the margin of the shell. The neck is as long as the shell, the head wide and lighter in color.

Natural history. This species is nervous and hard to approach. When it is grabbed, it is very lively and may even become irritable; it waves its long neck like a whip and tries to bite. It estivates for several weeks or months each year when the wetlands it inhabits dry up, but the populations in deeper and permanent water remain active throughout the year. The diet is varied, including crustaceans, small fish, algae and aquatic plants, larvae, and frogs. The long neck appears to be an adaptation to the mode of hunting; without moving the body, the turtle extends its neck slowly, like a snake, before surprising the prey and seizing and swallowing it. October and November are the preferred months for nesting. The species nests twice per year, the eggs measuring about 32 × 20 mm and being 9 to 16 in number per clutch. Incubation lasts about 200 days, but the delay is caused by the long wait for the rains; in captivity, eggs hatch in 70 to 100 days. Very thorough studies of this species have been undertaken by Gerald Kuchling at the University of Western Australia in Perth, with the objective of a better understanding of the reproductive cycle. He has

Common name

Oblong turtle

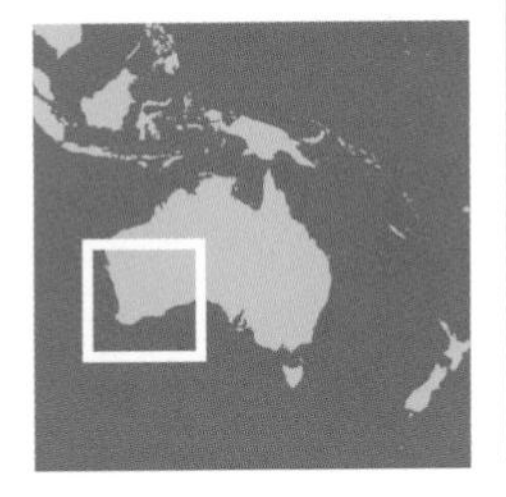

© B. Devaux

In this weird turtle the neck is longer than the carapace, and the latter is flat, elongated, and small. One wonders where the animal could find space to retract its head with such a huge neck.

© F. Bonin

This turtle has a serpentine appearance, with its tiny eyes and its long neck covered with fine tubercles. The throat is very expansive.

used ultrasound and transmitters located behind the animals' legs to visualize the follicular development and gain insight into the biology of the species.

Protection. This species is not hunted much and is rarely included in Western live collections. It remains extremely abundant and seems little affected by habitat alteration.

Chelodina parkeri Rhodin and Mittermeier, 1976

Distribution. This turtle is endemic to the island of New Guinea, in Lake Murray and near Balimo, as well as in the Aramia and Fly Rivers. Unlike some of its congeners, it is not found in the Australian continent.

Description. This species carries the name of Fred Parker, who has collected it in New Guinea. It is quite similar to *Chelodina siebenrocki,* but it is distinguished from all other *Chelodina* by its vermiculated, very bright white, yellow, or light green pattern on the top of the head and on the jaws. Maximum length is 267 mm (in females). The carapace is light brown, with vermiculated markings along the marginals. The plastron is moderately elongated and has a wide anterior lobe. In color, the plastron is yellow, without any special markings. The head is long, wide, and very flat, with a long, thick neck reaching 75% of the length of the carapace. Two tiny chin barbels are present. The males are smaller (about 150 mm), with a tail that is wider and flatter than that of the females. The juveniles and subadults are light brown, the top of the head and the mandibles being covered with little dots and yellow lines. The vermiculated adornment extends to the vertebral scutes.

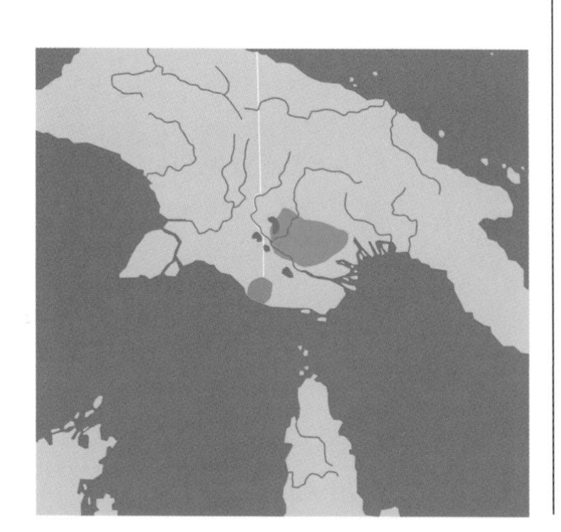

Common name

Parker's snake-necked turtle

© F. Bonin

Natural history. This turtle is found in great lakes and rivers with significant vegetation and well-shaded waters. It may also be found in periodically flooded wetlands, and it may make overland treks to reach permanent water. It often hides itself in the mud and seems to be mostly carnivorous.

The carapace of this species is striped and sculptured, as in the majority of *Chelodina* species before they reach maturity. Under natural conditions, some individuals become covered with algae and moss, which adhere to the rough carapace surface and provide excellent camouflage.

Chelodina reimanni

Philippen and Grossman, 1990

Common name

Reimann's snake-necked turtle

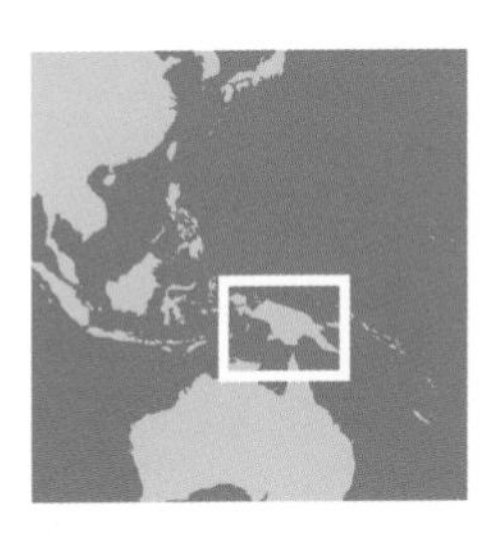

Distribution. This species is recorded only from a minuscule range in southeastern New Guinea, in the Merauke River.

Description. This turtle was discovered by Michael Reimann in 1972. It is quite distinct from other congeners like *C. rugosa* and *C. longicollis*. The head is larger and the neck shorter. The head and neck together attain only about 60% to 65% of the length of the plastron. The head is massive, wider and longer than in other *Chelodina* from New Guinea. The eyes are lower and located somewhat farther forward. It has a distinctive appearance that emphasizes its voluminous head. The carapace is oval, flatter than that of other *Chelodina,* and the scutes have deep, parallel grooves. The first vertebral scute has the same shape as the costal scutes. The holotype has a carapace length of 180 mm. The carapace is very dark brown to black in old females. The head and neck are dark gray to black, the underside of the head and neck being creamy yellow to white. The skin of the head is covered with very small tubercles. The plastron is unmarked and bright yellow to cream in color. The limbs are well webbed, the forelimbs having five claws and the posterior limbs four. The plastron of the hatchlings is bright red, a unique feature in the genus.

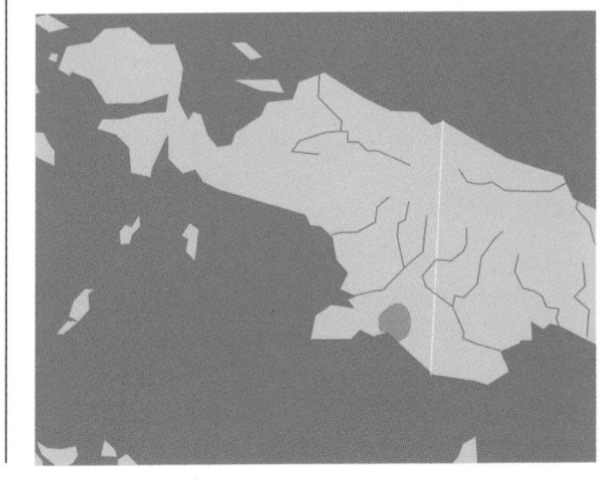

Natural history. The massive head of this species facilitates the consumption of crustaceans, mollusks, and large insects. The females lay 8 to 12 eggs, about 28 × 19 mm in size and 8 g in weight. Incubation lasts for about 62 days at 30°C.

Protection. There have been two episodes of international trade in the species, but no data are available on its status in the wild.

© J. Maran

With its huge head and eyes placed near the front, this turtle has a disturbing air—but it only seeks to dive from its basking place into the water. The carapace is marked with numerous grooves, which also seem to be engraved into the bony shell beneath.

Chelodina rugosa Ogilby, 1890

Common name

Northern snake-necked turtle

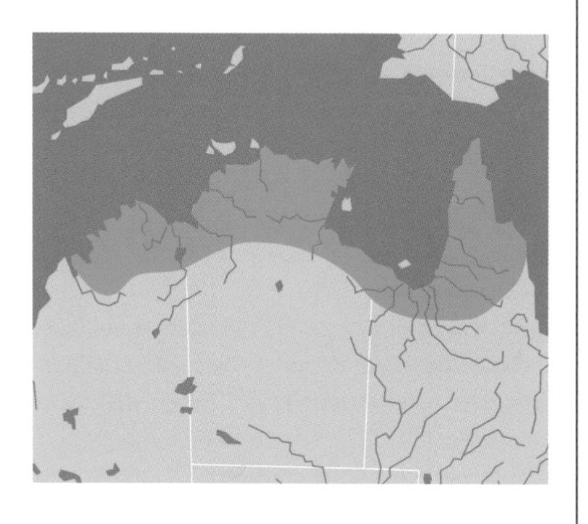

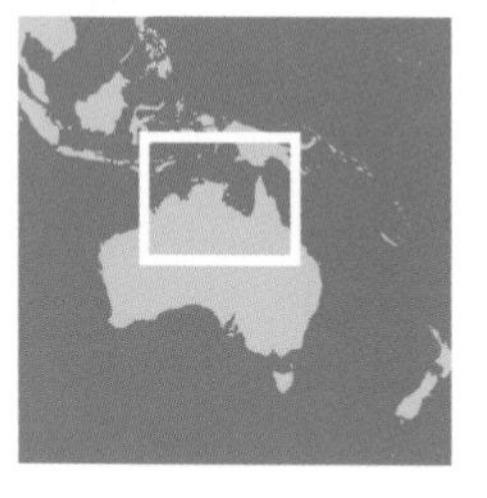

According to Cogger and Cann, this taxon includes *C. intergularis, C. siebenrocki, C. parkeri, C. reimanni*, and *C. kuchlingi*, although Bour recognizes *C. parkeri* (Rhodin and Mittermeier, 1976), *C. reimanni* (Philippen and Grossman, 1990), and *C. siebenrocki* (Werner, 1901) as distinct.

Distribution. This species group is found throughout humid areas of northern Australia, from the Kimberley District east of Broome to Cape York (Cooktown), as well as in New Guinea.

Description. The morphology of this species group is variable, according to locality, which would seem to legitimize the recognition of multiple forms, a topic on which the scientific community is still divided. We shall describe the features common to the group as a whole, while keeping the general denomination of *Chelodina rugosa*. The description is based primarily upon animals from the center of the range, near the Gulf of Carpentaria. The carapace has a uniquely rugose aspect, and the maximum size is probably about 350 mm. The carapace is somewhat elliptical and is slightly wider behind than in front. The color varies from light brown to light green and even bronze, steel gray, and black (in the eastern edge of the range), but it is not usually as dark as that of *C. oblonga*. The head is wide, with a strong depression behind the eyes and with smooth, shiny skin, often with little spots or darker striations on the light background of the cranium and the sides of the head. Two large barbels are sometimes present. The tubercles on the neck are very small and indistinct. The plastron is very narrow and occupies between one-half and one-third of the carapace. The plastron is generally yellow but may have irregular black or reddish spots.

Natural history. The general behavior of this species in its tropical environment is similar to that of *C. novaeguineae*. It rarely emerges from the water and has been little studied by naturalists since the 1990s. Rod Kennett discovered that this species had evolved a technique unique among

© F. Bonin

In this isolated region of the Kimberleys, south of Kalumburu, very large specimens of *Chelodina rugosa* are found. They are also unusually dark, and John Cann is of the opinion that they may be a distinctive local subspecies.

turtles: it lays its eggs underwater! After following the turtles and noting the results of their oviposition, Kennett placed transmitters in artificial eggs that were then replaced surgically in the oviducts of the animal. Seventeen turtles were equipped in this way, and it was possible to follow exactly the strange nesting behavior of *C. rugosa*. At the end of the wet season, when the lake or pond starts to dry up, the turtles nest in water about 20 cm deep, several meters from the edge. The eggs, usually about a dozen in number, have a thick, tough shell and measure about 35 × 26 mm. From the moment of deposition, development of the embryo is suspended. Several weeks later, the water retreats further and the clutch proceeds with normal "dry" incubation. Embryogenesis is thus delayed, and incubation takes three to four months. When the water returns, it contacts the eggs and stimulates the emergence of the hatchlings. This unique trait enables this species, alone among turtles, to adapt to an extremely demanding environment. Furthermore, the turtle can also nest in normal fashion on the margin of a river or lake, and the underwater nesting behavior is actually rarely observed. In other aspects, *C. rugosa* is lively and hunts a great variety of prey. The long, powerful neck and the wide head make it a fearsome predator upon crustaceans, fish, frogs, toads, and aquatic insects. It has even been seen catching snakes. During the hot dry season, it can migrate long distances in search of a water hole, or it may just estivate in the mud of dried-up ponds.

Protection. Hoofstock may damage nesting areas and destroy eggs. Some of the included taxa are hunted in New Guinea and sold to Western hobbyists. The status of populations in New Guinea is poorly understood and should be investigated.

Chelodina siebenrocki Werner, 1901

Distribution. This turtle seems to be endemic to Papua New Guinea, along the south coast, but it might also be found in West Papua and at Cape York in Australia.

Description. This species is named after Friedrich Siebenrock of the Vienna Natural History Museum. It is of medium size (about 300 mm for the females, maximum 380 mm), elliptical in outline, and somewhat wider posteriorly than anteriorly, with a rugose surface to the scutes. In older animals, especially females, a median groove may develop. The head is long, wide, and rather flat (generally similar to that of *C. rugosa*), and it has finely reticulated skin. The neck is long and thin, reaching 75% of the length of the carapace. Four or more chin barbels are present. The general coloration is dark brown to blackish on the dorsal surfaces and yellow to beige ventrally. The plastron is elongate and narrow, with an anterior lobe much narrower than the posterior one. It is cream in color, but sometimes the seams are outlined in brown. Males have a much longer and thicker tail than females do. The hatchlings are covered with numerous small black dots, irregularly arranged, on the lighter background of the animal, and the rugosity of the carapace is already evident.

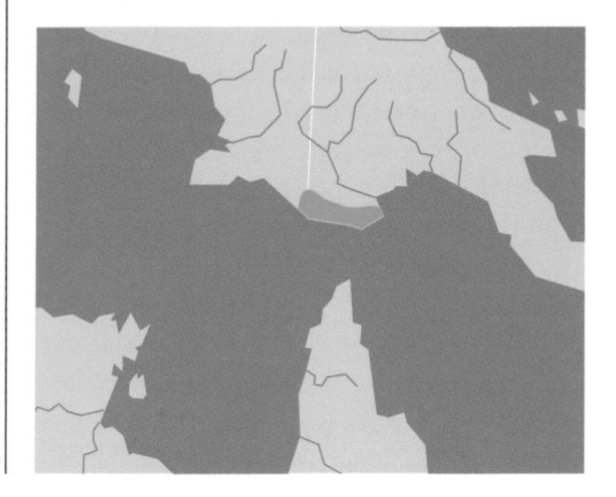

Siebenrock's turtle

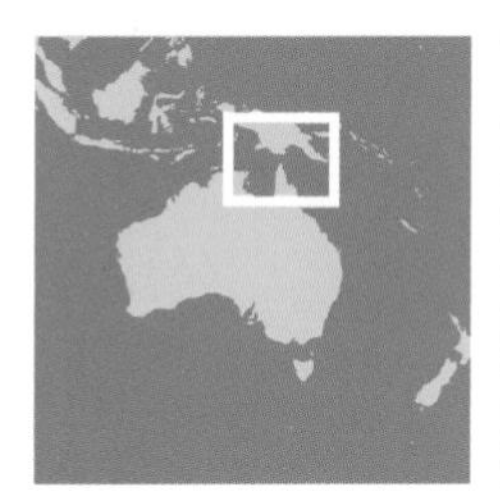

© J. Maran

This turtle resembles *C. oblonga* or *C. rugosa,* but the skin is finer and there are four barbels under the chin. The carapace, sometimes depressed in the midsection, is very long and narrow and strongly rugose.

Natural history. This turtle lives in wetlands along the coastal belt, as well as in lagoons and oxbow lakes, and is never seen basking. It lives in the mud and in detritus-laden bottom substrates of shallow waters. It is not aggressive and does not bite when picked up. This is a carnivorous species, and the long neck is instrumental in the capture of living food. Eggs are laid at the end of the wet season in May, and hatching occurs the following wet season, in November or December. Clutch size ranges from 4 to 17, and the eggs measure 35 × 28 mm. Incubation lasts from 187 to 272 days.

Protection. Exports from Papua New Guinea are controlled, but export from West Papua is frequent. The species still appears to be reasonably abundant. It breeds freely in captivity, and the captive population in the United States and Europe is very large.

Chelodina steindachneri Siebenrock, 1914

Distribution. This species occupies a vast area of Western Australia, where it is the only turtle species present. This hot, arid zone extends from the De Gray River in the north to the Irwin River in the south. This turtle is especially plentiful in the Murchison River and its tributaries.

Description. The vernacular names describe this species very well: it is round and flat. It is also the smallest member of the genus, the maximum length of females being 180 mm. The fact that it lives in an extremely arid habitat explains the reduced body size. The coloration is very light, of a beautiful, bright red-brown, with just a few dark lines or spots. A strong depression runs along the three mid-vertebral scutes. The marginals follow the overall line of the carapace, giving to the entire structure a very regular pattern. The plastron is very light in color, yellow to orange, with perhaps a few fine dark streaks along the seams, especially in young animals. Some specimens have an orange tint to the plastron, dotted with black. The neck is very long, beige to orange above and cream to yellowish below, with small, poorly developed

Common name

Flat-shelled turtle, dinner-plate tortoise

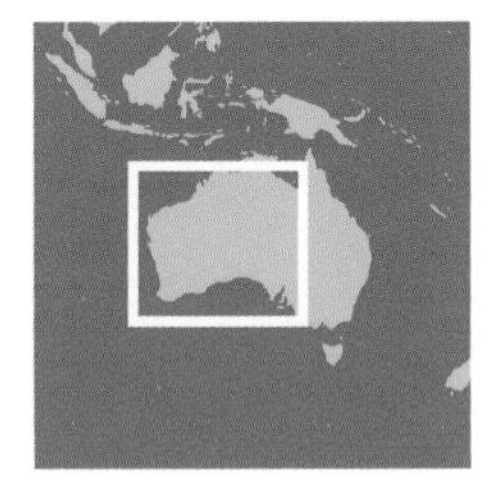

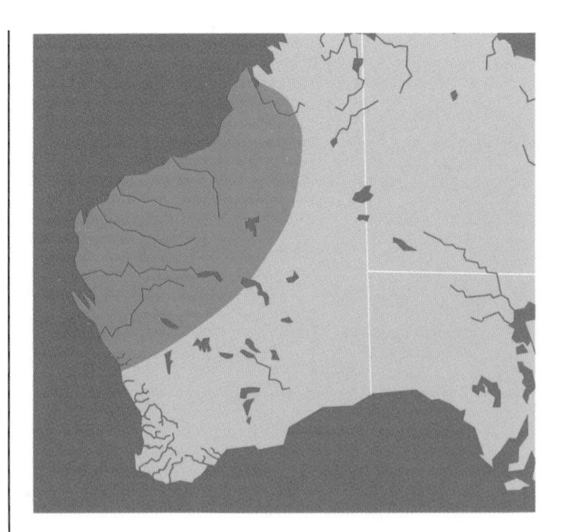

tubercles, squeezed closely together. The head widens behind the temples and is narrower in front, with a short mouth and large yellow eyes. The skin of the head is very smooth, the coloration being similar to that of the neck: brown to orange on the upper surface, and cream to yellow below. Males may show a slight plastral concavity, but the plastron may also be convex. In mature females, the carapace usually bears scratch marks made by the claws of the males during repeated copulatory attempts.

Natural history. *Chelodina steindachneri* is adapted to an unusually arid habitat. It has the ability to resist loss of body water and to recover the aqueous fraction of its urine, which enables it to survive under conditions of extreme dryness. Specimens are sometimes found in small ponds and isolated water holes, far from any major source of water, and one has no idea how they were able to reach such locations. According to J. Cann, they have probably been transported here and there by aborigines, who may carry them to remote locales as a sort of food reserve for hard times. The water holes where the turtles live dry up fast and remain dry for ten months or more. The temperature may reach 45°C, but the turtles survive, deeply buried in the dried mud. Their extraordinary estivation is similar to that of the African *Pelomedusa subrufa*. As soon as the first heavy rainfall strikes the ground in spring, the turtles come out and feed heavily, and the females nest immediately (about seven or eight eggs, 30 × 18 mm, once per year). Incubation may take months, as the eggs, with their embryos within, wait for the arrival of the rains that facilitate emergence (this may take up to a couple of years, according to Burbidge and Cann). This remarkable response to extreme survival conditions demonstrates the adaptability of the species.

Protection. This turtle is collected and consumed by aborigines, but it is not subject to any commercial pressure.

The shell of this species is flat and very oval, even tending toward perfect roundness. The size is small, and the head and neck are both relatively narrow and short. The turtle is mostly a rather light orange-brown, no doubt because it often lives in soil and mud of this color.

Chelus fimbriatus Schneider, 1783

Distribution. This turtle is only found in South America, throughout the Amazon Basin in the Guianas, northern Bolivia, western Peru, Ecuador, eastern Colombia, and Venezuela.

Description. This species could not be confused with any other. Its jagged carapace, russet brown in color, together with the lappets hanging from its neck make it look like a pile of dead leaves. Its crypsis and coloration are perfect for allowing it to disappear into the background of its typical environment of stagnant, shallow, muddy waters with deep mats of vegetative debris. The carapace may reach as much as 500 mm, and the weight up to 15 kg. The shell is blackish and brown, with some areas lighter-colored in beige to reddish. The shell is flat, with three rows of inflated areolae that together form lumpy longitudinal keels in older animals. The scutes are rugose, with clearly marked growth rings. Algae and other vegetation becomes attaches to the carapace, making it even more camouflaged. The plastron is narrow, with short bridges. There are no hinges, but there is a prominent anal notch. The head is unique among turtles: it has the shape of a flattened triangle, very wide behind the eyes, ending in a sort of breathing tube similar to that of some of the softshells. Numerous lappets of skin and tubercles adorn the head and neck, giving an organic mien to the animal. It so happens that the neck and head of the matamata have a precise resemblance to leaves of the mucka-mucka, an emergent aquatic plant with triangular leaves among which the matamata hides to ambush its prey. There are two barbels under the mouth, very well developed. The young are more colorful than the adults, possessing a light brown carapace with a dark line running the length of the median keel and an orange spot on the external border of each marginal. The plastron as well as the bridge and the lower surfaces of the marginals are pink, with dark borders. The head is brownish and has three darker stripes along the top, and three black stripes, separated by pink or reddish skin, run along the underside of the neck.

Natural history. This turtle essentially lives on the edges of wetlands or slow rivers and hunts its prey by ambush. It prefers quiet, muddy waters, rich in detritus and warm and shallow. When potential prey (small fish, tadpoles, frogs, etc.) approaches, the turtle's head swings to the side as the mouth opens, a strong sucking movement is set off by rapid depression of the hyoid, and the mouth closes upon the prey in the blink of an eye. Slow-motion cinematography at the University of Vienna has made it possible to analyze this unique motion, which seems to be a significant "evolutionary success." Furthermore, the matamata can hold its breath for several hours, absorbing oxygen through its pharyngeal and cloacal membranes. Except when actively feeding, this turtle is very lethargic, usually moves around by bottom-walking rather than swimming, and is almost never seen on land. Living in water that is always warm, it does not bask on emergent branches or logs, as do many other species. Courtship is rarely seen in the natural environment. The head of the male snaps forcibly toward the female, the mouth is opened and closed, and the limbs are extended as far as possible. The turtle then discards its lethargy and virtually leaps onto the female. The eggs, laid in eroding river cliffs, number between a dozen and 18; they are hard-shelled, nearly spherical, and quite small (34×37 mm on average). Several nestings may occur in a season, usually during November to December. The duration of incubation is quite variable, according to the humidity of the nest site and the exposure to sun, but the average is about 80 days.

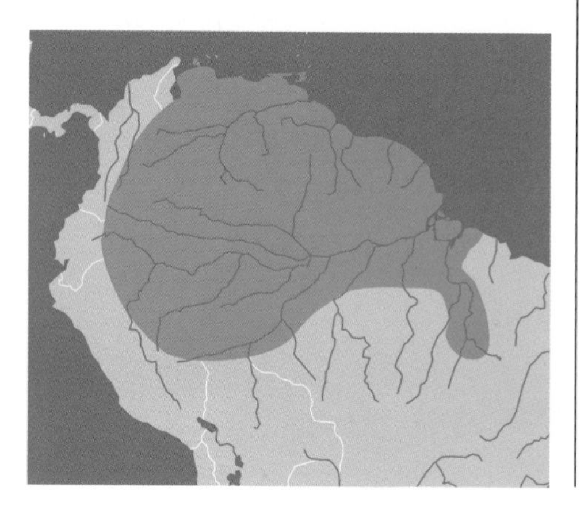

Common name

Matamata

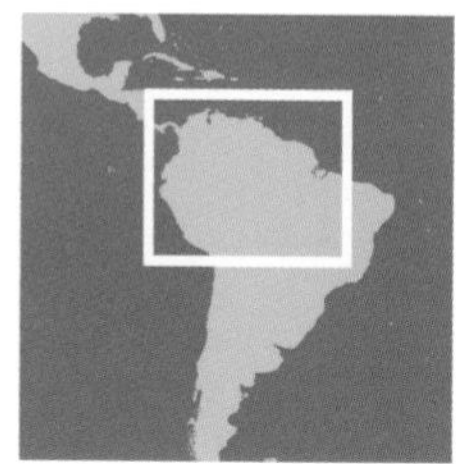

This strange turtle looks like an old piece of rotten bark, with its head like a leaf and the carapace studded with protrusions like an old pebble. These features provide for excellent concealment at the bottom of the water, as it lies in wait for potential prey.

Protection. The matamata is rarely utilized by local people; its appearance is simply too bizarre. Furthermore, the flesh is said to taste of mud and detritus and thus is simply not eaten. The Amerindians are put off by its awful appearance and disgusting smell. Unfortunately, its unusual appearance also makes it sought after by hobbyists, and it is caught illegally in French Guiana, even though there is nominally complete protection in this French department. The species is usually captured in bottom-raking seine nets, and some fishermen have reported catching as many as twenty 10 kg matamatas in a single pull of the net in French Guiana and elsewhere. Because of such actions, the species is becoming rare in certain areas that have been overexploited by commercial hunters. In Western menageries several years ago, hatchlings and juveniles were often seen for sale, although no captive rearing operation is known to us. This is a rather difficult species to keep in captivity, and breeding is rare. The large size and specialized behavior of this turtle make it difficult to raise successfully.

The most astonishing feature of this turtle is the long "breathing tube," which allows it to reach for air while keeping watch, 20 cm below the water surface.

Elseya bellii (Gray, 1844)

Distribution. The range is restricted to a small area of southeastern Australia, in the MacDonald, Namoi, and Gwydir Rivers. It is also found in Bald Rock Creek, a little farther north, but whether the two parts of the range interconnect is not currently known.

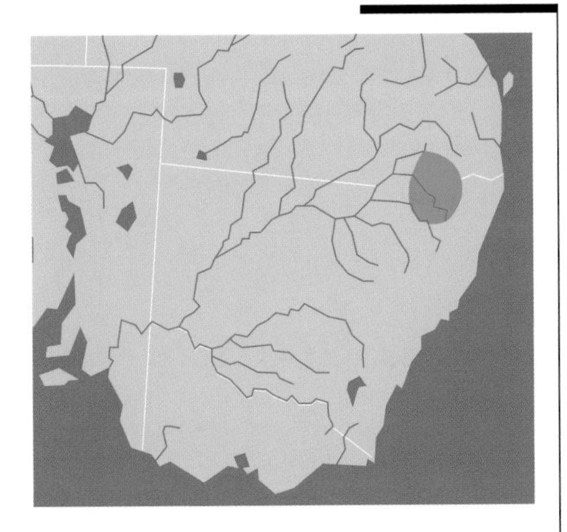

Common name

Bell's turtle

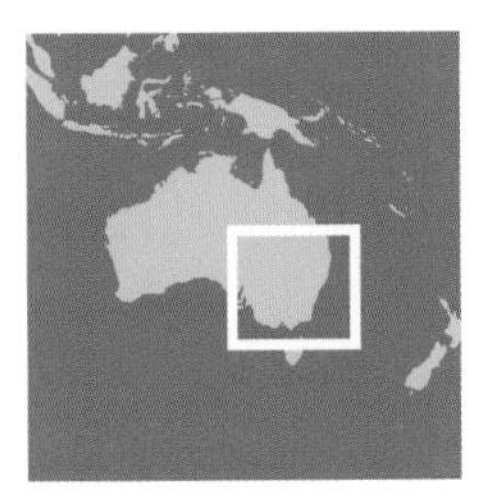

Description of the genus *Elseya.* These turtles have a very short neck. They are similar to *Emydura,* and their size is comparable: 200 to 350 mm, with a maximum of 400 mm for the Burnett River population. *Elseya* differs from *Emydura* in the following features: it sometimes lacks a nuchal scute; the carapace is more trapezoidal than oval and is anteriorly enlarged; the head is massive, with a short mouth and powerful jaws; the animals are lively and aggressive, as indicated by their common name "snapping turtles"; the intergular is narrow; on the top of the head, the skin is often dotted with large flat tubercles; the casque on the head is rough and gnarled; and there is no light yellow line along the neck.

Description of *E. bellii.* Initially described by Gray in 1844 and recently rediscovered in 1980 by Anders Rhodin, this species resembles *Elseya latisternum*. The carapace is very flattened and rounded and may reach 135 mm. The plastron is dirty gray, strongly marked with black seam lines and dark clouding; sometimes it is completely black. The entire periphery of the carapace is marked with a yellow band that extends to the underside of the marginals. The head is grayish to blackish, with a yellow line that originates just behind the mouth, at first very wide narrowing and fading out posteriorly on the neck. The neck is dotted with large, somewhat tapered protuberances. The young are almost discoidal in shape. They have serrated posterior marginals and a flattened carapace, brown with black spots. The head is wide and has a yellow spot behind each eye, the yellow line being very obvious behind the mouth.

The light line that extends from the mouth broadens at the base of the neck. The head is of medium size, with an expansion in the temporal area. The neck, as in all the *Elseya* species, is rather short.

Natural history. Fresh nests have been observed between October and mid-January, and the females lay between 8 and 23 eggs, around 32 × 21 mm in size. Under artificial conditions, incubation takes 80 days at 27°C. This is an omnivorous species, preferring aquatic insects, crayfish, and aquatic vegetation. In certain locations with dense populations, it has been noted that half of the turtles have eye diseases. Some have serious cataracts, and others are completely blind. As yet, it has not been possible to determine the cause of these infirmities.

Protection. This species is classified as endangered by the Turtle Conservation Fund.

Elseya branderhorsti (Ouwens, 1914)

Common name

Southern New Guinea snapping turtle

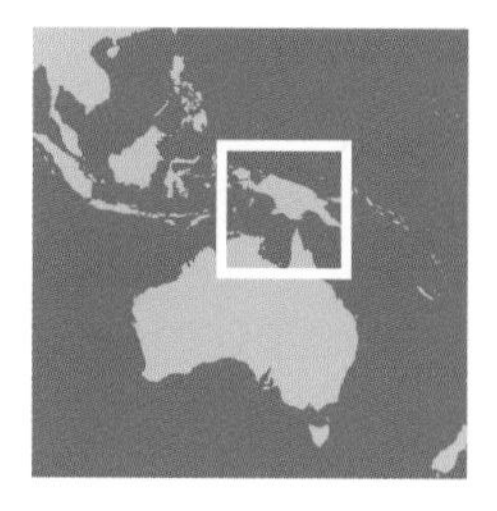

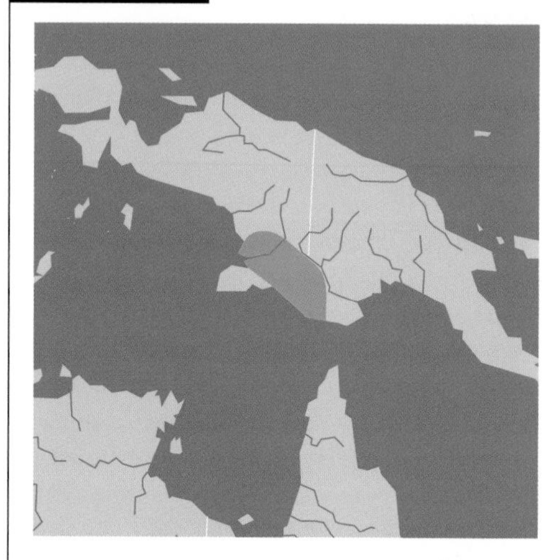

Distribution. This species is found in southern West Papua, from Timika to the Papua New Guinea border.

Description. Some specialists do not accept this species, but the recent literature differentiates it from other turtles of New Guinea. This is a rather large turtle, which may reach 50 cm in length, although the other local freshwater turtles do not exceed 30 cm. It is distinguished from other *Elseya* in the region by its black irides, a fine white spot behind each eye, and the absence of a nuchal scute. The carapace is very round in juveniles and becomes relatively elongate in adults. The young have tubercles on each vertebral scute. The neck

© B. Devaux

This is certainly the most placid *Elseya* species, and it rarely attempts to bite. A light, bright area starts below the snout, covering the throat and the base of the neck. The tubercles on the neck are very developed. Unfortunately, this species shows up more and more in animal dealerships. It is found in southern Papua New Guinea.

is dotted with small tubercles similar to those of *E. latisternum*. Two pairs of chin barbels are present. The posterior part of the anterior limbs is soft-skinned, without spines or tubercles. The carapace is olive to beige. The head is also olive to beige on the upper surface and cream below. A light line extends from the snout up to the base of the neck, passing below the tympanum and clearly separating the base of the neck from the head. The base of the jaws is pink or cream-colored. The plastron is white.

Natural history. These turtles occupy lowland swamps, as well as small rivers, and they prefer shady, quiet areas. The eggs are oval and about 55 × 33 mm in size.

Protection. Classified as vulnerable by IUCN, this species is still not protected, and one sees it offered for sale by Western animal dealers more and more frequently.

Elseya dentata (Gray, 1863)

Distribution. The overall distribution of this species breaks down into distinct zones, and the variation among isolated populations suggests that several ecotypes or subspecies may exist. One of the zones encompasses the Kimberley Plateau; a second is along the Barkly Plateau; and the last is along the east coast of Queensland, up to the Mary River.

Description. Great variation is evident in different parts of the range. The type specimen is from the Victoria River. It is a large species, females reaching up to 7.5 kg in weight and 350 mm in length. The carapace is thick and strong, with a pronounced tectiform profile. The top of the head bears a sort of thick casque or shield. The jaws are powerful, and the snout is well developed and projecting. Old animals often show macrocephaly, and along with the powerful jaws and the strong beak, they certainly appear aggressive, even dangerous. Watch out for these jaws! In certain old animals, the rear of the upper jaw extends downward and reinforces the aggressive mien. The general color is dark brown to steel gray, with small, indistinct lighter markings. Some specimens are yellowish, others are marked with radiating lines, and still others may be greenish or blackish. The top of the head may be brown to bronze. Lighter zones, often beige in color, may occur along the sides of the mouth and on the cranium. The eyes are dark brown and greenish. The underside of the neck is light-colored—either green, orangish, cream, yellow, or sometimes nearly white—up to the lower jaw. Two well-developed chin barbels are often present. The plastron is light, leaning toward yellow, but it darkens with age and may be almost black in old animals. The plastron is narrow, hardly one-third of the total length of the animal. The female is always distinctly bigger than the male. The young have a median keel and a blue-black carapace, or sometimes the carapace may be steel gray, brown, or beige, often with dark spots on the vertebrals and the costals. The margin is frequently serrated during the first years of life. The plastron is plain yellow. The head is brown to blackish, sometimes greenish. The subadults may have light lines on the sides of the head and the neck.

A related species, which lives in the Nicholson River in Queensland, is called *Elseya lavarackorum* by J. Cann. It shows two dark points on the iris, as do certain species of *Emydura*.

Natural history. These turtles are omnivorous. They chase their prey and can catch quite large

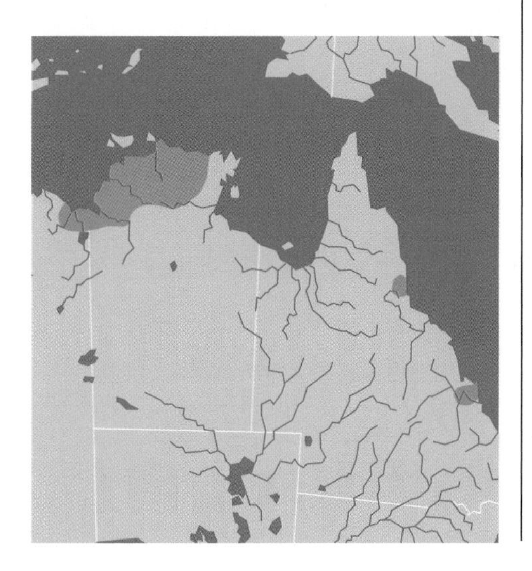

Common name

Northern snapping turtle

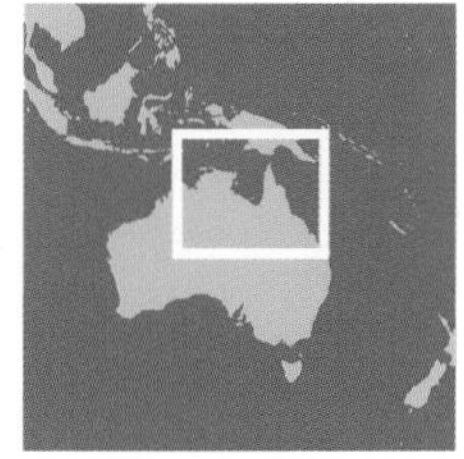

© F. Bonin

This *Elseya* looks like a tough customer, and is indeed an aggressive animal, with powerful jaws. The long limbs are webbed, allowing for great underwater speed in the pursuit of prey. This turtle is hardly ever found out of water.

fish, young birds, or small mammals that happen to come to the edge of the water, and they show equal appetite for aquatic plants, fruits that have fallen into the water, and *Pandanus* roots. The young are much more carnivorous than the adults; the latter ultimately become completely herbivorous. The speed of this turtle underwater is remarkable and has been reported by a number of observers; it is an excellent swimmer and has the ability to thrust its well-webbed limbs backward very rapidly, like a paddle, to advance rapidly through the water. Nesting has been observed from February to May, and incubation lasts for about 120 days. The large eggs (50 × 30 mm; 23 g) average about 10 per clutch. These turtles nest earlier in the season than other species, mostly in October, a trait made possible by their very warm, humid environment.

Protection. There is no solid information about the status of this species in its natural environment. It is not hunted much, and the populations still appear to be strong. The bellicose disposition of *E. dentata* limits its appeal as a pet.

Elseya georgesi Cann, 1997

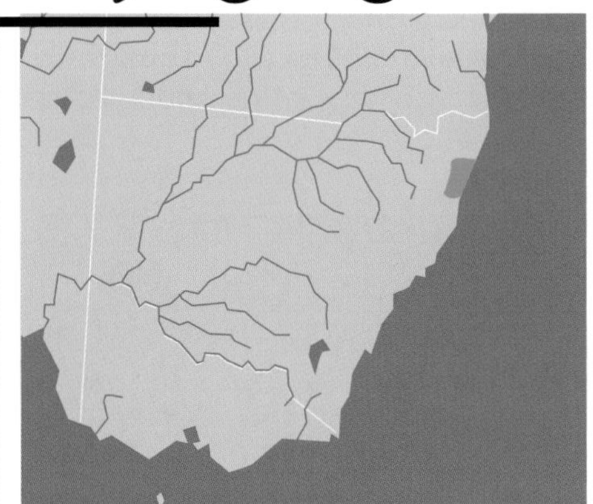

Common name

Bellinger river turtle

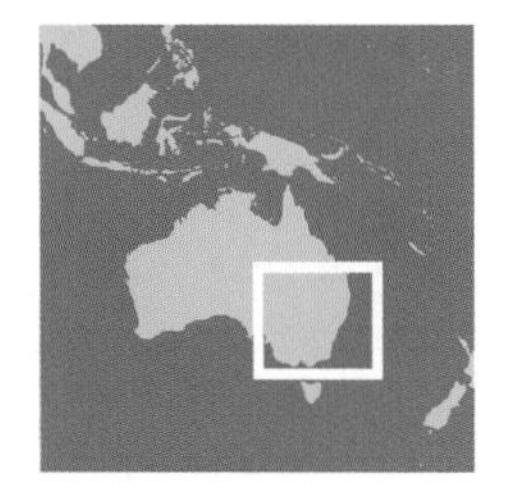

Distribution. This species is found only in the Bellinger River, near Thora, New South Wales.

Description. The species name refers to Arthur Georges, a renowned Australian turtle scientist. It resembles both *E. latisternum* and *E. purvisi,* but the yellow line along the neck is more diffuse and does not reach the angle of the jaws. There is no V-shaped marking under the head, just an indistinct yellow blotch. The plastron is yellowish, often marbled with black. There is no keel on the carapace, and indeed there is a slight depression along the midline of the shell. Sometimes the tail is marked with black and yellow stripes. The maximum length is 220 mm. The young are almost circular, with serrated margins and a vertebral keel; they are gray-bronze in color, and the yellow line along the head and neck is bright and well defined.

Natural history. A clutch may contain about a dozen eggs, and nesting has been observed between October and December. The diet varies with the season and includes both carnivorous and herbivorous components.

The yellow line along the side of the neck is very marked in this individual. The plastron and the underside of the marginals are light and yellowish, no doubt making for effective camouflage when viewed from below.

This turtle has an amused air, with its big, questioning eyes and its short mouth that seems to form a "smile." But be careful—from time to time, it may be quite aggressive.

Protection. The Bellinger River is an area of heavy cattle raising, which can disturb the turtles. The very limited natural range makes one worry that the total population could be small and endangered.

Elseya irwini Cann, 1997

Distribution. These turtles are found only in the basin of the Burdekin River in eastern Queensland, not far from the town of Ayr.

Description. This is a rather large turtle (322 mm), close to *Elseya dentata,* but its head is white, sometimes with a rosy pink snout. The eyes are very dark, with almost black irides, characteristics that easily distinguish the two species. The skull is expanded at the level of the tympana, and there is also a scooplike extension of the upper jaw. The plastron is yellow, with intensely black sutures and more or less regularly dispersed black spots. The young are virtually circular, with serrated margins, an almost white plastron, and often regular black spots on a brown or grayish background.

Natural history. This is an omnivorous species, eating snails, aquatic vegetation, fruits, and insects. The young are apparently more carnivorous. Nesting has been observed in September (a dozen eggs of quite large size, 45 × 35 mm). In one observed nest, the young hatched 111 days later.

This underwater photo by John Cann shows the wide "face" and astonished expression of this turtle, as well as its powerful limbs, with extensive webs that allow it to swim rapidly.

Common name

Irwin's turtle

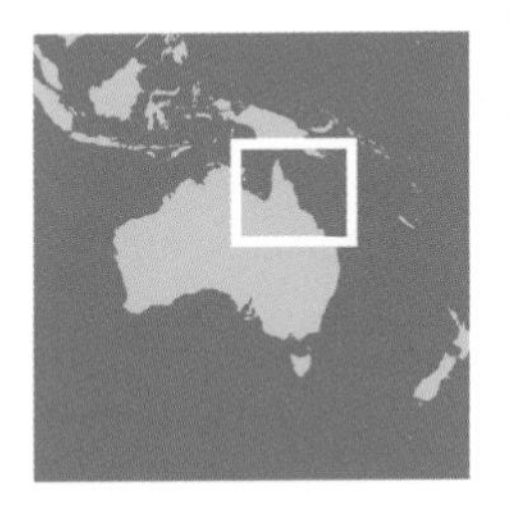

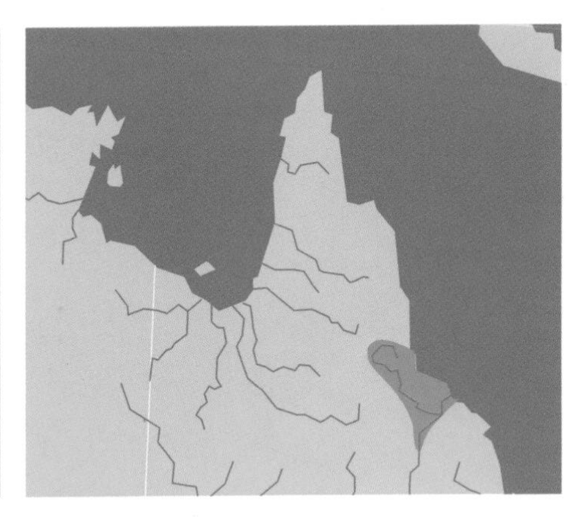

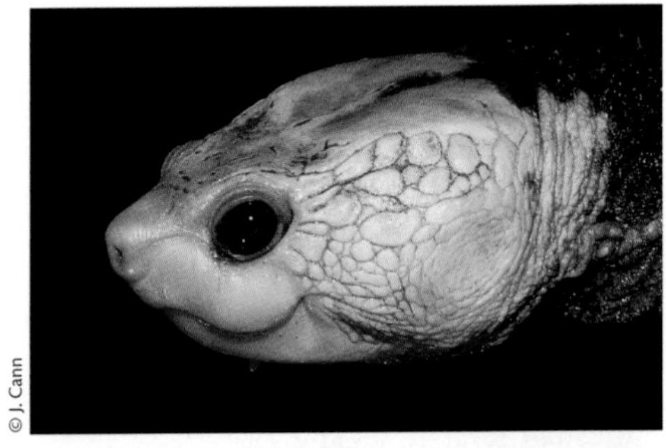

The upper jaw is very rounded, accentuating the "smiling face" that is seen among several Australian turtle species.

Elseya latisternum Gray, 1867

Distribution. The range is huge and discontinuous, in northern and eastern Australia. The species is known from the Alligator and Liverpool Rivers in Arnhem Land, as well as in the Gregory River and down to southern Queensland, as far as the Richmond River south of Brisbane. It abounds in the Flinders River and reaches 1,000 m altitude in the Atherton Tableland.

Description. This species is very similar to *E. dentata,* and the juveniles of the two can easily be confused. Adult *E. latisternum* can be distinguished as follows: the carapace is slightly constricted in the middle; the size is smaller (280 mm); the head is less enlarged, the jaws are less developed, and the gnarls on the surface of the skull are also less extreme. Macrocephaly is rare. The upper surface of the head is sometimes completely smooth and of an orange-brown that looks rather like metallic copper, and the snout is very prominent. In old animals, the carapace is often rather flattened, sometimes even depressed. Some individuals have serrated marginals toward the rear. There is often a yellow or whitish line along the side of the head and of the throat, extending to the shoulder. The plastron is yellow, but it changes rapidly with growth and becomes grayish, orange spotted with black, or dark gray. The intergular is as large as, or larger than, each gular scute. When disturbed, *E. latisternum* will emit a strong musky odor. The young are grayish, with a yellow line around the shell, serrated marginals, and a well-marked median keel. The head is gray or black. The lower jaw is yellow, and there is a line of the same color under the head.

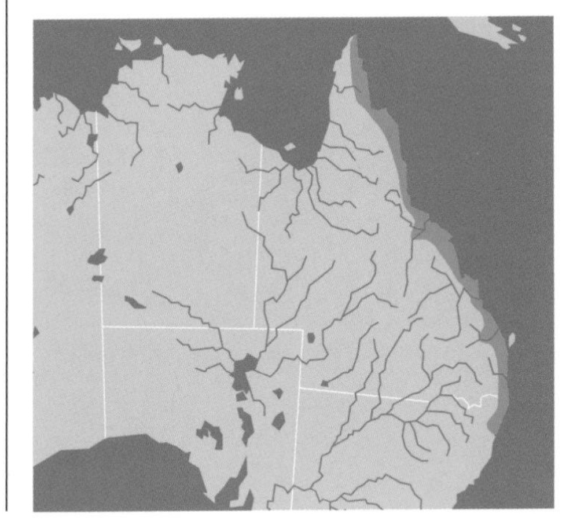

Common name

Saw-shelled turtle, serrated snapping turtle

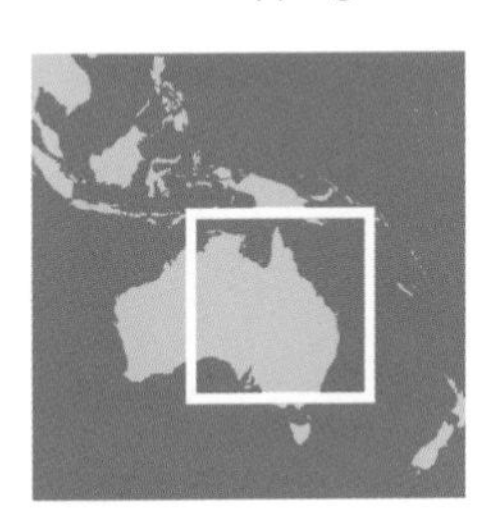

Natural history. Although less aggressive than *E. dentata,* this species is very lively and can bite with great force. It is omnivorous and eats fish, insects, crayfish, dead animals, toads (including *Bufo marinus*), and berries and fruits that fall to the ground or into the water. Along with *Chelodina novaeguineae,* this species utilizes the musk glands more frequently than any other Australian turtle. These odiferous emissions may also play a role in the reproductive cycle. Nesting takes place from September to January, several times per season, with clutches numbering 9 to 17 fairly large eggs

© G. Kuchling

With its blackish face, strong tubercles, and wide, inquisitive eyes, this is indeed an astonishing turtle. It is a powerful and rather aggressive species, whose jaws can inflict serious wounds.

Juveniles have a colorful plastron and a delicate head. The bright colors darken up quickly with growth.

© F. Bonin

(30 × 23 mm); in an entire season, as many as 50 eggs may be laid.

Protection. There is no evidence that natural populations of this species are depleted.

Elseya novaeguineae (Meyer, 1874)

Distribution. This species occurs only in New Guinea (West Papua and Papua New Guinea).

Description. This turtle could be a subspecies of *Elseya latisternum,* which it closely resembles. The length does not exceed 300 mm. The carapace is oval, flat, and brown, blackish, or sometimes tinged with olive and has very narrow vertebral scutes and very high costals. The marginals along the bridge are narrow. A nuchal may be present but is small. No keel is visible. Black spots are present on the upper part of each costal scute in young specimens. The wide, flat head, and the powerful jaws fully demonstrate the name "snapper." The top of the head is covered with a single large cornified scale. The sides of the head are somewhat tuberculated. The top is dark gray, and the underside yellow-white. The neck is dotted with tubercles of good size. The plastron is long, narrow, and yellow to cream-colored, with white seams and several dark spots on the edge of the scutes. The anal notch is very reduced. In the young, the posterior marginals are serrated, and there is a minor vertebral keel, which disappears rapidly with growth. The color of the young is chocolate brown, with a dark spot on each costal scute.

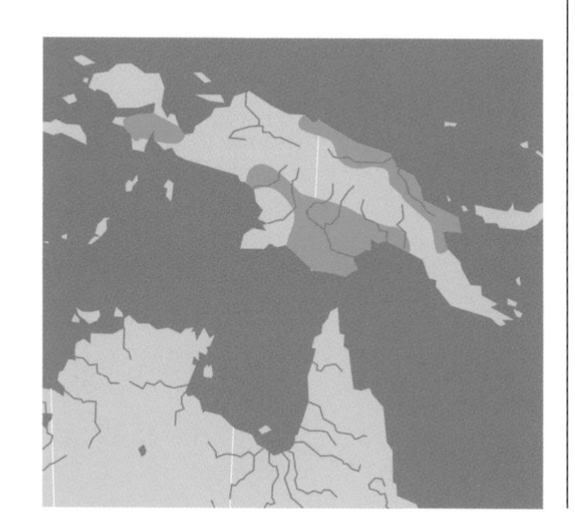

Common name

New Guinea snapping turtle, New Guinea snapper

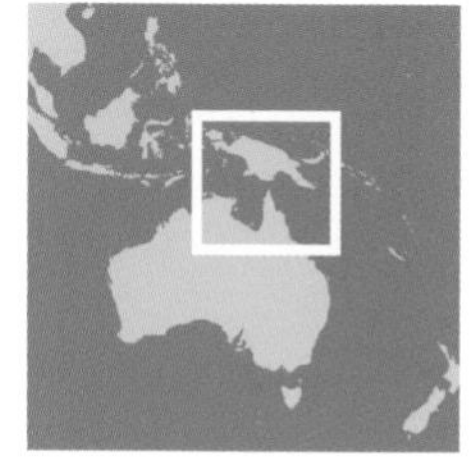

© F. Bonin

Natural history. This species is very widely distributed in New Guinea in a variety of aquatic habitats, including lakes and coastal swamps. The diet is primarily carnivorous. About six elliptical eggs, 55 × 33 mm in size, are laid per clutch.

Protection. These turtles are regularly caught for food by local people, but as far as is known, populations are in sound condition.

This turtle is much lighter in color than the other *Elseya* species. It has a smaller head and an elongate snout, is very light through the neck and limbs, and has a cream-colored plastron.

Elseya purvisi Wells and Wellington, 1985

Distribution. Confined to a very small range in the Manning River in New South Wales, about 25 km from the sea and from the town of Taree.

Description. This species was first described in an informal local Australian publication, and the description was confirmed by J. Legler and J. Bull in 1973. It is very close to *E. georgesi*, from which it is distinguished by a conspicuous detail of the head coloration: a striking yellow line that starts at the base of the neck, widens below the tympanum, and encircles the base of the head, up to the upper part of the upper mandible. Under the neck, a yellow V-shaped marking brightens the

This individual shows an especially bright light line running from the lower jaw to the base of the neck.

© J. Cann

gray background. The carapace is brown, continuing the coloration of the top of the head. There are two chin barbels. The plastron, the bridges, and the underside of the marginals are bright yellow, with brown spots and black seams in adults. The tail is equally marked with black and yellow lines. The juveniles are brightly colored, the head being set off by a brilliant yellow line. The plastron is bright yellow, with perhaps a touch of orange. The carapace is sometimes greenish, with a moderate keel.

Natural history. Nesting extends from February to April, and the number of eggs ranges from 7 to 23. They measure 32 × 23 mm, and the young weigh 5 g. The diet is varied but includes lots of aquatic insects.

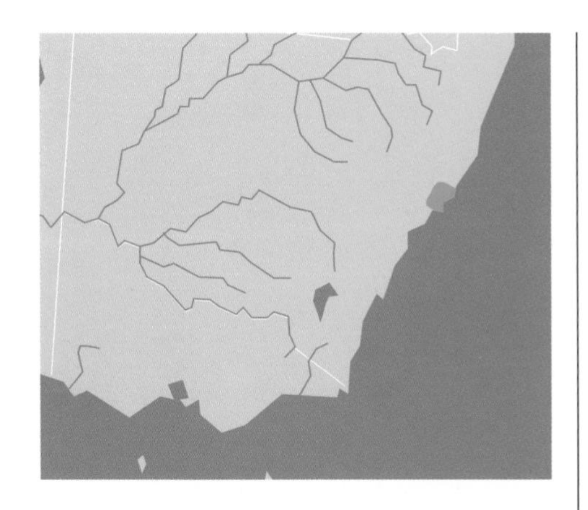

Protection. Turtles in general are protected in New South Wales, but this highly restricted species must be considered especially vulnerable.

Common name

Purvis's turtle

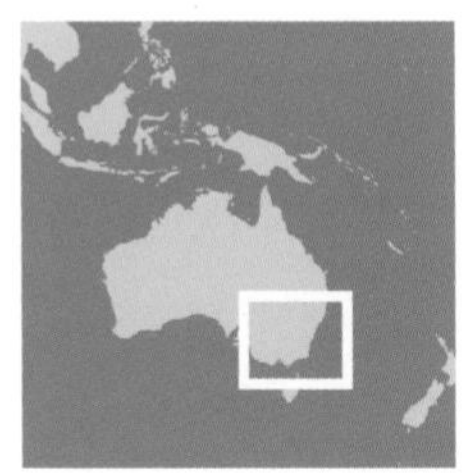

Elusor macrurus Cann and Legler, 1994

Distribution. This species has a very limited range on the Mary River in southeastern Queensland. It may occur throughout the length of this drainage, but up to now it is best known from the area south of Maryborough.

Description. This turtle has been known since 1961, at which time it was widely sold by certain Australian animal dealers, but it was confused with *Elseya latisternum*. It was not until 1994 that it was characterized and named by John Cann and John Legler. They described it as being large, up to 400 mm in males and 350 mm in females, and above all noteworthy for the very long tail of the males, sometimes attaining 70% of the length of the carapace, very wide at the base, laterally compressed, and with the cloaca located toward the tip. At the end of the tail, the vertebrae are long and heavy and form a sort of "bony mass" whose function is still unknown. In the females, the caudal appendage is less developed. The carapace is heavy, oval, very flat, somewhat depressed in males, and olive-brown, sometimes greenish or beige, or even tending toward black. The plastron is pearl gray or cream-colored, without additional pigmentation. The carapaces are frequently covered with moss and vegetative debris. The marginals bear a light line on the underside. The plastron is light, from yellowish white or gray-white to a dirty dark yellow. There are two mental barbels, very well developed, and the neck is covered with small tubercles. The features recall *Rheodytes leukops*, but in *Elusor macrurus* the eyes are dark, with a black iris (not yellow), and the line of the jaws is less "smiling." The young are almost circular, with the posterior marginals serrated and of a color somewhere between orange and dark gray. The seams are clearly marked with black. The head is very wide, with protruding eyes, and two fine yellow lines run along the length of the neck. The plastron is gray-white.

Natural history. This species is endemic to the basin of the Mary River, just as *Rheodytes leukops* is confined to the basin of the Fitzroy River, farther

© J. Cann

Note the gaping cloaca and the caudal thickening posterior to the cloaca.

Common name

Mary River turtle

This turtle has a slightly abnormal wide, flat carapace, beige in color; an elongate, narrow plastron; and a notably long, thick tail.

north. These two taxa, each the only representative of its genus, have a variety of points in common. There is a barrage at the mouth of the Mary River that keeps the sea from coming in and thus limits the salinity of the river water. This river is rich in turtles, having large numbers of both *Elseya dentata* and *Emydura krefftii*. The river is wide, calm, shallow, and warm, and *Elusor macrurus* is usually found in shallow areas 1.5 to 2 m deep. The species is omnivorous and eats algae as well as crustaceans, fish, frogs and toads, and freshwater shrimp, but it prefers shellfish and has been observed using its jaws to open gastropods from the side in order to feed upon the interior. Reproduction does not appear to differ in major points from that of other turtles in the same ecosystem. Nesting has been observed in late October on the sandbanks, 10 m or so from the water's edge. Incubation is rapid (eight weeks) and hatching occurs in December. The eggs (35 × 25 mm) number about 15, with a maximum of 25. The males can be quite aggressive. It captivity they are liable to fight and have to be kept separately. They are easily captured, as is *Rheodytes leukops*, because they are rather sluggish and show little fear when approached. This is a totally aquatic species that never comes out of the water except to nest. The cloacal bursae allow it to remain underwater for long periods of time. Underwater it can be very active, and the wide tail helps it to swim rapidly; John Cann considers it to be the fastest of the short-necked species.

Protection. This species is highly restricted and scarce; it apparently is much rarer than during the time of heavy exploitation of eggs and hatchlings for the pet trade. As a cloacal breather, it also is very vulnerable because of its sensitivity to siltation, water pollution, and depressed oxygen levels.

Emydura australis (Gray, 1841)

Common name

Australian big-headed turtle, northwest red-faced turtle

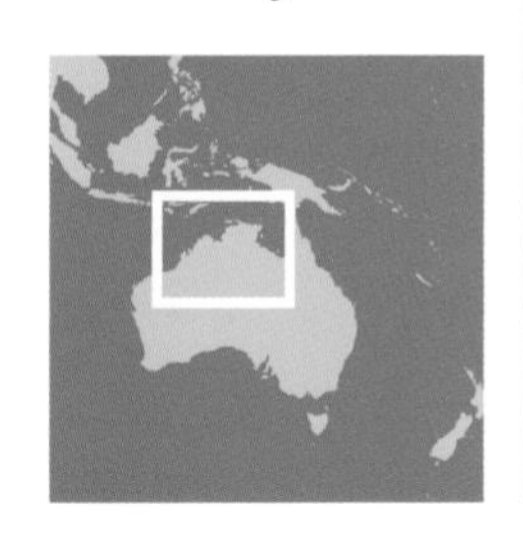

Distribution. The classification and distribution of this form are not agreed upon within the specialist community. According to J. Cann, it is found in the Fitzroy River and northward to the Princess Regent River in the Kimberleys (northwestern Australia).

Description of the genus *Emydura*. These pleurodires have a rather short neck and a shell length not exceeding 300 mm. Of rather stocky build, they have a somewhat large, casqued head. The tubercles on the neck are regular but small and poorly developed, and the chin barbels are so

© B. Devaux

In this juvenile specimen, the median keel and the lightly serrated posterior marginal scutes can be seen.

small as to be almost nonexistent. Certain species, such as *E. victoriae,* exhibit decided macrocephaly in advanced age. The skin at the level of the temples is very fine and smooth in almost all species of the genus. A white or yellow line along the side of the head and on the neck makes for quick recognition. A nuchal scute is often present. These turtles have five claws on each forelimb. At the front of the plastron, the gular scutes are completely separated by the intergular. The manner of moving through the water is noteworthy: these turtles walk slowly on the bottom of deep water, with careful placement of the feet and with head held out. Numerous new species within the vast genus *Emydura* remain to be described or recognized. Some are considered today to be subspecies of *macquarii,* while others, according to J. Cann, have been observed in recent years but still need to be studied and named.

Description of *E. australis.* Gray's name *australis* refers to the Southern Hemisphere distribution of this species. For a long time included with, or confused with, *E. m. macquarii,* this turtle is easily recognized by the red coloration on the head, a line extending from the mouth to the middle of the neck, and an elongated streak behind the eye, sometimes including the nose and the outline of the eyes. Sometimes the color tends toward orange or yellowish, but the elongate bar behind the eye is a reliable feature. Some individuals have very long tails. J. Cann has observed a male with a carapace 121 mm long and a tail 101 mm long. The maximum carapace length is 140 mm. The general form is wide and rounded behind and narrow in front, but both the carapace and plastron of individuals from the King Edward River are quite narrow. The plastron is yellowish, sometimes with grayish spots in old animals.

E. australis is a very pretty species, with a smiling face. Although it is small in size, its lively colors nevertheless make it one or the most attractive Australian turtles.

© B. Devaux

Emydura krefftii (Gray, 1871)

Distribution. This species occupies a rather narrow stretch of northeastern coastal Australia in Queensland, extending south to Fraser Island.

Description. This is the northern equivalent of *E. macquarii,* to which it is very similar. It is smaller (250 mm) and can be distinguished by the light cream marking along the neck, narrower and less contrasting than in *E. macquarii*. In general, *E. krefftii* is marked with a light area of cream or yellow behind the eye and extending to the tympanum. The carapace is also marked with deep, regular striations. The color is lighter than that of *E. macquarii,* tending toward chocolate beige or orange, with a light line encircling the carapace. The plastron is black, with yellow seams, especially in the center. The iris is black and surrounded by a conspicuous yellow ring. Juveniles have a modest keel on the carapace, and the borders are serrated. The carapace is grayish to

This turtle seems to be well protected, with its huge carapace, broad powerful head, temporal helmet, neck covered with tubercles, and strong limbs: truly a miniature army tank.

greenish, with the underside of the marginals yellow-orange. The light marking behind the eyes is bolder and brighter in the juveniles than in the adults.

Subspecies. The population of Fraser Island has been provisionally named by J. Cann as the Fraser Island short-necked turtle; no scientific name has been given as yet. It greatly resembles the typical form but is differentiated by the black coloration of the carapace; a very light plastron, pearl gray to whitish; and the absence of a light spot behind the eye. The snout is tapered and projecting. The black coloration of the animal may be due to the need to gain heat by basking, given that the lakes are rather cold. The marginals are very flat. This subspecies shows a tendency toward macrocephaly, which does not occur in the nominal race. The eggs measure about 35 × 20 mm, and one female is known to have laid 33 eggs in a season. Incubation lasts just 70 days. These turtles rarely travel between the various lakes on Fraser Island, and the population seems to be quite fragmented; the turtles in the three principal lakes show strikingly different eye color.

Natural history. This species is found offered for sale or present in collections with increasing frequency in the West, although the channels of the commerce remain obscure. As with all the turtles of Australia, exportation from the continent is prohibited.

Protection. These turtles have legal protection, and they also constitute the most abundant turtle species on Fraser Island.

Common name

Krefft's river turtle

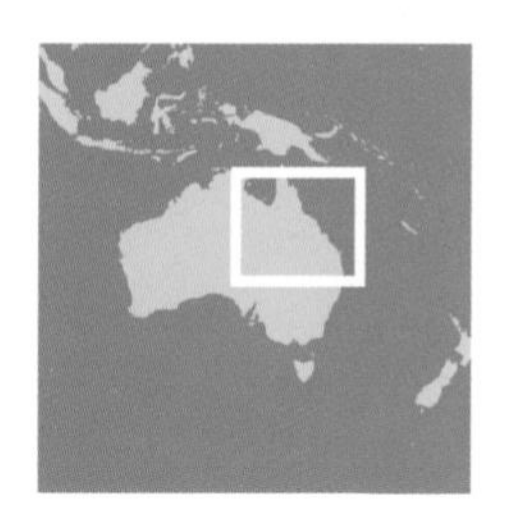

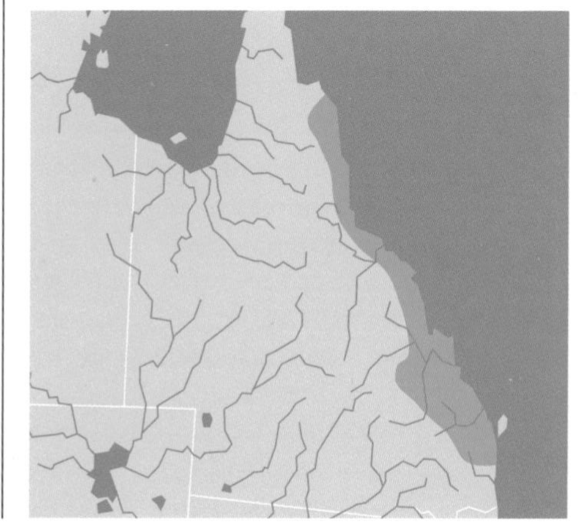

Emydura macquarii (Gray, 1841)

Distribution. The Murray River turtle is widely distributed in southeastern Australia, including the Murray-Darling system and the Great Dividing Range, without ever approaching the sea.

Description. The name *macquarii* derives from the name of a former governor of the colony of New South Wales. This turtle, the most widespread in the genus, measures about 240 mm (maximum 340 mm for females). The carapace is oval in adults, slightly domed, and wider behind than in front. The marginals extend almost horizontally, especially toward the rear. The shell is quite rugose in texture, with regular grooves and sculpturing. The color is beige to blackish gray, the sutures being dark. The plastron is long and narrow and totally yellow in color, with very long bridges. The intergular is quadrangular and completely separates the gular scutes. There is a moderate anal notch. The limbs are powerful, wide and extensively webbed, the posterior pair being especially long and strong. A series of soft, flexible scales runs along the length of the anterior limbs. The head is of medium size, with a rather elongated snout. Coloration is brown to beige on the upper surface and creamy yellow below. A yellow streak, expressed in most individuals, runs from the mouth to the base of the neck, but it never borders the tympanum, as it does in *E. signata*. The tubercles on the neck are very rounded but few in number and present only toward the rear of the head. The eyes, anteriorly located, are slightly protuberant, and there is a slight depression between the orbits. The horny beak is light yellow and extensively marked. The male is smaller than the female. The young are almost circular, with the rear marginals flaring, a slight median keel, and coloration ranging from ocher to yellowish, spotted with beige. There is sometimes a yellow spot behind the eye.

Subspecies. Five subspecies of *Emydura macquarii* are recognized.

E. m. binjing (Cann, 1998): The name comes from the aboriginal word used for turtles in the region, where it occurs. The subspecies occurs in the Clarence River and its tributaries. It does not exceed 235 mm. The carapace is dark brown, striated with yellow. There may be a modest vertebral keel, and in the juveniles the keel is strongly tuberculate. The iris is greenish yellow. The plastron is orange, with dark areas. The light line along the side of the neck also tends toward orange. Each forelimb is boldly marked with a large yellow-orange spot on the anterior face.

This animal is not an exceptional specimen. Many Australian turtles become covered with algae and moss, tightly bound to the striations in the carapace and camouflaging them perfectly on the muddy bottoms of the streams where they live.

The eyes are small but bright, and a broad light-colored stripe extends from the angle of the jaw.

E. m. dharra (Cann, 1998): This subspecies is found in the Macleay and Hastings Rivers, not far from Kempsey. The adult females are about 185 mm long. The carapace is dark brown and has a moderate median keel. The iris is orange-yellow. The yellow line is rather indistinct and commences just behind the mouth and disappears halfway along the neck. The plastron is grayish to beige. The intergulars are very wide and form a large V.

E. m. dharuk (Cann, 1998): *Daruk* is another name for local turtles. This subspecies occurs in the Hawkesbury River, above Sydney. It may reach 240 mm. The coloration is much lighter than that of *E. m. gunabarra,* beige to brown with strong streaking, and this subspecies has no keel. The skin is light gray. The light line is rather long and continues to the beginning of the neck. The iris is yellow-green. The plastron is light, yellowish to gray, orange, or beige. The young are almost perfectly circular, with a serrated margin and of a slate-gray color. The eggs usually weigh just 2 g, with a maximum of 4 g. Incubation is very short: 52 to 53 days at 27°C to 28°C.

E. m. gunabarra (Cann, 1998): *Gunabarra* is one of the names that the aborigines use for local turtles. This subspecies occurs in the Hunter River near Newcastle and reaches 280 mm. This is a rather dark animal, brown-black to gray-black with light streaks, and it has no real keel. The iris is a rather dark yellow-green. The light line commences behind the nostrils but thins out and disappears in the middle of the neck. The plastron is grayish with yellow areas. The young are almost black and have a median keel.

E. m. signata (Ahl, 1932): This subspecies is found only in the Brisbane River. The young are brown and have a rather pronounced keel, and

the adults have a shell with a more of less circular outline. The lateral marginals are lightly reverted in the young as well as in the adults. The adults are a medium brown, with a striated texture. The iris tends toward orange or greenish. The front lobe of the plastron is rounded. The maximum length seems to be 300 mm.

Natural history. This species is very lively, without doubt the most energetic of the *Emydura* genus. The rather eclectic diet includes frogs, insects, and fish, as well as algae and aquatic plants. Nesting occurs in November to December, and on average 18 rather large eggs (36 × 22 mm) are laid per clutch. Nests of up to 38 eggs have been recorded, as well as a total productivity of 84 eggs for a season. The population density of this turtle in certain places, including the Murray River, may be very high; some observers have counted up to 1,000 hatchlings in one season.

Protection. This species is abundant in nature. It is heavily collected and is present in many wildlife parks and zoos. This is also one of the Australian species that is most frequently found in Western collections.

Common name

Murray River turtle

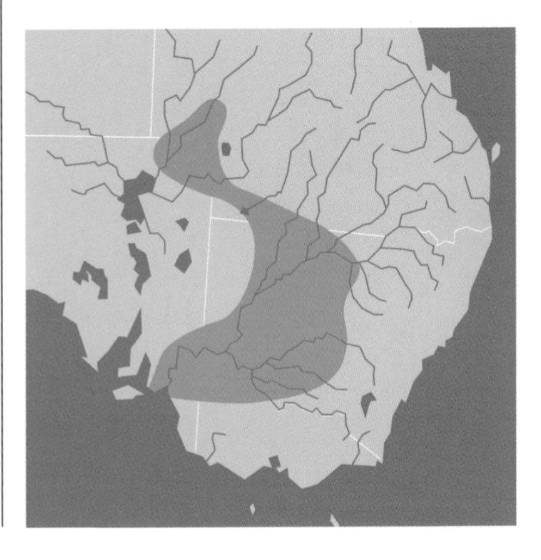

Emydura subglobosa (Krefft, 1876)

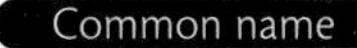

Red-bellied short-necked turtle, painted sideneck turtle

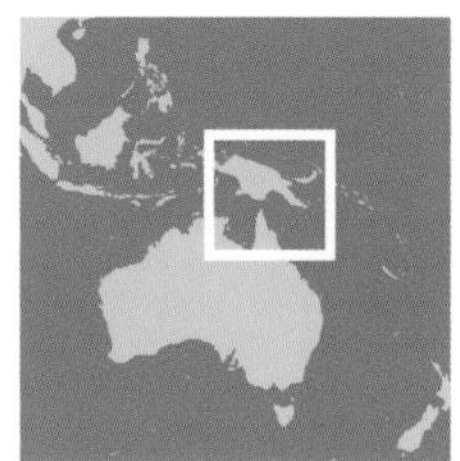

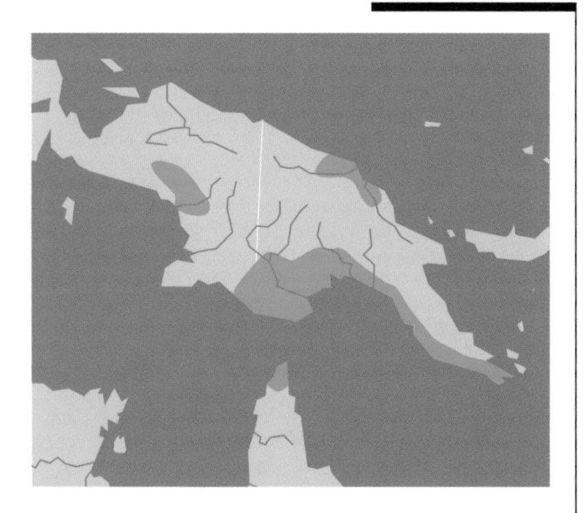

Distribution. This turtle occupies a wide territory in New Guinea, from the western province of West Papua up to the southeast at Port Moresby, in the Kemp Welch River. In Australia it has been found only in the Jardine River at the northern tip of Cape York, where it was discovered by Harold Cogger in 1975. It lives in very tropical, very wet areas in the rain forests of New Guinea.

Description. Considered to be one of the most attractive turtles in Australia and New Guinea, this species is much sought after by collectors. It is easy to identify, and its common name "painted sideneck turtle" characterizes it well. A wide, light yellow band passes from the nostrils to the base of the tympanum, sometimes with a touch of pink on the neck. Another line passes from the snout to the temples, just above the eyes. The general color tends toward light brown, reddish, or bronze. The undersides of the feet are red, as is the border of the carapace. The scutes of the shell are sometimes underlined with a lighter color. The length does not exceed 260 mm. The general shape is oval, bulging in the middle and with a "skirt," or margin, extending out posteriorly. The young are particularly brightly colored. They have a great, shiny V on the snout, a yellow-orange line around the carapace and under the marginals, and a reddish plastron, against a velvety black body. They have a moderate vertebral keel and lightly festooned marginals.

Natural history. These turtles live in warm, very rainy environments, and it is difficult to see them in the wild. They seems to be mainly carnivorous. The females lay mostly in September, depositing 10 or so eggs on average. One of their main predators in northern Australia is the freshwater crocodile, *Crocodylus johnsoni*.

Here the typical features of *Emydura* can be seen: short snout, rounded jaw, small nose, and large eyes, giving the physiognomy a smiling, amused appearance that is quite characteristic. Note the circular, conspicuous tympanum and the light bar that extends from the snout to the back of the head.

Protection. This is the *Emydura* species most commonly seen in European collections, popular because of its bright coloration and also because it comes from New Guinea, where it is heavily collected. It seems likely that the populations suffer from this level of collection as well as from habitat destruction.

Emydura tanybaraga Cann, 1997

Common name

Tanybaraga turtle

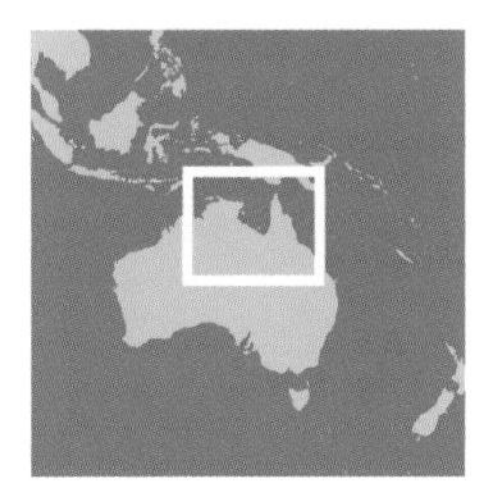

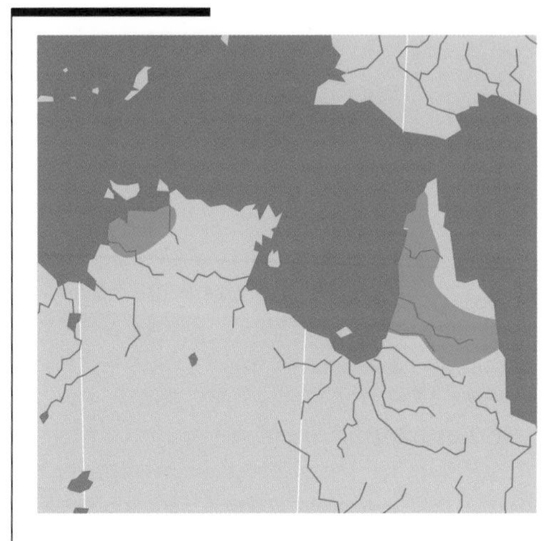

Distribution. At the present time this species is known only from the Daly and South Alligator Rivers in northwestern Australia and from the Mitchell River in northern Queensland.

Description. The name comes from the aborigine word *tanybar-arrga.* In the western part of its range, it share the Daly River with *E. victoriae* and *E. worrelli*. These species are very similar in appearance and morphology but differ principally in coloration. *Emydura tanybaraga* is identified primarily by its yellow head; from the snout to the base of the neck, a paramedian line divides the head into the dark gray of the upper surface and the bright yellow of the underside. A yellow line also starts at the snout, passes above the eye, and becomes wider above the tympanum. The iris is traversed by the same color; a black band across the center separates the two yellow areas.

The carapace is in the shape of a very elongated oval, widened behind and narrow in front, of a brown to orange-beige color, with more or less obvious brown vermiculations. The maximum

This turtle is very light-colored, with a carapace widened toward the rear, a short neck, and light markings between the eyes and behind the jaw.

size seems to be about 260 mm. The plastron is of a light whitish yellow color, sometimes with irregular reddish spots. The barbels may be distinct or almost invisible. The young have a dorsal keel and serrated marginals. They are brown to black, sometimes greenish with black spots. A yellow line encircles the carapace. The eye is very bright and is traversed by a black bar. The snout is quite short.

Natural history. According to Legler, the females lay from August to mid-November, usually producing about 15 eggs of around 32 × 26 mm in size.

Protection. Throughout the range, populations seem to be strong, but we really do not know any details of the present status.

Emydura victoriae (Gray, 1842)

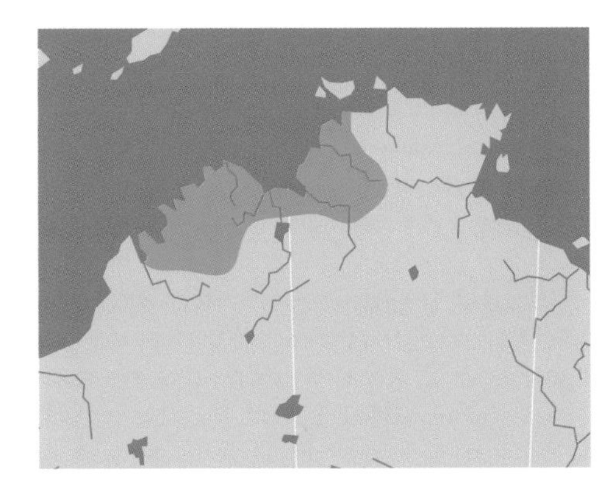

Distribution. This species is quite widely distributed in northwestern Australia, around Darwin and especially in the Victoria River.

Description. This is a reasonably large turtle (300 mm) with a neck of medium length. The overall shape is oval, with a slight posterior depression, especially in the young. The marginals are greatly flared and flat toward the back. The nuchal scute is not always present. The general coloration is not very contrasting, with tones of brownish black and sometimes light brown streaks. The plastron is light in color, whitish and sometimes spotted with pink or salmon. The head is enlarged behind the eyes but does not carry a casque, or helmet, as do certain other *Emydura,* except in old specimens. By way of contrast, on the Bullo River, specimens of this turtle may be found that show a major degree of macrocephaly, but again only in the oldest animals. The head is gray to brown, with an isolated light spot behind the eyes, and a pale neck. The skin of the temporal region is very fine. The young are beige to light brown, with a well-defined keel and feebly serrated margins. The head is lighter, grayish with rosy streaks, and the irides are bright yellow.

Natural history. This species is poorly known. It is shy and similar in lifestyle to *Chelodina rugosa,* which occupies the same habitat. The females nest in November, and the young emerge two and a half months later, although irregularity of rainfall may cause great variation in the timing of emergence. The turtle is primarily carnivorous, and its size might lead one to believe that it could overcome fairly large prey. But this is not the case; the species is not very active and is distinctly

Common name

Victoria short-necked turtle

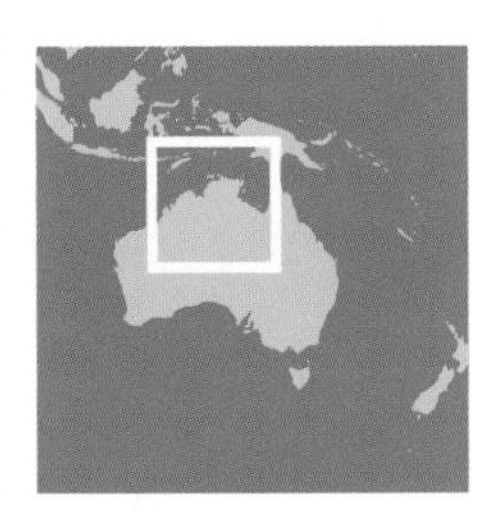

The head is small; the patch behind the eye, discrete. The carapace is often colonized by algae and moss.

unaggressive, which makes one think that it is content with sluggish or dead prey (amphibians, including tadpoles; invertebrates; etc.).

Protection. This species is still poorly known and rarely present in live collections in Australia or elsewhere.

Emydura worrelli (Wells and Wellington, 1985)

Common name

Worrell's short-necked turtle

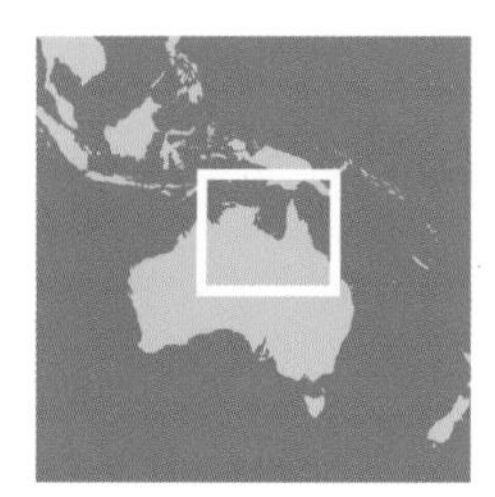

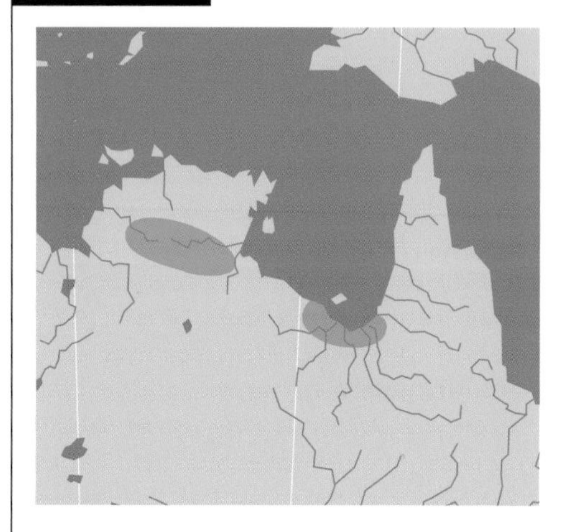

Distribution. These turtles apparently occupy a huge territory in the north of Australia, from the Katherine River (Northern Territory) to the Saxby River, 1,200 km to the east, in Queensland.

Description. The name is a patronymic for Eric Worrell, a Gosford herpetologist. The species is often subsumed within, or confounded with, *E. australis* or even with *E. victoriae* (sometimes by Worrell himself). The carapace is oval, very wide behind, and slate gray to dark beige or blackish. A yellow line starts at the snout, widens along the jaws and throat, then disappears around the middle of the neck, where it breaks up. Under the neck there may be grayish marbling on the yellow background. Another bright, light line passes from the upper surface of the orbit straight to the back of the head. Sometimes this line takes on orange tones; it is reddish in the young. The iris is not yellow and black, as in *E. tanybaraga,* but greenish yellow, often darkening with age. The plastron is rather wide in the anterior and yellowish in color. The length is at least 240 mm in females. Certain old animals become a uniform anthracite gray. The young have a wrinkled appearance,

This is a dark, almost black species, with just a light line behind the eyes and a rather light throat. The shell of this individual shows deep rugosity along the seams and the marginals eroded at the sides and rear.

without serrated marginals. Brown or grayish in color, they have a reddish line behind the eye and a yellow neck.

Natural history. This species feeds upon snails and mussels, as well as *Pandanus* fruits and various vegetative species. Nesting takes place between August and mid-October.

Protection. This species is probably sporadically consumed by aborigines, but there is no reason to believe that populations are endangered.

Hydromedusa maximiliani (Mikan, 1820)

Distribution. This species is endemic to Brazil and has a restricted range on the eastern coast in the states of Espírito Santo, Minas Gerais, Rio de Janeiro, and São Paulo, as well as the island of São Sebastião.

Description. The name *maximiliani* refers to the former emperor of Brazil, but the English name is a more descriptive one for this long-necked turtle. The carapace is oval and small, with parallel sides and a median depression. Each vertebral scute has a slight median keel on its posterior part, these together forming a series of pointed projections from the first vertebral scute to the beginning of the fifth. The keels disappear with age. The cervical scute, very wide, is situated behind the anterior pair of marginals, and its corners are oriented toward the front, rather than posteriorly, as in *H. tectifera*. The shell is very dark brown to dark gray, with an extremely smooth surface. The plastron, the bridges, and the undersides of the marginals are yellow to light brown, shaded with brown vermiculations. The anterior plastral lobe is quite rounded anteriorly and a little more narrowed toward the back. The head is small and the snout somewhat projecting. The upper jaw shows neither a hook nor a notch. The upper side is covered with very smooth skin, but behind the eye the scales are well developed. An excrescence of skin (actually a muscular fold) is visible externally on each side at the corner of the mouth.

Natural history. Some authors have called this species diurnal, but it seems to be active primarily on rainy days during the spring and summer (September to March). One food study, conducted in the Carlos Botero State Park, has shown that 65% of the diet is aquatic crustaceans, and 14% insect larvae. This turtle also eats spiders, cockroaches, and termites that have fallen into the water, as well as tadpoles and small fish. P. Pritchard thinks that the long neck, shaped by evolutionary selection, is a manifestation of its adaptation for rapid, agile prey. In the Carlos Botero park, high population densities may be reached—193.5 turtles per hectare. Predators upon this species include jaguars, cougars, raccoons, and possibly giant otters. At breeding time, the females appear to go to the places where males are to be found rather than the reverse. Nesting occurs in December and January, and hatching in September and October, with recorded incubation times of 250 and 300 days. One observed female laid three eggs 40 × 25 mm in size. The young are surprisingly large, about 50 mm long.

Protection. The range of this turtle, close to the coast in the densely inhabited Atlantic forest area, is threatened by human-caused habitat degradation. This species is listed as threatened and vulnerable at both national and international levels.

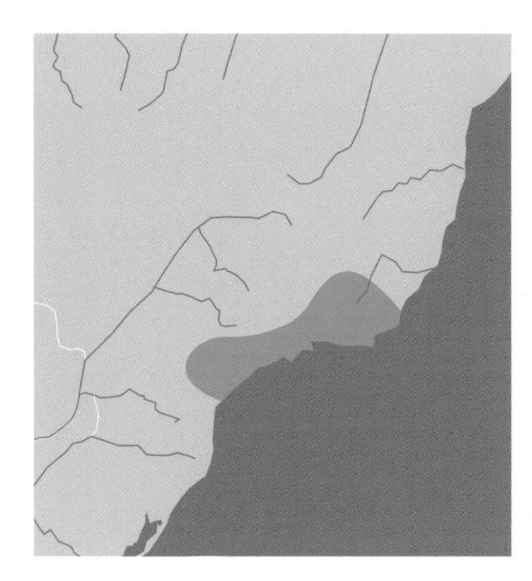

Common name

Brazilian snake-necked turtle

Hydromedusa tectifera Cope, 1869

Distribution. This species lives farther south than *H. maximiliani* and in a much wider area: from the southeast of Brazil (São Paulo, Rio Grande do Sul, Paraná) into Uruguay and as far as the northeast of Argentina (Rio de la Plata). Its presence in Paraguay has sometimes been suggested but needs to be confirmed .

Description. The maximum length of this species is about 300 mm. The carapace is ovoid and somewhat wider posteriorly and is characterized by rough-surfaced scutes, each of which is raised into a conical tubercle toward the rear. The unusual appearance is most developed in the young and, together with the presence of encrusting algae on the scutes, enhances the crypsis of the turtle in its habitat. With age, this sculpturing smooths over, and only a median keel may remain in the oldest animals, from the rear of the first vertebral to the beginning of the fifth. The carapace is chocolate-colored to very dark brown. The scutes are marked with black dots and dashes. The plastron, the anterior lobe of which is rounded in front and much wider than the posterior lobe, has a deep anal notch forming an angle of about 90 degrees. It is light in color, yellowish to beige, and irregularly spotted with dark brown, and the males have a deep femoral concavity. Certain individuals will develop very darkly pigmented seams on the plastron, which otherwise is little marked, whereas others have light seams delimiting the scutes. The underside of the marginals is equally light, with darker spots, or may include a black triangle covering half of the scute. Certain males may have a reddish color, particularly those from the area of Misiones, Argentina. The reddish ocher substrates in this zone may explain this particular detail of coloration. The head is of medium size, has a short, pointed snout, and is covered with small irregular scales. The upper jaw is neither hooked nor notched, and this species, in contrast to *H. maximiliani,* does not have a muscular fold at the corner of the mouth. The head is dark gray, sometimes brownish, and is underlined by a long, thick light stripe with black borders, which originates at the upper lip and goes as far as the base of the neck, running right across the tympanum. The underside of the neck also has discontinuous light lines, often in the form of a V,following the mandible. The neck of this snakeneck turtle is very long, perhaps equal to the length of the carapace. The feet are olive-gray above, with light areas below. There are four claws on each foot, as well as extensive webbing of the digits. In males, the carapace is lower behind than in front.

Natural history. This is a shy species that one never see basking at the edge of the water, and its habitat embraces swamps, lakes, ponds, irrigation canals, and muddy streams. It remains motionless on the bottom of the water, waiting for moving prey. The juveniles, with their algal camouflage, hide for long periods of time, and aquatic vegetation seems to be an important part of their environment. In coastal regions, this species tolerates slightly saline waters perfectly well. The diet is mostly carnivorous, largely composed of snails but also of mussels, insects, crabs, shrimp, amphibians, tadpoles, and small fish. It has the capacity to suck prey into the throat with a powerful inrush of water, as some other species may do also. Using its jaws, it is able to extract a snail from its shell. It may be seen leaving the water under cover of night, but it will race back to its aquatic medium when disturbed. In the southern part of its range, it hibernates in June, July, and August, in the muddy bottom or in a hole dug in the bank. Courtship is remarkable. The male bites the female on her forelimbs and neck and ends up mounting her. Coupling occurs in November to January. Incubation lasts for 70 days. The eggs are elliptical, white, and hard-shelled, measuring about 35 × 21 mm. The young, very rugose and very dark, measure 30 mm.

Common name

South American snake-necked turtle

Protection. The species has been observed being attacked and eaten by a giant otter in São Paulo State, Brazil. Even though some humid areas within its range are threatened by intensive agriculture, it does not seem to be in danger except in the biggest cities. It has been observed that this species in not pollution-sensitive. Isolated populations should be studied thoroughly; it is thought that subspecies may exist.

The species has a pointed snout, large eyes placed well forward, and a very long neck covered with nodosities.

© J. Maran

Mesoclemmys (Phrynops) gibba

(Schweigger, 1812)

Mesoclemmys is a recently revived genus. Its name comes from the Greek *mesos,* meaning "medium," and *klemmys,* meaning "turtle." The "medium turtle" is an intermediate between *Hydraspis* and *Platemys,* according to Gray.

Distribution. The range is very wide and is apparently divided into two blocks. One comprises eastern Venezuela, the island of Trinidad, the Guianas, and northeastern Brazil. The other, separated from the first by the llanos of Venezuela and the Guainía Plateau in Colombia, includes the southern part of the Orinoco Basin, eastern Ecuador, eastern Peru, and runs as far as the northern point of Bolivia and the entire western edge of Brazil.

Description. This is a small species, 160 mm long and 115 mm wide (with an exceptional record from Venezuela of 230 mm). The carapace is oval, slightly wider posteriorly. It has a high profile with a median keel, giving it the French name *bossue.* The marginal scutes have upcurved edges that form a gutter on each side, and there is a supracaudal notch. The shell is smooth in adults, somewhat rugose in juveniles. The general color is very dark brown to pure black. Some individuals may be lighter, possibly for environmental reasons. The plastron is large and oval and is wider

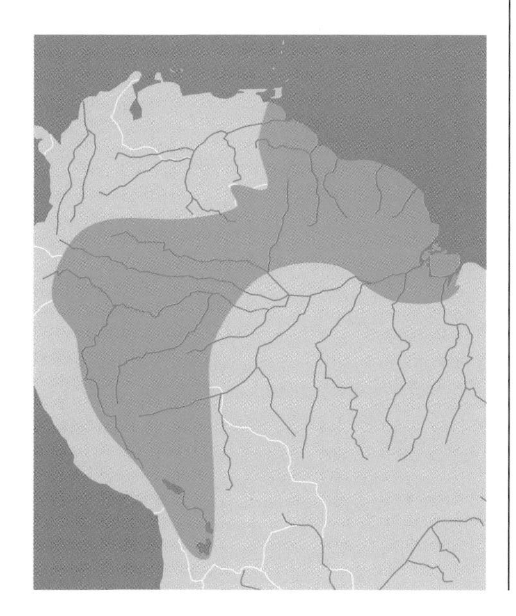

Common name

Gibba turtle

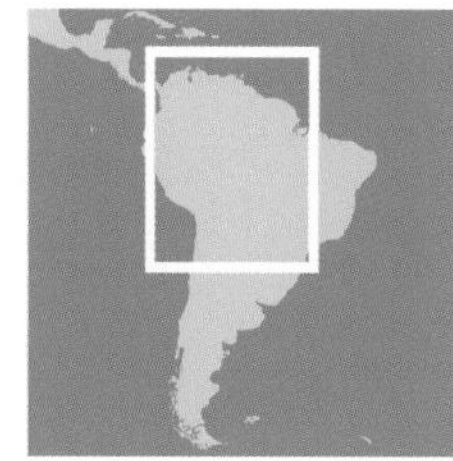

The deeply rugose carapace of this small species forms a rooflike shape. The head is bicolored, with light markings on an almost black background.

© R. Bour

and more rounded anteriorly. It has a medium anal notch. It may be light yellow with darker spots but is often almost totally black, with a fine light border. The head is rather small, the snout pointed and short. The top of the head is somewhat flattened and covered with small, irregular scales. In color it is dark brown to black, lightening at the base of the neck. There are also notable individual differences, certain animals being entirely black. The jaws have fine dark lines, and two small barbels are present on the chin. The feet are extensively webbed, with four claws on each forelimb and five on each of the hind limbs.

Natural history. This shy species is difficult to observe and largely crepuscular in nature. During the day, it remains hidden under debris at the bottom of the water. In the rainy season it may sometimes be found on land, especially at night. During drought periods, it buries itself in the ground or in mud. It never basks in the sun. It can emit a foul-smelling fluid, presumably to discourage potential predators. It is mainly found in forest pools, swamps, slow creeks, and shallow puddles. In diet it is mostly carnivorous (insects, larvae, tadpoles, crustaceans), but fecal analysis in French Guiana has shown that it will also ingest filamentous algae and the seeds and flowers of aquatic plants. The females sometimes deposit their eggs right on top of the ground; the eggs measure about 42 × 31 mm and are white, elongate, and hard-shelled. Incubation takes 150 to 200 days. Nesting seems to occur mainly at the onset of the dry season, thus timing the hatching to result at the beginning of the wet season. The hatchlings remain hidden in leaf litter for many months.

Protection. Even though this species is very small, it appears to be consumed by humans regularly. During the 1980s, it was offered for sale in terrarium magazines in the United States (P. Pritchard). Despite its inconspicuous lifestyle, it is much sought by collectors. In French Guiana, export is prohibited.

The genus *Phrynops* formerly comprised a dozen species, but after the revision of William McCord, William Lamar, and Mehdi Joseph-Ouni (2001), it now includes just four. The name *Phrynops* (phryne—toad and ops—face or head in Greek) describes them well.

Phrynops geoffroanus (Schweigger, 1812)

Distribution. This species has a very wide distribution in South America: southeastern Venezuela, eastern Colombia, eastern Peru and Ecuador, northeastern Bolivia and possibly Paraguay, the extreme north of Argentina, a large part of Brazil (but not in the lower Amazon Basin), the three Guianas, and the southern part of the Orinoco Basin.

Description. This is a large species, measuring up to 350 mm. The carapace is oval, very flat, and widened toward the rear, without a median keel. Older animals may have a slight median depression. The shell is dark brown, shading to anthracite gray or even to black. There is a yellow band around the periphery of the shell. The scutes are rugose and usually show evident growth annuli. The plastron is large and is rounded toward the front; the two lobes are of equal size. The anal notch is very large, and the bridges well developed. The plastron and the bridges are quite colorful in younger animals, with dark irregular spots and markings on a background of orange-yellow, red, or even pink. This coloration fades with age, becoming cream to yellowish, modestly shaded with gray black. The head is wide and large and has a pointed, protuberant snout. The jaws are not notched or hooked, and there are two barbels on the chin. The upper surface of the head is covered with small, irregular scales, dark gray to olive in color. The sides of the head are barred with distinct black lines. The first starts at the nostrils and passes across the eye and the tympanum, continuing along the neck. The second, following the neck, runs from the upper jaw to the shoulder. The skin in general is cream to yellowish. In some individuals the neck bears small soft tubercles. The underside of the head, like the jaws, is yellow or cream, marked with dark lines that continue the patterning of the plastron. The underside of the limbs is similar, but the soles of the feet are dark, almost black. The hind limbs are equipped with five claws, and each forelimb with four claws. There is extensive webbing of the digits. Sexual dimorphism is only modestly developed, the males having a less expanded posterior lobe and a very slight plastral concavity.

Natural history. This species prefers small slow-flowing streams, lagoons, and oxbow lakes along large rivers. It is diurnal but may sometimes be seen at night in shallow water. It has the habit of basking in the sun on banks or fallen trees and snags, with up to 15 individual sharing the same site. This is a shy species that cannot usually be approached within 50 m before the flight reflex takes over, and the animal drops into the water. At night it sleeps among submerged logs and branches. In Venezuela, numerous turtles have mutilations of the feet due to piranha bites (Medem, 1960). The diet is strongly carnivorous and includes small fish, aquatic insects, snails, and certain vegetative species. The females lay their eggs in rudimentary nests, several meters from the edge of the water when the water is at a high level during the rainy season. But in the southern zone the rainy season may extend for a longer time. The nests are about 15 cm deep, and the eggs are hard-shelled, white, and spherical, 25 to

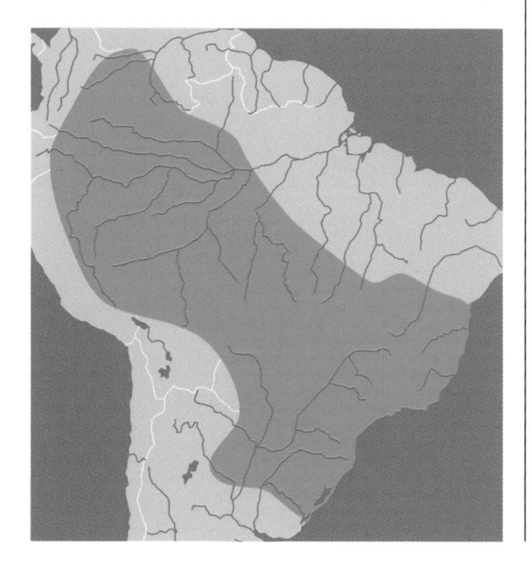

Common name

Geoffroy's side-necked turtle

© F. Bonin

The carapace of this species is rather elongated and flat on top. The neck is long and thin, with a serpentine snout, lightened by the pale lines that extend to the base of the neck.

31 mm in diameter. Clutches contain between 10 and 20 eggs on average, 30 on occasion. Incubation lasts for about 120 days, and the hatchlings, 38 to 48 mm in length, weigh 8 to 14 g. They have rugose carapaces, with a very minor median keel.

Protection. This species is consumed by certain ethnic groups, although the flesh is said to promote a type of urticaria. The turtles are caught with harpoons or seized by hand, and the eggs are also eaten. As a general observation, the flesh of *P. geoffroanus* is not of high repute in gourmet circles, having a rather fishy taste. On the other hand, collectors haul them away in numbers. There is a breeding center in São Paulo, Brazil, with the objective of both promoting studies of the species and conserving it in that country.

Phrynops hilarii (Duméril and Bibron, 1835)

Common name

Hilaire's side-necked turtle, spotted-bellied side-necked turtle

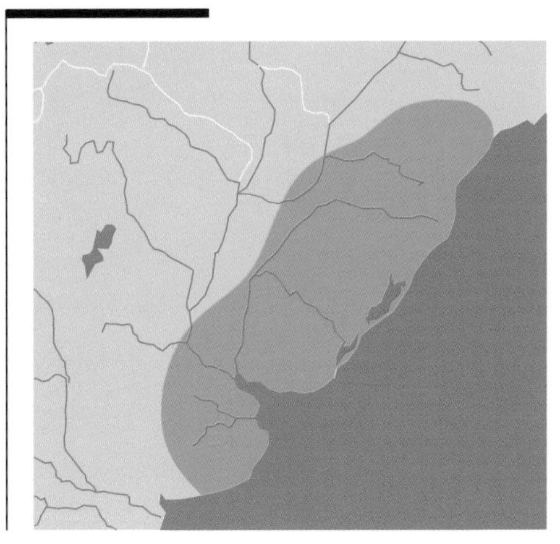

Distribution. This species is found in southern Brazil in the provinces of Rio Grande do Sul, Santa Catarina, and Paraná; in Uruguay; and in northern Argentina in the provinces of Córdoba and Santiago del Estero. It may be present in southern Paraguay, and hints of its existence in Bolivia need investigation.

Description. This is certainly the largest member of the genus, reaching as much as 400 mm. It is also the most elegant of the group, with its wide, dark gray shell, spotted with a multitude of white dots. The shell is flat, slightly flattened along the vertebral scutes. The general shape is oval, but

some individuals have a parallel-sided carapace. The marginal scutes are very wide, except for the very narrow nuchal. The scales are sometimes rugose, but with age they tend to become smooth. The outline of the marginals is always delineated by a light line. The well-developed plastron is widest at the level of the anterior lobe. The anal notch has an angle of about 90 degrees. The bridges and the undersides of the marginals are yellow to light beige, with irregular dark spots. This coloration is especially bold in the young and is maintained by most adults. In other adults, however, the pattern disappears. The head is large, flat, and widened at the level of the tympana. The snout is short and pointed. The top of the head is mid-gray, somewhat lighter than the carapace. The scales of the head are small and irregular. The eyes are protuberant and remind one of those of a frog. The iris is cream or bright yellow, traversed by a black line that extends from the nostrils, running the length of the head and passing over the tympanum. The top of the neck, covered with small dispersed tubercles, is gray, like the surface of the head. The underside is light yellow to cream, sometimes relieved by small black dots. The jaws have neither hook nor notch and are gray in their upper section and lighter below. There are two small chin barbels. The limbs are often bordered by two light lines. They carry five strong claws posteriorly and four claws anteriorly. The digital webbing is well developed and black, like the soles of the feet. Sexual dimorphism is slight, but the plastron of males is somewhat concave and the carapace flatter than that of the females.

Natural history. This turtle prefers sluggish streams, lagoons, weed-choked swamps, and rivers with a muddy or sandy bottom. It is not aggressive toward its conspecifics, and group basking is often observed. During the hot season, the animals may be seen floating sleepily on the surface or napping on branches. During the winter, they hibernate at the bottom of the water, but when the days become warmer, they resume activity. The diet is almost completely carnivorous (insects, mollusks, fish, amphibians, and even small mammals). They hunt in the water during the day and do not seem to be active at night. Courtship occurs in the classical fashion: the male sniffs the cloacal region of the female, bites her on the hind limbs, and climbs on top of her carapace while biting her neck and the crown of her head. Copulation is observed in September to October in the northern part of the range and in January to early February farther south. The eggs are spherical and white, with a hard, smooth shell. Clutches number up to 23 eggs, and the diameter ranges from 27 to 37 mm. Incubation lasts for 70 to 140 days. The newly hatched young have a slight median keel and a bright brown color, with a dark spot on each scute.

Protection. The large size and spectacular coloration make this animal a prize for collectors. In Argentina and Brazil, it is eaten, as are its eggs, which are boiled or cooked with potatoes to make an omelet. On the other hand, in Uruguay, the local people consider the turtle to be useless and its meat poisonous. The natural habitat is subject to widespread alteration and destruction. As of the time of this writing, the species does not appear on any protected list.

This beat-up animal, bearing the scars of the years, still shows signs of vigor. One often see turtles on banks or tree trunks, beside slow-flowing rivers, sunning themselves hour after hour. But they dive off at the slightest noise.

Phrynops tuberosus (Peters, 1870)

finely outlined in yellow. The head is gray, with a thick longitudinal cream band that starts at the nostrils and ends at the shoulders. The light line is bordered above by a black line starting at the orbit and another along the jaws.

Natural history. This species is often observed basking on dry land. The diet is mainly carnivorous: fish, insects, aquatic invertebrates. Reproduction and habits are similar to those of *P. geoffroanus*.

Protection. This species is not extensively hunted, and populations are largely intact.

This turtle has an almost rectangular carapace, with a vertebral depression in place of the usual median keel. The head is broad, covered with enlarged scales, and protected by a temporal helmet.

Common name

Peters's side-necked turtle

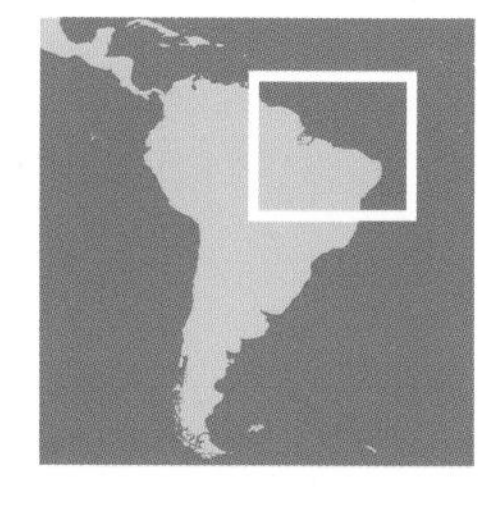

Distribution. Formerly a subspecies of *Phrynops geoffroanus,* this turtle occupies the southeastern part of Venezuela and is also found in Guyana, Suriname, French Guiana, and the entire northeast of Brazil as far as the region of Salvador.

Description. This sister species of *P. geoffroanus* resembles the latter closely. Some individuals have a median keel and are marked on each side by a series of outgrowths on the costal scutes, forming a discontinuous line. The carapace is dark brown, lightened by a reticulum of mahogany-colored streaks and dots. The edge of the shell is

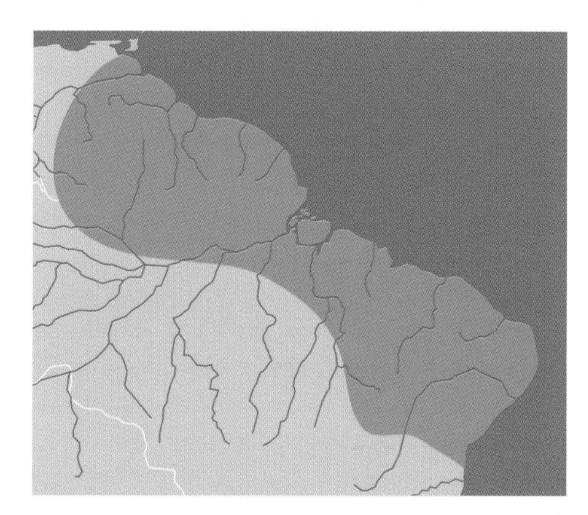

Phrynops williamsi
Rhodin and Mittermeier, 1983

Distribution. The range is limited to the south of Brazil (Santa Catarina and Rio Grande do Sul), the north of Uruguay (Uruguay and Cuareim Rivers), the north of Argentina, and the Rio Paraná Basin in southern Paraguay.

Description. This species in named after Ernest E. Williams, the former curator of reptiles at the Museum of Comparative Zoology at Harvard (Massachusetts). It is close to both *P. geoffroanus* and *P. hilarii*. The carapace is oval, widened at the level of the seventh and eighth marginals, and very flat, without keel or median depression, but certain individuals have a modest elevation of the carapace at the fifth vertebral scute. The scutes are slightly rugose, but growth rings are not evident in the adult. The general color is very dark brown, almost black, with fine light ocher reticu-

lations at the areola of each scute. These lines may have a radiating pattern or may be parallel. The marginals, which are slightly serrated, are underlined by an orange-yellow border, especially in the juveniles. These features disappear with age. The plastron is rather wide, a little wider at the level of the anterior lobe, and has a feebly developed anal notch. In color it ranges from white or light gray to very pale yellow, sometimes immaculate but in other specimens covered with diffuse spots. The head is not very wide, and the jaws are neither hooked not notched. The iris is yellow, and the chin bears two light-colored conical barbels. The skin on the top of the head is incompletely divided into small irregular scales and presents a smooth appearance, as is also the case with the underside of the neck. The species can be identified by the coloration of the head: the upper side is very dark, almost black, except for a wide light line that runs from the nostrils to the base of the neck. The band is underlain by a black band that originates above the upper jaw and also runs to the base of the neck. A third line, black in color, appears on the throat, where the ground color is pale yellow or even reddish. The line, in the form of a horseshoe, has its rami directed posteriorly. The neck, doubtless the shortest of any *Phrynops*, is decorated below with large black irregular spots. The limbs are powerful, the upper surface being of a dark brown to black and the underside reddish or yellowish, spotted with black. There is strong digital webbing. There are five claws on each forelimb, and four on each hind limb. The internal aspect of the palms is black, bordered with orange yellow. The soles of the feet are black. The females may reach 350 mm, although the biggest males are not longer than 200 mm. The juveniles and the subadults often have contrasting markings on the plastron, pale yellow spotted with brown.

Natural history. This highly aquatic turtle is not found at altitudes above about 500 m. It occurs in small forest rivers with clear water, rapid current, and pebble-covered bottom. The claws are strong and allow the turtle to climb across emergent rocks when it seeks basking sites. The diet is primarily carnivorous; the flattened form of the crushing plates of the jaws suggest that the main diet is mollusks. Copulation, which occurs in the water, has rarely been observed. Nesting apparently occurs in November and December (9 eggs on average, ovoid and white, with hard shells and measuring 32×34 mm. The hatchlings appear in March to April and have a brown carapace about 35 mm in length, and each scute is decorated with a large dark blotch. The plastron of the hatchling is sometimes very light gray, with a central dark line. The lines on the head are similar to those of adults but are more likely to take on shades of red or orange.

Protection. Populations are probably intact. This species is rarely eaten by humans, and it does not appear to be particularly sensitive to pollution.

Common name

Williams's South American side-necked turtle

This species is decorated with a pattern of beige, pointillist vermiculations on a dark background. The head is adorned with long, light-colored bands that extend from the snout to the top of the head and run from there to the beginning of the neck.

Platemys platycephala (Schneider, 1792)

Distribution. This species has a very wide range throughout the central part of South America, extending from the southern Orinoco drainage in Venezuela to the Amazon Basin in Brazil and including the three Guianas. It is also present in southern Colombia, eastern Ecuador, and northern Peru. In French Guiana, the species occurs in the coastal zone, as well as in the upper Maroni and the upper Oyapock Rivers.

Description. This is today the only species in the genus *Platemys*. As a result of the systematic revision of Ernst (1983), this taxon was separated from all the other former *Platemys* species, which were placed in the genus *Acanthochelys*. Its name describes it perfectly: it is doubly flat, not just in the carapace but also in the head. It is a small turtle, the females not exceeding 165 mm. The elliptical carapace has a strong median depression, bordered by two keels on the upper edges of the costal scutes. The lateral edge of the carapace is rolled up slightly, forming two shallow gutters. The scutes are smooth in adults, but in younger animals the growth annuli are clearly visible. The carapace is orange-brown or mahogany, sometimes light brown or orange, with somewhat extensive black areas toward the front. The plastron is almost entirely black, with just an orange or yellow-green border. The bridges are yellow, with black bars, as are the undersides of the marginals. The plastron is well developed, with a slightly raised anterior lobe that is somewhat wider and longer than the posterior lobe. The head is rather small, flat, and heart-shaped, with a single large scale covering the whole thing, from the point of the mouth to the beginning of the neck. The sides of the head are covered with polygonal granular scales. The top of the head is always light yellow-orange, and a darker area may originate between the eyes, extending to the occipital region and the remainder of the neck and limbs. The neck is entirely covered with well-developed, pointed, conical tubercles, possibly offering some protection when the animal withdraws its head. There are two minuscule barbels under the chin. The jaws are narrow and weak and neither hooked nor notched. The limbs have the usual complement of four claws on each hind limb and five on each forelimb and are dark brown, with light spots here and there, and covered with large, imbricate scales. The digital webbing is onlymodestly developed. This species has two special features: the male has a hook at the base of the foot, doubtless for holding on to the female during copulation; and the female has a slightly flexible suture across the plastron, which can open up at the moment of oviposition. A similar suture is undetectable in males.

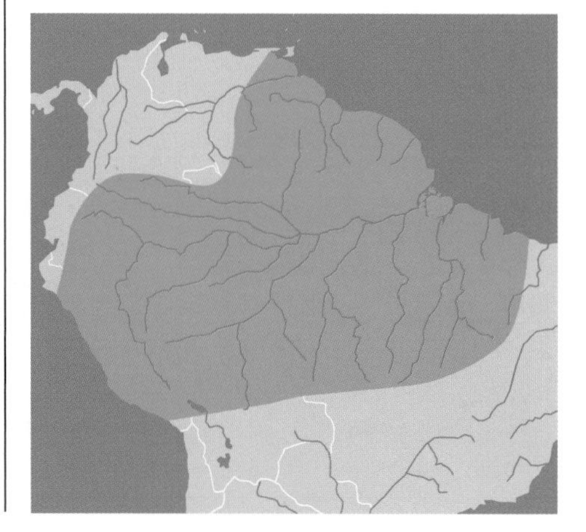

Common name

Twist-necked turtle

Subspecies. Two subspecies are recognized: *P. p. platycephala* (described above) and *P. p. melanonota* (Ernst, 1983), the western twist-necked turtle. The latter lives in the Amazon system, in the basins of the Santiago and Cenepa Rivers in Peru, and in the Napo and Curaray Rivers in Ecuador. It has a much darker appearance, being almost entirely dark brown or even black (*melanonota* comes from the Greek *melania,* or "black," and *notos,* meaning "back"). Only scant light areas of yellow-orange may sometimes be seen in the vertebral depression.

Natural history. The behavior of this turtle is strongly influenced by seasonal climatic variations. It lives in the substrate of humid forests and often moves around after tropical showers. It is never found in large rivers. It prefers lakes, creeks, or muddy puddles, and its crypsis is almost perfect. One doesn't see the little orange head; one sees only a heap of dead leaves. During the dry season, it buries itself in the ground and does not emerge until the next rain shower. It may also be active for part of the night. Many individuals are parasitized by leeches. J. Fretey (1981) found a turtle near Mana in French Guiana that had 81 leeches on

© D. Madec

This turtle is quite flat, as its name indicates, with a deep hollow along the top of the carapace and a similarly flat, small head. The orange to brown coloration allows it, at will, to remain in deep waters without being spotted.

its soft parts, although other specimens from the same site did not have any. The diet is opportunistic but includes amphibian eggs as a favored item. This is a peaceful, shy animal, totally nonaggressive, although some authors have described fights between males during the mating season. Coupling takes place from March to November and may occur on land or in the water, at dusk or even in the middle of the night. The male shakes his head and neck violently and bites the female on her extremities. He also rubs his chin barbels against the head of the female and may spray her with a jet of water from the nostrils or the mouth. Oviposition takes place during the dry season. The female lays a single, enormous egg at a time but may lay a second one a few weeks later. The egg is 45 to 50 mm long and 25 to 26 mm wide. The white shell is hard and seems to be unaffected by the prevailing humidity. Incubation takes 130 to 170 days. The hatchlings measure 60 × 46 mm and weigh about 20 g. Their coloration is similar to that of the adults, but in general the tones are brighter.

Protection. This species is rarely consumed on account of its small size, and it is not particularly sought by collectors because it does not do very well in captivity. In France, since May 15, 1986, it has been on the list of protected species in French Guiana, and capture is totally prohibited, as are hunting or trading throughout this overseas department of France.

Pseudemydura umbrina Siebenrock, 1901

Distribution. Populations of this turtle, considered to be the rarest in the world, are very reduced and are found on the Ellen Brook Nature Reserve east of Perth, on the Mogumber Nature Reserve north of Perth, and in a breeding colony at Perth Zoo.

Description. This is the smallest of the Australian turtles, not exceeding 150 mm in carapace length. The neck is quite short. The following features distinguish this species: gray in color (hence the specific name *umbrina*), carapace strongly flattened, head elongate and with an enlarged temporal scale or casque, plastron with gray sutures against a cream background and with small dark spots scattered about, and large well-spaced tubercles on the neck. The gular scutes are well separated by the intergular, which also separates the humerals. The young, similar to the adults but a little lighter in color, grow extremely quickly during their first few years, certainly having the fastest growth rate known in the Chelidae. They may reach 120 mm after three years, although these growth data were obtained with a captive colony, not in the wild.

Natural history. At the time of its rediscovery in 1953, nothing was known of the biology of this species. The information acquired at Perth Zoo by Andrew Burbidge and later by Gerald Kuchling's team made possible a better understanding of its reproduction, feeding, and ecological details. It is extremely adapted for its environment, where a drastic seasonal climatic cycle governs its activities completely. This is a freshwater turtle for part of the year, when it is immersed in the seasonal swamp, but it estivates "high and dry" during the arid season. This harsh and contrasting annual regimen doubtless explains both the rarity and the small size of this turtle. The species shows little sexual dimorphism. The females are only slightly bigger than the males. The best breeding results in captivity are found when the sexes have access to each other for only a brief period. Nesting occurs in November, with three to five eggs laid per clutch. On occasion, just a single egg may be laid. The nesting site is situated under a bush and in a shady zone, because the ambient temperature is very high at this time of the year. Incubation lasts for several months, and emergence occurs once the waters return and inundate the site. The rainfall pattern has total control over the existence of this species, which is forced to compress its annual activities of nourishment and reproduction into just a few months, when there is water in the swamp. When the water recedes and disappears, the turtle seeks out a shaded site, with much leaf litter and debris, where it digs a deep hole to find the degree of humidity that it needs. Once thoroughly buried, it will estivate for several months, several tens of meters away from the dried-up swamp. When the rains return, at the end of the dry season, it leaves its hiding place for a brief new cycle of activity.

In the rearing facility at Perth Zoo, the eggs are placed in an incubator at 29°C. After three months, this temperature is lowered to 24°C for three more months. Then the temperature is lowered still further to 19°C, and the eggs hatch. The hatching is induced by the change of temperature, which is designed to mimic the cycle in nature. It has also been discovered that these turtles dig their nest in a unique fashion, using the forelimbs. The female starts to dig a sort of gutter with its anterior limbs, somewhat as if it were burying itself for the summer season. It continues digging in this way until a depth of about 10 cm is reached, at which point she will turn around and adopt a near-vertical posture of oviposition. Laying of the eggs is quick, after which the turtle leaves the hole and closes off the nest with all four limbs. It is not clear if this behavior is atavistic, a vestige of ancient schemes, or if, by contrast, it is a modern innovation that this species alone has been able to develop. This turtle is carnivorous, eating crustaceans, insects, and amphibians. Under captive conditions, it is fed on insect larvae and other invertebrates raised specially for this purpose.

Protection. The first specimen was discovered in 1839 by Ludwig Preiss and deposited in the Vienna Natural History Museum, where it was finally named by Siebenrock in 1901. No other specimens were known in the wild. It was not until 1953 that a student at the University of Perth rediscovered the species in the wild and named this "novelty" *Emydura inspectata*. A little later, Williams demonstrated that *E. inspectata* was actually *P. umbrina*. Several additional animals were located in the suburbs of Perth, and these became the first breeding colony at Perth Zoo. Gerald Kuchling became the manager of this program, which was led by the University of Perth and the Department of Conservation and Land Management (CALM) of Western Australia. A decade ago

Common name

Western swamp turtle

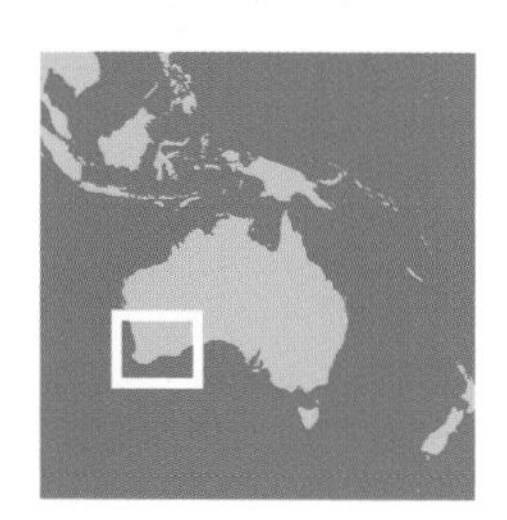

© F. Bonin

The coloration of this small species ranges from black to dark gray (hence the name *umbrina*). The head is somewhat rounded but flat, and the limbs are well webbed, wide and powerful, and provided with strong claws. This is a semiaquatic animal, able to dig in the mud at the bottom of lakes and also to swim rapidly.

the surviving population was estimated to number about 50 individuals. The extreme rarity of the species seemed to be due to progressively worsening dry seasons and predation by exotic species, including the European fox. Another introduced species, the rabbit, was equally harmful, in that its habit of burrowing constantly caused it to destroy the turtles' buried eggs.

The action program has been conducted on two fronts: the captive breeding effort at Perth Zoo and the improvement of a natural site close to Perth, the Ellen Brook Nature Reserve. The drying up of the swamps has been stopped by increasing the amount of water artificially, and high electric fences have been installed to control foxes. The rabbits have been controlled by means of poisoned food with a view to limiting their increase. In 2004 the zoo facility housed 200 juveniles and adults, the latter including 16 males and 14 females. Each year, 40 or so three-year-old turtles, about 100 mm in length, are released in the reserves, most of them with radio transmitters. A second reserve was created in 2001, the Mogumber Nature Reserve near Lake Wannamal, where about 20 adults have been released, also radio-equipped. The seasonal swamps at this second site have a natural regimen that seems to be appropriate for *Pseudemydura,* but there is no barrier to prevent access by foxes, which need to be controlled by farmers in the region. In 2004 the total population of the species was about 450 specimens—200 in captivity and 250 in nature. It is still too soon to declare victory, but this conservation program is a pilot project whose lessons are numerous and of interest to anyone who is attempting turtle conservation.

Protection. This species is classified as endangered by the Turtle Conservation Fund.

© G. Kuchling

In the lakes near Perth, it is not easy to observe this species, which is always shy and spends part of the year under moist vegetation or in the mud of inundated pastures. They may be seen in the Perth Zoo, which has financed part of the protection program.

Ranacephala (Phrynops) hogei

(Mertens, 1967)

The face of this turtle has a decidedly froglike appearance, as its name indicates. It is one of the least-studied species of *Phrynops,* and it is able to conceal itself with great efficacy within its habitat.

Distribution. This turtle is endemic to a small area of southeastern Brazil, in the coastal watersheds of the Rio Paraíba do Sul (states of Rio de Janeiro and Minas Gerais) and in the basin of the Rio Itapemirim (state of Espírito Santo). The populations are reduced and apparently isolated from each other.

Description. The name derives from the Latin *rana* (frog) and the Greek *kephale* (head). This species is still very poorly studied and seems to be ignored both by local people and by scientists. Its carapace may reach a length of 350 mm. The carapace is oval, slightly domed, and keelless and has a slight depression between the second, third, and fourth vertebral scutes. The marginals are somewhat rolled up to form lateral gutters. The overall color is a uniform brown, and the scutes do not have any actual markings or even rugosities. The plastron is large, with the anterior lobe larger than the posterior. The anal notch is very deep. The plastron is entirely yellow, except in certain individuals in which it may have brown or gray speckles. The head is especially small, a remarkable difference from other *Phrynops*. The snout is pointed and projecting, and there are two strong barbels on the chin. The head is bicolored—dark gray-brown on the top and yellow to whitish below. The females have a dark red line running along the side of the head, separating the differently colored areas. The jaws are yellow with gray at the joint, and they are neither hooked nor notched. The eye is brown, finely outlined with yellow. The feet are dark above and yellow-orange below.

Natural history. This species is not especially aquatic. It is found in lakes and small watercourses, in regions below 500 m altitude. Diet is unknown, as are details of reproduction.

Protection. Even though the species has not been studied in situ, its populations are surely reduced, for the habitat has been heavily deforested and subject to progressive aridification. It would certainly appear to be an endangered species. Various authorities have correctly identified it as endangered—IUCN included the species in its 1982 *Red Data Book* for amphibians and reptiles, and the Turtle Conservation Fund included it in a list of the 48 most endangered turtle species in the world (2002). A captive reproduction program and the restoration of habitat would both seem to be vital steps toward the survival of this species.

Common name

Hoge's side-necked turtle

Rheodytes leukops
Legler and Cann, 1980

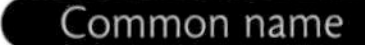

Fitzroy River turtle

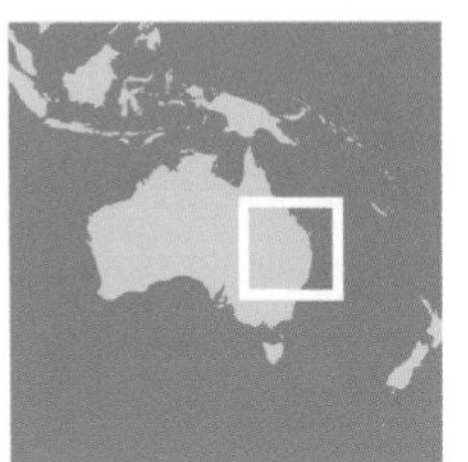

Distribution. This turtle has a minuscule range in the eastern part of Australia, in the Fitzroy River a few kilometers from the sea. It was not known to science until J. Legler and J. Cann discovered it in 1976.

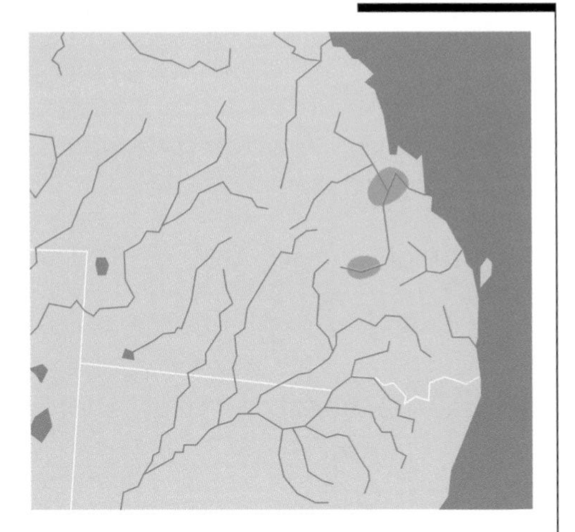

Description. *Rheodytes* is different from all other Australian turtles and indeed merits a genus of its own. This is a short-necked pleurodire, not more than 210 mm in length. One of its characteristics is the stark white iris, which, along with the pronounced elevation of the posterior part of the mouth, gives this turtle a strange "smiling" physiognomy. It is slow-moving and unaggressive, walking on land with relaxed deliberation. It walks the same way underwater. J. Cann has observed that a specimen placed on a rock will stay there for a long time, with head held out, like a statue. The tubercles of the neck are particularly large and prominent, although relatively small in number. *Rheodytes leukops* is brown, with orange or bronze tints and irregular darker areas. The plastron is long, narrow, and cream-colored or pale yellow. The intergular scute is narrower than either of the gulars. The monotypic genus is close to *Emydura* but differs in the following characteristics: the third and fourth costals are in contact with the sixth and eighth marginals, respectively; the forelimbs have five claws; a single pair of barbels is present on the chin; and there is a wide white ring over the iris. Australian turtles often use their cloacal bursae to extract oxygen from the ambient water, but J. Cann has observed that in this species the cloaca remains open for long periods of time, much longer than in other species.

Natural history. This species is still poorly known, but its populations are reduced. It appears to lay about 16 eggs—not bad for a turtle less than 210 mm long. The food consists mainly of insect larvae. Its sheer lethargy leads one to be-

© F. Bonin

This strange turtle has large white eyes and a wide, smiling mouth. But the surprising thing is its behavior. On the ground it moves very slowly, like a mechanical toy, but once back in the (preferably muddy) water, it reveals its considerable swimming prowess.

© F. Bonin

The great development of the temporal region and the jaws indicates that this species feeds on mollusks.

lieve that it would not be very successful at catching fish or other nimble prey. It is also somewhat surprising that it can survive in an environment where there are numerous predators, including large eels, catfish, and even saltwater crocodiles. Its only hope for survival would appear to lie in its habit of hiding among the rocks at the bottom of this fast-flowing river and remaining immobile. Early observations suggested that this was a carnivorous species, but in reality it may be equally an herbivore, often living in areas of dense *Vallisneria* grass, which it consumes in quantities. Perhaps it should best be considered as opportunistic.

Protection. In view of the remoteness of the habitat and the small number of individuals observed, the rarity of the species implies the need for monitoring and for studies as to how to best protect it. Along with *Elusor macrurus* and *Pseudemydura umbrina,* it is one of the rarest and strangest turtles of Australia and one whose conservation is far from guaranteed. All three of these species need highly specific survival plans that prevent overcollection (indeed, *any* collection) and provide environmental protection.

Rhinemys (Phrynops) rufipes (Spix, 1824)

Common name

Red-headed sideneck turtle

Distribution. The range is limited to the upper course of the Amazon in western Brazil, extending to southeastern Colombia and possibly the northeastern frontier of Peru.

Description. This species is sufficiently distinct from other *Phrynops* that it has become the only species in the revived genus *Rhinemys*. Basically, as indicated by both its Latin name and its vernacular, it has bright red coloration on the head and neck, unlike other *Phrynops*. The head is rather wide and flattened on top. The snout is pointed and protuberant. The temporal regions and the cheeks are covered with irregular scales. The skin of the top of the head is marked with a dark band. Another band, equally dark, is present on each side of the head, running from the nostrils to the base of the neck and traversing the orbit. Two small barbels are present on the chin. The legs and neck are equally red, augmented with irregular lines and spots. The oval carapace may reach a length of 260 mm and is strongly keeled. From the front, the appearance is that of a roof with a top angle of about 90 degrees, making the species instantly recognizable. The carapace is dark brown. The marginals are lightly serrated, sometimes with a light border. The plastron and bridges are yellow, darker or lighter from one individual to another but never with markings or spots. The plastron is strongly concave in the males, without doubt an accommodation to the high keel on the female. The anterior lobe is larger than the posterior. The anal notch is strong, and the limbs are quite powerful, well webbed, and equipped with five claws in front and four behind.

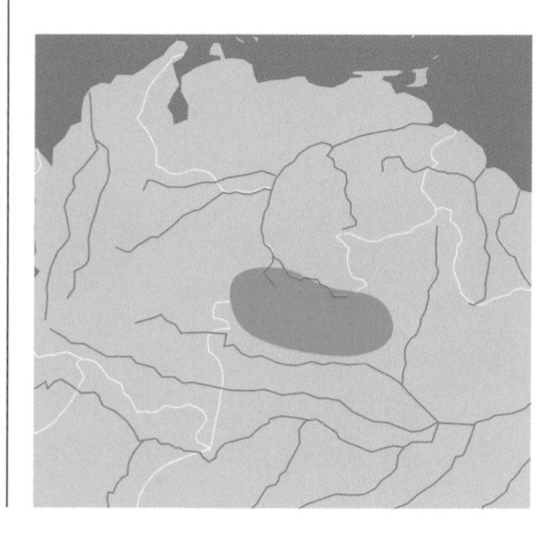

Natural history. Populations of this species appear to be reduced and fragmented. Formerly considered to be a great rarity, it has recently been found in the forest reserve of Adolphe Duke in quite good numbers. It lives in creeks and forest streams, fast or slow-flowing, where the water may be clear or turbid. On rare occasions it has been seen basking. It patrols the bottom in the search for food, but its diet is essentially omnivorous, and it seems fond of palm fruits that fall into the water. Sexual maturity of females is achieved at an age of 6 to 10 years. In Colombia, nesting has been observed from December to February but may also occur from the end of June to early August. The eggs number 3 to 12 per clutch and are white, spherical, and about 40 mm in diameter. The young have a light brown carapace and a strong keel. The marginals are strongly serrated. The red mask of the head and the coloration of the limbs are especially vivid in the juveniles but fade with time, and in old females the coloration is more brown than red.

Protection. No doubt the species is hunted and even caught on hook and line by local people, but the worst threats come from deforestation of the region. It is also sought after by hobbyists, mainly because of its beautiful colors, but it does very poorly in captivity in stagnant water, and its skin quickly becomes covered with outbreaks of fungus. It is listed by IUCN as endangered.

PELOMEDUSIDAE Cope, 1868

These pleurodires are the most ancient of living turtles, having appeared about 120 million years ago. They are found only in the Southern Hemisphere, especially in Africa and Madagascar. There are two subfamilies: the African Pelomedusinae, with emarginated skull roofs, and the South American and Madagascan Podocneminae, with extensively roofed-over skulls. The five genera are Pelomedusa, Pelusios, Podocnemis, Erymnochelys, *and* Peltocephalus.

Subfamily Pelomedusinae Cope, 1868

Pelomedusa subrufa (Lacepède, 1788)

Common name

Common African helmeted turtle

Distribution. This species is the most widespread turtle in Africa, occupying the entire continent south of the Sahara. It also occurs in western and southern Madagascar and in Yemen, presumably carried there recently by human agency.

Description. *Pelomedusa subrufa* differs from *Pelusios* by having a poorly ossified unhinged plastron and five claws on the forelimbs. Turtles of this species are typically small (200 mm maximum), but individuals may reach as much as 330 mm in South Africa (P. Pritchard, pers. comm.). The laterally retractile neck is very short. The shell is flattened and is smoothly oval in profile. In the juveniles, there may be light serrations around the posterior rim of the carapace. The coloration of adults is variable but is uniformly beige to greenish, with the shade depending on the local environment. The head is wide and flat, capped by a

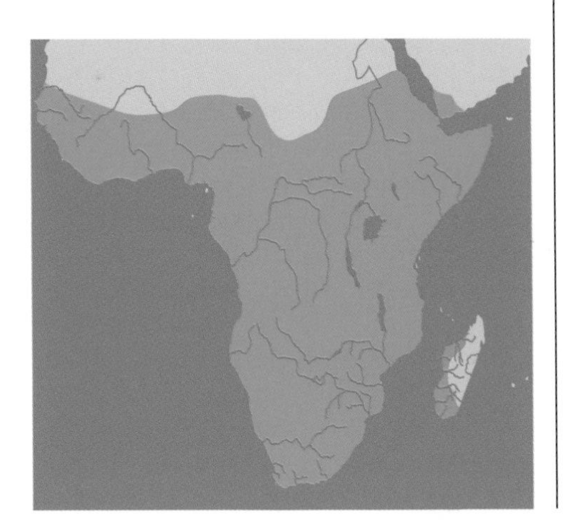

This turtle is quite hard to see in its natural environment. Brown to beige in color, it hides at the bottom of creeks or in the mud, where it estivates for long periods. This is an ancient turtle species, which today still occupies a major part of the African continent.

© B. Devaux

sort of "helmet" (hence the vernacular name) that is formed by two large supraorbital scales combined with the two temporal scales and a single large frontal scale. The color of the head usually mirrors that of the carapace: beige, brown, or olive, with perhaps some light or dark speckling on top, and a light-colored throat with two small yellow barbels. The plastron is yellowish to cream in color but may sometimes be darker, with the seams almost black. The plastron has no concavity, and is sharply cut off in front, posteriorly ending in a large anal notch. The bridges are quite narrow and are yellow to brown. The undersides of the marginals are yellow, but traces of dark triangles follow the seam lines, especially in *P. s. nigra.* The tail is the same color as the rest of the animal—that is, brown to olive above and cream to yellowish below. The feet are moderately webbed and have just two phalanges in each digit.

Subspecies. Three subspecies are recognized. They differ primarily in the arrangement of the scales of the plastron. *P. s. subrufa* (Lacepède, 1788): In the nominal subspecies, the eastern helmeted turtle, the pectoral scutes touch each other along the plastral midline. This subspecies occupies the largest territory in eastern Africa, from Sudan to Ghana and southward to Western Cape Province, as well as Madagascar (presumably introduced). *P. s. nigra* (Gray, 1863): The black-spotted helmeted turtle is found only in South Africa and in KwaZulu-Natal, from the Free State to Kimberley. The pectoral scutes make contact in the midline of the plastron, the latter being black to dark brown, with dark triangles on the light surface of the marginals and dark areas on the top of the head. Maximum size is 330 mm. *P. s. olivacea* (Schweigger, 1812): The western helmeted turtle is found to the north in Ethiopia and the Sudan, west to Nigeria and Cameroon. The pectoral scutes are widely separated, and the color tends toward olive.

Natural history. Extremely specialized for a demanding environment where heavy rainfall gives way to extreme drought, this species is found in swamps, ponds, and water holes in lowland areas and at altitudes up to 3,100 m. During the rainy season, it travels from swamp to swamp, and when these waters dry up, it buries itself deeply in the mud and estivates for months; during that time, the soil forms a hard crust under which a modest level of humidity may persist. Such behavior is also typical of some of the Australian sidenecks and contrasts with that of the related genus *Pelusios,* which needs permanent water. During the rainy season, these turtles often share their water holes with both humans and ungulates, without apparent inconvenience to themselves. They also like to bask on banks and emergent branches. They are opportunistic in diet but by preference carnivorous. The environment is lean, and they consume anything that comes their way: insects, crustaceans, fish, and amphibians, as well as terrestrial species such as snails, earthworms, small reptiles, young birds, and young mammals. They often hunt in packs, not hesitating to seize a sick bird or mammal larger than themselves, which they pull down underwater before dismembering it. They feed happily on carcasses, even very large ones. They have been filmed "cleaning" rhinoceroses, delicately extracting the parasites lodged in the skin of these huge mammals. There seems to be a kind of hierarchy within the group of turtles employed in this way, with the dominant individuals selecting the largest animals to clean.

Copulation occurs in spring or autumn, according to latitude. The male, with its neck held out, pursues the female and sniffs at her cloacal orifice. He then bites her on the tail and the hind feet and keeps a hold on her with his strong claws. The male energetically swings his head from front to back, blowing water out from his nostrils (Ernst, pers. comm.). Clutch size is quite large, up to 42 eggs per female (16 is normal), but apparently only one clutch is laid annually. The nest is quite deep and is flask shaped. The eggs average 38 × 22 mm and have a mucus-covered external membrane that protects them from desiccation. Hatching occurs 80 days later. The hatchlings are olive to blackish and about 30 mm long.

Protection. Ethnic Africans catch and consume the helmeted turtle quite widely. The turtles may also be sold to motorists by children standing beside the highway at the start of the wet season, when the turtles emerge from their mud cocoons and start to walk about. But they are also dug up with spades by villagers who are aware of their estivation sites. Nevertheless, the modest size of these turtles does not make them a cherished item of food, and the collection is not systematic nor does it lead to significant commerce. Popuations still seem to be in good shape throughout the range of the species.

Pelusios adansonii (Schweigger, 1812)

Distribution. This species is the northernmost member of the genus. It occupies a wide expanse of the Sahel-Sudan region, extending north as far as Mali and Senegal, through Niger, Nigeria, northern Cameroon, Chad, and the Central African Republic, and as far west as Sudan.

Description of the genus *Pelusios*. The Latin name *Pelusios* means "mud" or "clay," and these turtles indeed love to bury themselves in mud and to find refuge and food therein. They differ from *Pelomedusa* by having a hinge across the front lobe of the plastron. They do not undergo prolonged estivation in dried mud during the dry season; instead, they need to find a wet or humid place to survive. The head is wide and flat, with a seemingly "smiling" face. Color varies with environment, and the carapace often takes on the color of the substrate. Although collected in significant numbers by native people, populations of these turtles remain in good shape because they reproduce rapidly and the growth of the subadults is also rapid. Nevertheless, certain species in isolated places or with reduced populations need to be watched.

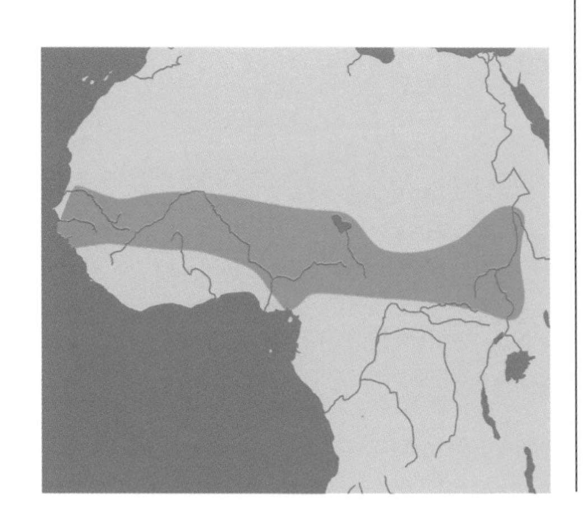

Common name

Adanson's mud turtle

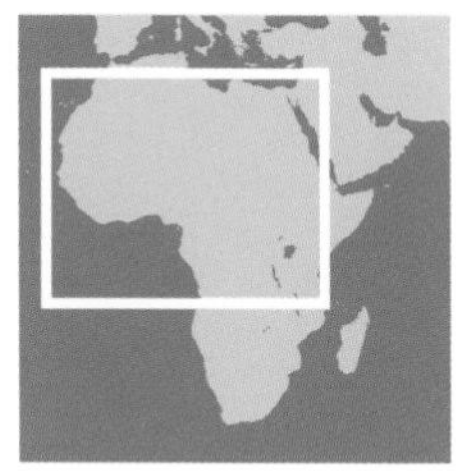

© B. Devaux

Nearly all the *Pelusios* species have a rounded, pebble-like form, which allows them to bury themselves during the dry season. They are gray to brown and have no nuchal scute.

The head of this species is wide, with well-differentiated tympana and with small scales on the top of the head. The snout is rounded.

© B. Devaux

Description of *P. adansonii*. Adanson gave his name not only to this turtle species but also to the genus of the baobab tree, after his studies in Senegal. This turtle is of medium size, about 220 mm maximum. It has an elliptical carapace, wider behind than in front, with a flattened vertebral area and a minor keel present on the first four vertebral scutes. There is no nuchal scute. The posterior marginals are not notched. The carapace is gray-brown, and the scutes are sometimes adorned with dark radiating lines or irregular black spots. The plastron is smaller than the carapace and cannot close the shell openings completely. The long, mobile anterior lobe is rounded in front, whereas the posterior lobe is narrow and rigid. The anal notch is well developed. There are no axillary or inguinal scales. The plastron and the bridges are yellow, either somewhat shaded or practically immaculate. Just the seams and sometimes a few growth rings are slightly darker. The head is strong and wide, ending in a short, rounded snout. The upper jaw is straight-edged, with neither hook nor notch. There are two quite large barbels on the chin. The head is gray or dark brown with yellow vermiculations. It is lighter, becoming pale yellow, underneath, as is the neck. In some individuals, a yellowish line adorns the side of the head, from the orbit to the tympanic region. The jaws are light yellow. The feet are medium sized, well clawed, and only slightly webbed.

Natural history. This turtle is found in rivers and calm waters, large ponds, flooded savannas, and all more or less permanent wet places within its range. During the dry season, it will spend several months under the mud at the bottom of the pond. Its diet is omnivorous, but mostly carnivorous; recorded prey include larvae and other invertebrates, amphibians, and fish, as well as carrion and vegetation. It nests several times per season (four or even five times). A clutch includes about 15 eggs, which are soft-shelled and measure about 30 × 18 mm on average.

Protection. This species is caught and consumed by local people. It also suffers stress from the arid environment, with the ever-present Sahara never far away. In Senegal, good populations exist in the Lac de Guiers, although they seem to be absent in the wetlands of Djoudj.

Pelusios bechuanicus FitzSimons, 1932

Distribution. Found in the center of Africa and in Angola, northeastern Namibia, northern Botswana (including the Okavango Delta), Zimbabwe, and Zambia.

Description. The Latin name comes from the former name of Botswana, "Bechuanaland." This species is without doubt the largest of the *Pelusios,* certain females attaining 330 mm (300 mm in males). The carapace is oval and elongate, with a pronounced dome, and is evenly rounded at the marginals, giving the animal the appearance of a smooth rock or huge pebble. The marginals are not serrated, and there is no nuchal scute. The carapace is very dark, often almost black, and lightens up to yellow or orange only at the sides. The vertebral scutes may have a very slight median keel, as a vestige of the shell form of the hatchling. The five vertebral scutes are longer than wide. The supracaudal scute is split. The plastron is well developed and can close the shell completely; the front lobe is rounded, articulates by means of a hinge, and is quite mobile. The black plastron is sometimes touched with a lighter color near the center. The bridges are dark, and there is no axillary scute. The head is wide, ending in a slightly pointed snout. The upper jaw is slightly cusped, although some individuals have completely smooth-edged jaws. There are usually three, sometimes two, barbels on the chin. The head is very dark, decorated lightly with a reticulated pattern of yellow or white. The feet are medium in size, each armed with five claws. On the forelimbs, the third claw is the most developed, but on the hind foot the second claw is the longest. Males have a slight plastral concavity, visible only toward the rear of the abdominals and on the femorals. The young have a vertebral keel that disappears rapidly with growth.

Natural history. This turtle is found in clear, deep, calm waters, in rivers, and in vegetation-choked sloughs. It is often seen in the water-

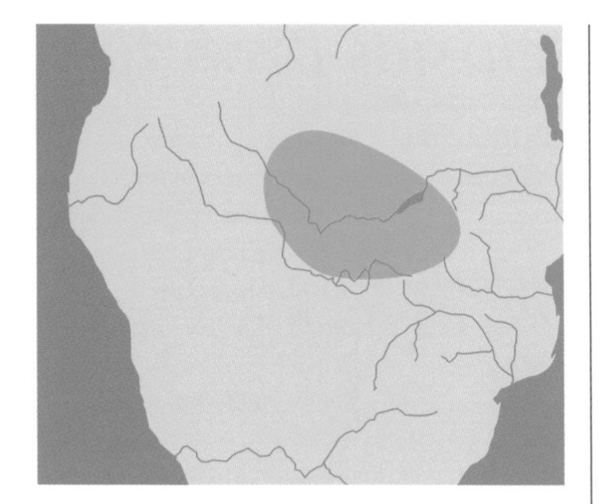

courses crossing the Okavango Delta, in Linyanti, and in the Zambezi River. This shy species is difficult to observe, except when it is looking for temporary aquatic refuges during the dry season. It is also a frequent prey of the formidable shoe-billed stork *(Balaeniceps rex)*. It is a carnivore, eating fish and invertebrates. The breeding season occurs during the southern summer. Clutches are quite large: 20 to 50 eggs, with fine, flexible, elliptical shells, averaging 36 × 22 mm.

Protection. The species is protected in the reserves of Okavango, Moremi, and Chone, as well as in the Zambezi National Park and Victoria Falls.

Okavango mud turtle

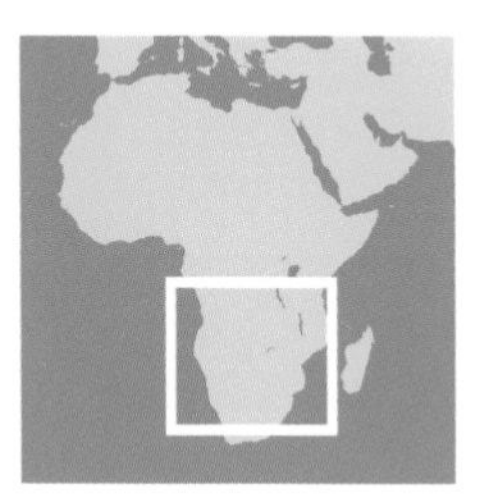

© B. Devaux

The carapace of this species is rather elongate, very smooth, and almost black in color, while the head is decorated with light yellowish markings and stripes.

Pelusios broadleyi Bour, 1986

Distribution. This species occurs only in a very small area in northwestern Kenya (Marsabit District) on the southeastern corner of Lake Turkana.

Description. Turtles of this species are small, not exceeding 155 mm. *P. broadleyi* is close to *P. adansonii,* and its carapace is elliptical, widening toward the middle. There is a median keel on the carapace, most developed on the third and fourth vertebral scutes, that reaches a point toward the rear of vertebral 4. The marginals are not serrated. A very small nuchal scute is sometimes present. The overall color is gray-brown, each scute being marked with fine dark lines and radiating dots. The plastron is dark brown to black and is almost completely rigid, without a functional hinge. The anterior lobe is round and well developed, although the posterior lobe is parallel-sided and narrowed at the level of the femoral and anal scutes. The plastron does not completely close the shell. There is a strong anal notch. The pectorals and abdominals contribute to the rigidity of the bridges, in the absence of axillaries and inguinals. The head is wide and large, with a short, slightly projecting snout. The upper jaw, lighter in color, is spotted and barred with black. The jaw is neither hooked nor notched. The frontal scale is very large. The head is basically brown, adorned with a light reticulated pattern on the top. The remainder of the skin is grayish, sometimes yellow-brown, becoming lighter below. There are two barbels on the chin. The forelimbs have a series of wide, enlarged scales on their anterior face. Juveniles have a very marked keel, which is lost rapidly with growth. They have a yellow plastron with dark spots on the bridge. This coloration darkens with age, but in some cases the plastron retains a number of yellow spots throughout life.

Common name

Lake Turkana mud turtle

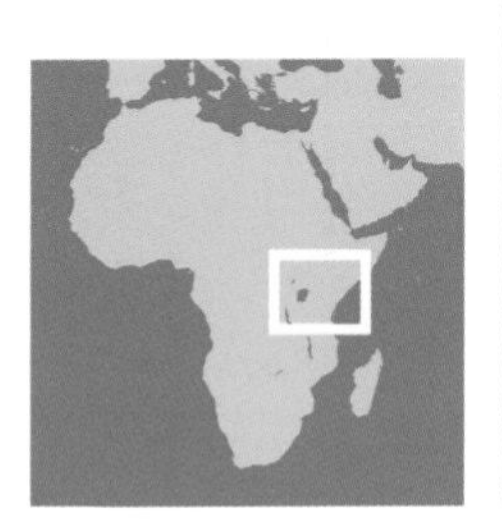

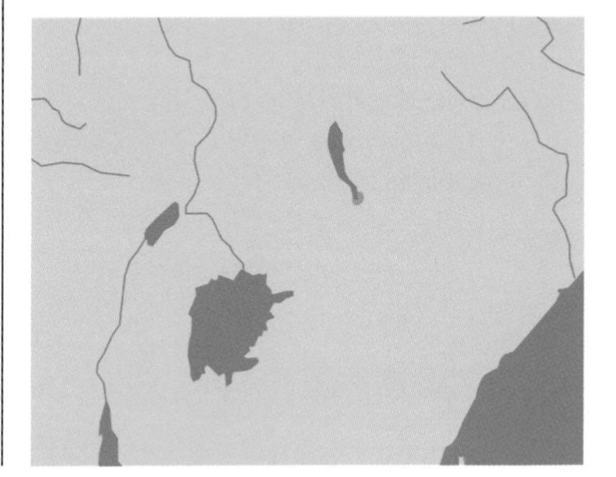

Natural history. This turtle lives in streams that flow into Lake Turkana. The young measure 25 mm at hatching.

Protection. Within its extremely limited range, the populations seem to be abundant, but the species is collected for food, and its rarity is a warning that should not be ignored. The status needs to be evaluated rapidly and protective measures inaugurated as necessary.

Pelusios carinatus Laurent, 1956

Distribution. This species is found only in the equatorial region of West Africa, in the main basin of the Congo River in the Democratic Republic of the Congo, and in southeastern Gabon, in the region of Franceville.

Description. This species may reach a carapace length of 265 mm. The tectiform carapace has a keel that is slight on vertebral 1 but much higher on vertebrals 2 to 4, with promontories on the posterior parts of the last two. The posterior marginals are serrated and uplifted. The overall shape is oval, expanded toward the middle of the carapace. The plastron has a rounded anterior lobe, narrowing posteriorly and ending with an anal notch. The bridges are short and lack axillary scutes. The plastron is yellow, with a black peripheral band, especially on the anterior lobe. The head is of medium size, relatively delicate for a *Pelusios,* rounded and terminating in a short, rounded snout. In color the head is gray or brown, adorned with a yellowish reticulated design. The upper jaw,

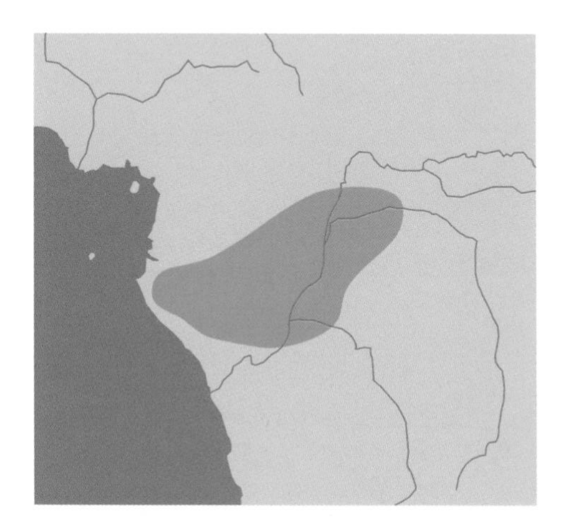

forth into flooded savannas during the rainy season. Mostly carnivorous or opportunistic, it eats fish, amphibians, aquatic mollusks, insects, and other invertebrates, as well as fallen fruits, seeds, and certain vegetables. The females lay 6 to 12 eggs.

Protection. These turtles are widely caught with fishing lines or enmeshed in nets extended across streams and rivers. The species appears to be still abundant, but the status of populations should be examined to determine whether conservation steps are necessary.

Common name

African keeled mud turtle

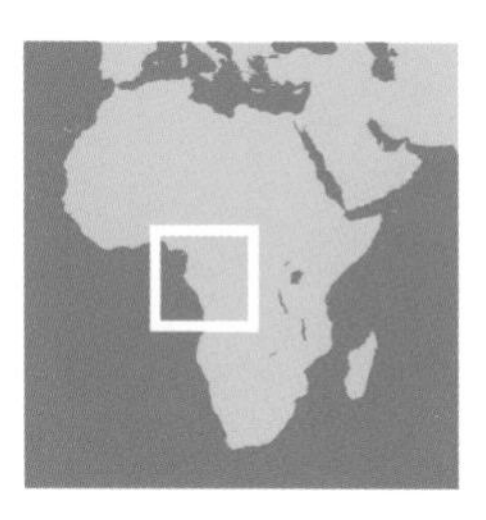

lacking hook or notches, is slightly lighter and bears dark vertical streaks. There are two barbels on the chin. The neck is rather long and gray, similar to the top of the head. This coloration may lighten up to yellowish on the underside of the head and neck. The limbs are of medium size, and the forelimbs have wide, enlarged transverse scales. The juveniles have a keel even stronger than that of the adults, and their vertebral scutes are generally wider then long, the reverse of the situation in adults.

Natural history. This species lives in large rivers, streams, lakes, and swamps. It may venture

This young *P. carinatus* still has an obvious median keel and a brightly colored head. With age, the head pattern fades.

© J. Maran

Pelusios castaneus (Schweigger, 1812)

Distribution. This species is found in a wide coastal band in West Africa, from Senegal east to the Democratic Republic of the Congo, as well as on the island of São Tomé. The species has been introduced to Guadeloupe (French West Indies), where it lives in wetlands between Le Moule, Sainte-Anne, and Saint-François, on Grande-Terre.

Description. The scientific name comes from the Greek *castaneus,* which refers to the chestnut brown coloration. The species may reach a length of 285 mm and is of oval form, widened behind. The carapace is round and fairly steep, resembling a large pebble or stone. Some individuals have slightly flattened or even depressed vertebrals, with a trace of a median keel. The general color is olive-brown, often with reddish touches, but most animals have a very dark gray, almost black shell. The marginals are not upcurved into gutters. The plastron is large and almost completely closes the

© J. Maran

The whole turtle is brown, but the tint varies from mahogany to blackish among individuals. The head is also brown and is dotted with cream to whitish spots.

This view of the species shows the light-colored carapace, but the tints always remain mahogany or burnt sienna.

© F. Bonin

anterior carapace opening. The rounded anterior lobe is articulated, whereas the posterior lobe is constricted. The plastron is mainly dark yellow, but often outlined with black along a narrow peripheral band. Some individuals have a completely dark brown plastron (Breuil, 2002). The anal notch is well marked. The bridges are wide, with axillary scutes. The head is medium sized, flat, rounded, and with a prominent snout. The head is brown, sometimes reddish, and has a light reticulated pattern. The jaws are vertically streaked with dark pigment. The upper jaw may show two small serrations or notches. There are two barbels on the chin. The limbs are yellow-gray, darker on top. The anterior limbs have large transverse scales.

Natural history. This turtle is found in wet, shallow environments with abundant vegetation, including swamps, inundated brushy savannas, and lagoons situated between the coast and the rain forest. It does not favor actual forests. It may be active by day or night and can be seen patrolling the bottom of the water in search of food. The diet is opportunistic, including fish, amphibians, snails, and insects, as well as seeds, fruits, algae, and various aquatic plants. In some regions, the species estivates in the mud during the dry season. This is not an aggressive species; it never tries to bite, but it may excrete a nauseating, oily yellow liquid. The females lay up to 25 soft-shelled eggs, which are white and oblong and measure about 35×25 mm. The young are almost black and remain hidden under vegetation or in the puddles where they reside. In Guadeloupe, nesting occurs in February and March, and the eggs number 6 to 20, approximately. The species probably came to the Antilles around the same time as slaves were imported from Africa.

Protection. This species, together with the sympatric *P. niger,* is heavily harvested for food. It also suffers from progressive drying of its habitats, as occurs near Dakar, Senegal, in the Nyayes region. On São Tomé, stuffed specimens are offered to tourists as souvenirs.

Common name

West African mud turtle

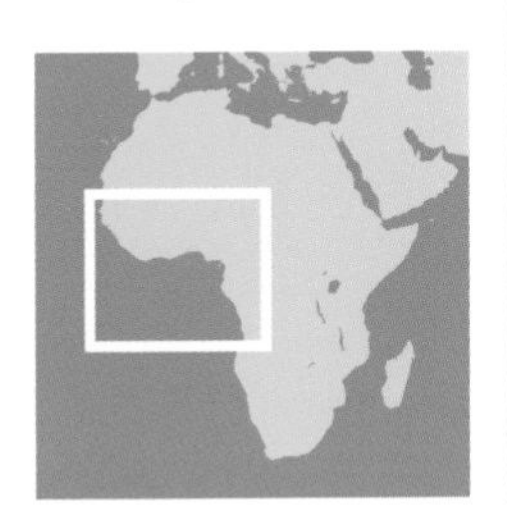

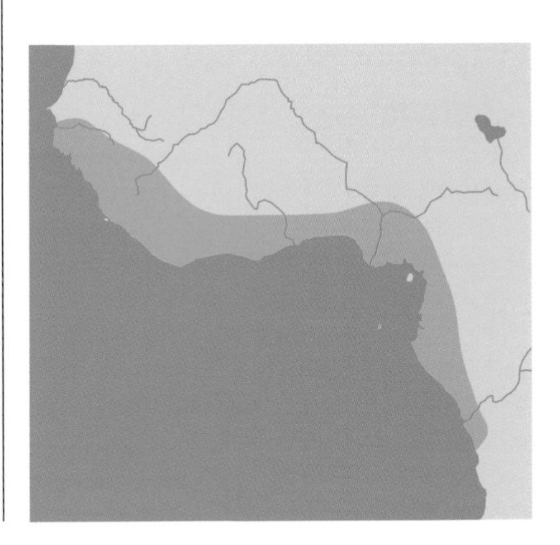

Pelusios castanoides Hewitt, 1931

Distribution. Found in East Africa, Malawi, Mozambique, Swaziland, and north to the city of Durban in South Africa, this species is also found in the Seychelles Islands and in Madagascar.

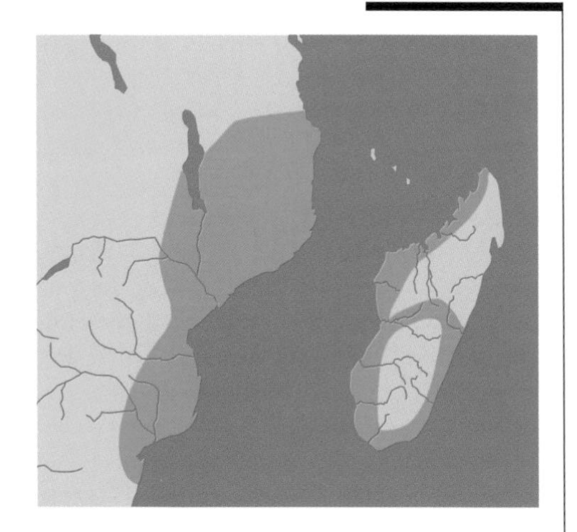

East African yellow-bellied mud turtle

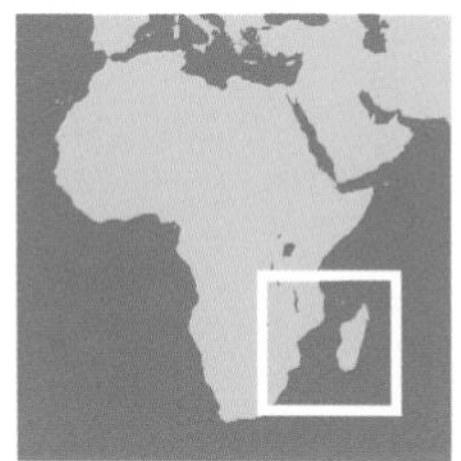

Description. The Latin name is based on the turtle's chestnut color *(castaneus* in Greek). This species may reach a carapace length of 220 mm and weigh up to 900 g. The carapace is domed, oval in outline, and widened posteriorly. The marginals are not serrated and may be slightly upcurved in the vicinity of the posterior limbs. The five vertebral scutes have a light median keel, principally on the rear of the fourth and on the fifth. The supracaudal is divided in two, and there is no nuchal scute. The carapace is beige, yellowish, brown, or more or less dark, sometimes nuanced with reddish. The plastron is large and covers the entire opening of the carapace. The anterior lobe is connected by a fully functional hinge and is rounded. The posterior lobe is lightly constricted at the level of the junction between the abdominals and femorals. There are no axillary scutes. The plastron and the bridges are yellow, sometimes with darker touches. The head is medium in size, wide, and flattened on top, terminating in a blunt, rounded snout. The upper jaw is sometimes bicuspid, but more often not. There are two chin barbels. The head is brown, with a yellowish vermiculated pattern. The eye is black, circled with silver. The limbs are light yellowish, as is the rest of the skin, and each limb is armed with five claws but little webbing. The posterior lobe of the plastron of the female is often wider than that of the males.

Subspecies. *P. c. castanoides* (described above): This subspecies is from the African continent. *P. c. intergularis* (Bour, 1983): In this subspecies, from

© J. Maran

This individual has a complex head pattern, wrought in light colors on a blackish background. Other specimens may have a more uniform coloration, ranging from beige to blackish.

the Seychelles, the intergular scute is pentagonal and separated from the gulars by parallel sides, whereas this seam is diagonal in the nominal race. Old individuals of *P. c. intergularis* may become very light in color, with just a few irregular dark spots.

Natural history. This species lives in shallow, muddy ponds and swamps. These habitats are often seasonal, and the turtles estivate for long periods during the dry season, beneath the hardened ground. The species is carnivorous, feeding principally on snails, as well as other invertebrates, fish, and aquatic vegetation. Nesting has been observed after the appearance of the first rains and continues through the southern summer. Five to 25 eggs are laid, elliptical in shape and measuring, on average, 32 × 22 mm.

Protection. *Pelusios castanoides* is protected only in formal reserves, such as the Greater St. Lucia Wetland Park, the Mdumu Game Reserve, the Sodwana Bay National Park, and the Lake McIlwaine National Park, and it is frequently caught for food. In the Seychelles, it is subject to tourism development of the wetlands and invasion of the swamps by water hyacinths. The populations are in decline, and the species is restricted to the Mare Anglaise on Mahé and Anse Kerlan on Praslin. Captive reproduction on Silhouette Island may ultimately reinforce the natural populations (Gerlach, 2002).

Pelusios chapini Laurent, 1965

Distribution. This turtle occurs in equatorial central Africa, from southeastern Gabon to the Democratic Republic of the Congo, Uganda, the Central African Republic, and Cameroon.

Description. Formerly a subspecies of *P. castaneus*, this species is now elevated to the rank of species. It can measure as much as 380 mm. The carapace is narrow, in the form of a dome slightly flattened around the second and third vertebral scutes, and without a keel. The marginals are cut away above the four limbs and are reduced at the sides. The carapace is black. The wide plastron completely covers the shell openings. The anterior lobe is rather short and is anteriorly rounded. A flexible hinge connects the anterior lobe to the midpart of the plastron, which is larger. There is a deep anal notch. The plastron is black, somewhat relieved with lighter coloration along the median seam and the hinge. The bridges and the undersides of the marginals are black. The head is rather wide and ends in a round, short snout. The top of the head and the neck are dark brown, sometimes tinted with greenish. The underside is lighter. The mandibles are whitish. The feet are quite strong, dark above and lighter below. In juveniles both the carapace and the plastron are black, and each scute is adorned with lighter, radiating streaks. The carapace forms a roof shape, with a modest keel that is best developed on the second and third vertebral scutes. The head is dark but has a lighter vermiculated pattern. The jaws are light, as are those of the adults.

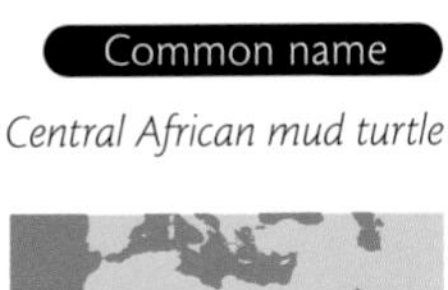

Central African mud turtle

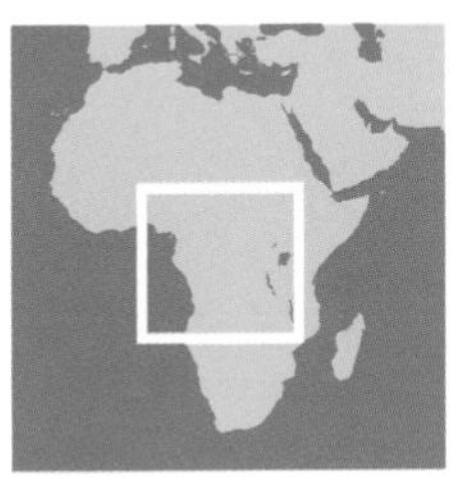

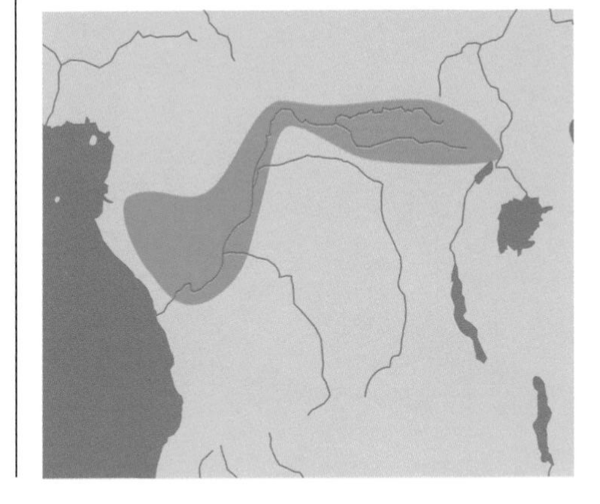

Natural history. This turtle occurs in brushy savanna, where it lives in lakes, streams, and rivers, both large and small. During the rainy season, it moves out into the flooded savannas; J. Maran (2002) noticed that it was even found in water-filled ditches in urban areas. It is omnivorous and

The carapace of this species is narrow and somewhat elongate. The head is sometimes greenish to dark brown on top and lighter below. The snout is rather short.

opportunistic, consuming aquatic vegetation and fruits that fall into the water, as well as insects, crustaceans, fish, amphibians, and dead animals. Reproduction is probably similar to that of *P. castaneus*.

Protection. Because of its large size, this species is heavily harvested for food, as are *P. niger* and *P. gabonensis*. It is rare in Gabon, and the remainder of the range is quite fragmented.

Pelusios cupulatta Bour and Maran, 2003

Distribution. This is a newly described species, discovered by J. Maran. The distribution is as yet incompletely known. It is present in the southwestern forests of Ivory Coast (Taï Forest, Mano River, Mani, Dodo River, and the Grand Berebi at San Pedro). It may be present in Assinie, as well as in Ghana, and there are a few individuals recorded from Nigeria.

Note the flattening on vertebrals 2, 3 and 4. The upper jaw is hooked. The head is gray on top and somewhat yellowish below, with discrete dark spots.

Description. The name comes from the Corsican word *cupulatta,* which means a "light blow" and also "turtle" and is the name of the Corsican association that financed the J. Maran's research. The species does not exceed a length of 214 mm (for the male). The carapace is elliptical, slightly flattened on the second to fourth vertebrals. The median keel is discontinuous and is really visible only toward the rear of each of the five vertebrals, especially on the fourth. The color is tobacco

© J. Maran

Dorsal and ventral views of this new species, recently discovered and described.

Common name

Cupulatta's mud turtle

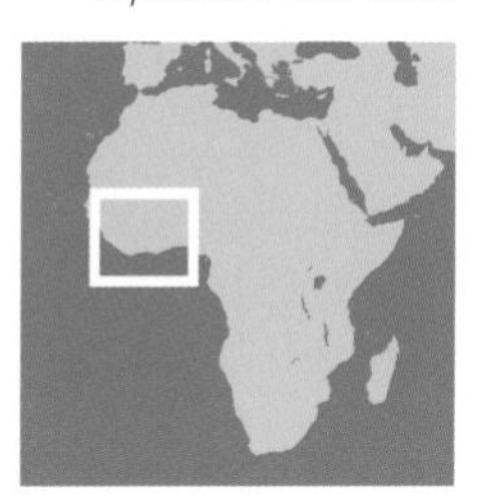

brown, with a very dark (almost black) median line running down the carapace and composed of a succession of small dark radiating spots. The line fades out on the first two marginals. The plastron is large and entirely covers the opening of the carapace. The anterior lobe is short and rounded in front, while the posterior one is very wide and large and ends in a well-defined anal notch. The plastral hinge passes straight across and is well defined. In color, the plastron is dark, sometimes with light spotting. The head is wide, flat, and modestly depressed along the top and ends in a rather long, feebly pointed snout. There are two chin barbels. The beak of the upper jaw is hooked and is bordered on each side by a deep notch. The head is grayish to yellowish above and on the sides, with small points and short dark lines. The iris is silver gray. The feet are gray, more or less dark, on the upper surfaces and whitish below. Numerous wide, rectangular scales cover the forelimbs. The shell of the juvenile is almost round, café au lait in color, and covered with a constellation of asperities that render it rugose. The keel is well developed and overlain by a dark streak that extends onto each vertebral scute, forming a series of squares connected by their corners. Each costal is equally spotted.

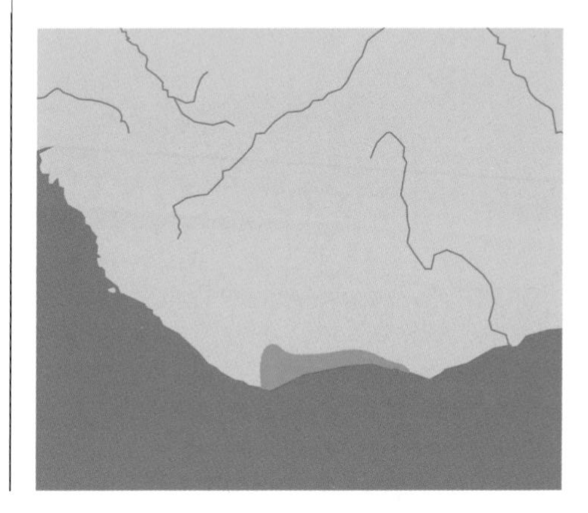

Natural history. This turtle occupies environments identical to those of *P. gabonensis:* creeks, calm watercourses, forested rivers, and flooded forests. No information on its biology is available.

Protection. This species features on the menu of local people, who catch it in various ways. Numerous midden remains have been found, which help in developing range maps for the species.

Pelusios gabonensis (Duméril, 1856)

Distribution. This species is found in a very large area of West Africa, including Ivory Coast, and in northern Angola, Cabinda, and the Democratic Republic of the Congo, up to the extreme northwest of Tanzania and Uganda. Its existence is uncertain in Liberia, Ghana, Togo, Benin, Nigeria, and Equatorial Guinea.

Common name

African forest turtle

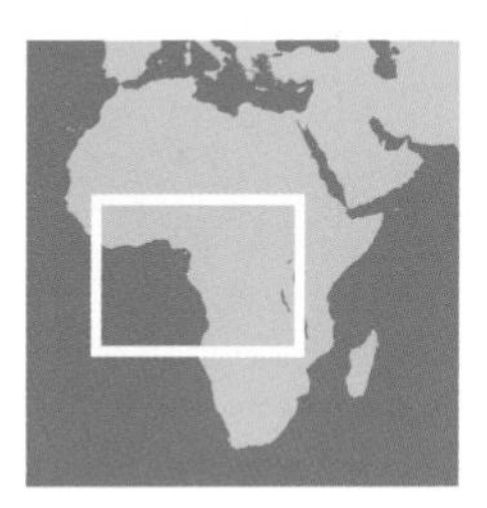

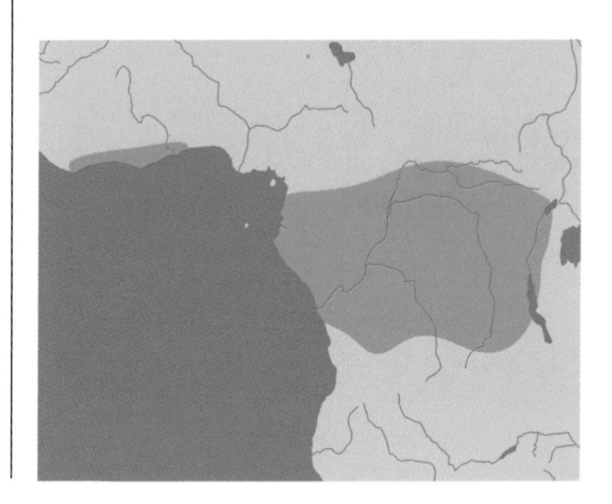

Description. This is a large turtle, measuring up to 330 mm. The shell is dorsally flattened and has a slight median keel, which is more defined in the young than in the adults. The carapace is chocolate brown, light or dark, with a broad, almost black median line that follows the keel and broadens at the first marginals. This line continues on the neck and skull, where it forms a Y-shaped mark. The head is brown and rather small for a *Pelusios* and ends in a pointed snout. The upper jaw has a modest anterior hook with a denticle on each side.

© J. Maran

The carapace is somewhat elongate and is brown to blackish. The snout is rather pointed, and the head dark. The plastron is black.

The jaws are generally lighter than the skin of the head, and they have no dark spots or stripes. The eye is completely black. There are two barbels on the chin. The plastron is large and nearly closes the anterior openings of the carapace. It is almost entirely black, although the seams may be outlined with cream or white. The plastron is connected to the carapace by a band of cartilage rather than by bony bridges as in other *Pelusios*. The front lobe is mobile and long, and the plastron ends in a deep anal notch. The feet are brown, rather fine, and well armed with five claws on each, but only slightly webbed. The hatchlings have a carapace that is almost circular and is light chocolate in color, with rugose scales covered with small rough points. A strong median keel is outlined by a dark band that widens on the first vertebral. The plastron is dark, almost black, and only the seams are lighter. The underside of the marginals is beige, like the skin. The hatchling already shows the black spot on the top of the head.

Natural history. This turtle occurs in all the types of water bodies within its gross range: rivers, streams, flooded forests, creeks, and so on, with a preference for acid waters (pH 5.5 to 6; Maran, 2002). The adults seem to prefer rivers, while the young remain in shallow puddles. The species is omnivorous and consumes small fish, aquatic invertebrates, amphibians and their larvae, mollusks, and vegetation, such as cylinders of manioc, which the villagers macerate in the shallow water. The females lay about a dozen eggs, about 35 × 25 mm in size.

Protection. Because of its large size, this *Pelusios* is heavily collected for food, as are its sympatric congeners *P. chapini* and *P. niger*.

©R. Amann

Even nowadays, some people can still find new species. An example is Jerome Maran, a young naturalist about 30 years old, self-taught, and crazy about Africa and turtles. He has made expeditions to Gabon and Ivory Coast and found two previously undescribed species of turtles. One carries his own name, *Pelusios marani* (Bour, 2000), and the other has received the name of the wildlife center that finances his expeditions, *Pelusios cupulatta* (Bour, 2003). A great traveler who is very curious about the world of turtles, Jerome Maran has also found new turtle species in South America that will soon be described and named. A fascination with this group of animals—combined with a taste for adventure and research, a knowledge of biology, and a love of fieldwork—makes these wonderful discoveries possible, even today.

Pelusios marani Bour, 2000

Common name

Maran's sideneck turtle

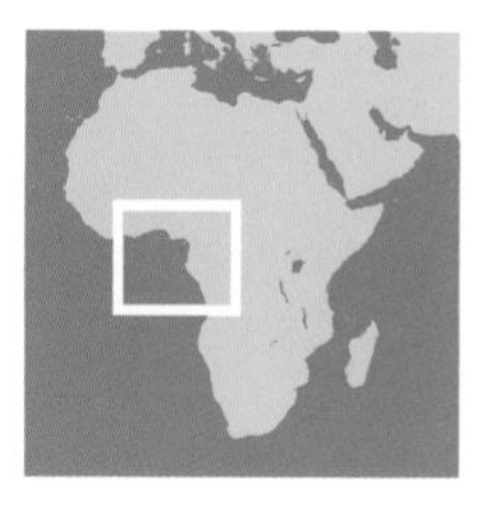

Distribution. *Pelusios marani* has been observed in the dense forests of Gabon, near Yombi. But the discovery of a carapace at Roungassa, 260 km from Yombi, poses the question of its presence farther east. The range needs to be studied more.

Description. The turtle may attain a length of 275 mm, but certain authors, noting the rapid captive growth, postulate a maximum of about 350 mm. The coloration is very contrasting, with a nearly black carapace and a bright yellow plastron. The carapace has an elliptical profile, with modestly serrated margins. The posterior marginals are curved upward above the hind limbs. The carapace is rather flat, but a light median keel is sometimes apparent on the vertebral scutes, which are wider than long—a unique feature within the genus *Pelusios*. The plastron is large, with a rounded anterior lobe and a barely functional transverse hinge. There is a small anal notch. The plastron and bridges are almost immaculate yellow. This coloration provides immediate differentiation from the sympatric congeners *P. chapini* and *P. gabonensis*. The head is wide and is extended anteriorly by the long, pointed snout. There are two small barbels on the chin. The upper surfaces of the head and neck are very dark gray, almost black, sometimes lightened with yellow. The underside is golden yellow. The tympanic region is grayish, touched with yellow. The upper jaw is rounded and shows two small denticulations. The feet are very dark above, light yellow below. The males have a slight midplastral concavity, but it is not easy to tell the sexes apart.

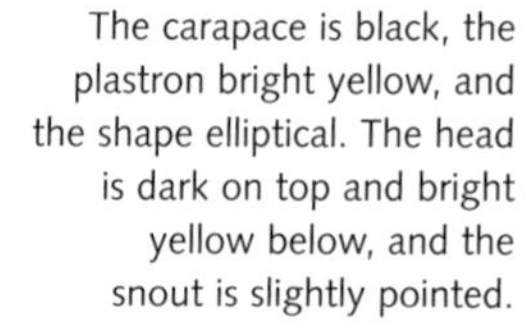

The carapace is black, the plastron bright yellow, and the shape elliptical. The head is dark on top and bright yellow below, and the snout is slightly pointed.

© J. Maran

Natural history. This species occurs in dense primary forest and is found in swamps and marshes and in water holes along rivers and creeks. Sometimes it is seen basking on emergent logs, but it is wary and dives off quickly. During the dry season (July to October), it estivates. It apparently feeds on aquatic insects, amphibians, fish, and carrion. The jaw structure is not suggestive of any special adaptation toward a diet of crustaceans and mollusks. When it is first caught, it may threaten with open mouth or excrete a nauseating oily yellow fluid. Reproductive details are unknown, but one female laid a single egg, 45 × 21 mm in size.

Protection. This species is occasionally used for food, and indeed the first discovery was made by J. Maran in a village midden. A local fisherman specialized in the capture of this turtle, using traps placed in the flooded channels connecting swamps. This species also suffers from habitat destruction and deforestation.

Pelusios nanus Laurent, 1956

Distribution. This species is found in southern Africa, north to the Zambezi, and in the Congo (Democratic Republic of the Congo) and as far as Angola.

Description. This turtle used to be considered a subspecies of *P. adansonii*, but it has emerged as a full species. This is the smallest member of the genus, not exceeding 120 mm in shell length—hence its name. The carapace is oval and flattened but sometimes presents a slight median keel. There is no nuchal scute. The marginals are unserrated, except for a modest notch on the supracaudal scute. The nuchal in juveniles is very eroded anteriorly and is wider than long, but in adults it becomes longer than wide. The carapace is brown, sometimes decorated with dark lines. The plastron is large, with a rather long anterior lobe, and the plastral hinge is poorly developed. The axillaries and inguinals are absent. The plastron is yellow, with a black margin. The bridges are also black. The head is of medium size and has a short, rounded snout. The upper jaw is lightly notched, and two chin barbels are present. The anterior limbs do not have enlarged transverse scales.

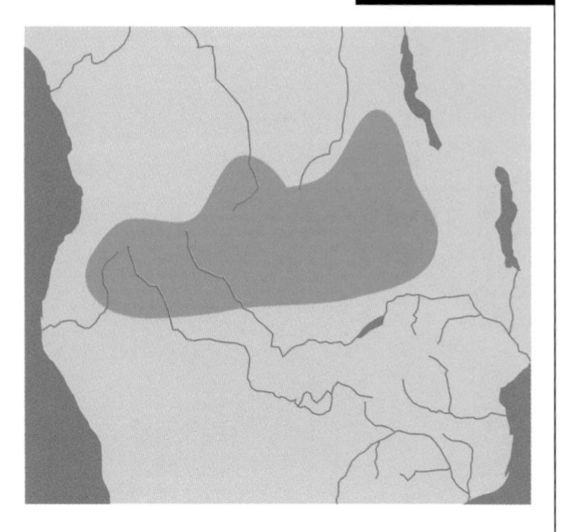

Natural history. This turtle is found in watercourses and the wet zones of warm savannas. Its reproduction and life history are similar to those of *P. adansonii*.

Protection. These turtles are very small and are not overexploited by humans, as far as we know.

Common name

African dwarf mud turtle

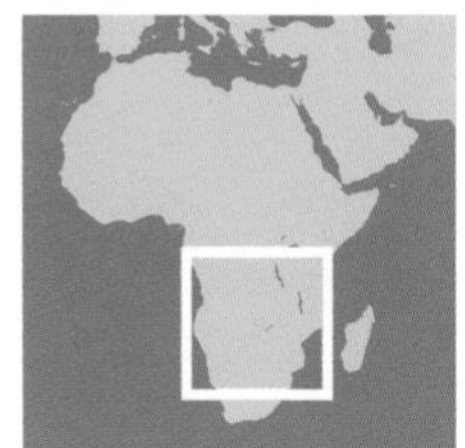

Pelusios niger (Duméril and Bibron, 1835)

Distribution. This species occurs in equatorial Africa, along a wide coastal band from Ivory Coast to northern Angola.

Description. This is a large species, with some individuals reaching as much as 350 mm and 4.5 kg. The females do not exceed 280 mm. The carapace, uniformly black, is slightly domed and a little flat on top, and there is a slight median keel. On the other hand, some examples are a lighter red-brown, with dark streaks on the scutes. The marginals are not serrated but are very slightly raised above the hind limbs. There is no nuchal scute. The plastron is of medium size and does not completely close the shell opening. The anterior lobe is short and rounded, and the hinge is straight and fully functional. There is a moderate anal notch. The plastron is very dark, often black, with lighter sutures. The bridges include neither axillary nor inguinal scutes and are the same color as the

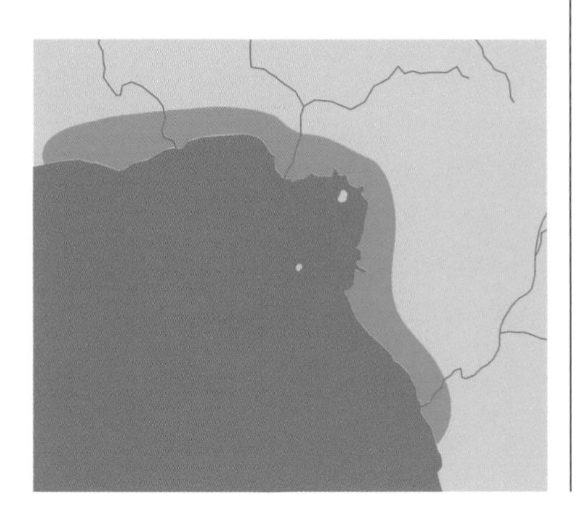

Common name

West African black turtle

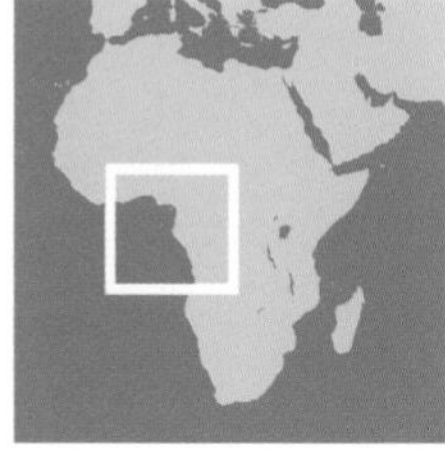

This species of the genus *Pelusios* has a well-developed beak. The carapace is black and is less flattened and more domed than in other members of the genus.

plastron. The head is fairly large, terminating in a pointed, hooked snout, characteristic of the species. The head is yellow, with a complex reticulated pattern of dark spots. The upper jaw is strongly hooked anteriorly, light yellow in color, and vertically streaked with black. The tympanic region is visible as a wide, plain yellow area. Two small chin barbels are present. The limbs are strong, lightly webbed, and armed with strong claws. The forelimbs each bear three or four rows of enlarged, transverse scales. The young fall into two distinct color varieties. Usually, they are very dark gray, with a strongly keeled, tectiform carapace and a dark head. The plastron is dark, with light seams. Other specimens from Nigeria have a carapace the color of burnt bread, with irregular dark spots on the scutes, or the carapace bears a strong black median line against a yellow background, the plastron also being dark with light seams. These may correspond to an undescribed subspecies.

Natural history. This species lives in humid environments, is often found on land as well as in the water, and inhabits lagoons, lakes, marshes, swamps, and flooded savannas. The diet is very catholic: water snails, mollusks, and fish, as well as seeds and aquatic vegetation, including water hyacinths. The females lay 6 to 10 eggs per clutch; they are oblong, soft-shelled, and rosy in color, and they measure 48 × 24 mm.

Protection. Like other large turtle species, this *Pelusios* is widely caught for food. Juveniles from Nigeria are sought after by hobbyists, because of the striking markings. Considerable export occurs, and in the year 2000 Togo sold 3,000 specimens, of which 500 were reported as wild-caught.

Pelusios rhodesianus Hewitt, 1927

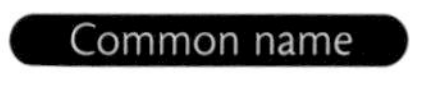

Variable mud turtle

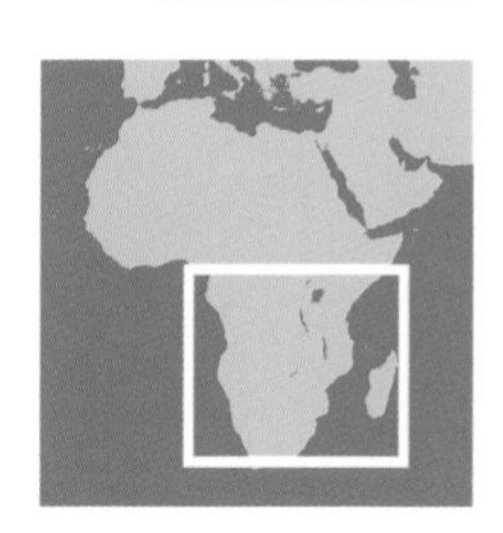

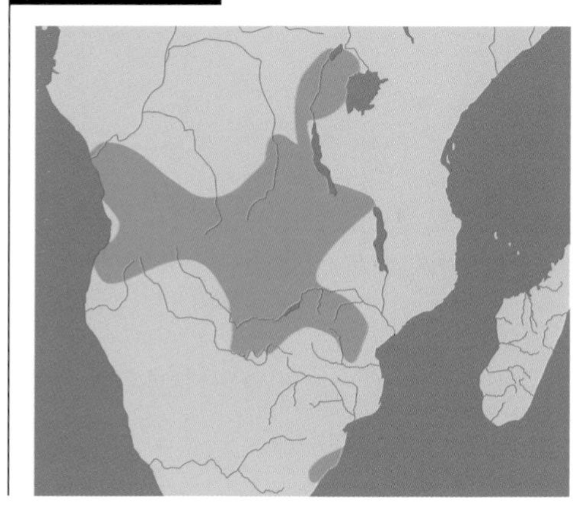

Distribution. This species occupies a wide area of central and East Africa, from the southern part of the Democratic Republic of the Congo to Angola, Uganda, Zimbabwe, Rwanda, southern Tanzania, eastern Mozambique, northern Botswana, and Zambia. There is an isolated population in South Africa, in KwaZulu-Natal and also around Durban.

Description. This is a medium-sized species, reaching about 250 mm. The females are larger than the males. The carapace is oval, somewhat elongated, domed, and widened behind. Some-

times there is a slight flattening of the top of the shell. A modest keel is visible at times, often reduced to a pointed projection on vertebral 4. The marginals are not serrated, and there is no nuchal scute. The supracaudal is divided. The carapace is dark gray or totally black. The large plastron is able to close the shell opening almost completely. The anal notch is deep. The plastron is completely black, but certain individuals have light spots, especially along the midline seam, although in some cases, notably some Zimbabwe populations, the entire plastron may be yellow (Broadley, 1981). The bridges and the undersides of the marginals are black. There are no axillary scutes. The bridges are strong but do not include the pectoral scutes. The head is small and flat, extended by a slightly pointed snout. The upper jaw is bicuspid, and two chin barbels are present. The head is dark brown to black on the top, with yellow sides in animals from the southern part of the range, although in others the head is decorated with yellow vermiculation. The remainder of the skin is light brown to yellow. The males have a carapace widest at the level of the ninth marginal scute, while in females it is widest at the level of the seventh or eighth marginal.

© B. Devaux

The carapace is often domed, but it may be more flattened in certain individuals. The head is rather small, with light streaks and dots on a blackish background.

Natural history. This turtle is found in calm waters of large rivers, swamps, and marshes, as well as in artificial impoundments. It is primarily carnivorous, but some individuals also eat aquatic plants. It is most active during the rainy season and is often found walking about on dry land. Nesting takes place with the first rains of September but may continue through the southern summer. The eggs number 10 to 15 per nest and are white, soft-shelled, and elliptical, averaging 35 × 22 mm. The young hatch in December or January.

Protection. Beyond the protection measures offered by existing reserves, the species receives no specific protection except in South Africa, where it is included in the *Red Data Book*.

Pelusios sinuatus (Smith, 1838)

Distribution. This species has a huge distribution in the eastern parts of Africa, including Somalia, Ethiopia, Kenya, Uganda, Rwanda, Burundi, Tanzania, Zambia, Malawi, Botswana, northeastern South Africa, and into Swaziland.

Description. The specific name reflects the characteristic sinuous rear margin of the carapace. This is the largest member of the genus, with some individuals reaching 485 mm and weighing 9 kg. The shell is somewhat domed and is often flattened or even slightly depressed along the vertebral scutes. There may be a light keel, often reduced to single points on each of vertebrals 1 to 4. The shell is especially widened toward the rear, behind the plastral hinge. There is no nuchal scute, and the supracaudal is divided. The color is generally gray, always more or less dark, and some-

© F. Bonin

This species may reach a fair size and a weight of 5 kg or more. The head is blackish, with a few light markings and the snout is moderately pointed.

times actually black. In some individuals the seams may appear light. The growth annuli on the scutes are visible for many years. The plastron is large but does not completely close the shell opening. The anterior lobe is short and rounded, and the hinge is well developed. The wide anal notch is angled in males and rounded in females. There is a reduced axillary scute on each bridge. The bridges are black, as are the undersides of the marginals. The plastron is yellow to orange and has a black periphery with wavy borders, similar to the sinuosity of the border of the carapace. The head is wide and flattened, with a lightly pointed snout. The upper jaw is unicuspid or even bicuspid, and there are two chin barbels. A very large frontal scale covers the top of the head. The head is gray to brown on top, sometimes presenting a yellowish appearance. The chin and the underside of the neck are yellow. The remainder of the skin of the neck and limbs is light to dark grayish.

Natural history. This is a very aquatic species, confined to permanent waters in rivers and lakes in the coastal savannas and to an altitude of about 1,500 m. It is often seen basking on emergent trunks and logs. The diet is primarily carnivorous: fish, amphibians and their eggs and larvae, insects, snails, and worms, as well as fallen fruit and aquatic vegetation. The species is also known to feed on ticks attached to large mammals that venture into water holes. It is thought never to estivate. Nesting occurs after the first summer rains, in October to January. The clutches include 7 to 25 elliptical eggs with parchmentlike shells, about 43×25 mm in size. Incubation is brief, sometimes as little as 48 days. The juveniles have a very denticulate shell, with a strong keel. The vertebral scutes are wider than long, the reverse of the situation in adults. The growth rings give a rugose aspect to the shell, and lighter radiating lines may be present.

Common name

East African serrated mud turtle

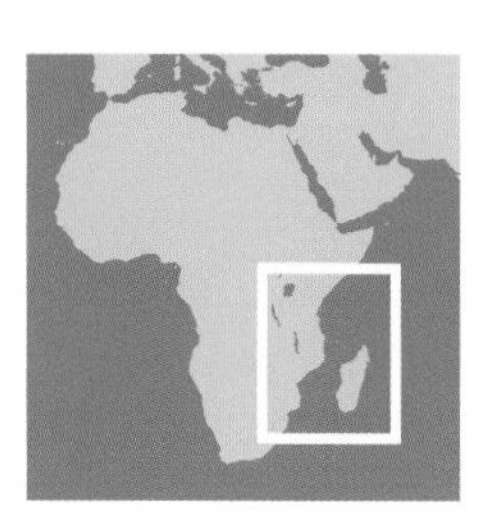

Protection. This species is widely eaten by local people and represents a good quantity of meat. Juveniles are eaten by herons and stilts, as well as by monitor lizards. The adults may be crushed and swallowed by Nile crocodiles. *P. sinuatus* is not included in any protective laws or regulations. There is some international trade in the species, but in general this turtle remains widespread and fairly common.

Pelusios subniger (Lacepède, 1788)

Distribution. This species has a wide range in southeastern Africa, including Rwanda, Burundi, Tanzania, Mozambique, Zimbabwe, extreme northeastern South Africa, northeastern Namibia, Zambia, and southeastern Democratic Republic of the Congo. It also occurs in eastern Madagascar and in the Seychelles, doubtless having been introduced by humans to these regions.

Description. The name *Pelusios* means "mud" or "clay" in Latin, and *niger* means "black." Carapace length reaches about 200 mm, and the females are larger than the males. The shell is rounded, keelless, and unserrated. The supracaudal scute is divided, and there is no nuchal scute. The marginals are straight, nearly vertical, and unnotched. Coloration varies according to habitat but ranges from dark to light earth tones such as olive-beige and can be plain black. The plastral hinge is very well developed and functional. The anterior lobe is rounded and wider than the posterior, which includes an anal notch. The plastron as a whole is guitar-shaped and provides complete closure of the openings of the carapace. In color it is more or less dark, with black seams. The bridges are beige to light brown or gray, and axillary scutes are lacking. The head is wide and rounded, ending with a short, rounded snout that has a simple unicuspid beak. The color of the head ranges from gray to brown or even black above, with the jaws a somewhat lighter yellow to cream. The chin bears two well-developed barbels. The limbs are dark above, lighter below, and modestly webbed but armed with strong claws. The young hatch with a tectiform carapace and a yellow head with extensive scattered or regular dark spots.

Subspecies. *P. s. subniger* (described above): This race occupies the entire African and Madagascar range of the species. *P. s. parietalis* Bour, 1983: Found only in the Seychelles, this subspecies is differentiated by its cephalic scale patterns.

Natural history. *Pelusios subniger* occupies many types of aquatic and moist habitats: swamps, lakes, ponds, rivers, and temporary wetlands. It is mostly nocturnal, but it walks on land after rains and sometimes basks on dry land. The diet is omnivorous and opportunistic: crabs, worms, insects, amphibians, fish, and even dead animals. Vegetation is also consumed. During the dry season, it estivates in the mud and awaits the return of the rains to resume activity, but a severe drought may kill it. The reproductive period extends from February to March (southern summer). From 8 to 12 eggs are laid on average, and there are several

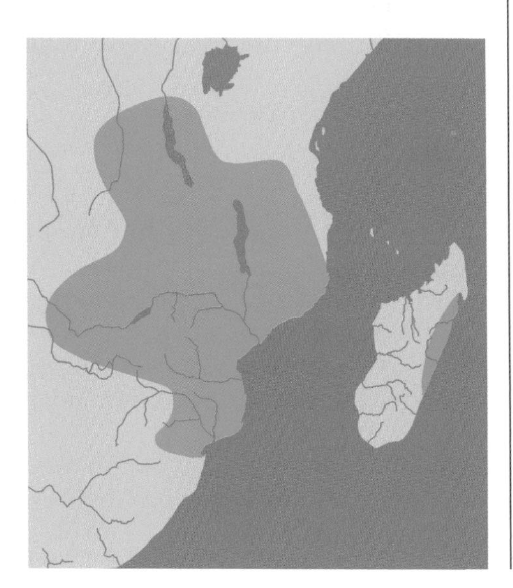

Common name

East African black mud turtle

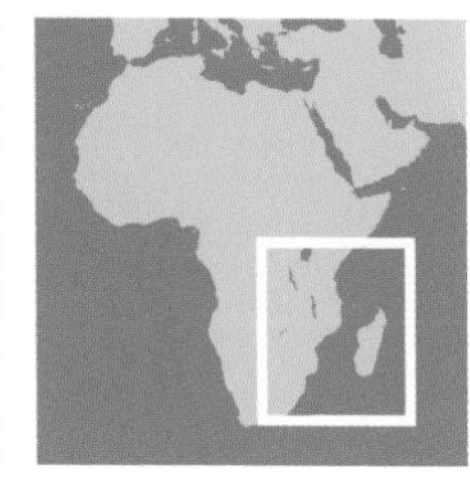

This turtle is mostly black and has no nuchal scute. The carapace is rounded and elongate. The snout is short and rounded.

© J. Maran

nestings each season. The eggs are fragile and measure 36 × 21 mm. Incubation time depends on the start of the rains and varies from 90 to 120 days on average.

Protection. These turtles are constantly gathered, fished for, trapped, and consumed by local people, and many populations of this species are threatened in certain regions of Africa and Madagascar. Some years ago they were also caught to make tourist souvenirs from specimens that were dried and stuffed. The species also bears the brunt of habitat destruction, hotel construction, and the drying up of wetlands. It is additionally subject to demand for domestic pets in some urban areas and is exported to Western countries. The Seychelles subspecies has only very small, isolated populations that have been subject to such stresses as the extension of the international airport, the creation of golf courses, and hotel construction. At present there are no more than 200 Seychelles individuals left (Gerlach, 2000). Today these turtles are studied and protected on Silhouette Island by the Nature Protection Trust of Seychelles. In Africa the species is not officially protected anywhere, and it enjoys some degree of security only in national parks, of which Kruger National Park in South Africa is the outstanding example.

Pelusios upembae Broadley, 1981

Distribution. The range is a very small area in the southeast of the Democratic Republic of the Congo in the Upemba National Park, in the Fungwe and Lualaba Rivers and their tributaries.

Description. This form was originally proposed as a subspecies of *P. bechuanicus* but was recently elevated to full species status (Bour, 1983) because it was considered closer to *P. rhodesianus*. It is a medium-sized species, reaching 230 mm. The carapace is elongated and oval, dorsally flattened, and sometimes has a light median keel. The carapace is widened posteriorly, from the center on back, and the marginals are not serrated. Vertebrals 1, 2, 4, and 5 are wider than long. The first vertebral is widened anteriorly, and the fifth is widened toward the rear. The color of the carapace ranges from dark brown to black. The plastron has a fully functional hinge and can completely close the shell opening. The intergular is wider than long. The anterior lobe is wide and rounded, and the posterior lobe is curved up at the level of the abdominofemoral seam. The anal notch is very marked. The plastron is black and spotted or dotted with yellow, especially along the midline seam. The head is wide and ends in a short, rounded snout. The upper jaw is smooth-edged, and two chin barbels are present. The head may be uniform brown or may be vermiculated, with a light yellow design. The skin of the neck and of the throat is finely granulated and lighter than the top surface.

Natural history. This species inhabits deep rivers, and its diet is certainly carnivorous. Its ecology remains virtually unknown, although it is probably similar to that of *P. bechuanicus* and *P. rhodesianus*.

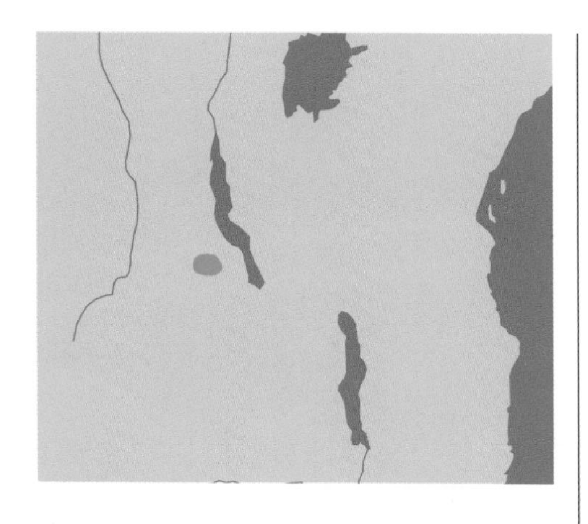

Protection. This species is certainly collected for food by local people. Moreover, this region is politically unstable, so undertaking detailed studies on its status is difficult.

Common name

Upemba mud turtle

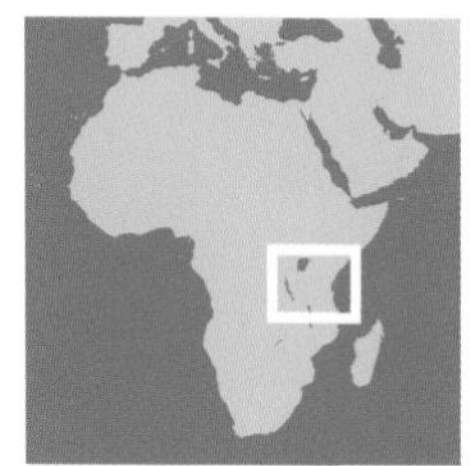

Pelusios williamsi Laurent, 1965

Distribution. This turtle has a rather limited distribution, in the upper regions of the White Nile, including Lakes Victoria, Edward, and Albert, to the extreme east of the Democratic Republic of the Congo, in Uganda, Kenya and Tanzania.

Description. This species is named after Ernest E. Williams, chelonian authority at Harvard University. It does not exceed 250 mm. The carapace is oval, slightly expanded toward the rear. It is somewhat flattened on top, maybe even a little depressed, but it always has a slight median keel, reduced to several blunt points. The marginals are not serrated. The carapace is generally dark brown to black. The large plastron almost covers the opening of the carapace but is constricted at the level of the seam between the abdominals and femorals. The anterior lobe is short and rounded. The anal notch is well developed, and the bridges are wide and lack axillary scutes. The head is wide, with a slightly projecting snout, and of a gray color without any special markings. The upper jaw is bicuspid, and two barbels are present. The feet are gray above and yellowish below, and a wide lamella traverses the anterior limbs.

Subspecies. Three subspecies are recognized. *P. w. williamsi* (described above): This race is found in the upper course of the Nile and in Lake Victoria. The plastron is black, with light seams. The posterior lobe is equal to or shorter than the

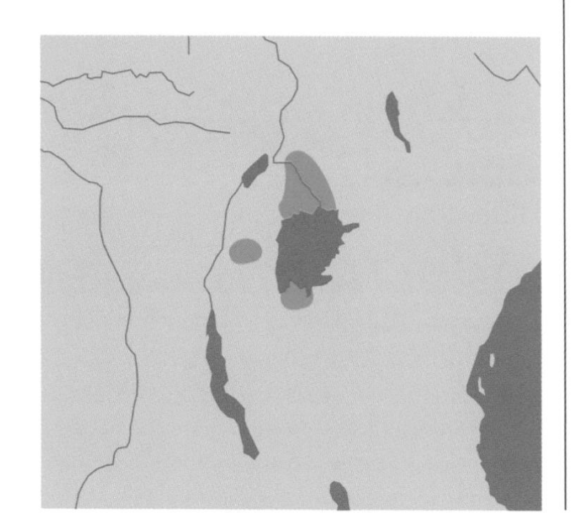

Common name

Lake Victoria mud turtle

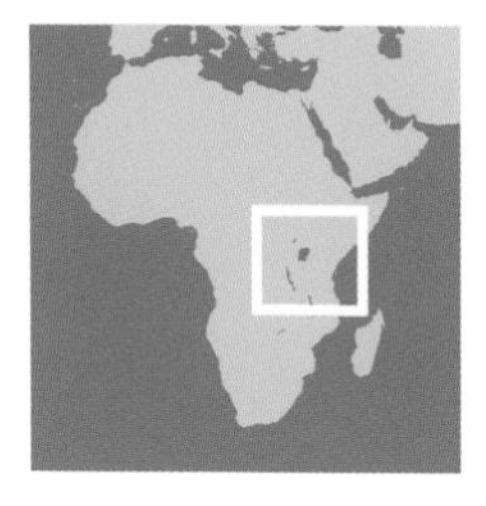

The elongate carapace is somewhat domed, has a light median keel, and is almost blackish. The head is also dark, without light markings.

anterior lobe. The intergular scute represents about half the length of the anterior lobe. *P. w. laurenti* (Bour, 1984): Laurent's pelusios is found only in Lake Victoria, one locality being the island of Ukerewe, situated in the southern part of the lake in Tanzanian territory. This turtle's posterior lobe is longer than the anterior, and the intergular is more than half the length of the anterior lobe. The first vertebral is widened and T-shaped. *P. w. lutescens* (Laurent, 1965): The Lake Albert pelusios occurs indeed in Lake Albert, as well as in the Semliki River and as far north as Lake Edward. The plastron is entirely yellow, with some degree of gray and brown spotting. The posterior lobe is as long as or longer than the anterior lobe. The intergular is less than half the length of the anterior lobe.

Natural history. This species occurs only in the deep waters of rivers and great lakes of the African Rift Valley. Its diet is primarily carnivorous.

Protection. This turtle is rather rare but is little exploited. Its biology is unknown, and its status should be evaluated to determine whether conservation measures are needed.

Subfamily Podocneminae Cope, 1812

Podocnemis cayennensis (Schweigger, 1812)

Distribution. This turtle occupies a large area of South America, usually in the far interior, although in Brazil and French Guiana, it actually comes close to the sea. It occurs in Guyana, Suriname, French Guiana, Venezuela, Colombia, Ecuador, Peru, and northern Bolivia and throughout the Amazon Basin in Brazil.

Description of the genus *Podocnemis*. *Podos knemis*, in Greek, refers to the large scales

on the limbs of turtles in the *Podocnemis* genus. This is the most ancient genus among living turtles and also one of the most widespread, having occurred in South America since the end of the Cretaceous, with African fossils from the Eocene and the Pleistocene. The skull is well ossified, and the neck is fairly long. Some of the species, especially *P. expansa* and *P. vogli,* show a remarkable tendency to aggregate in large numbers. The former gather for nesting purposes, and the latter form high densities as their water holes dry up. The genus is characterized, among other features, by the groove on the front of the head, passing from the nostrils to between the eyes, as well as by the enlarged scales of the anterior limbs and the four claws on the forelimbs.

© J. Maran

This young *P. cayennensis* displays bright colors: a yellow line around the carapace and bright head markings on a black background.

Description of *P. cayennensis.* Often called *Podocnemis unifilis,* this large species may reach a length of 476 mm. Adults have an oval carapace without widening of the posterior flanks, as in several other *Podocnemis* species, and with a smooth posterior border. A low keel may be present on the third and fourth vertebrals. The color is brown to greenish gray. The plastron is cream, with yellowish spots and scattered black pigmentation. The coloration of the head is very specific, with large yellow spots around the tympana, as well as above and below the nostrils. The head is relatively small and short, the nostrils forming a small, black, slightly projecting snout. The young have a well-developed median keel and a yellow line around the carapace. They lie in the sun and look rather splendid, the bright yellow spots contrasting with the black background color of the head.

Natural history. This very aquatic species is rarely observed in a thermoregulatory posture. It occupies major rivers and large lakes, but where it is sympatric with *P. expansa,* it leaves the main watercourses to the latter and occupies the more tranquil backwaters. It is primarily herbivorous and enjoys aquatic plants, including water hyacinths, and fallen fruit, but it feeds equally on mollusks and aquatic insects. It lays from January to March in the north and from November to February in the south, in small groups of nesting females. The eggs are hard-shelled and elongate, numbering up to 28 per clutch. Average egg size is 22.4 × 39 mm. The females almost always nest by night, sometimes a good distance from the water. Typically, there are two clutches per season.

Protection. Like *P. expansa,* this species is heavily collected for food and may even be somewhat tastier than the latter, according to Mittermeier (1978). It is widely sold in Brazil, and when a given area is cleaned out, the hunters switch to *P. sextuberculata*. The animals are caught on beaches when they attempt to nest or are captured with traps and nets. Formerly they were sold in large numbers for the pet trade, but the countries of origin and CITES have now stopped this export trade.

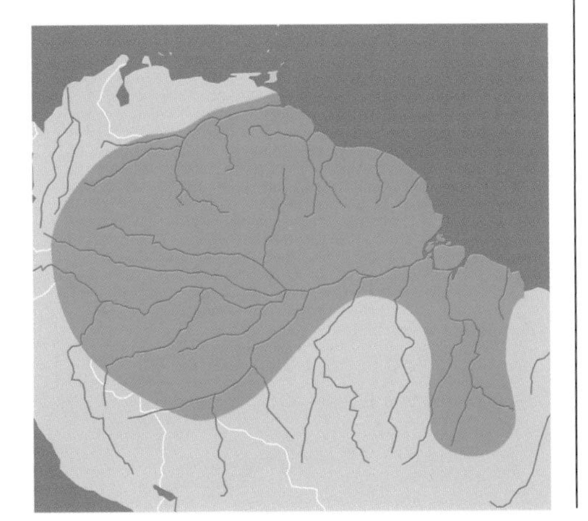

Common name

Yellow-spotted river turtle

Podocnemis erythrocephala (Spix, 1824)

Common name

Red-headed river turtle

Distribution. This species has a restricted distribution, essentially confined to the Rio Negro and its tributaries in Brazil, the extreme south of Venezuela, and the eastern extreme of Colombia.

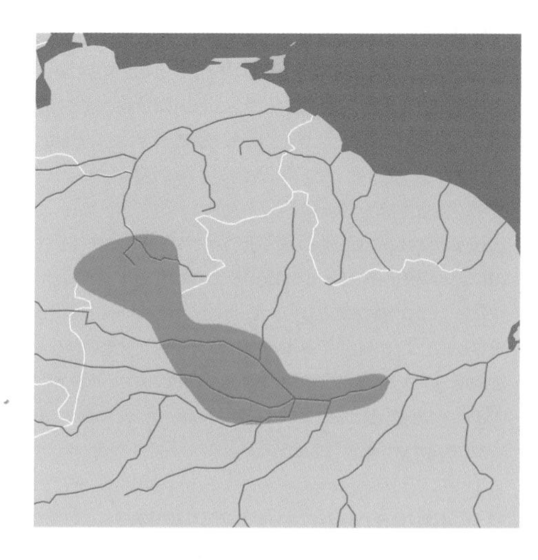

Description. This small turtle (320 mm maximum) has a blackish carapace that is widened toward the rear. The name derives from the broad red band that passes across the crown of the head, although the red markings disappear in mature females. The carapace is somewhat domed, with the highest point at the second vertebral. The head is elongate and has a projecting snout. The three enlarged scales on each forelimb reflect the generic name *Podocnemis*. The tail, like that of *P. vogli,* bears nine whitish scales behind the cloaca. The young have large orange or red markings on the nose, the tympana, and the top of the head.

Natural history. This species is confined to quiet waters, swamps, and turbid zones. The small size may be an adaptation or response to an oligotrophic environment and the limitations of the ecosystem (Pritchard and Trebbau, 1984). The adults are primarily herbivorous, feeding on aquatic vegetation, algae, and fallen fruit, but they may be caught on hook and line with fish or meat as bait. They also feed on fine floating particles, using a suction system known as neustophagia. This is accomplished by opening the jaws at the surface of the water, creating a depression into which organic detritus will flow. This species nests from late autumn to early November, laying 5 to 13 white, elongate, hard-shelled eggs, about 40 × 28 mm in size.

Protection. Although this is a small species, it is exploited on the Río Negro for both meat and eggs. It may be caught by harpoon, by net, or on fishing line or collected on land on the nesting grounds. Formerly exported to Europe and North America, it is now protected by law in Brazil and Venezuela, greatly decreasing its collection for the pet trade. Nevertheless, it is considered vulnerable, and its populations are much smaller than in former years.

© F. Bonin

Note the deep gutter between the eyes and the wide, smooth scales on the top of the head.

Podocnemis expansa (Schweigger, 1812)

Distribution. This species has a wide range in all of northwestern Brazil, the southern half of Venezuela, eastern Colombia, a small part of Bolivia, and Peru. In the Guianas it is known only from the Upper Essequibo River in Guyana.

Description. This is the largest of the pleurodires, with a maximum weight of 90 kg for females (up to 890 mm in length) and 60 kg for males (550 mm). The name *expansa* refers to the posterior widening of the carapace, which is wide, flat, and gray-brown or blackish. The young are beige to greenish gray, with a light yellow line around the shell, expanded marginal scutes, and a rather large head. The front of the head has a noticeable gutter or groove from the nostrils to the rear of the eyes, which is outlined with yellow lines in the young, and there are also two yellow spots in the temporal region.

Natural history. The large size and powerful limbs of this species enable it to colonize major deep rivers with strong currents, including the Orinoco, the Amazon, and their tributaries. This turtle is herbivorous, eating fruit, flowers, aquatic plants, and especially a sort of freshwater sponge called *pica pica,* which proliferates on trees that have fallen in the river. During the dry season, *P. expansa* may migrate in search of new areas for food and nesting. The reproduction is noteworthy, this species being a socially gregarious one during the nesting season, and one that basks at the edge of the nesting sandbanks for long periods of time as the eggs ripen within. Formerly, thousands of turtles would assemble for basking, especially on sandbanks in the Rio Meta. These fantastic gatherings, called arribadas and otherwise known only among the marine turtles *(Lepidochelys),* were described by the first explorers of these regions and also inspired certain passage of Jules Verne's *Superbe Orénoque.* At the end of the dry season, the females collect near the sandbanks. There they are fertilized by the males, which have also migrated. Then the females leave the water and bask in the sun for long periods of time, which bring the eggs to a mature state, ready for laying. Nesting occurs several weeks after the basking episode, nearly always at night and a good distance from the water so that the eggs will not be inundated as the floodwaters return. The eggs are almost spherical (42 × 40 mm), but a few of them may be oversized (50–60 mm). The latter are called *huevos de manteca* ("butter eggs") and are not viable. Incubation is short, about 45 days, allowing for emergence before the waters rise, inundating the sandbanks. The young hatch in April to May and exit the next during the night, presumably to minimize attack by predatory birds. On the other hand, in the river itself, caimans and carnivorous fish consume a large proportion of the hatchlings.

Protection. The heavy exploitation to which this species has been subject is well documented. From the first arrival of the European colonists, published works have condemned the excessive exploitation of the eggs by Amerindians, then by missionaries and later arrivals (Humboldt, 1814). Today the population level is very low, a result of centuries of overexploitation. Even at Santa Maria on the Venezuelan Orinoco, just a few hundred females gather each season to nest, although two centuries ago there were tens of thousands. Today the species is protected in most of the major nesting zones, but illegal take continues. Programs that protect nests and the young, as at Santa Ma-

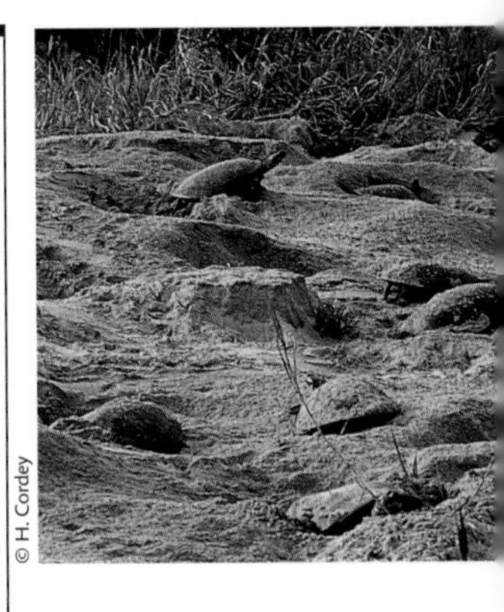
© H. Cordey

A century ago, nesting groups of thousands of turtles on sandbanks in the Orinoco inspired the writer Jules Verne. Today the emergences are more limited, but even so, dozens of turtles still get together for annual group nestings on sandbanks in the middle of the river.

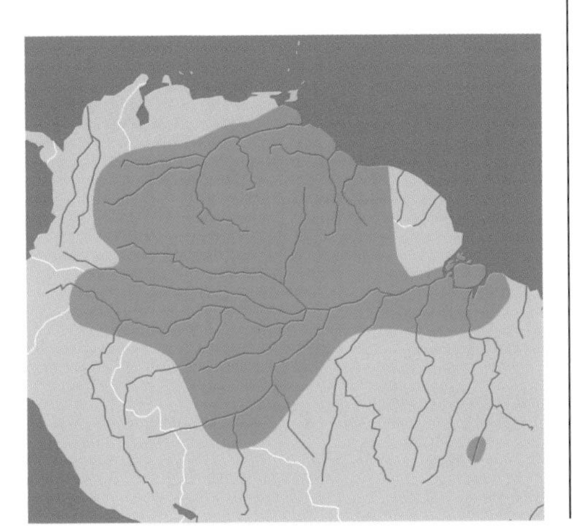

Common name

South American river turtle, arrau sideneck

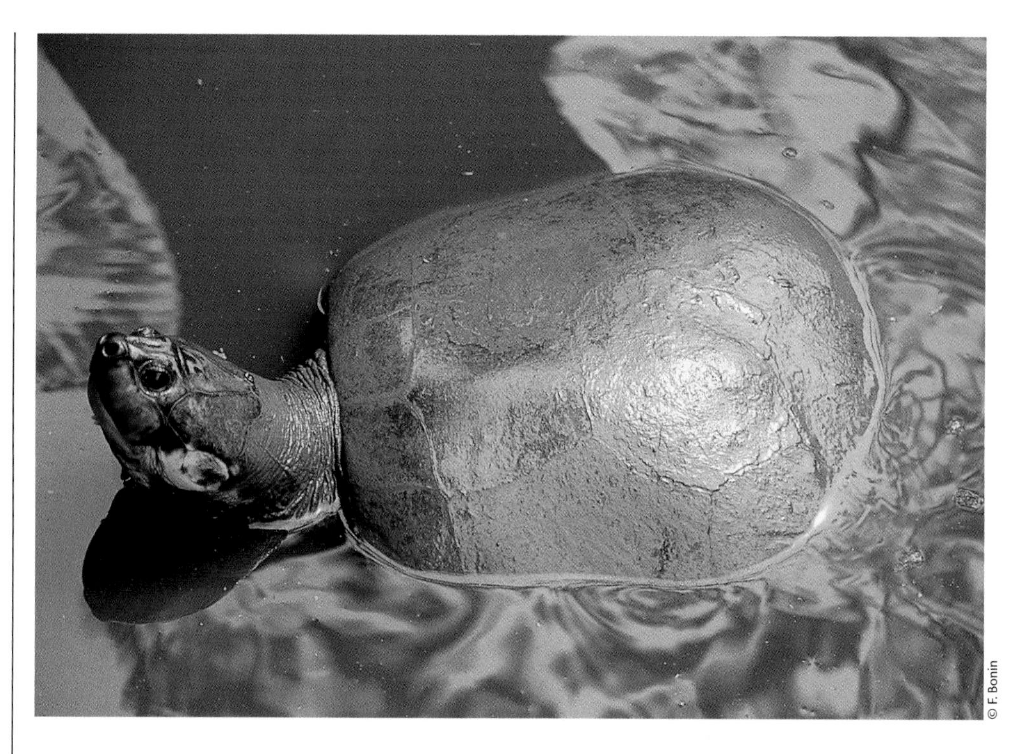

This species may reach a weight of 90 kg. The carapace is massive and bronze to greenish. The head is wide, strong, and expanded in the temporal area, with a short snout and the eyes placed high and well forward.

ria, should ensure the safety of at least a good nucleus of the population. Release of numerous young turtles is undertaken every year in the central part of the Orinoco. Almost everywhere else, the species continues to decline as a result of human activities.

Podocnemis lewyana Duméril, 1852

Distribution. This species if found only in Colombia, in the Magdalena and Sinu Rivers.

Description. The name drives from M. Léwy, a French banker who discovered the species in Venezuela. It has a flattened, wide carapace, with neither serrations nor median keel, and is up to 400 mm in length. The fifth vertebral scute is lightly notched. The color of the shell ranges from olive-brown to blackish. The plastron is smaller than the opening of the shell, with the anterior lobe shorter than the posterior one, and is also olive-brown to blackish. The head is rather small, with a groove between the nostrils and the eyes. There are two barbels under the chin. The snout is long, and the upper jaw somewhat rounded. The nostrils form at the tip of the snout, and the eyes are placed high and well forward on the head. A light band on the side of the head extends from the snout to the back of the tympanum. The juveniles have a gray to brown shell with a light peripheral line, a modest vertebral keel, and serrations on the posterior marginals.

Common name

Magdalena River turtle

Natural history. This species keeps a low profile in swamps, lakes, and slow-flowing rivers, even high up in the Cordillera Oriental. It feeds by neustophagia, as do *P. erythrocephala* and *P. vogli,* but the young are more vegetarian. A clutch includes 15 to 30 eggs, each about 34 × 40 mm in size.

Protection. This turtle suffers from many human activities, including direct consumption and water pollution. In the Magdalena River, it seems to be rare. It is classified as in Appendix 2 of CITES and is considered endangered by the Turtle Trust Fund.

The head of this species is relatively small, with a somewhat elongated snout. The carapace is black, and the head is lightened by a number of light streaks and bands.

© P. Pritchard

Podocnemis sextuberculata Cornalia, 1849

Distribution. This species is found in the Amazon Basin in Brazil, Peru, and Colombia.

Description. The name refers to the six tubercles, three on each side of the plastron, which are clearly visible in juveniles but disappear with age. The carapace does not exceed 330 mm. It is oval and wider behind than in front. A low keel may persist on vertebrals 2 and 3. The carapace is smooth, with tones of gray, brown, or blackish. The plastron is relatively narrow and grayish, with an anterior lobe that is wide and rounded in front.

© F. Bonin

This turtle has a medium-sized head, with a rather short snout. There are light markings on the dark background of the head. The gutter between the eyes does not reach the nostrils.

The head is wide and gray and has several light spots. The nose is projecting, and a deep groove is visible between the eyes but does not reach as far as the nostrils. There are two small barbels on the chin, which is beige or light gray. The young display clear markings on the head, and their shells are beige to grayish.

Natural history. This species lives in large bodies of water and is mainly carnivorous, eating fish. Nesting has been observed in July in the Amazon Basin, in November and December in the Putumayo, and in October in the Caquetá. The nests are excavated on the borders of lakes and in riverbanks during the dry season, before the water rises. The eggs, oval and 40 × 28 mm in size, number 8 to 13 per clutch.

Protection. The status of this species is poorly known. It is sometimes caught for food, as are the eggs, but the population trends are unknown.

Common name

Six-tubercled river turtle

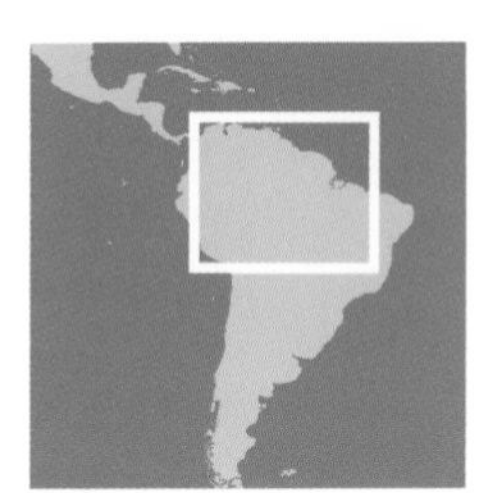

Podocnemis vogli Müller, 1935

Distribution. This turtle is a specialized form found in the llanos (Orinoco floodplain) of Venezuela (the states of Apure, Barinas, Bolivar, Cojedes, and Guárico) and Colombia (the department of Meta).

Description. The name derives from one Cornelius Vogl, a German missionary who collected numerous species in Venezuela. Members of this species are small, not exceeding 380 mm in carapace length and usually much shorter. The carapace is slightly domed, without a median keel, with very smooth scutes, and beige to brown in color, perhaps with bronze or reddish in some old individuals. The shell and bones are very thin. The plastron is large yellow-beige, with black dots. The head is wide and "smiling," has a projecting snout and pronounced temporal swellings, and is sometimes orange-yellow in old animals. The entire animal shows the same coloration as the dry mud environment of the llanos during the dry season. The tail has 12 to 16 whitish scales on each side. The males are recognized by a greenish ring that surrounds the pupil, the ring being wider than that of the females. The juveniles have a brown shell, with black seams and a yellow line along the marginals and with a slight keel.

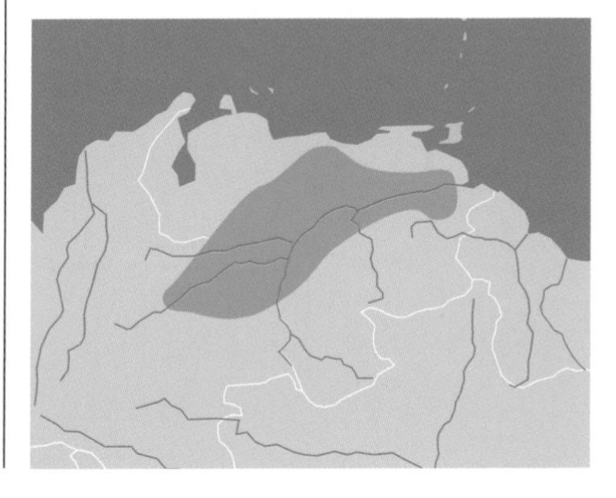

Natural history. This species is extremely gregarious, and the populations concentrate in the remaining pools of water during the dry season, with thousands of individuals sometimes found in a small space. They sometimes estivate in holes dug in the river banks, but they continue to be active if even a very small amount of water remains. During the humid season, they take advantage of the enormous flooded prairies that extend from the Orinoco Delta to the source of the Apure River. In the llanos, this is the dominant turtle, but it

Common name

Savanna side-necked turtle

© F. Bonin

During the dry season, individuals of *P. vogli* may assemble by the hundreds in the scarce water bodies, which they share with caimans. The beige coloration is identical to that of the mud of the water holes.

shares its habitat with occasional individuals of *P. cayennensis*. It is mainly herbivorous, but in the pools where large populations exist, it will feast on any shreds of organic material left by the caimans. In reality it is an opportunist, and during the dry season it will eat anything that goes past its door. Nesting occurs earlier than in *P. expansa,* doubtless because of the rapid inundation of the llanos in May to June. Oviposition is mainly observed in November, December, and January, sometimes several hundred meters from water. The eggs number 7 to 13 (maximum 20), have a well-calcified shell, and measure about 26 × 40 mm. After 120 days of incubation, hatching takes place from late April to early May, just before the arrival of the heavy rains. The eggs are heavily predated by numerous animals during the dry season, including birds and reptiles. The young measure about 35 mm and are smaller than those of *P. expansa,* a trait that probably adds to the danger of predation.

Protection. This turtle is collected in quantity by the inhabitants of the llanos, more so because it may be the only turtle species available. Piles of carapaces may be seen behind ranches and village huts in the llanos, testifying to heavy exploitation. But although the take is considerable, the species does not appear to be in danger. Nevertheless, the exploitation of this species could easily spread or be displaced to other turtles and could ultimately constitute a threat to all the *Podocnemis* species.

© F. Bonin

The head is wide, with well-defined scales. This is a calm species that never demonstrates any form of aggression.

Erymnochelys madagascariensis
(Grandidier, 1867)

Distribution. This species is confined to the western part of Madagascar, occurring in the Betsiboka, Mahalavy (including Lake Kinkony), Tsiribihina, Mangoky, and other rivers and farther south, in Lakes Bemamba, Masama, and Befotaka.

Description. This is the only pleurodire in Madagascar (excluding *Pelusios* and *Pelomedusa,* which are presumed to be early human introductions). It is a very large species, with a maximum length of 480 mm and a weight that may exceed

© B. Devaux

This powerful beast may reach a weight of a dozen kilograms. The head is robust and plated with thick shields, and the temporal scales are large and thick. The upper beak, hooked and sharp, is a formidable weapon. This is an aggressive, carnivorous species, but it lives quietly at the bottom of lakes and swamps.

Common name

Madagascar big-headed turtle

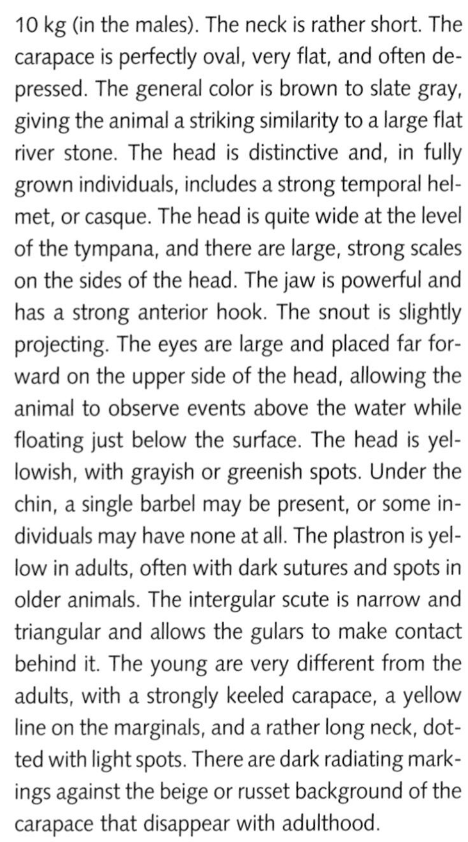

10 kg (in the males). The neck is rather short. The carapace is perfectly oval, very flat, and often depressed. The general color is brown to slate gray, giving the animal a striking similarity to a large flat river stone. The head is distinctive and, in fully grown individuals, includes a strong temporal helmet, or casque. The head is quite wide at the level of the tympana, and there are large, strong scales on the sides of the head. The jaw is powerful and has a strong anterior hook. The snout is slightly projecting. The eyes are large and placed far forward on the upper side of the head, allowing the animal to observe events above the water while floating just below the surface. The head is yellowish, with grayish or greenish spots. Under the chin, a single barbel may be present, or some individuals may have none at all. The plastron is yellow in adults, often with dark sutures and spots in older animals. The intergular scute is narrow and triangular and allows the gulars to make contact behind it. The young are very different from the adults, with a strongly keeled carapace, a yellow line on the marginals, and a rather long neck, dotted with light spots. There are dark radiating markings against the beige or russet background of the carapace that disappear with adulthood.

Natural history. This turtle keeps a low profile in its natural environment, hiding in the mud bottom of lakes or side branches of rivers. It is never seen basking and can navigate through fast waters to reach quieter areas; it is a powerful animal and an excellent swimmer. It is also an aggressive predator, feeding on fish, amphibians, birds, and so on. When picked up, it will defend itself vigorously, and the jaws are quite dangerous, especially in the large males. There are few morphological differences between the sexes, but the plastron is slightly concave in some males. The males are also larger than the females and have a slightly thicker tail, a cloaca situated toward the tip of the tail, a wider and more massive head, a more conspicuous beak, and a more aggressive disposition. The behavior of the males is generally more energetic than that of the females. Sexual maturity may develop precociously at just 5 or 6 years of age. Nesting usually occurs in October to November, nearly always at night; sometimes the season will be extended into January. July nestings have also been witnessed, which is surprising given that July is the coldest month of the year, when the waters are low and the weather dry. Embryogenesis may be postponed until the end of the year in the case of these atypical nestings. Nests contain 16 to 22 eggs, with a maximum of 24, but there are several nestings per season, and each large female may produce 60 eggs per year. Hatching occurs in December to January, at the beginning of the rainy season.

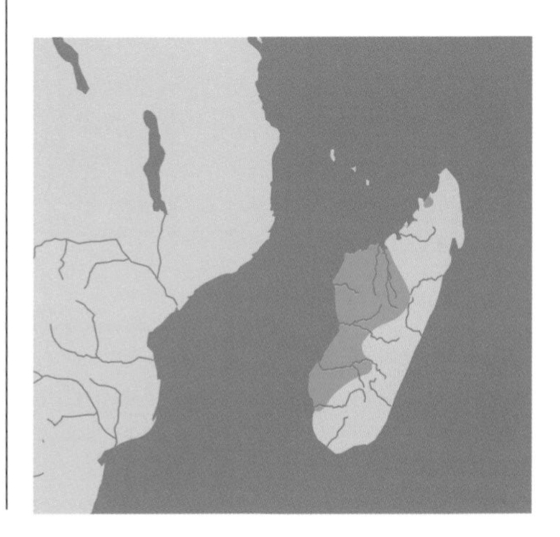

Protection. As we have learned from the work of Gerald Kuchling and the Chelonian Captive Breeding Centre in Ampijoroa, founded by the Durrell Wildlife Conservation Trust, this species is disappearing rapidly in most of the lakes where it was common a century ago. It is greatly favored for food by the Malagasy people and is fished intensively. Certain large lakes are now totally bereft of these turtles. Fires, deforestation, and habitat degradation impact heavily on it also. On the other hand, it is not the object of organized commercial collection and trade. A major feat of conservation has been initiated at Ampijoroa to reverse its gradual disappearance through an urgent education campaign. Since 1999, a breeding population has been established at the Ampijoroa center, with the intention of releasing young turtles back into the natural environment. This program is led by Don Reid. With full authorization, seven males and three females were collected from the wild. Twenty-three hatchings took place in 1999, and there have been about 100 per year since then. The breeding process requires that the males be put together for a brief period. This rivalry and resulting combat allows for good hormonal development, which gets the overall breeding process started. The real trick is to protect the young from rats and other nocturnal predators. After a period of rapid growth, the subadults are released (and monitored) in a lake near the center of Ampijoroa; later they are released in other suitable sites in

western Madagascar. But the true salvation of this species will not occur until widespread awareness programs for the villagers have been instituted, along with instigation of complete protection in some lakes, tight fishing quotas in other lakes, and negotiations with villagers to switch to alternative resources, such as eggs, chickens, pigs, fish, and so on, to replace turtle meat.

Ethnozoology. The réré is well known to the people of western Madagascar. It has always been consumed and is part of the folklore of local cultures, often represented on pottery, engravings, and mural paintings. It is feared and respected, because of its strength and aggressive nature, but unfortunately the flesh is much appreciated also. Today the local people admit that it is becoming rarer and rarer, but they still continue to catch every one they can.

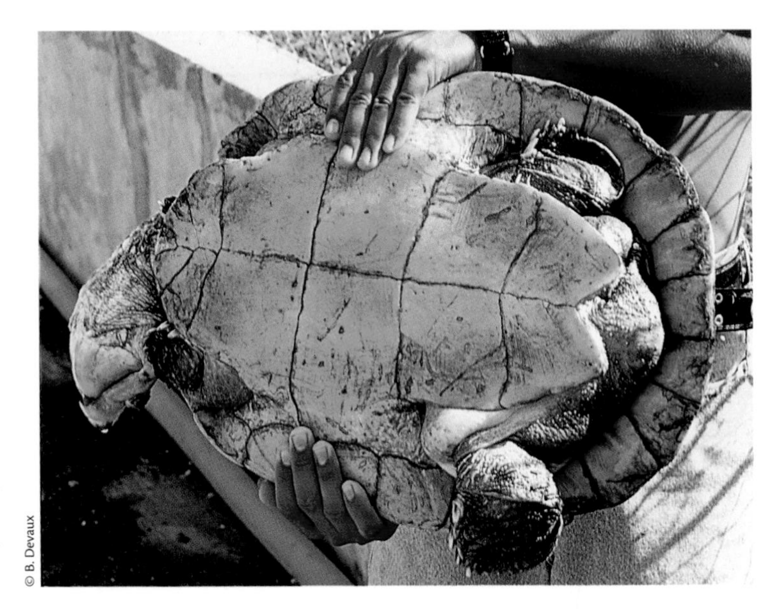
© B. Devaux

The plastron is wide and completely yellow, with several black streaks and spots.

Peltocephalus dumerilianus

(Schweigger, 1812)

Distribution. This species has a wide range that extends from the Orinoco Basin to the Amazon Basin in Venezuela, in eastern Colombia and Ecuador, northeastern Peru, northern Brazil toward the mouth of the Amazon, and southeastern French Guiana.

Description. The generic name derives from the Greek *pelte,* which means "shield," and *kephale,* which means "head." The specific name refers to the famous herpetologist André-Marie-Constant Duméril (1774–1860). This is a massive turtle, up to 480 mm in length and weighing up to 15 kg. The carapace is oval and strongly domed, and the rear marginals are often serrated. The carapace is brown, dark gray, or black. The scales are very smooth in adults. The plastron is yellow, cream, or light brown, sometimes with dark areas. The anterior lobe is wide and large, although the posterior of the plastron is narrowed and parallel-sided. The anal notch is strong. The head, very striking in this species, is huge, triangular, and covered with large scales, giving the appearance of a helmet. The beak is powerful and strongly hooked. There is a single barbel on the chin. The color of the head is dark, but the underside may be relatively light or reddish, like the underside of the neck, in some individuals. The tympanic region is always very light, sometimes almost white. The feet and the underside of the neck are equally dark but seem to be rather gracile in view of the overall powerful aspect of the animal. The feet are

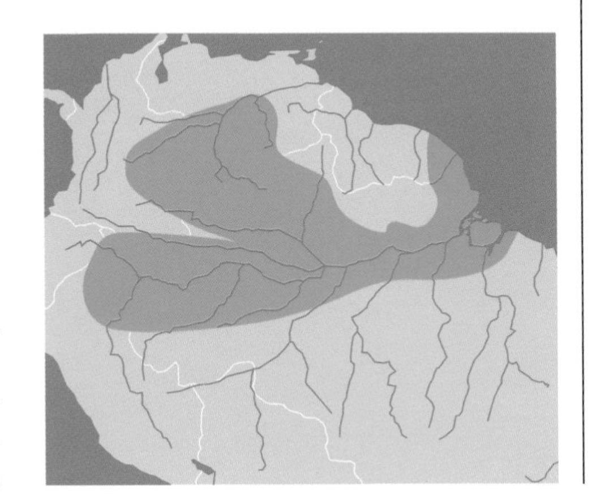

Common name

Big-headed Amazon river turtle

© T. Frétey

This massive species has an all-covering shell and a broad, strong head with a temporal helmet. The beak is powerful, and the animal is a formidable predator.

webbed. The males are somewhat larger than the females.

Natural history. This turtle lives in rivers, streams, and deep watercourses, both clear and turbid, as well as in flooded savannas and swamps. The diet in the natural environment is not well known—this species is difficult to observe. Apparently it is omnivorous: one stomach yielded seeds and various vegetative species, whereas others had fed on dead fish. This is an aggressive animal, seeking to bite painfully when picked up. Nesting occurs from August to December. The females lay between 12 and 24 eggs in a hole 20 cm deep. The eggs are elongate, with soft shells, and measure 55 × 35 mm on average. Incubation lasts for 125 days. Then the young hatch with a carapace length of 45 to 60 mm and are light brown with black sutures. They have a median keel, well marked on the third through fifth vertebrals. Even the hatchlings are irascible, and they also emit a nauseating musk when threatened.

Protection. This species is included in Appendix 2 of CITES. In French Guiana it is completely protected (decree of May 15, 1986) but is still often sought for food because of its large size. The eggs are rarely consumed, mainly because they are difficult to find. In French Guiana this turtle has been observed only rarely but has been recorded in the commune of Matoury and in the swamps of Kaw, where one specimen 290 mm long had a head width of 65 mm.

© T. Frétey

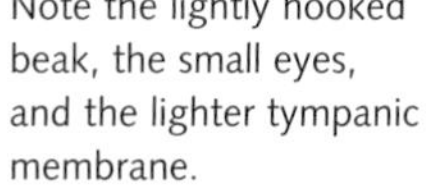

Note the lightly hooked beak, the small eyes, and the lighter tympanic membrane.

Cryptodira

The infraorder Cryptodira (Cope, 1868) is more recent in origin than the Pleurodira, and it includes the large family Emydidae, which appeared about 40 million years ago. It is characterized by the neck, which is retracted vertically as it is raised into the interior of the carapace.

Chelydridae Gray, 1831

This family includes two species of large American turtles with various areas of distribution. They have rough shells, a very elongate, crocodilian-like tail, a reduced cruciform plastron, and a powerful head with crushing jaws.

Chelydra serpentina (Linnaeus, 1758)

Distribution. This species is widely distributed from southern Canada (Nova Scotia, New Brunswick, Quebec, Alberta) down to the southern United States (East Coast and central).

Description. This is a massive turtle, which may attain an extreme length of 550 mm and a weight of 35 kg. The carapace has three strong keels, forming pointed projections on each scute, although in old individuals the shell is much smoother. The hind marginals are strongly denticulate. The color is generally brown to olive or tending toward black. The scutes may be impressed with radiating lines, which are most obvious in young specimens. The plastron is very reduced, allowing the limbs, neck, and tail to move freely. The bridges are very narrow. The coloration is yellowish, from light to dark. The head is large and wide, and the temporal regions as well as the rear of the skull are covered with wide, flat, overlapping scales. The snout is conical and pointed, and the upper jaw is hooked. The jaws are powerful and well developed, and it is these that give the snapping turtle its common name. The jaws are light in color, streaked with fine dark lines. The eyes are rather small and situated high up on the head. There are two chin barbels. Various warts and tubercles, often somewhat pointed, are visible almost all over the neck, the limbs, and the tail; the crocodile-like tail is long and has three rows of heavy pointed scales on the upper surface. The limbs are strong, giving the animal a robust, vigorous aspect. The feet are well webbed and have impressive claws. The overall effect is that of a primitive, aggressive crocodile. The young are very small (about 30 mm) and already have three well-defined, rugose keels. Gray or brown in color, the young have a yellow ocellus on the upper side of each marginal

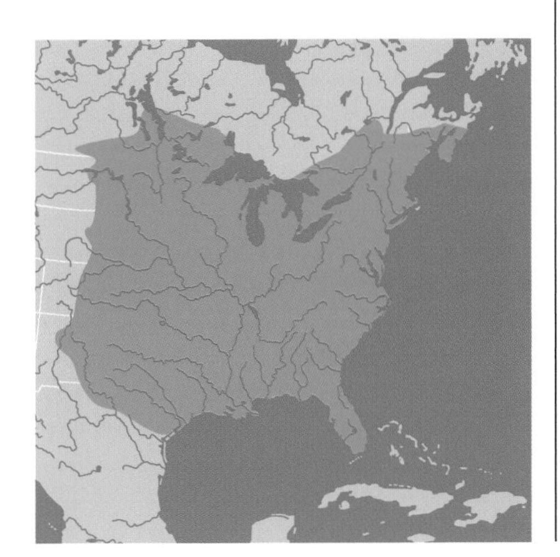

Common name

Common snapping turtle

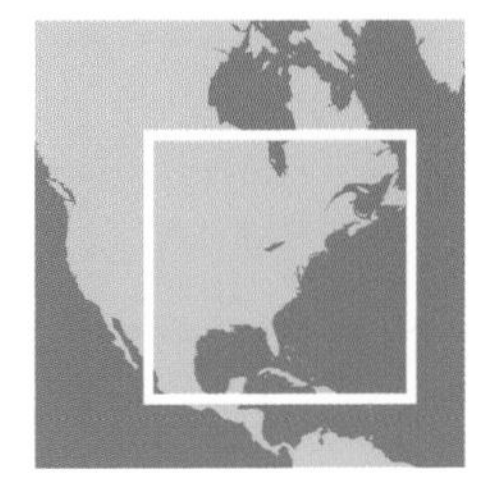

© B. Devaux

This turtle, about 10 years old, has a massive head, a strong beak with sharp cutting edges, and powerful limbs with large claws. The entire animal exudes strength and predatory prowess.

Subspecies. Two subspecies are recognized. *C. s. serpentina* (described above): The common snapping turtle is found from southern Canada to northern Florida. *C. s. osceola* (Stejneger, 1918): The Florida snapping turtle is found only in Florida. The greater number of tubercles on its neck distinguish this subspecies from *C. s. serpentina.* The function of these tubercles, which are strongly pointed, is poorly understood, but perhaps they play some role in cutaneous respiration.

Natural history. This turtle, blessed with remarkable adaptability, is found in all aquatic environments within its range, even in blackish water or highly polluted places. It needs muddy bottoms and enough detritus to allow it to hide and lie in ambush for prey as it rests in concealment. It passes almost all of its time in this fashion, under roots and submerged stumps, but from time to time it rises slowly to the surface, where its eyes and nostrils may emerge. It is omnivorous, with a carnivorous preference, and consumes insects, crustaceans, mollusks, worms, fish, amphibians, snakes, other turtles, birds (which it drowns by seizing them from below as they swim by), and small mammals, as well as plants, algae, and freshwater sponges. This is a truly voracious animal. Schmidt and Inger (1957) cited a case in which Indians were able to locate the bodies of drowning victims after a cyclone by using snapping turtles attached to a long cord or leash. The long, mobile neck allows the turtle to project its head rapidly to the front or the side, causing potential injury to someone trying to hold it. Although this is a massive animal, it has great agility, and its sudden movements are unpredictable. It may be active by night in the southern part of its range, and it hibernates in the north. Copulation occurs from April to November. The couple swim facing each other, but courtship is quite varied. Nesting occurs from May to September. The nest is large and provided with a narrow entrance and a wide chamber below. The clutch ranges from 10 to 83 eggs (about 30 on average), spherical, hard, and white-shelled, with a diameter of about 32 mm. Incubation lasts for 55 to 125 days, according to environmental conditions, and hatching occurs in late August up to October.

Protection. This species is extremely adaptable and is not rare in its natural environment; Iverson (1983) reported densities of 59 turtles per hectare in a pond in Tennessee. Nevertheless, it is widely captured by humans for two reasons: some are appalled by its appearance and presumed dangerousness, whereas others propose to eat it. Formerly, American Indians used it quite widely in soups and broths, as in Louisiana and Florida. Even today it is still sold commercially in Florida, in cans bearing an image of a green turtle. The eggs are sometimes fried, because the white does not harden on boiling but remains gelatinous. In Florida, fields may be adorned with thousands of carapaces of *C. serpentina,* and the stuffed animals are sold in boutiques as souvenirs. Urban household demand thus has endangered numerous populations. These turtles are also exported to oriental countries, where the buyers of the little 20 g hatchlings are unaware that they may grow up to become large and dangerous. When this happens, the turtles may be dropped off in local waterways, where they may profoundly impact the local environment. Ecologically, this is a top predator, and its flesh and fat may accumulate toxic products at the top of the food chain, which may cause many mortalities. The young are also subject to predation from many directions, and the adults may carry large numbers of leeches. Finally, both naturalists and nonnaturalists may not hesitate to destroy an animal considered a dangerous nuisance, like an alligator. Protection of this species is thus not always well received by the public. The result is that populations are diminishing from north to south, and some regions have already seen their last snapping turtle.

Chelydra acutirostris Peters, 1862

Distribution. This turtle occurs from northern Honduras as far as the Gulf of Guayaquil (Ecuador), including Nicaragua, Costa Rica, and Panama.

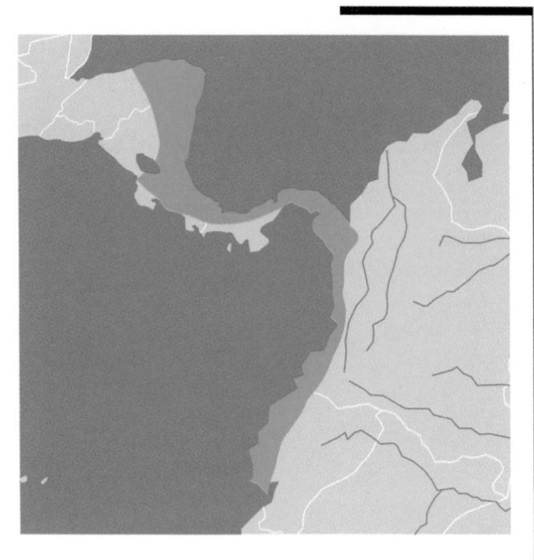

Common name

South American snapping turtle

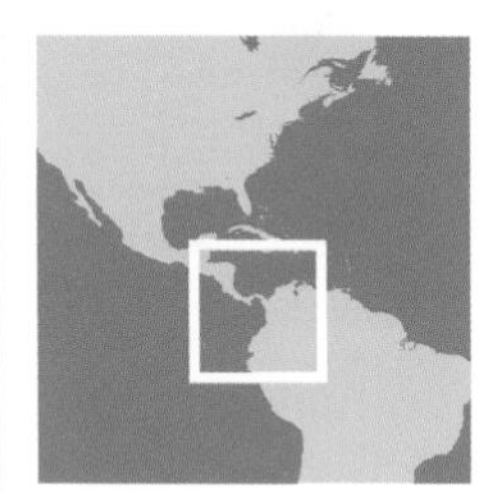

Description. Formerly considered a subspecies of *Chelydra serpentina,* this turtle is now recognized as a species in its own right. It differs little from *C. serpentina,* but the plastron is bigger and represents more than 40% of the area of the carapace. The anterior margin of the third vertebral scute is shorter than 25% of the width of the carapace. The neck is covered with small round tubercles. The head has a wide, light brown band along each side, bordered above and below by fine dark lines.

Natural history. The behavior and activities are the same as those of the common snapping turtle, and one finds them in the same kinds of places, up to 1,200 m above sea level.

Protection. This species often falls victim to caimans, and it may be locally consumed in soups and *estofados* (stews). The eggs are prized and are often compared to the eggs of sea turtles, which are protected, whereas those of the snapping turtle are not. It is unfortunate that in Costa Rica the juveniles are sold as "baby living fossil turtles" in boutiques, and they end up as "terrarium animals" in the homes of hobbyists.

Chelydra rossignonii (Bocourt, 1868)

Common name

Central American snapping turtle

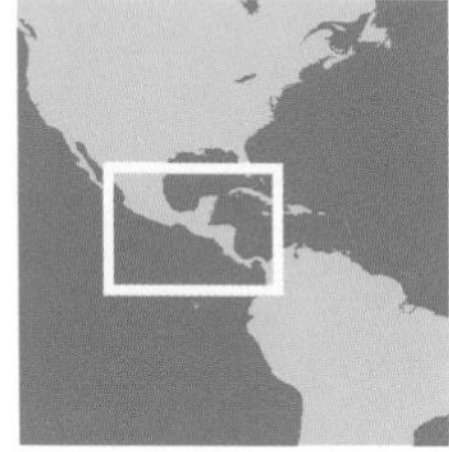

Distribution. This species is found in the Gulf drainages of Mexico (Veracruz, Tabasco, Campeche, Chiapas) and Caribbean lowlands of Belize, Guatemala, and northeastern Honduras.

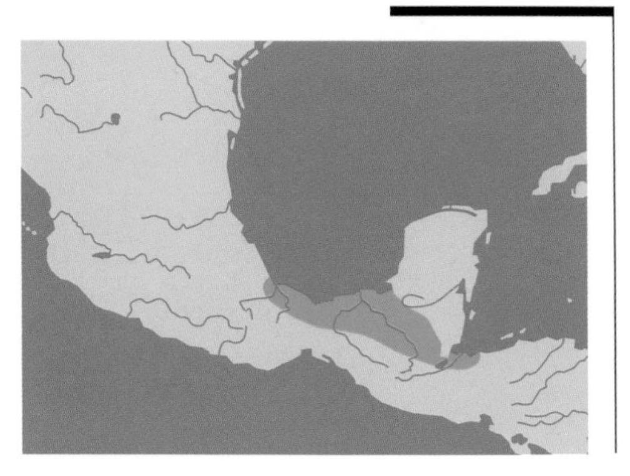

Description. This former subspecies of *C. serpentina* now is recognized as a full species. It differs only modestly from its congeners. The anterior lobe of the plastron is longer than 40% of the length of the carapace, and the anterior face of the third vertebral scutes is longer than

The carapace has longitudinal keels. The head is smaller, and the pointed neck tubercles are less developed than in *C. serpentina*.

25% of the width of the carapace. The neck is adorned with conical, pointed tubercles.

Natural history. The habits of this species are identical to those of *C. serpentina*. According to P. Pritchard (1979), this species would be closer to *C. s. osceola* than to *C. s. serpentina,* and he thinks that the population restricted to Florida and that of Mexico and Central America were formerly contiguous, before the glacial periods of the Pleistocene separated them. The pointed tubercles on the neck in both turtles may be an adaptation to poorer oxygenation in these warmer waters.

Protection. Both the turtles themselves and their eggs are heavily collected for food. In Veracruz, Mexico (at least), the wild populations are now quite depleted.

Macrochelys temminckii (Harlan, 1835)

Distribution. This species occupies all the river systems that drain into the Gulf of Mexico (USA), including the whole Mississippi Valley, as far up as Kansas, Iowa, and Illinois. It is also present in the southeast in Georgia, northern Florida, Alabama, Mississippi, Louisiana, and the eastern part of Texas.

Description. This is the only species in the genus, and it is certainly the largest freshwater turtle, at least in North America. Some individuals may weigh as much as 113 kg and with carapace length of 800 mm, not to mention the famous "Beast of 'Busco," captured in 1948, which allegedly weighed 227 kg. The carapace is rough and very strongly serrated, especially along the rear marginals. There are three strong keels, made up of ridges that rise to a point on each scute of the carapace, and they persist even in old animals. This turtle has 23 marginal scutes, and 3 to 8 supplementary scutes, known as supramarginals, are situated between the pleurals and the marginals on each side. The carapace is dark brown, dark gray, or black. Often the shell becomes covered with algae, which exaggerates the archaic aspect of the animal and also allows it to conceal itself in aquatic vegetation. The plastron is very reduced, is cross-shaped, and has very narrow bridges. The plastron as a whole is marbled with gray or is evenly dark. The head is wide, with a pointed snout, and the upper jaw is equipped with a huge, dangerous hook. On the floor of the mouth is a red, soft, mobile worm-shaped structure, unique among turtles, that the alligator snapper uses to entice prey into its mouth. The head has large

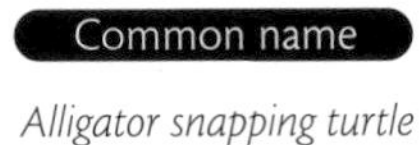

Alligator snapping turtle

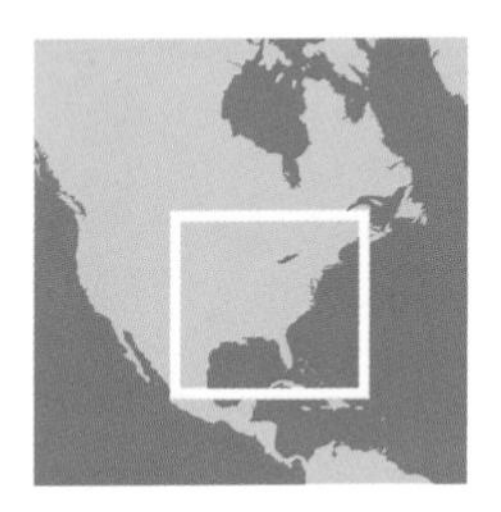

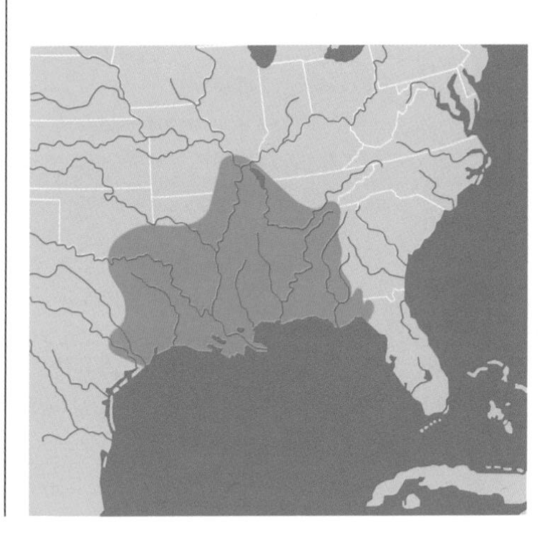

© B. Devaux

This turtle, photographed in an aquarium at the Bronx Zoo, weighs more than 100 kg. The powerful head and sharp beak bear witness to an aggressive nature. The least strike of the jaws may leave deep wounds, which rapidly become infected.

scales on the top and sides. The neck and chin are covered with conical tubercles and lappets, which contribute to the overall camouflage of the animal. The head is too big to be completely retracted within the shell. The skin is grayish, brownish, or even yellow and often has dark spots, particularly along the head. The limbs are powerful, webbed, and armed with long, strong claws. The tail is very long and "crocodilian," equal in length to the carapace, and has three rows of tubercles along its dorsal surface. The underside is covered with numerous small scales. This is an animal of truly antediluvian mien, is dangerous to handle, and has a devastating bite and rapid "snap." In the young, the carapace is heavily rugose, with three keels, and the head is covered with lappets of skin. There is a ring of fleshy tubercles around the eye, as in the adults. The jaws already function well, and the hatchlings are just a bad-tempered as their parents.

Natural history. This is a highly aquatic species, never leaving the water except to nest. It lives in deep rivers, canals, lakes, bayous, and other places that may have connections with running water and where it may find a mud bottom. It can tolerate brackish water, as is testified by the occasional discovery of specimens bearing external barnacles. It likes to rest immobile, mouth wide open, with the lure on the tongue leaping and somersaulting, waiting for the arrival of prey. The diet is actually quite omnivorous and varied: fish, crustaceans, mollusks, snakes, small alligators, seeds, and fruits that fall in the water. Even acorns, in some quantity, have been recovered from stomachs of alligator snappers. In Florida, mating occurs from February to April. The male is larger than the female, and he bites her brutally on the head and neck before rapidly proceeding with copulation. There is usually only one nesting per year, the nest being hidden on the riverbank. Nesting always occurs by day, and the eggs, around 40 in number, are spherical, 35 to 51 mm in diameter. Incubation lasts 100 to 140 days. The young hatch in September and October.

Protection. Natural predators are rare, although the eggs are destroyed by raccoons. This species has been heavily exploited, especially for human consumption (meat and soup). A single dealer in Louisiana, during the years 1984–1986, handled a total of 17,117 kg of alligator snapper meat. Protective measures need to be undertaken at the federal level, and some states, including Florida, have already outlawed all purchase, sale, or export of the species. The image of this formidable predator does not prevent local people from appreciating the meat of the alligator snapper, and the Cajun people of Louisiana have made it a major cultural item. These turtles are also sold in some European pet shops, and one shudders at the thought of the danger these "pets" may pose when they get big. The powerful jaws can sever digits in a moment, and children could be severely injured by one of these monsters at liberty. In practice, when the turtles get too big for the home, they are quietly liberated in the wild. They have been found in lakes or ponds in Germany and France and even in the sea (in the Mediterranean near the island of Port-Cros).

© F. Bonin

This turtle has a unique feature—namely, the small red wormlike filament in front of the glottis, which attracts prey right into the cavernous mouth.

PLATYSTERNIDAE Gray, 1869

Cette famille est représentée par une seule espèce asiatique, à la dossière très plate, à la tête de grande taille armée de mandibules puissantes et non rétractiles, au crâne large et massif, à la queue aussi longue que le corps.

Platysternon megacephalum Gray, 1931

This family is represented by a single species, found in Asia. It has a very flat carapace, a huge nonretractile head armed with powerful jaws, a wide skull, and a tail fully as long as the carapace.

Distribution. This species is widely distributed in Southeast Asia, from central China to Hong Kong, northern Vietnam, Laos, Myanmar, and northern Thailand.

Description. This turtle is immediately recognizable by its enormous flat-topped head, half the width of the carapace of the animal and too big to retract within the carapace. The top of the head is covered with a single huge scute or shield, which forms a sort of helmet with laterally descending arms in front of the tympana. The jaws are powerful, and there is a heavily hooked beak. The shell length does not exceed 185 mm. It is elongated and extremely flat, sometimes has a slight median keel, and it has moderately serrated marginals. It may be yellowish or olive-brown, with dark spots. Each scute carries very clear annual growth rings. The plastron is unhinged and well developed but is attached to the carapace only by cartilaginous bridges. There is a deep anal notch. The plastron is yellow to light brown, decorated with darker spots. The head is brown to olive, marked with orange-yellow spots or with red or pink ones in some subspecies. The limbs are strong, well armed, and adapted more for walking and climbing than for swimming. The tail is as long as the carapace, thick at the base and becoming more pointed toward the end. When disturbed, the turtle may secrete a foul-smelling musk. The young hatch with brighter colors than the adults. The carapace of the young is often keeled. If disturbed, they have the habit of emitting audible squeaks.

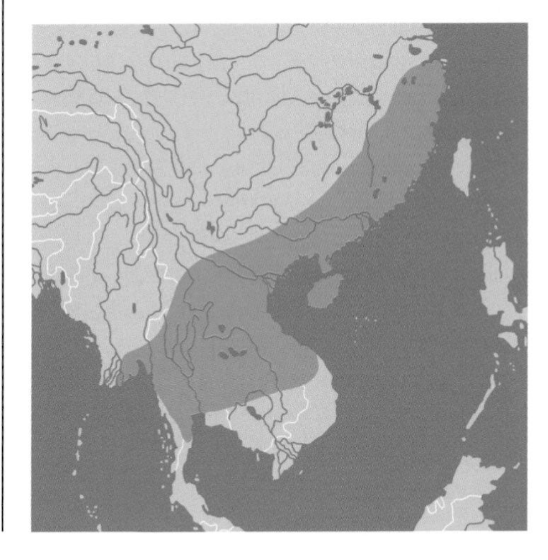

Common name

Big-headed turtle

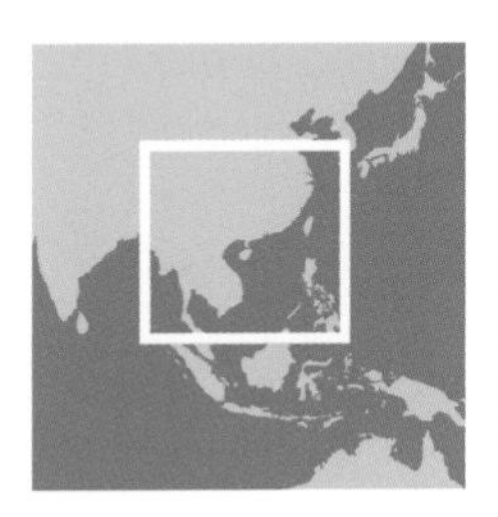

Subspecies. Five subspecies are recognized today. *P. m. megacephalum* (described above): The Chinese big-headed turtle occurs in southern China and has a plain plastron, a carapace with discrete keels, and no indication of growth rings on the scutes. The marginals are only slightly serrated, and those along the bridge area curve upward. The cephalic scale is very developed and reaches quite low behind the orbits. *P. m. peguense* (Gray, 1870): The Burmese big-headed turtle is found from western Vietnam to the southern part of Burma. The plastron is decorated with dark areas along the sutures. The carapace has a feeble median keel, and sometimes two lateral keels, and has clear growth annuli and serrated marginals behind. The jaws are immaculate, and the upper jaw is strongly developed. *P. m. shiui* (Ernst and McCord, 1987): The Vietnamese big-headed turtle is found in northern Vietnam. The head, neck, carapace, upper surface of the limbs, and underside of the tail are heavily dotted with yellow, orange, and pink. The carapace is smooth, and the upper edge of the marginals is not notched. There

© B. Devaux

A very flat carapace, a long and strong tail, and a head armed with a hooked beak render this creature suitable for adornment of the upper reaches of Notre Dame.

are neither growth annuli nor keels. The head casque is reasonably developed but does not extend behind the orbits. The upper jaw has a very strong hook. *P. m. tristernalis* (Schleich and Gruber, 1984): Found only in the province of Yunnan, China, the Yunnan big-headed turtle has three additional scutes at the junction of the humerals and gulars. *P. m. vogeli* (Wermuth, 1969): Found only in northwestern Thailand, the Thai big-headed turtle has the same plastral pattern as *P. m. peguense,* but the upper jaw is small and narrow, without a heavy beak. The carapace is smooth, and the marginals are not serrated.

Natural history. Big-headed turtles live in rivers and waterfalls at high altitude and with fast currents and a rocky bottom, and during the day they hide among the rocks. They prefer a cool temperature of about 12°C to 17°C. They are almost never seen basking. They are mostly nocturnal and take advantage of darkness to hunt mollusks, crustaceans, and fish, which they dismember with their powerful jaws. This species is not a great swimmer, but its unusually morphology allows it to pursue a unique means of getting about: climbing. Thanks to its powerful head and robust jaws, it is able to haul itself over rocks and to climb violent waterfalls, aided by its long tail, upon which it leans. Is it this behavior that has favored the development of the huge head, or did the big head allow it to develop this method of climbing? This is an aggressive species, and it should be handled carefully. Reproduction is poorly studied. The females are said to lay just one or two white eggs, ellipsoidal in form and 37 × 22 mm in size.

Protection. This species is seriously threatened throughout its entire range. It is widely caught for human consumption, particularly in China, where it is rapidly disappearing. It is also threatened by the construction of hydroelectric dams and by deforestation. Once very common in certain regions, it has become rare and is on the path toward extinction. In Vietnam it is also hunted for export markets and is listed as one of the 48 most threatened species in the world by the Turtle Conservation Fund Action Plan. It is also collected and exported for hobbyists. There is some urgency to establish measures of protection that will reverse the negative trends.

TRIONYCHIDAE Fitzinger, 1826

The carapace of these turtles is covered with continuous leathery skin, which is laid over papillose areas of the carapace and plastral bones known as callosities. There are only three claws on each limb. The overall distribution parallels that of the Emydidae. The genera are divided into two subfamilies.

Subfamily Cyclanorbinae Lydekker, 1889

Cyclanorbis senegalensis (Duméril and Bibron, 1835)

Common name

Senegal flap-shelled turtle

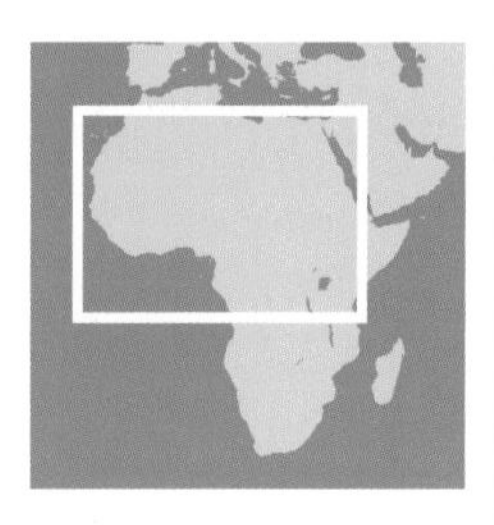

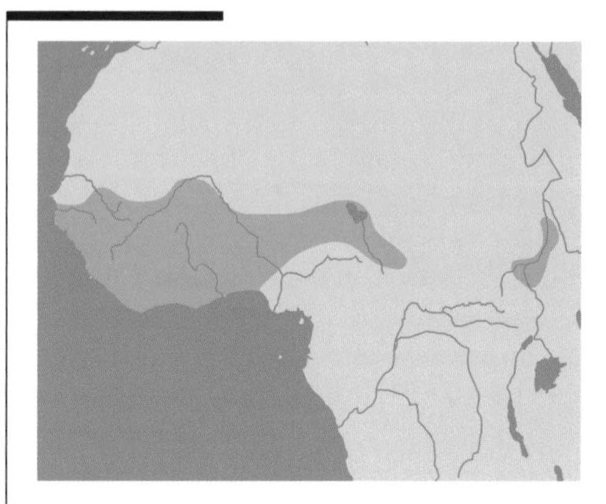

Distribution. The species is best known from Senegal, but it may be found from southern Sudan to northern Cameroon, Ghana, and Gabon. It may also occur in Chad, the Gambia, Togo, and Ivory Coast.

Description. *Cyclanorbis senegalensis* is usually thought to be much smaller than its congener *C. elegans,* but certain individuals may attain as much as 600 mm in carapace length. The shell is slightly more domed and is oval, slightly granulated in texture, and grayish to green, with numerous dark dots that fade out with age. The carapace is wide posteriorly, with a series of thin peripheral bones and a notch above the tail. The plastron is gray-white and has seven to nine well-developed callosities in adults, as well as two semilunar valves that can close up to cover the hind limbs. The head is medium-sized, has a very short snout, and is grayish to greenish on top, often with a marbled, cream-colored design below. The remainder of the body is light gray. The forelimbs have five or six well-developed cutaneous lamellae. The tail is thicker and longer in the males than in the females. The young are more colorful, with

© J. Maran

This turtle is very rounded, flat, and smooth and is almost impossible to grasp at the bottom of waterways. The neck is moderately long, and the dark greenish to brown head is decorated with small light spots.

numerous tuberculate dark lines on a cream to greenish background, and they have a slight median keel on the carapace. The periphery of the dorsal disk bears a light line, whitish to light green in color. The head is beige to grayish, with a marbled effect, and is white below. The plastron is almost yellowish, spotted with gray.

Natural history. This species lives in sympatry with *C. elegans* in a part of its overall distribution (Sudan, Togo, Chad). It is elusive and rarely spotted in its typical habitats of swamps and marshes, as well as oxbow lakes and other stagnant water bodies. It feeds on organic detritus, fallen fruit, alevins, dead animals, and anything else that passes its portal. It nests several hundred meters from its pond or residence, sometime in March or April. Several nests may occur within a season, with about a dozen elongate eggs each time. The eggs measure 45 × 25 mm on average.

Protection. This species is still consumed in the villages of Casamance and Sudan. It may be caught in the wetlands on hooks or by hand. It is difficult to estimate the size of populations, but the species seems to be common in southern Senegal and in the countries of West Africa.

© B. Devaux

The well-developed snout allows this turtle to see and to respire while remaining underwater.

Cyclanorbis elegans (Gray, 1869)

Distribution. This species has a wide but fragmented and discontinuous range in Ghana, Togo, Benin, Nigeria, Central African Republic, southern Chad, and southern Sudan.

Description. This is a large species with a rounded shell that may measure 600 mm in length. The color is olive to brown, dotted with a myriad of yellowish or greenish spots along the flanks of the carapace. There is a line of small tubercles and a slight longitudinal keel in the young, but these disappear with growth. Larger tubercles on the front of the carapace above the neck remain throughout life. The plastron is yellowish, with dark spots. Two crescent-shaped large callosities protect the viscera, and two smaller ones may be present at the rear of the plastron. The head is small, with a pointed snout in the form of a double tube. The head is brown above, with yellow or light green vermiculated pigmentation. The chin and neck are lighter and spotted with numerous small yellow dots. The forelimbs have four transverse lamellae in the shape of a crescent. All limbs are well webbed. The skin is brown above, lighter below. The juveniles, spotted with yellow dots, are very colorful. The plastron has no femoral valves.

Natural history. This species occurs in slow rivers and wetlands, where it is often sympatric with *C. senegalensis*. Its life history and diet are

Common name

Nubian flap-shelled turtle

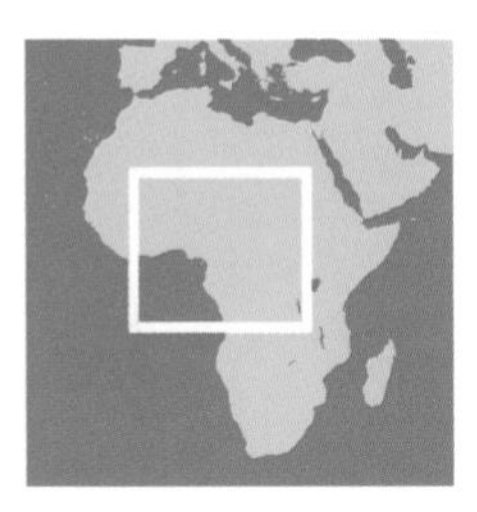

© K. Dodd

This species is more colorful than the preceding one, and the vermiculated effect on the greenish to brown background is evident. The snout is rather short.

poorly known but are presumably close to those of the other *Cyclanorbis*.

Protection. Widely eaten by local people throughout its range, the populations of this species appear to be much smaller than those of *C. senegalensis*.

Cycloderma frenatum Peters, 1854

Distribution. This is an East African species, found from Tanzania as far as the Save River in Mozambique, passing through southern Zambia and Malawi. Its occurrence in Zimbabwe needs to be confirmed.

Description. The name comes from the Greek *kiklos* (circle) and *derma* (skin), and from the Latin *frenum* (bridle), in reference to the rounded, skin-covered carapace and the parallel lines along the neck that look like a bridle. This is a large turtle, which may measure 560 mm and weigh 14 kg. The color is greenish to brown, usually rather dark, with a scattering of darker dots and spots. The rather large plastron is whitish or pink, and it develops a gray, vermiculated appearance in certain females. There are seven plastral callosities, which develop progressively as the animal grows and ages. The head is elongate and is slightly flattened on top. The eyes are bulging and placed on the very top of the head. The head ends in a pointed greenish or grayish snout, with a dark line connecting the orbits and five longitudinal lines ending at the base of the neck. These lines, usually very straight but sometimes interrupted, tend to disappear with age. Old animals just have white spots on the head and neck. The chin and the base of the neck are pale, sometimes with darker spots. The limbs are gray and have four or five crescent-shaped scales on each anterior one, and all are well webbed and equipped with three claws. In the young, the carapace bears a slight median keel and often has small rugose tubercles. In color it is green or gray, with a white margin. The plastron is yellowish to whitish, the only markings being a dark zone around the umbilicus and a smaller one in the anal region.

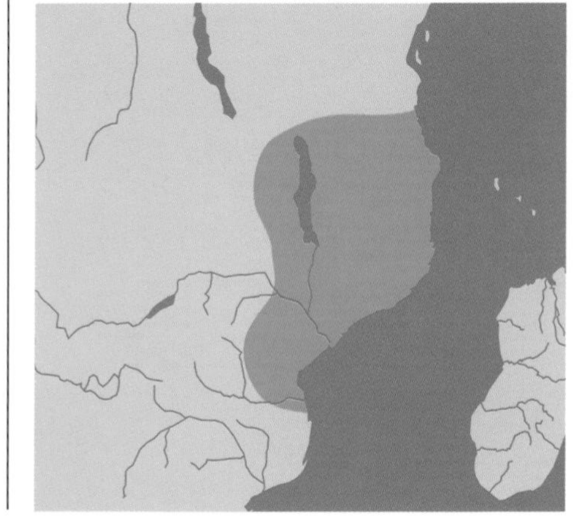

Natural history. This species occurs in rivers, streams, and lakes and also in stagnant swamps. It is carnivorous in nature, the diet including a large number of snails and bivalve mollusks. It can bury itself deeply in the mud, where it searches for buried mussels. The species in general is shy and nervous, never seeking to bite when handled. It is an excellent swimmer and also runs rather fast on dry land; it is fully capable of disappearing in a flash as it buries its neck and limbs in the sand or the mud. It spends long periods of time totally buried underwater. The clutches, which are laid from December to April, during the southern summer, include 15 to 25 eggs. They have a hard, spherical shell, sometimes slightly prolate, and measure about 32 × 33 mm. There may be several clutches per season. The young hatch in December to February of the following year.

Protection. This species is poorly studied in the field, and the status of populations is unknown. It is heavily utilized for food by local people.

Common name

Zambezi flap-shelled turtle

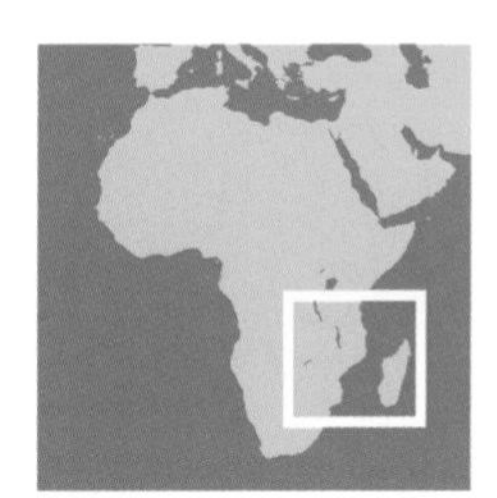

Cycloderma aubryi (Duméril, 1856)

Distribution. This turtle is found only in the western part of central Africa, in Gabon, in the extreme east of the Democratic Republic of the Congo, and in Cabinda (a separated enclave of Angola). Its occurrence in the south of the Central African Republic needs to be confirmed.

© J. Maran

This species has a very long neck, dark in color and marked with a black line running from the eye as far as the carapace. The head is rather flattened, small, and extended into a pronounced snout. The eyes are well developed and placed very high.

Description. This is a large turtle, measuring up to 610 mm and weighing up to 18 kg. The carapace is oval, somewhat extended in the nuchal region, smooth, unkeeled in adults, and red-brown in color, with a very dark median stripe. The plastron is yellowish, with dark spots here and there. There are seven very large, granular plastral callosities. The skull, flat or even slightly concave, ends anteriorly in a bony tubular snout. The head is brownish to reddish, sometimes very dark, and is decorated with five fine longitudinal lines, one along the middle of the skull, reaching to the base of the neck, two others from the top of the orbits to the back of the head, and two more starting at the nostrils and extending through the orbits to the base of the neck. These dark lines on the carapace and the head, which may be outlined with fine white borders in some individuals, are unique within the Trionychidae, but, curiously, they occur in *Pelusios niger* and *P. gabonensis,* which live in sympatry with *C. aubryi* (P. Pritchard, 1979). The chin and the throat are yellow, with scattered small dark spots. The upper jaw has a flexible angular "lip" on each side and completely covers the lower jaw. The limbs are strong and each forelimb has six or seven crescent-shaped scales on the anterior surface. The males are larger than the females and have a flat plastron, whereas that of the females is convex. When threatened, this turtle excretes a foul-smelling musk. The young are extremely colorful. The carapace and head are orange or light red, and the fine dark lines are very clearly indicated. The carapace has a keel and a dark median stripe. A pattern of concentric lines and small tubercles is evident on the dorsal side. The limbs are almost black, and the black spots on the plastron are striking, although the plastral callosities have not yet appeared.

Natural history. These turtles live in watercourses in humid tropical forests and seem to enjoy the muddy or sandy bottoms of lagoons and rivers with turbid waters. They also occur in inundated forests or in isolated pools during the dry season. They are sometimes seen basking on fallen trunks, but they are also active at night, when they feed. The diet is omnivorous and includes freshwater crabs and small crustaceans, as well as fish, which it may seize by means of powerful hyoidal sucking action, as is also practiced by *Chelus fimbriatus*. The species is often heavily parasitized by leeches, even on the inside of the mouth, during the dry season. The eggs, 15 to 35 in number, are almost spherical and have white, hard shells. They measure about 33 × 390 mm and are deposited in a shallow nest not far from the water. There may be two nestings per season. Hatchling emergence occurs during the major wet season, between March and April.

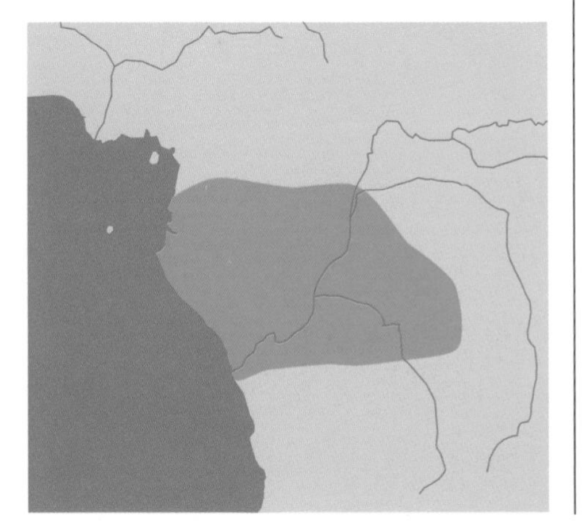

Common name

Aubry's flapshell turtle

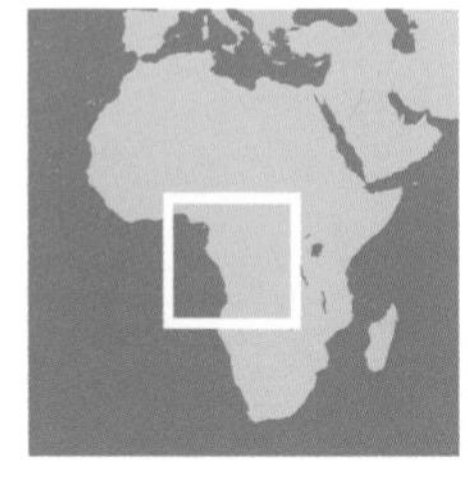

Protection. This species is hunted for its meat and is not protected. Hunters know how to extract the musk glands so that the meat will not take on a bad taste.

Lissemys punctata (Lacepède, 1788)

Distribution. This species is found throughout the Indian subcontinent, in India, Pakistan, Nepal, Myanmar, and Bangladesh, as well as in the Andaman Islands, where they have been introduced by humans.

Description. This is a small soft-shelled turtle, measuring up to 270 mm and weighing up to 7 kg. The carapace is oval, somewhat domed, and dark brown or dark green, but with great variation in color according to locality. Sometimes it is boldly adorned with green, yellow, or even brown markings (hence the name *punctata*), but other specimens are immaculate. The plastron is cream-colored, without any markings. The anterior lobe can bend upward to close the shell opening completely and to protect the head and forelimbs. There are seven callosities on the plastron. The head is grayish to green above, pale green below. Young individuals have three dark parallel lines running the length of the head.

Subspecies. Currently, two subspecies are recognized. The third, *L. p scutata,* is now raised to full species status. *L. p. punctata* (described above): The southern flapshell turtle occurs in India, from south of the Ganges to the southern tip of India and into Sri Lanka (with the exception of the east coast) and Bangladesh, in the Khulna District. *L. p. andersoni* (Webb, 1980): The northern flapshell turtle is found in the Ganges Basin, the Brahmaputra and the Indus in northern India, as well as in eastern Pakistan, southeastern Bangladesh, and northeastern Myanmar. The shell is more olive in color, often with a light peripheral line.

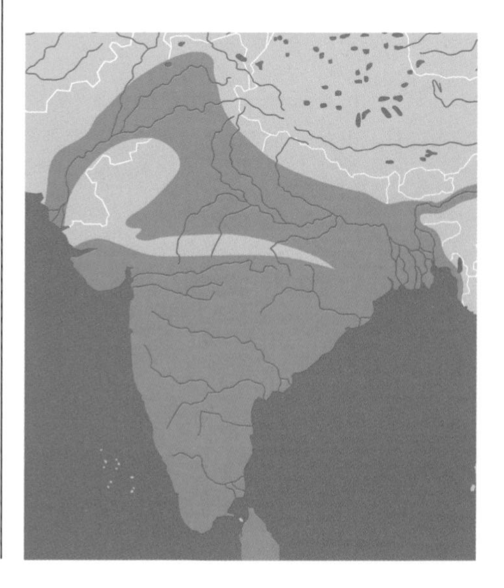

Indian flapshell turtle

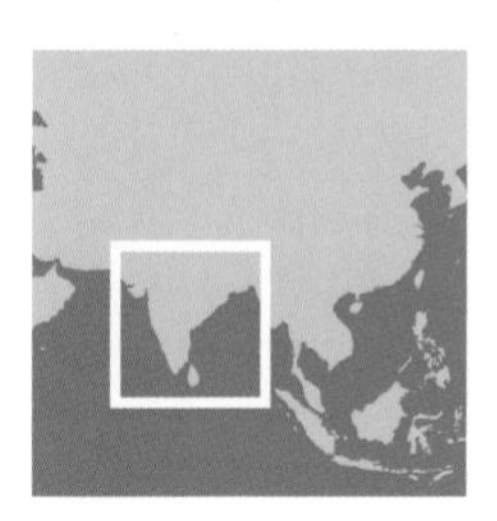

Natural history. This turtle lives in calm, deep waters in streams, rivers, lakes, swamps, wetlands, and irrigation canals, with either a muddy bottom or a sandy one. It is often seen basking on banks or on floating vegetation. It is very nervous and dives at the slightest disturbance. It is not aggressive and does not bite when picked up, but it can excrete a nauseating yellow fluid. During the dry season, it buries itself in the mud to protect itself from predators and desiccation and awaits the return of the rains before emerging. The diet is omnivorous and includes frogs, tadpoles, fish, crustaceans, snails, worms, and aquatic vegetation. It also consumes carrion, and cases of cannibalism have been observed. The flat jaw surfaces allow it to smash the shells of mollusks. In general, it is mostly herbivorous when it emerges from hibernation, which occurs from November to February, and then become carnivorous, rediscovering herbivory before the following hibernation. Courtship takes place in April. The male swims above the female, caressing her carapace with his chin. Then they face each other and butt heads. The female allows herself to drop toward the bottom, and the male establishes a grip on her carapace whereupon copulation ensues. The eggs are laid in shallow nests covered with vegetative debris. Numbering from 3 to 14, the eggs are spherical and white, with brittle shells, and they average 33 × 25 mm in size. Several nestings may occur within a season. Incubation lasts for 270 days. Emergence occurs just before the rains, which give

the young the opportunity of feeding on mosquito larvae, which are numerous at this time.

Protection. The principal predators of the eggs and the young are otters, mongooses, jackals, monitor lizards, and vultures. In addition, the species is heavily harvested for human consumption and is highly valued by local ethnic groups for its nutritive value. It plays a vital role in Vedic religious rituals, in the course of which it is sacrificed in large numbers. Most temples in the region maintain populations of this species in their ponds or tanks. It is also sold in certain city markets such as those of Calcutta. It has been proposed that "farms" for this species be created, as has been done for another soft-shelled turtle, *Pelodiscus sinensis,* but these have not yet come into being. Even though the species is classified in Appendix 2 of CITES and is protected by national laws, major stresses on the populations remain, and the numbers have gone down significantly.

© I. Das

This species has a short snout and often has light markings on the head and neck. Certain old individuals are darker; the young have lighter lines on the head and neck.

Lissemys scutata (Peters, 1868)

Distribution. This turtle is found only in Myanmar, in the Ayeyarwady, Sittang, and Salween River systems and their tributaries.

Description. Longtime listed as a subspecies of *L. punctata,* this turtle is now considered to be a valid full species. Certainly the two species look rather alike, but there are several important differences. The carapace of *L. scutata* is olive-brown, rather strongly domed and decorated with numerous dark spots in juveniles, giving way to a brown, reticulated texture in adults. The head is also olive-brown but bears two dark, well-defined lines extending backward, the first between the orbits and the second on each side of the posterior edge of the orbits, extending to the neck. In the plastron, the entoplastral callosity is very large and ultimately comes into contact with the hyo-hypoplastral callosities. The overall size is somewhat smaller than that of *L. punctata*.

Natural history. This species frequents the slow-flowing sections of streams, as well as irrigation canals, swamps, and wetlands. In diet it is omnivorous, eating fish, amphibians, carrion, and aquatic vegetation.

Protection. Information on the status of this turtle in Myanmar is scarce, but it is known that the species is heavily collected both for food and for export, the evidence for the latter being the large numbers offered for sale in China. Certain

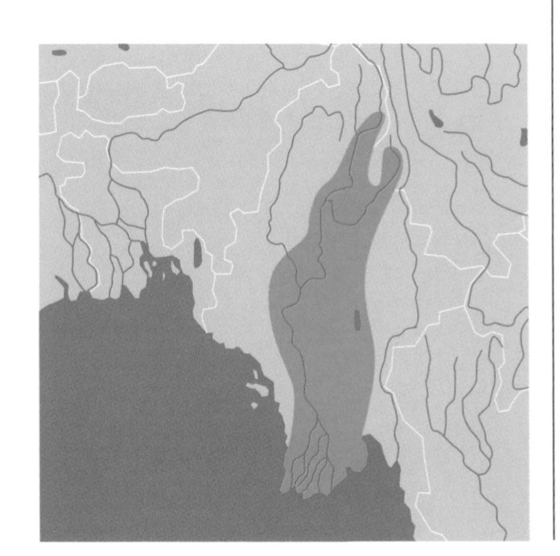

Common name

Myanmar flap-shelled turtle

© J. Maran

The carapace is brown, rather domed, and quite smooth. The young have a number of dark spots on the gray or beige background; in the adults the spots are scattered and form a reticulated pattern, but sometimes the entire carapace becomes black. The head is brown, with one or two dark lines on the side.

oral traditions allude to the danger of eating the meat of this turtle, which can cause skin diseases in women—and these traditions, for better or worse, help protect the species. Furthermore, the legal penalties are severe; turtle hunters run the risk of two years in jail. But the commerce continues, the law is rarely applied, and it is probable that this species, endemic to the single nation of Myanmar, is severely threatened.

Subfamily Trionychinae Fitzinger, 1826

Amyda cartilaginea (Boddaert, 1770)

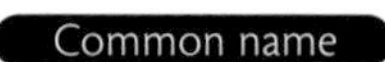
Common name

Asiatic softshell turtle

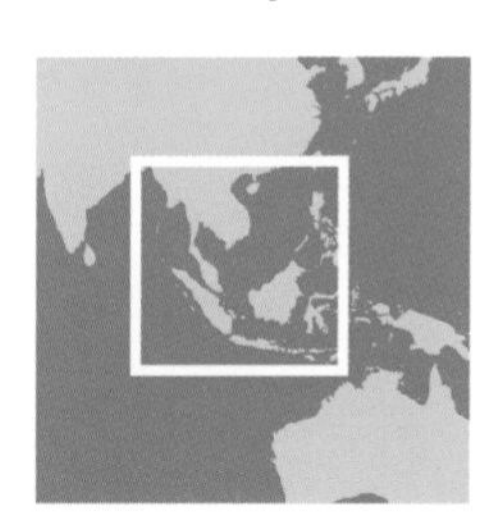

Distribution. The range is of this species is very wide and also discontinuous. The species is found from eastern India (Mizoram, Ngengpui) and in the Indian states between Bangladesh and Myanmar, as well as in Thailand, southern Laos, Cambodia, and southern Vietnam to the Malay Peninsula, Borneo, Sumatra, and Java.

Description. This is a large turtle with an oval carapace, which may reach a length of 800 mm (in 1987, a specimen of 202 kg was captured in the Chao Phya River in Thailand). It is olive-gray to greenish brown in color, with numerous yellow spots and, in young specimens, a black carapace. This adornment fades out with age and is replaced by broad, radiating black bands. Some adults are immaculate olive green. The young have several linear series of small tubercles on the leathery disk. These eventually disappear. The adults keep only the dense patch of tubercles in the nuchal area. The plastron is white or sometimes grayish and

has five roughened callosities. The head, neck, and limbs are olive green, with numerous yellow spots, and in addition there may be orange or pinkish spots on the top and sides of the head, behind the eyes, in very young specimens. Very old animals may only retain a cross-shaped figure on the top of the head. The snout is long and tubular, and the orbits are located high on the head. The young are very colorful. The carapace and head are dark green, with numerous small yellow spots, and the plastron is pure white or slightly grayish.

Natural history. This turtle is found in a wide variety of aquatic habitats, ranging from highland waterfalls to slow, turbid creeks, as well as in irrigation canals, lakes, and wetlands. The diet is primarily carnivorous and includes fish, crustaceans, aquatic insects, and other invertebrates. This is a burrowing species that hides itself under the mud or the sand for long hours, with only the tip of the snout showing, but during the night it may emerge from the water, presumably to feed. It is somewhat aggressive and will try to bite. Three or four nestings may occur annually, with 6 to 10 eggs per clutch (maximum 30). The eggs are thin-shelled and spherical, with a diameter of 21 to 33 mm. Incubation lasts 60 to 145 days.

Protection. Although the meat of this species is less sought after than that of certain other softshells, this turtle is still heavily caught and consumed. In Vietnam, for example, it has become rare, but even so, it still appears on the menu of some restaurants. In Indonesia, where the human population is mostly Muslim and turtles are not consumed, it is sent to the markets of China instead. In Malaysia, the eggs and meat are both heavily exploited, and the species has become very rare. This turtle also is stressed by the trade in traditional medicines. Furthermore, its habitats are very damaged and altered by humans, and pollution thus adds to the stress of overharvest. Populations are becoming more and more reduced, and it is vital that comprehensive conservation of the species be initiated.

This rather young specimen, photographed in Malaysia, has very clear, light decoration, although old animals are darker. The snout is thin and elongate. Altogether this is an aggressive creature.

Apalone spinifera (Lesueur, 1726)

Distribution. The six subspecies occupy a vast range, from Mexico into Canada, including much of the USA. The species occurs from Vermont, Ontario and Quebec, the Dakotas, and Montana to the Gulf States and New Mexico and into the states of Tamaulipas, Nuevo León, Coahuila, and Chihuahua in Mexico.

Description. The species of *Apalone* were formerly included in *Trionyx*. The carapace of the present species, which does not exceed 550 m (smaller in males), is rounded and very flat, rugose in texture, and sometimes with pointed excrescences, tubercles, or small spines (hence the name), which may run together at the front of the carapace, above the neck. The carapace is olive to brown, with numerous ocelli and small black spots, and sometimes a dark line that follows the border of the dorsal disk. This species appears to be able to alter the coloration of the carapace according to its environment. The plastron is white or yellowish and immaculate. Large callosities are pres-

© F. Bonin

Young animals sometimes have remarkable and unusual coloration. With age, the carapace becomes dark, with a light border. The neck is ornamented with light spots or streaks, and the snout is fine and elongate. The spines are visible at the front of the carapace and are most obvious in young specimens.

© F. Bonin

Numerous subspecies exist, and only young specimens can be identified with ease. This specimen appears to be a *hartwegi* from the southwestern United States.

ent on the hyo-, hypo-, and xiphiplastra but are not evident on the epiplastra and entoplastron. The head is rather small and pointed and tapers anteriorly into a long, tubular snout ending in large nostrils. The color is brown, olive, or gray, covered with dark dots. There are two light bars, bordered by fine black lines; one bar extends from the rear of the orbit and passes along the neck, and the other extends from the corners of the mouth. The lips are yellow and spotted with dark dots, and they conceal sharp jaws that can cause serious wounds. The limbs are powerful and thoroughly webbed, and each has three claws. This is an irascible species that will attack other turtles of its own or other species.

Subspecies. Six subspecies are recognized.

A. s. spinifera (described above): The Eastern spiny softshell occurs east of the Mississippi, from Vermont to extreme southeastern Canada and into Wisconsin, North Carolina, Virginia, and Tennessee. The disk bears large black ocelli and a single dark peripheral line.

A. s. aspera (Agassiz, 1857): The Gulf Coast spiny softshell is found from North Carolina to Mississippi, Louisiana, and Florida. The disk, often very light in color, is frequently marked with more or less discontinuous dark lines outlining the border. The postorbital and postlabial lines sometimes reconnect.

A. s. emoryi (Agassiz, 1857): The southern spiny softshell occurs in the basins of the Rio Grande and the Pecos, in Texas and New Mexico, as well as the Colorado River basin in California, Utah, Nevada, and Arizona and then into Mexico in the states of Chihuahua, Coahuila, Nuevo León, and Tamaulipas. The carapace is bordered with a light line that becomes wider toward the rear. A curved line joins the orbits, and the postocular lines are reduced to a spot on the side of the head.

A. s. guadalupensis (Webb, 1962): The Guadalupe spiny softshell is restricted to the drainage basins of the Guadalupe, San Antonio, and Nueces Rivers in southern Texas. The dark carapace is covered with small light tubercles and decorated with black ocelli.

A. s. hartwegi (Conant and Goin, 1948): The western spiny softshell is found to the west of the Mississippi, in Minnesota, the Dakotas, Montana, and into Louisiana, Oklahoma, and New Mexico. The carapace is covered with dark dots of variable size. There is a dark band around the margins and along the sides, as well as on the rear of the carapace. Black lines run along each digit of the forelimbs.

A. s. pallida (Webb, 1962): The pallid spiny softshell occurs in the basin of the Red River, which reaches the Gulf of Mexico near Galveston, Texas. The carapace is very pale and dotted with small white tubercles on the posterior half.

Natural history. This turtle frequents creeks and rivers, but it also occurs in other types of water bodies, including irrigation canals, lakes, and

bayous, as long as the bottoms are sandy or muddy. It spends all its time in the water, burrowing in the mud, floating at the surface, or dug into the sand or other bottom substrate to ambush its prey. It is active by day and by night and sleeps down in the mud or wedged on a branch in the current. It does not avoid brackish waters and occasionally may be seen basking on banks or on submerged wood. A cutaneous respiratory system allows it to obtain part of its oxygen requirements from the water itself. This turtle is almost completely carnivorous, and it eats everything it can catch—fish, tadpoles, frogs, mollusks, and aquatic insects—and also carrion. It seems to be particularly fond of crayfish. Some vegetation may also be consumed. Copulation takes place in April and May, and nesting from May to August. There may be two nestings per season. The nests are sometimes a good distance from the water, and the clutch numbers from 4 to 39 eggs. Incubation lasts 50 to 95 days, according to latitude. In contrast to nearly all other turtles, the sex of the individual does not depend upon the temperature at which the egg was incubated. The young are marked similarly to the adults.

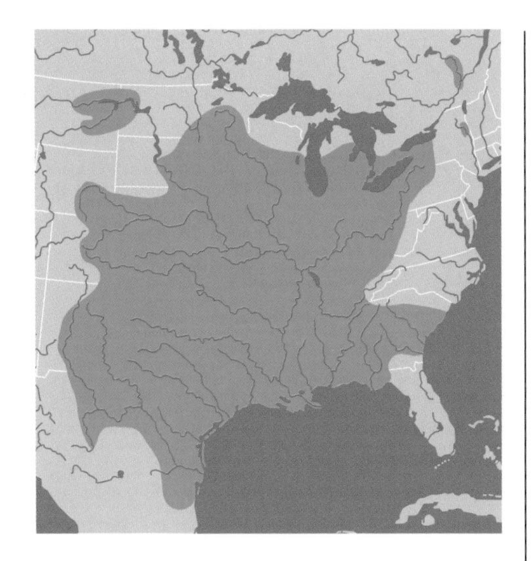

Protection. This species does not appear to have been depleted by commercial harvest, but certain populations may have been reduced by habitat destruction and pollution. The most endangered subspecies are those with the smallest ranges.

Common name

Spiny softshell turtle

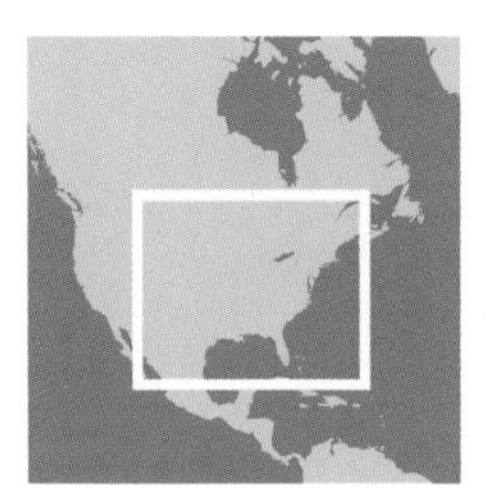

Apalone ater (Webb and Legler, 1960)

Distribution. This species has an extremely restricted range in the basin of Cuatro Ciénegas, in the state of Coahuila in northern Mexico.

Description. The validity of this species remains controversial among herpetologists. It was formerly considered a subspecies of *A. spinifera,* but one school of thought now considers it to be a full species. The carapace has a slightly granulated texture, like sandpaper but sometimes smoother, and there are no tubercles present on the anterior margin. The posterior margin is irregularly serrated. The color is dark gray but may actually be black (hence the name *ater*). Males have a light line around the border of the disk. The head is large, and the tubular snout is larger in the male than in the female. The skin is very dark gray, with small whitish dots adorning the rear of the cranium.

Natural history. This turtle seems to be very rare and lives only in *pozas* (freshwater springs) in Cuatro Ciénegas, which it shares with other species of turtle, including *Terrapene coahuila.* These springs are now connected by irrigation canals to creeks outside the valley, and the species as seen today, may be a hybrid with *A. spinifera,* the two having been brought together by these hydrological alterations. The upshot is that, during numer-

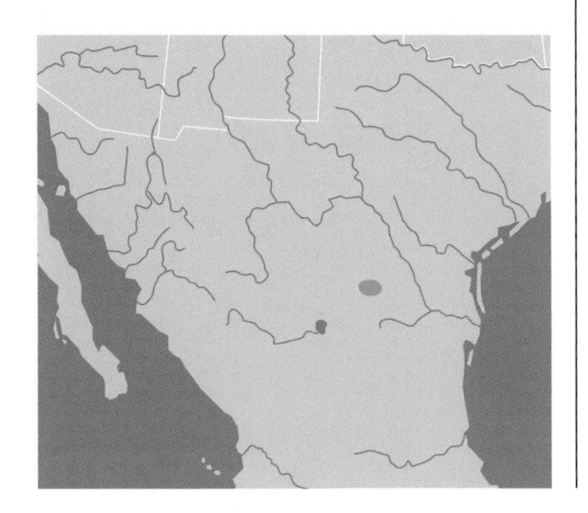

Common name

Black spiny softshell

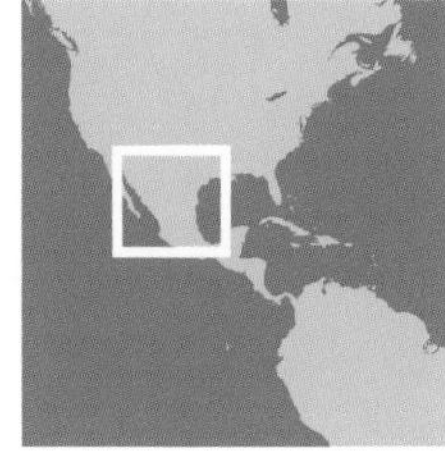

ous investigations at the site, the species has not been observed recently, and the ecology and biology have not been described.

Protection. The Cuatro Ciénegas Basin is legally protected and is the object of permanent vigilance by local environmental teams. The two species *A. ater* and *T. coahuila* have been placed in Appendix 1 of CITES, and the softshell is placed by the Turtle Conservation Fund on its list of the 24 most endangered turtle species. All in all, deciding upon its exact status and evaluating its complete biology and ecological needs remain to be done.

Apalone ferox (Schneider, 1783)

Distribution. This species is found only in the southeastern United States, from South Carolina through part of Georgia and into the Florida peninsula, excluding the Keys. It reaches southern Alabama as far as Baldwin County and Mobile Bay.

Description. This is a rather large softshell, which may reach 600 mm in length. The carapace is flat and has a thickened edge, decked with a series of wide, short tubercles, most numerous in the crescent-shaped anterior area, above the neck and the feet. Often there are also longitudinal lines of small tubercles running the entire length of the disk. The disk is oval, without a real keel, and is grayish or olive to brown, often adorned with darker spots. The plastron is white, with somewhat of a gray tinge. The hyo-, hypo- and xiphiplastral callosities are conspicuous; others are absent or at most poorly developed. The head and the limbs are gray or brown, with lighter marbling and reticulations. There is also a reddish or yellowish line on the side of the head, extending from the eye to the corner of the mouth. The jaws are powerful, and the alveolar surface of the upper jaw expands considerably in some old individuals. The nose is tubular, rather long, and is sharply cut off at the tip. The four limbs are powerful and very strongly webbed. The males are smaller than the females and do not exceed 330 mm. The young have a rather dark carapace, with an orange or yellow line around the edge and with large ocelli, bordered by orange-yellow rings. The head and limbs are dark, with broken light lines, one of which forms a V on the head, from the corner of each eye to the point of the snout. This is a quite aggressive turtle, and it may turn and bite the hand of the motorist who innocently tries to pick it up to help it off a dangerous highway.

Common name

Florida softshell turtle

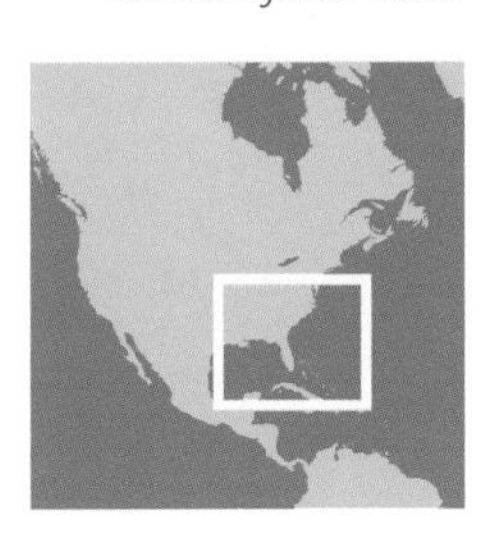

Natural history. The Florida softshell lives in calm, quiet backwaters of rivers, swamps, marshes, and lakes, preferring muddy or sandy bottoms. In some areas it lives sympatrically with *A. spinifera*, but the latter normally prefers running waters. Sometimes this species finds itself out in the ocean, and it can regain the shore without difficulty. It often spends long periods of time partially buried, with just the head visible and extended, but it can remain underwater for a long time, thanks to its pharyngeal, cloacal, and cutaneous respiration. It also tolerates surprisingly warm waters, of up to 40°C. It is often seen basking on banks or on branches and snags and may also travel overland, sometimes with fatal results when it attempts to cross highways. It is basically carnivorous, feeding on snails, bivalve mollusks, crayfish, other invertebrates, fish, frogs, turtles, snakes, and occasionally aquatic birds. Some old animals will specialize in snails and bivalves, and their jaw surfaces become broadened accordingly. They also feed upon carrion and vegetation. Nesting takes place from mid-March to July in Florida and from June to July further north. Some females may choose to nest actually within alligator nests, thus taking advantage of the watchful eyes of the mother gator to

© F. Bonin

The head of this species is quite gracile, with a very elongated snout. The dark lines marking the head and neck are most attractive. The carapace bears large tubercles, but certain individuals have a smooth, dark carapace.

chase off predators. Clutches consist of 4 to 24 eggs, with thin, brittle shells 24 × 33 mm in diameter. Two to six nestings are possible within a season.

Protection. Numerous predators feast upon the young of this species, including large fish, birds, certain turtles (e.g., *Chelydra serpentina*), and many kinds of mammals. In the past, all of these stresses were surpassed by the numbers eaten by Indians and early colonists of Florida. Today there are breeding centers in southern Florida, whose production is largely earmarked for Chinese markets. It is affected also by cultivation and the drying of habitats, as well as pollution and highway mortality. It is considered rare in many regions.

Apalone mutica (Lesueur, 1827)

Distribution. In the United States, this turtle occupies the basin of the Ohio River in Ohio, Indiana, and Illinois; the Upper Mississippi Basin in Minnesota to Wisconsin; and from the Missouri River, to extreme northwestern Florida and the eastern half of Texas. There is an isolated population in extreme eastern New Mexico. Today it seems to be no longer present in Pennsylvania.

Description. This is a turtle of medium size (360 mm for the females, 180 mm for the males), with a round, flat, leathery disk, lacking any excrescence, rugosity, or tubercle. In color it is olive to orange-brown, dotted with small black spots that vary greatly in size and number. The edge of the disk is often outlined with a light zone, set off toward the middle by a line of dots so close together that they appear to form a single dark line. The plastron is immaculate, white or sometimes grayish, and the callosities (xiphi-, hypo-, epi-, and entoplastral) are clearly visible through the skin. The head, neck, and limbs are orange to olive above, white to gray below. A light line, quite wide and bordered with black on each side, runs

Common name

Smooth softshell turtle

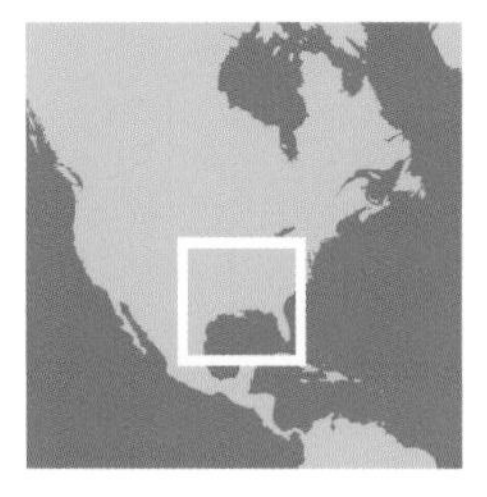

the length of the head, passing through the orbit and continuing along the neck. The snout is very long, and the tubular nose is cut off anteriorly at an oblique angle (hence the name *mutica*), with two small nostrils visible at the tip. The jaws are strong. The grayish limbs never show any distinctive coloration, except in a few individuals that have a light dusting of dark spots on the skin, but each limb is provided with four large cornified scales. The limbs are powerful, well adapted for swimming, and strongly webbed. The females have bigger claws than the males on the anterior limbs, and longer claws on the hind limbs.

Subspecies. Two subspecies are recognized. *A. m. mutica* (described above): The common smooth softshell occurs in Ohio, Minnesota, South Dakota, Tennessee, Louisiana, Oklahoma, Texas, and New Mexico. The young have dark dots and short lines on the carapace, as well as a light line, finely bordered with black, on the snout and the temples. Some specimens from the Colorado River have none of these adornments. *A. m. calvata* (Webb, 1959): The Gulf Coast smooth softshell is found only in the coastal plains of Florida, Alabama, and Mississippi as far as the Pearl River. The young, which have large dark ocelli on the shell, have no line on the dorsal surface of the snout nor on the upper side of the limbs. Furthermore, the light line that runs along the snout is bordered by wider dark lines than in the other subspecies.

The snout is obliquely cut, which gives this turtle its scientific name *mutica*. The head is olive to beige on top and light below and is separated by a more or distinct less light line that commences behind the eye.

© J. Maran

Natural history. This turtle, or at least the subspecies *A. m. calvata,* lives in large streams and rivers with strong current. The nominal race is also found in lakes, swamps, and shallow marshes with muddy or sandy bottoms. Females are rarely seen basking, although males may emerge onto the banks. During the winter, this turtle hibernates in the mud or clay, but during the summer season it is diurnal, burying itself in the mud at night. It can remain underwater for long periods, facilitated by its partially cutaneous respiration. It is an excellent, rapid swimmer and can swim against the current for kilometers. In diet it is mainly carnivorous and is about 75% insectivorous. In addition, it eats worms, snails, mollusks, crayfish, amphibians, small birds, and small mammals, as well as fish. But it can also subsist upon algae, seeds, and fruits that have fallen into the water, and it is capable of hunting on land—or at least the males are. Courtship proceeds at the end of hibernation. The male sniffs the cloaca of the female, and before long he starts to bite. The males become aggressive very easily. Copulation occurs in shallow water, and nesting occurs from May to July, about 30 m from the water's edge. From 3 to 33 eggs may be deposited in each clutch, with two or three clutches per season. The eggs are white and spherical, measuring 20 to 23 mm in diameter, and have a thick shell. Incubation lasts for about 65 to 77 days. The sex of the hatchling is independent of the temperature of incubation.

Protection. Adults have just one predator, the alligator, although their great speed usually allows them to get away. These turtles are aggressive animals, especially the females, which can give painful bites to those who seek to pick them up. In some places the species suffers from disturbance of the aquatic habitats where it lives, as well as from pollution of the larger rivers, to which it is especially vulnerable because of its cutaneous respiration.

Aspideretes gangeticus (Cuvier, 1824)

Distribution. This species is found in the northern part of the Indian subcontinent, in the drainage basins of the Ganges, Indus, Mahanadi, and Brahmaputra Rivers in Pakistan, India, Bangladesh, and southern Nepal.

Description. This is a large turtle, up to 700 mm in length, with a round or somewhat oval carapace, modestly domed, olive to light green, and having a vermiculated dark figure outlined in yellow on the central disk. It is smooth in texture and without tubercles. The plastron is white to cream, and the well-developed callosities show clearly through the skin. The head is wide, and the snout rather short and ending in a pair of prominent nostrils. The jaws are underlined by fleshy excrescences, very characteristic of the species. The tomial surfaces are flat and rugose. The head is greenish, sometimes adorned with dark markings, and has a pattern of oblique black lines in the form of a forward-directed Y. Sometimes these lines are broken or incomplete. A line on each side connects the edge of the orbit to the angle of the jaw. The top of the cranium also bears a line that starts between the orbits and continues to the nuchal area. The jaws are yellow; the chin and throat, cream to almost white. The limbs are strongly webbed, each bearing three claws, and they are generally unspotted green in color, like the rest of the animal. The hatchlings have several longitudinal rows of small tubercles on the dorsal disk. They are light green, decorated with a pattern of dark reticulations that often form large, yellow-bordered ocelli. No subspecies are currently recognized, but the overall distribution is so wide that it incorporates populations with variant morphology and color pattern.

Natural history. This species lives in great rivers, with muddy, turbid waters, but is also found in lakes, swamps, and irrigation canals. It is fond of burying itself in substrates with muddy or sandy bottoms but also may be seen sunning on riverbanks or floating on the surface of the water. The diet is omnivorous and opportunistic and includes insects, mollusks, other invertebrates, fish, frogs, mollusks, and even young aquatic birds, which it catches by seizing their feet. It also eats carrion, as well as members of its own species (eggs, juveniles, subadults), and consumes aquatic vegetation, seeds, and fruits that fall into the water. It is very active and can bite fiercely while forcefully pulling its head back, which leaves deep V-shaped injuries in the skin. Courtship coincides with the start of the rainy season in April. It is said that the males can give forth raucous cries. Nesting takes place almost throughout the year, but mostly in July and August. The clutches number 8 to 32 eggs. The white, spherical, hard-shelled eggs measure 30 × 33 mm in diameter. Incubation lasts 217 to 287 days.

Protection. This large turtle is extensively exploited for meat. In Bangladesh, the populations

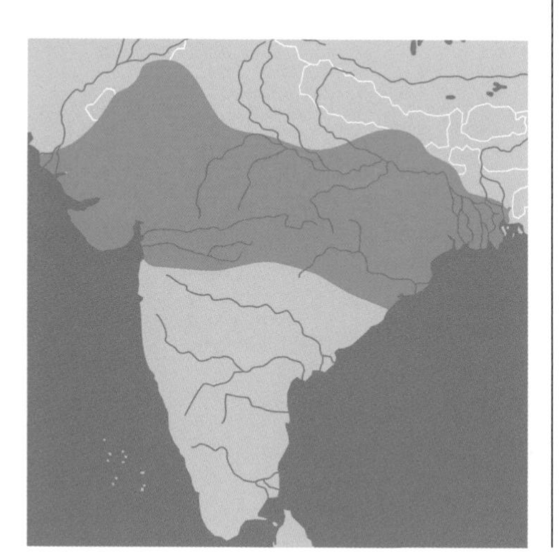

© I. Das

The soft-shelled turtles have extensive "pads" on their jaws, facilitating the seizure of prey when they are at the bottom of the water.

Common name

Indian softshell turtle

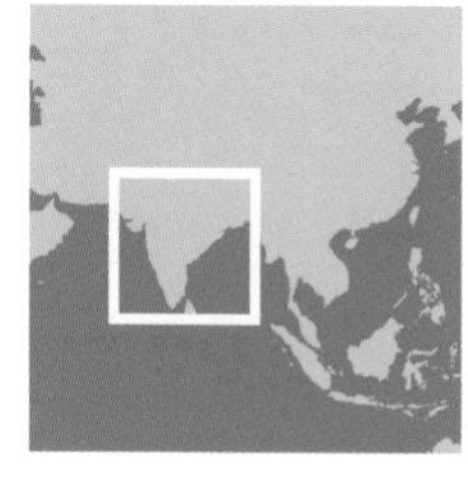

In one of the tributaries of the Ganges, these turtles are numerous, and they sun themselves in small groups not far from the water's edge.

© F. Bonin

have diminished rapidly through overexploitation. Nevertheless, the species still occupies all the great river systems, but in general it is restricted to the least accessible stretches. Nearly all of the turtles caught are exported to China. The eggs are gathered in great numbers for local consumption. In India, although officially protected, the species is also heavily exploited, but in somewhat more clandestine fashion. In Pakistan, the Muslim faith limits the exploitation of the animal. Since the late 1980s, a breeding and rearing station near Benares has produced 60,000 turtles, which were released into the Ganges with the objective of clearing the river of organic trash. The program was stopped in 2002 because the turtle population was then considered adequate (Rao, 2004). The species is classified in Appendix 1 of CITES but is still caught and exported to Hong Kong, Malaysia, South Korea, and Japan. Populations are dropping dramatically.

Aspideretes hurum (Gray, 1831)

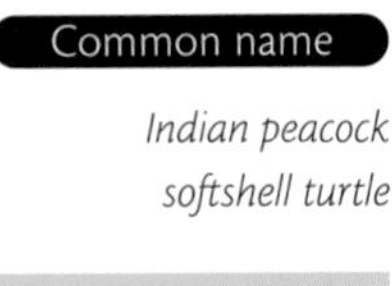

Common name

Indian peacock softshell turtle

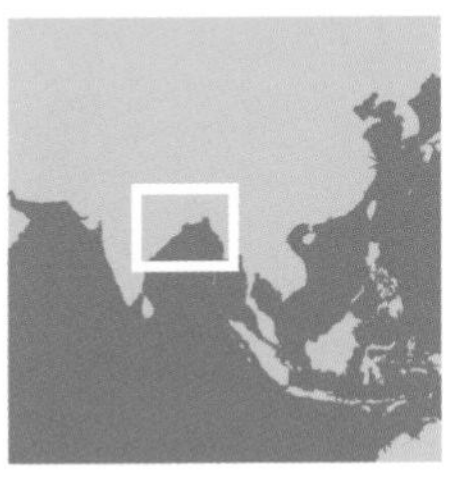

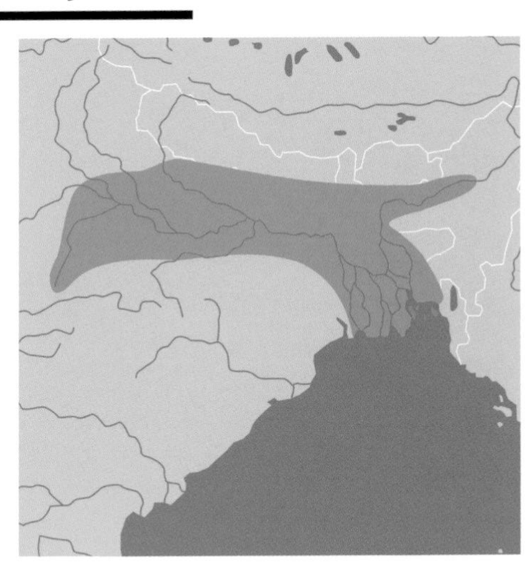

Distribution. This turtle is found in southern Nepal (at altitudes below 300 m), in central and eastern India (Assam, Bihar, Madhya Pradesh, Maharashtra, Rajasthan, Uttar Pradesh, Bengal, and Orissa), and in Bangladesh (Khula, Natore, Dhaka, Noakhali, Chittagong, Sylhet).

Description. The name *hurum* derives from an Indian vernacular for the species (I. Das). It is a large turtle, reaching 600 mm in length. The oval carapace is flat or perhaps slightly domed. It is beautifully marked, and the English name compares it to a peacock. The background color is dark green, but the disk bears four (or even six) large ocelli. The center of each ocellus is a black disk

surrounded by concentric yellow circles, and the entire group of ocelli are encircled by a light greenish belt. The remainder of the shell may bear irregular greenish markings. The edge of the carapace is adorned with a multitude of yellow dots reminiscent of the Milky Way, enhancing the decor of this already flamboyant species. Possibly the ocelli have the role of momentarily distracting or frightening predators, as they flash what seem to be huge, menacing yellow eyes (P. Pritchard). Nevertheless, they tend to fade with age, finally even disappearing. Above the neck is a group of soft tubercles, vestiges of those present in juveniles. The plastron is beige or gray and includes five well-developed callosities. The head is of medium size, is rather wide, and bears a strongly conical snout, wide at the base, sharply cut off anteriorly, and slightly downturned. The skin of the head and the neck is dark green, decorated with two yellow, black-bordered ocelli near the tympana, and with yellow or orange spots on the dorsal side of the snout. The remainder of the skin and the limbs is dotted with small yellow points. All of these decorative touches fade with age.

Subspecies. No subspecies are currently recognized, but considerable differences do exist between populations in the various river systems and regions that the species inhabits. For example, there is a melanistic variety in Bangladesh, locally called *bukum,* that has a wide black plastral band.

Natural history. This species is mostly nocturnal, living on riverbeds but also in lakes with muddy or sandy bottoms where it can hide. The diet is generally omnivorous; stomach contents have revealed the presence of snails, but in temple ponds this turtle will accept cooked rice and even pancakes. Fishermen catch it with various baits, including mussels, shrimp, and fish. Less aggressive than *A. gangeticus,* it is still dangerous and can extend and retract the head vigorously when biting. Courtship takes place in winter. The males are very violent, and they bite the females on the limbs and on the neck. The males can also emit loud sounds during copulation. The eggs are spherical, brittle, and about 30 mm in diameter. The nests include 20 to 30 eggs. Up to three nestings may occur within a season, and incubation lasts for 180 days.

Protection. Along with *A. gangeticus* and *Lissemys punctata,* this is one of the three most heavily exploited turtles species in its range. Some authorities have estimated traffic of more than 10,000 tons of this species from 1985 to 1992, exported to countries of southeast Asia (Japan, China, Singapore, and Hong Kong). In Bangladesh, the species is caught by means of long, baited lines placed across the rivers. The flesh is also sought for alleged medicinal values. Although listed in Appendix 1 of CITES and included in the protected lists in India and Bangladesh, this species is in danger of rapid extirpation. It is vital that the take be restrained and export stopped; currently exportation is estimated at 60 to 80 tons per week.

© I. Das

This species has a light-colored head, marked with dark streaks and reticulations. The snout is rather short, and the carapace is somewhat domed, with decorative ocelli and irregular markings that fade with age.

Aspideretes leithii (Gray, 1872)

© I. Das

The head is marked with yellow ocelli, but they disappear as the turtle grows. The snout is short. The head is greenish on top and white below, and dark lines pass from the eye to the base of the neck.

Distribution. Endemic to India, this species occurs in the central part of the peninsula, in the provinces of Madhya Pradesh, Maharashtra, Karnataka, Andhra Pradesh, Orissa, and Kerala, principally in the Moyar, Godavari, and Bhavani Rivers.

Description. This is a rather large turtle, which may reach 600 mm and 30 kg. The shell is oval, gray to olive-green, and ornamented with four to six yellow ocelli bordered by concentric designs that evoke images of huge, menacing eyes. These designs disappear with age. Warty tubercles are present on the nuchal area of the carapace, and some individuals retain the traces of the lines of tubercles present along the carapaces of juveniles, especially toward the back. The head is of medium size and has a short, fine conical snout. The head and neck are greenish, with extensive black lines of varying clarity; the first extends from the corner of the eye to the neck, and the others extend downward from the tympanic region to the sides of the neck. The corner of the mouth has a characteristic yellow spot. The remainder of the skin is greenish, usually rather dark on top, creamy or whitish below. The young show similar colors but are more brightly marked, and the carapace has several lines of small tubercles. The limbs are strong, very well webbed, and thoroughly adapted for swimming.

Natural history. This species lives in rivers and creeks with muddy bottoms and turbid waters, as well as in swamps and canals, and it is often seen in artificial "tanks" attached to religious sites. It is a shy species, almost never seen basking on land. In many of the environments in central and southern India, it seems to replace the more northern species *A. gangeticus.* The diet is carnivorous: bivalve mollusks, snails, fish, and crustaceans. Stomach analysis has also revealed the presence of small stones. In the temple tank at Kotipalle, the turtles seem happy with their captive diet of hibiscus flowers and bananas. The nesting season extends from January to June. The eggs are spherical, with a diameter of 30 mm, and there are several successive nestings per season.

Protection. This species is heavily harvested for local consumption and sold throughout the country, although it apparently is less often exported to China than *A. gangeticus.* Several sanctuaries exist, some in the Godavari River, which should allow at least nuclear populations to persist. Although protected by national law in India, the species appears to be disappearing fast, but in-depth studies will be necessary to define the status accurately.

Common name

Leith's softshell turtle

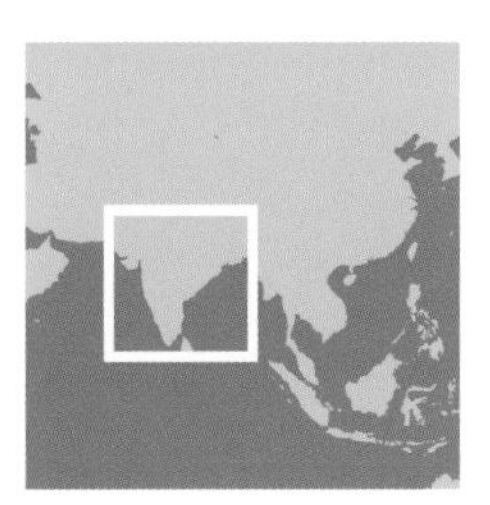

Aspideretes nigricans (Anderson, 1875)

Distribution. Formerly known only from the tanks of the temple of Hazrat Sultan Bayazid Bostami at Chittagong, Bangladesh, the species has now been identified in the wild in the Brahmaputra River of northeastern India.

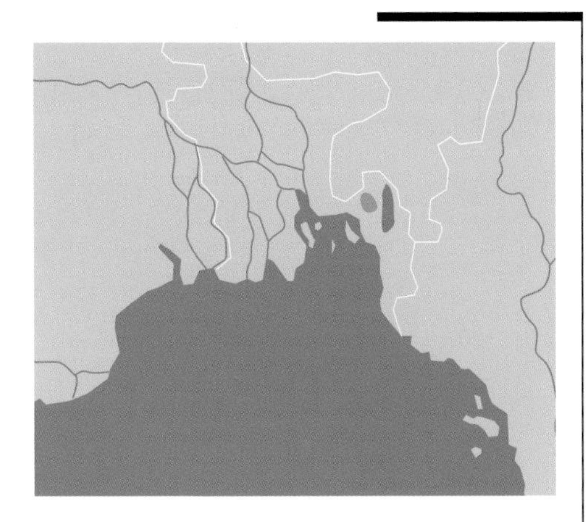

Common name

Black softshell turtle

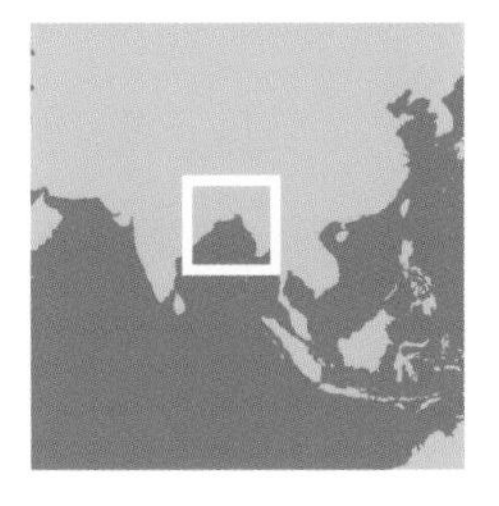

Description. This is a large animal, reaching a carapace length of 910 mm and a weight of up to 57 kg. It is very similar to *A. gangeticus* but differs from it in having a bony midline ridge on the lower jaw, well developed at least from subadult size onward. The coloration varies greatly, depending on the age of the animal. The young are quite colorful, with yellow ocelli on a dark brown background. The adults are usually plainer in color, almost black, but certain individuals are more copper-colored, with irregular dark spots. The plastron is gray, spotted with black, and the five callosities are large and well developed. The head has no light lines, in contrast to that of *A. gangeticus,* but there are small light spots and marbling on the upper surface of the head. These lines fade out with age. The snout is elongate, and the lips are light, almost white, in most individuals. The limbs are often outlined by a light border on the posterior edge.

Natural history. The ecology of the species, until recently thought to be confined to the temple tank, is still poorly known (Praschag and Gemel, 2002). In the wild these turtles live in sympatry with *Pangshura sylhetensis* and *Cyclemys oldhami,* in clear-water, slow-flowing rivers with sand or pebble bottoms. The preferred temperature is 22°C to 23°C. The population density in the Brahmaputra appears to be quite high. The eggs are small, numerous (up to 38), white, and spherical. There are several nestings per year. These turtles bask on the riverbanks when the water is cool, but they pass most of their time on sandy bottoms, lying in wait to ambush prey. In the temple tanks, they eat all the material offered them by human visitors and seem to prefer bananas. But in the wild, they eat only fish, meat, and shellfish, which they crush with the bony ridge on the lower jaw.

Protection. This species was formerly thought to be confined to the Chittagong temple tank. There conservation planning presents a challenge, made worse by an episode of poisoning that occurred a few years ago, which obliged the guardians to transfer the turtles to other tanks elsewhere. But the large, newly identified populations within the Brahmaputra itself seem to assure that the species has some kind of future, even though it is subject to quite heavy fishing pressure. The Muslims revere this species, which they believe brings them into close spiritual contact with the temple saint, Bayazid Bostami.

© P. Pritchard

This photo, taken at the temple pond at Nasirabad, shows the uniform dark appearance of the species. Other turtles, from the Brahmaputra River, have a more colorful appearance, especially the younger animals.

Chitra chitra Nutaphand, 1986

Common name

Southeast Asian narrow-headed softshell turtle

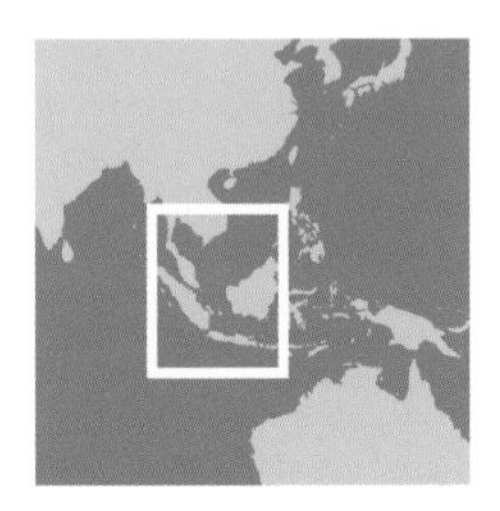

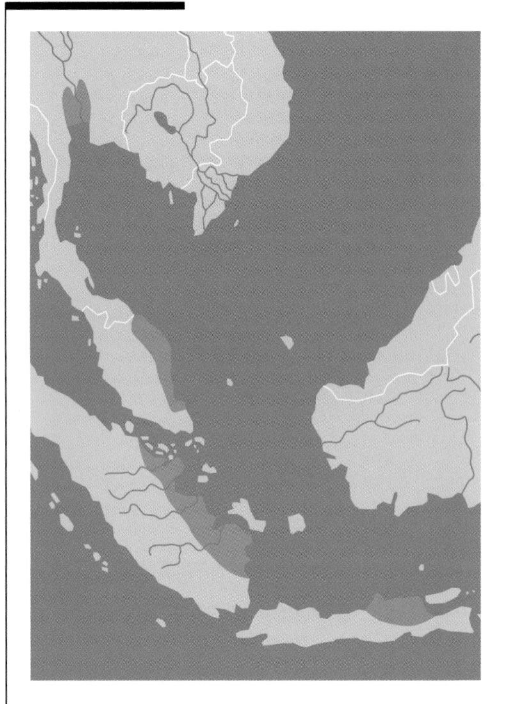

Distribution. The two subspecies live within small, isolated ranges, one in southwestern Thailand and southeastern peninsular Malaysia, and the other in Sumatra and Java.

Description. The word *chitra* means "picture" in Hindi. This is without doubt one of the largest of the freshwater turtles, attaining 1,100 mm and 100 kg (the record is 152 kg). The carapace is flat, and the limbs strong, but the head is bizarrely undersized and finds itself at the extreme end of a long neck with a very thick base, where the skin can bunch up in the manner of a sock being removed from a foot. The carapace does not really have a defined border above the neck, but rather the one grades smoothly into the other, giving the animal a massive, streamlined appearance. The skin of the limbs and the neck bears numerous light yellow decorative markings, which are displayed against a greenish background and are often bordered with fine dark lines. The carapace is also marked with an amazing design that makes one think of a piece of complex electronic circuitry. The minute head, with a tiny, short snout also bears light lines. The eyes, placed at the very front of the head, are also extraordinarily small.

Subspecies. *C. c. chitra* (described above): The nominal subspecies is found in southwestern Thailand as well as the southeast of the Malaysian Peninsula. It is olive-green in color, with wide yellowish lines giving it a light vermiculated aspect. Old animals may be plain colored, but their original pattern can still be perceived. The actual markings vary greatly from one specimen to another. The edge of the disk is marked with a light line, which continues to the tip of the snout and, together with two other lines, forms a V outlining the widened part of the neck. One or two transverse light lines mark the distance between the orbits. The underside of the neck is marked with dark spots. The plastron is immaculate and is yellowish, as are the undersides of the limbs. *C. c. javanensis* (McCord and Pritchard, 2002): This subspecies is found in the coastal plain of northern Sumatra (Indonesia), especially in the region of Lampung. It has also been found in eastern Java, in Probolinggo District. The dorsal disk is much darker and the vermiculations are less visible, often reduced to just a few spots. A light vertebral line is often present. As they age, individuals may become very dark. The head is decorated with complex lines and carries a marking in the form of an X, of which the two anterior points start at the border of the orbits and the other two are extended posteriorly to the beginning of the neck. A light triangular figure defines the outline of the snout.

Natural history. The biology in nature is still unknown, but it is probably similar to that of *C. indica.*

Protection. The legal protection that is supposed to benefit *Chitra indica* has not been extended to the new subspecies, which run the risk of exposure to exploitation and commerce. As with all softshelled turtles, these too are constantly hunted, widely consumed, and exported to eastern Asian markets. It is vital to protect *Chitra chitra,* which the Turtle Conservation Fund has placed on its list of the 24 most endangered turtles in the world. The fragility of the species has already been underscored during the summer of 2004. An epidemic was declared in the river basin where the turtles were found, probably caused by pesticides coming from upstream. The animals were caught in an emergency rescue operation, and the basin was completely disinfected. But the entire population remains very vulnerable. It is essential that two or three reserve populations in other, nearby basins be established.

© A. Cadi

This is the largest *Chitra* specimen found in Nepal in recent years. It weighed 80 kg. These turtles live in the mud at the bottom of waterways and are rarely visible.

Chitra indica (Gray, 1831)

Distribution. The wide range of this species extends from Pakistan, in the basin of the Indus, to the south and also northeast in the Jhelum River and the Sutlej, then throughout the north and center of the Indian Peninsula as far as the Cauvery River. The species also occurs in southern Nepal and in the basins of the Ganges and the Brahmaputra, into Bangladesh.

Description. This is an extremely large turtle, measuring up to 1,150 mm and with a weight of up to 120 kg. The shell is rather flat, olive to grayish in color, with a reticulated pattern and spots similar to the skin of a leopard. Light lines, sometimes finely outlined with black, extend from the tip of the snout to a point on the carapace where they form a V. The head is very small, and the neck is strangely conical and quite thick at the shoulders. The head grades directly into the carapace, with no overhang at the base of the neck. The eyes are tiny and are located close to the snout, which itself is very short. Two light lines extend from the posterior border of each orbit and reach to the middle of the narrow, long head. Several similar lines adorn the neck, the middle one starting with a Y just behind the base of the skull. The chin is almost immaculate. The plastron is cream-colored or slightly pink; four or five callosities may be detected through the surface skin. The limbs are strong, olive or gray in color, and have five to seven large scaly transverse lamellae on the side of each one. The young hatch with a gaudy appearance. The background is rather dark olive-green, overlain by a mélange of yellow lines, finely bordered with black on the head and continuing on the carapace. The shell has a slight median keel and numerous small rugose tubercles, which eventually disappear.

Natural history. This turtle occurs in major rivers with sandy substrates and turbid waters. It is

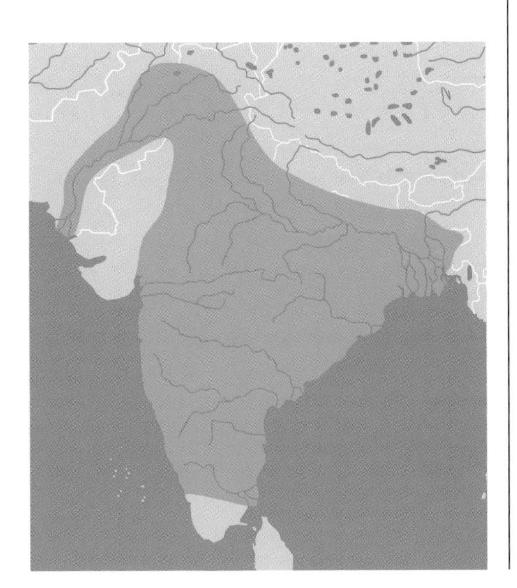

Common name

Indian narrow-headed giant softshell turtle

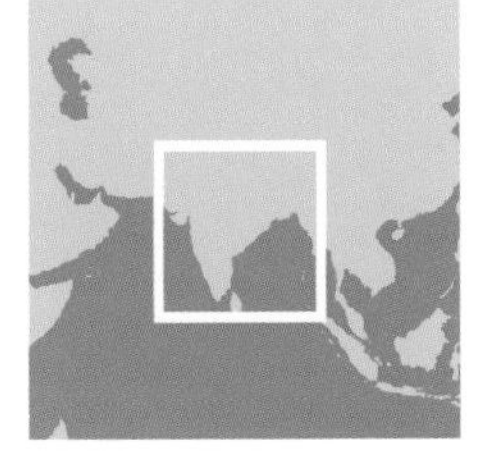

This softshell turtle rarely comes out of the water. The bright colors and designs are perhaps signals for potential predators underwater. Old animals are beige or brown and do not have the bright markings. The head is minuscule, with a short snout.

extremely aquatic and is never seen on land except when the females come ashore to nest. Ill at ease on terra firma, it walks with slow deliberation. In the water, however, it is fast and can spend long minutes on the bottom without coming up for air, thanks to its cutaneous respiration. It may attack large prey that come close to the bank, including goats. The usual diet consists of mollusks, fish, crabs, shrimp, and sometimes quantities of aquatic plants or vegetables that have fallen into the water. When hungry, the turtle buries itself in the bottom substrate, with only the minuscule eyes and nostrils showing. Smaller prey are simply sucked in and swallowed; larger ones may be cut with the jaws. The species is active during daylight hours and remains hidden on the bottom, where its coloration perfectly mimics the play of sunlight on the sand. This is a very aggressive animal; striking violently with its head if it is disturbed, it can inflict terrible bites. It is also able to emit a foul-smelling musk. Nesting takes place almost throughout the year, with a peak in August and September. The eggs are placed in a nest that may be a long way from the water; the nest itself is flask-shaped and may be 30 cm deep. Clutch size ranges from 60 to 178, and the eggs are small, spherical, and about 34 mm in diameter. Incubation takes 40 to 70 days.

Protection. The meat of this turtle is much appreciated by local people, and a few years ago it was common to see such products displayed in public markets. Today the meat is rarely seen for sale, which makes one think the turtle populations may be disappearing fast. In India, it is now nearly extinct in the Ganges. Sanctuaries have been declared, including ones at Varanasi and the Chambal River, as well as at Satkosia, in Orissa Province. There is little information available about populations of the species in Pakistan, and CITES had not yet listed this species. However, it is included by the Turtle Conservation Fund on its list of the 48 most endangered turtle species.

Chitra vandijki McCord and Pritchard, 2002

Common name

Myanmar narrow-headed softshell turtle

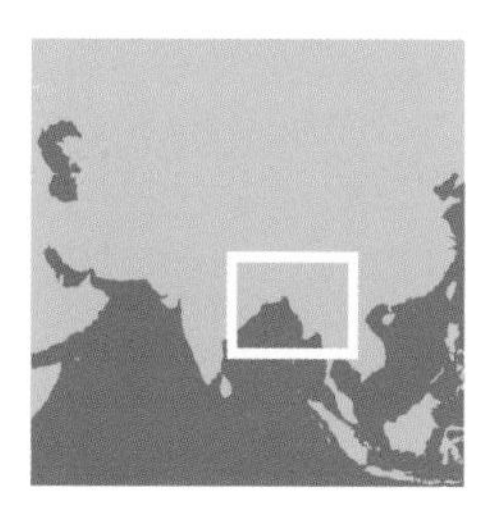

Distribution. This newly described species is confined to central Myanmar, in the basin of the Ayeyarwady (formerly Irrawaddy), Myitkyina, Yangon, and Bhamo Rivers.

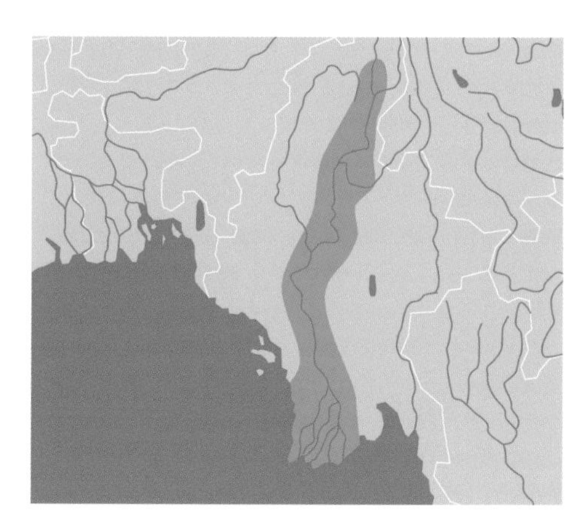

Description. The holotype, 220 mm in length, was found in the Ruili market in Yunnan, China, and later was found in nature by Steven Platt, in the River Ayeyarwady. The name honors Peter Paul van Dijk, for his conservation work on the soft-shelled turtles of this area of the world. It is very similar to two other *Chitra* taxa but differs in the arrangement of the V marking, which starts at

the point of the snout, extends onto the carapace, and does not form the bell-shaped design seen in the other species. There are three or four transverse cornified lamellae on the edge of each forelimb. The periphery of the dorsal disk is not marked with a light band. The head has just two lines on the top of the cranium, and two others extending back from the orbits, dropping down to form a right angle at the tympanic region. A single light line joins the anteriors of the orbits, but there is no arrangement of light lines defining the outlines of the snout, as in other *Chitra*.

Natural history. The natural history of this species is presumably similar to that of the others in this genus.

Protection. This turtle, like all Burmese chelonians, is collected for food, and its presence in a market in China suggests that it is also exported. Studies are needed to define its status better and to draw up protective measures.

This species is much smaller than *C. indica,* and its pattern is more similar to that of *C. chitra.* The V-shaped marking on the front of the carapace is easily seen.

Dogania subplana (Geoffroy Saint-Hilaire, 1809)

Distribution. Despite its name, this species is by no means confined to Malaysia. It occurs in southeast Myanmar, Thailand, the Malay Peninsula, Sumatra, western Java, Borneo (Sarawak, Sabah, Brunei), and Singapore.

Description. This turtle has an oval, very flat shell that may reach 350 mm in length. The coloration is dark olive to dark brown, with a narrow black line along the midline. Two or three pairs of black ocelli, bordered with a fine line of yellow dots, are visible in young individuals. With maturity, the carapace becomes dark brown, with just a few lighter streaks and radiating lines. Several longitudinal rows of small tubercles are present in juveniles and one or two of them may persist into adulthood, running together in the nuchal region. The plastron is whitish, sometimes cream or pale gray. There are four plastral callosities, often hardly visible exteriorly. The plastron is very wide anteriorly, narrower behind—not wide enough to protect the hind limbs. The head is quite large and has a long, narrow snout and breathing "tube." The head is dark brown, with yellow and black markings. A black line extends from the point of the snout, traversing the cranium and passing between the orbits. Another, short line runs along the side of the snout on each side, also passing between the orbits. Adults retain a reddish, or sometimes pink or orange, streak behind each eye. The chin has a dark, reticulated

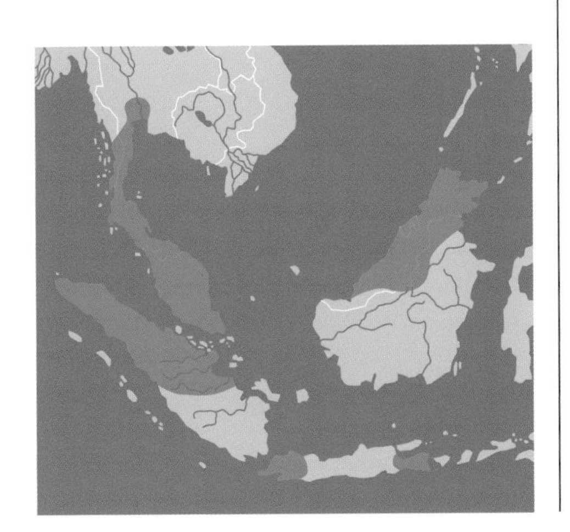

Common name

Malayan softshell turtle

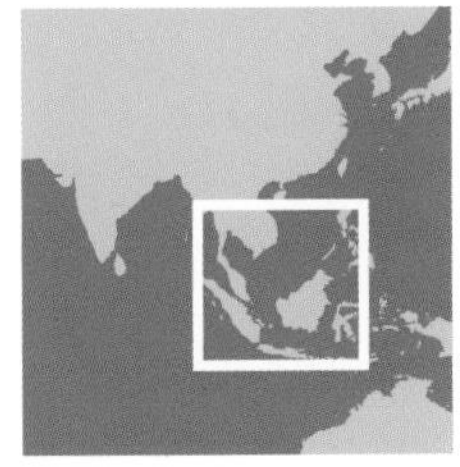

© I. Das

This moraylike head suggests a species found at great depths. The head and neck are very dark, with callosities and fine markings on the entire skin. Only the eyes, which are quite light, relieve this disturbing appearance. The snout is short.

appearance, sometimes becoming almost black. The remainder of the skin of the limbs is olive to brown, sometimes quite dark, with numerous small yellow dots. This species has a certain amount of mobility between the sutures in adults, which facilitates the retraction of the large head and may allow the turtle to slip between irregular rocks when it is hiding underwater (P. Pritchard, 1993).

Natural history. This species prefers clear, fast-flowing, shallow waters, with abundant rapids and waterfalls, in forested environments. The small size is an advantage in small, shallow aquatic environments. The diet is omnivorous—fish, crustaceans, insects—but the large size of the head suggests some adaptation toward hard-shelled prey, such as snails and bivalve mollusks. Algae and fallen fruit are also eaten. The clutch size is small (3 to 7), and the spherical eggs are 22 to 31 mm in diameter.

Protection. The flesh of this turtle may be consumed on a local basis, but there is also substantial export to China, where it is the second commonest turtle seen in restaurants, after *Pelodiscus sinensis.* Muslim people, who do not eat turtles but are willing to collect them for sale, are paid two to three euros per kilo, live weight. The eggs are also collected and eaten. Destruction of forest environments is an additional threat to this species. Furthermore, in the West, *Dogania* has become a species of interest to hobbyists, and Indonesia officially exported 13,500 live specimens in 1998. The introduction of the Chinese softshell *Pelodiscus* into environments inhabited by *Dogania* adds to the latter's problems.

Nilssonia formosa (Gray, 1869)

Common name

Burmese peacock softshell turtle

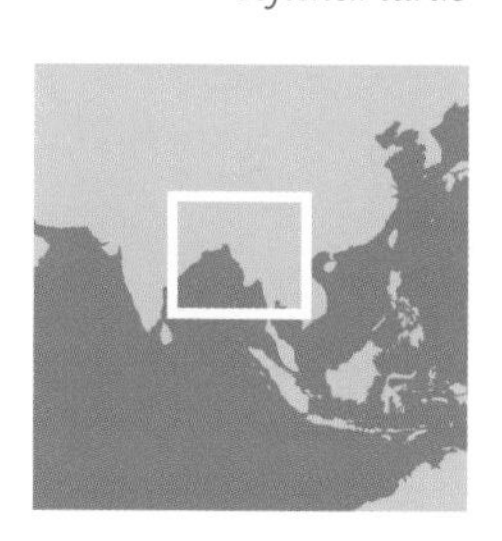

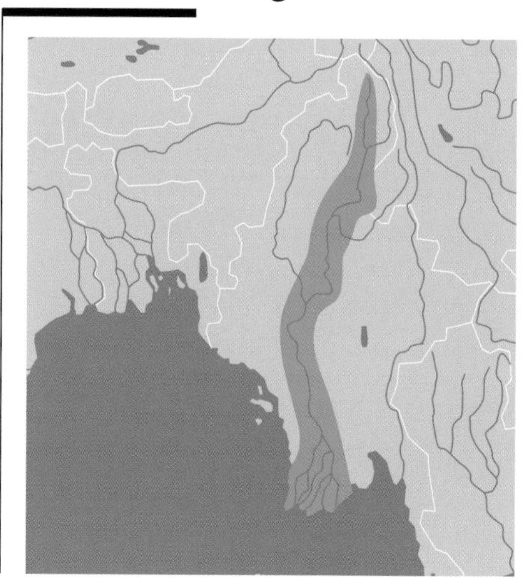

Distribution. This species is endemic to Myanmar, in the Ayeyarwady River, but it is possible that it could be found in Thai territory, and in 1912 Nelson Annandale indeed recorded it from the area of the border with China.

Description. This species does not exceed 400 mm. It is rounded and flat, and the flexible part around the edge of the carapace is very wide. The general coloration is green or olive, with fine yellow lines that form a reticulate figure. Four large black ocelli decorate the shell and, recalling the patterning of the tail of a peacock, give rise to the English name. Some individuals are very light in color; others are beige. The ocelli are always at least somewhat darker. Several rows of small tubercles are present on the dorsal disk of young specimens, but these—as well as the bright colors—disappear with age. The adults retain an area of large tubercles on the anterior border of the leathery carapace, and some individuals retain an indistinct dorsal keel. The head is of medium size; the neck is strong and conical. The nose and snout are long, pointed, and somewhat inclined downward. The coloration of the head picks up on the coloration of the carapace, perhaps with larger yellow spots, with dark bordering along the sides of the head and in the tympanic region, as well as at the corners of the mouth and perhaps at the tip of the chin. The plastron is cream to whitish and has four callosities. The anterior part of the plastron is very developed, in contrast to the hind lobe.

© I. Das

Natural history. Few observations have been made on this turtle in the wild; usually it is seen only in temple tanks or ponds. The animals there are considered sacred, and it is practically impossible—certainly strictly forbidden—to manipulate or study them there. Some of the specimens seen have serious deformities, presumably because of the genetic inbreeding and isolation of the stock. The faithful feed the turtles with balls of cooked rice.

Protection. This is a rare and poorly known turtle species, still consumed by local people and sold in market stalls. It is also caught for illegal export to China. The extent of these fishery losses is hard to evaluate. Some serious hobbyists in Europe have live specimens. It is included in the IUCN *Red Data Book* and listed as one of the 48 most endangered turtle species by the Turtle Conservation Fund. It is also protected by WARPA law in Thailand. Protection that involves both market surveys and field study should be initiated as quickly as possible, under the aegis of the Mandalay Zoo.

The shell is very flat, sometimes depressed in the middle, and olive or blackish in color, with yellow dots. The head and neck continue the design of the shell, sometimes with more spots and markings of a yellowish color. The snout is rather long.

Oscaria (Rafetus) swinhoei Gray, 1873

Distribution. The distribution and the status of this species are very poorly known. It was first reported from China, in the Shanghai River and in Tai Hu Lake, but it was also found in Vietnam, in Hoan Kiem Lake in the center of Hanoi, as well as in the Red River. The investigations undertaken by P. Pritchard during recent years have not yielded any living specimens in the wild, but one live animal was found in the Beijing Zoo. At the present time, the sole evidence for the presence of the species in either Vietnam or China is in the form of skulls and bones kept by fishermen.

Description. This species was found near Shanghai by M. Swinhoe. It has been synonymized with *Pelochelys maculatus.* It could be that

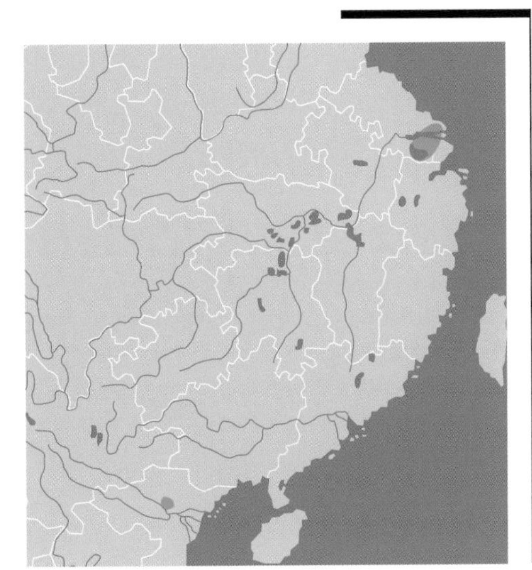

Common name

Shanghai softshell turtle

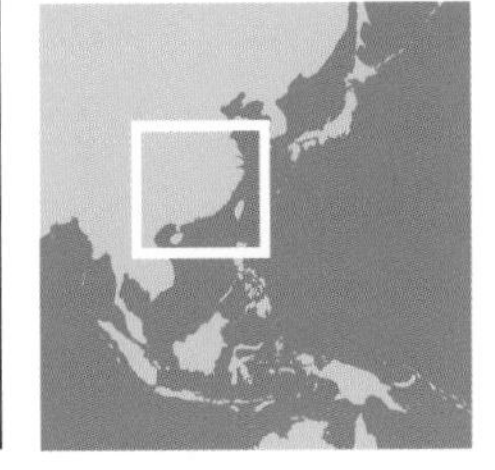

This amazing photo, taken in Hanoi, Vietnam, shows a huge softshell, *Oscaria swinhoei,* resident in Hoan Kiem Lake.

© J.H. Harding

both species are really the same thing, or possibly a new species occurs in Vietnam and in Hoan Kiem Lake (for which the name would have to be *Rafetus (Oscaria) hoankiemensis*). Estimates of the size of the species are quite variable. Ermi Zhao considers it to be a true giant, 1,800 mm long and 200 kg in weight, but P. Pritchard considers it to be much smaller than this. This is an oblong, very flat species, with the shell somewhat depressed along the midline. Grayish to brownish in color, the head, snout, and upper surfaces of the limbs are marbled with yellow spots. There are eight pairs of costal bones, the eighth pair small and meeting on the midline. The plastron is grayish to yellowish and has just two rather imperfectly developed callosities. The head is of medium size, with a short snout, and olive or grayish in color, with a pale yellow chin. The sides of the neck are yellow, with olive vermiculation. The tail is very short and is conical in shape.

Natural history. The behavior of this species is not well known. No juvenile has been observed. One adult was photographed in Hoan Kiem Lake a few years ago, but the turtles in the lake, of very large size, may be the species *Pelochelys bibroni.*

Palea steindachneri (Siebenrock, 1906)

Distribution. Found in China (Yunnan, Guizhou, Guangdong, Guangxi, Hainan Island), as well as in North Vietnam (Hanoi, Hue, Da Nang), this species has been introduced into United States territory, in the islands of Kauai and Oahu in the Hawaiian archipelago, as well as to the island of Mauritius (in the Indian Ocean) by Asian immigrants.

Description. The carapace length may reach 430 mm. The carapace, oval in shape, has an area of large tubercles in the nuchal areas. An additional, crescent-shaped area farther back is covered with small rugose excrescences. The color is brown to olive, with any pattern. The plastron is uniformly grayish, sometimes cream or yellowish, without markings. Five callosities are visible. The head is of medium size, with a long, narrow pointed snout. The overall color is brown-olive, with several fine black lines accompanied by small dark points curving around the orbit and following the curve of the cranium. A thick, pale yellow line commences behind the eyes and runs the length of the head and neck, fading out toward the end. There is a yellow spot in the angle of the jaw. These markings are bright and clear in juveniles, but they fade out with age. The character that always defines this species is the area of large, warty, coalescing tubercles on both sides and on top of the neck (hence the vernacular name). The limbs are strongly webbed, the front ones with several transverse horny lamellae. Males are smaller than females. At hatching, the carapace of the young is round and brown to olive, spotted here and there with black dots, and it bears several longitudinal rows of small tubercles. The pattern of the head, described above, is very bold.

Natural history. This species lives in mountainous regions, where it is found in running streams, rivers, and lakes, up to an altitude of 1,500 m. The diet is carnivorous: mollusks, fish, and crustaceans. The nesting season is from April to September, and the clutch size is three to eight. The eggs are spherical, brittle-shelled, and about 22 mm in diameter.

Protection. In China, the species benefits to a certain degree from official protection (Zhao, 1998), but populations are dropping dangerously because of illegal collection. Habitat destruction and pollution, as well as the construction of small hydroelectric facilities, add to the challenges. In Vietnam the species is still found on the menu of some restaurants, but it is rarely consumed by the general population. It is one of the species undergoing rehabilitation at the Cuc Phuong Turtle Conservation Center, with captive reproduction and liberation into natural environments. It is one of the 48 most endangered species listed by the Turtle Conservation Fund.

Common name

Wattle-necked softshell turtle

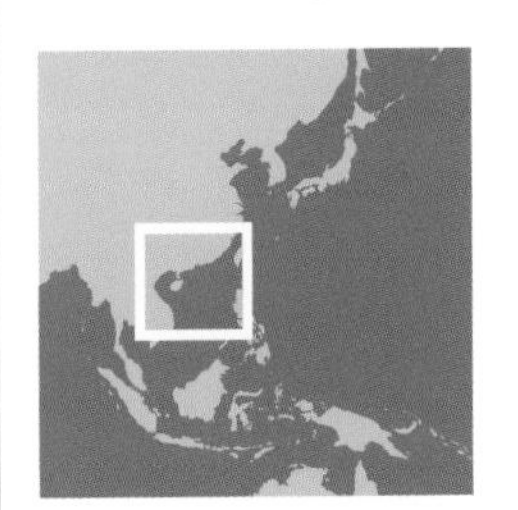

© F. Bonin

This species, found in rivers in China, lives in the mud and in warm waters. The coloration is a uniform grayish to greenish. The snout is long, and the head is wide and elongated. There are tubercles on the carapace above the head.

Pelochelys bibroni (Owen, 1853)

Distribution. This species occupies the southern and northwestern parts of the island of New Guinea, including the region of Uta, in Indonesia, and the river basins of West Papua, farther south in Indonesia. In Papua New Guinea, it occurs in the Palau, Digul, Fly, Kikori, and Putari Rivers, as well as in Lake Murray, in the Strickland River, and, farther east, in the Port Moresby region.

Description. This is one of the largest turtles in the world, measuring up to 1,000 mm. The carapace is very flat and broadens considerably above the limbs and the tail, leaving the dorsal bony disk round and distinct. This is certainly the most strikingly patterned of all the softshell turtles. The background is dark brown, marked with several wide, radiating, bright yellow lines that all originate at the center of the shell. Around the edges of the dorsal disk, there are numerous additional light markings. The neck is very thick and is conical in shape, and the skin grades smoothly from the neck to the shell, without forming a distinct edge.

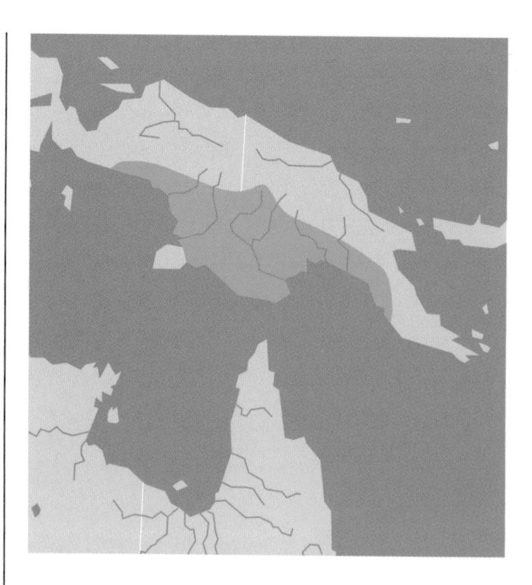

Common name

Bibron's frog-faced giant softshell turtle

The neck has a median line, with two additional pairs of S-shaped lines that traverse the edges of the shell above the anterior limbs. This pattern is quite similar to that of the softshell *Chitra.* The head is small and round, the eyes being situated on the upper surface toward the front of the skull. The iris is golden, with a fine black transverse bar. Large rolls of fat may be seen on the upper surface and along the upper jaw. The snout is tubular and very short.

Natural history. This species prefers lowland rivers and estuaries but may be found considerable distances upstream; it adjusts well to saline conditions in deltas and large estuaries. The diet seems to be almost totally carnivorous—fish, crabs, mollusks—but some vegetation is perhaps ingested with the prey. This is a voracious animal and will attack large prey. However, it is not aggressive, if one excludes the habit of striking with the head when it is picked up. Nesting usually occurs in the dry season, in the month of September, but may also occur at other times of the year. This species has been seen nesting at the same time and on the same beaches as *Carettochelys insculpta.* There are usually 22 to 45 eggs in a clutch, but on occasion as many as 100. They average 30 mm in diameter. Clutches have been found in the nests of crocodiles, which may be an effective way to avoid nest predation.

Protection. This species is known to be consumed by saltwater crocodiles, but the main stress is hunting by humankind, for both the flesh and the eggs. Tribal masks may be made from the bony carapaces. No real commercial hunt or distribution has been reported, and the species is protected by law in Papua New Guinea, although this is not the case in the Indonesian part of the island. It remains listed as vulnerable by IUCN, and the Turtle Conservation Fund is planning conservation programs.

The young *P. bibroni* below has a long, slender neck. The underlying ossifications and the wide, flexible marginal area can be easily seen.

© F. Bonin

Older animals have a strange, pearlike shape, with a tiny head, a short snout, and a wide neck with a bridled design. Some of these turtles may reach a weight of 120 kg and a length of more than a meter.

© B. Devaux

Pelochelys cantorii Gray, 1864

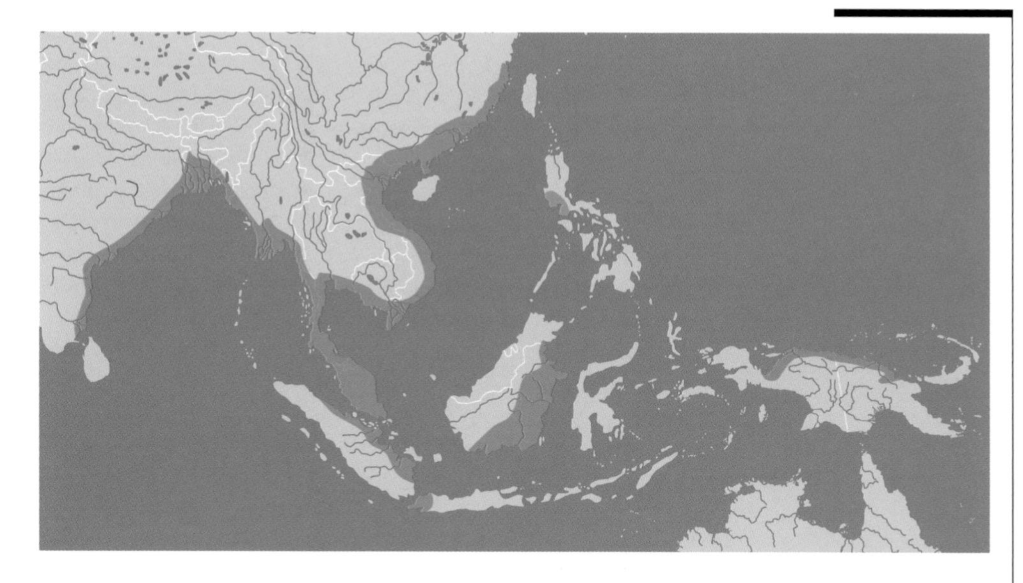

Common name

Cantor's frog-faced giant softshell turtle

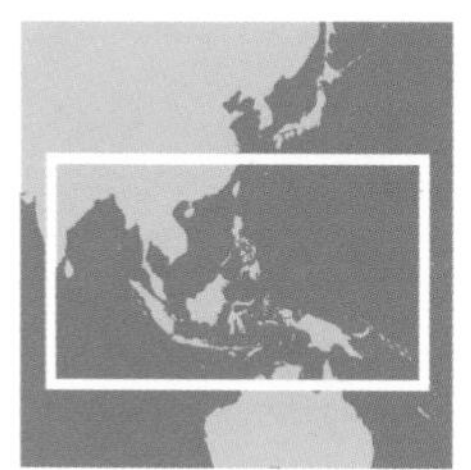

Distribution. Formerly included within the species *P. bibroni,* this turtle has a wide range that extends from southern India (Madurai) as far as Bangladesh, Myanmar, Thailand, the Malay Peninsula, Cambodia, and Vietnam and into Laos, with a northern extension into China to the region of Fuzhou. It also occurs, on the island of Luzon in the Philippines, and south and east to Borneo, eastern Sumatra, and in northern Java in Indonesia. The easternmost population is in the Sepik region of Papua New Guinea, where it is separated from *P. bibroni* by the mountain range that divides the island from east to west.

Description. The name of the species is derived from the Danish naturalist Theodore Cantor, who worked with the British East India Company. This is sometimes reported to be the largest of the freshwater turtles, reaching 1,300 mm and 200 kg, but actual measured voucher specimens are about half of this length. The carapace is very rounded and flat, olive-green to a more or less dark brown, and decorated with small dark dots or yellow spots, with perhaps some dark lines radiating from the center of the shell. The head is extremely small and rounded, slightly flattened on the upper surface of the skull, and with the eyes placed very high. The nose is tubular and quite short (shorter than that of *P. bibroni*), and the pattern of the head is continuous with that of the carapace. The neck is strong, conical, and pleated, and its skin passes smoothly onto that of the carapace, without edge or overhang. The plastron is cream, with four or five callosities, but without femoral excrescences. The limbs are strong, well webbed, olive to brownish in color, and cream below. The females are larger than the males. The young are born with a vivid color pattern, a modest keel, and numerous tubercles on the carapace, all of which disappear with the passage of years. The distribution of this species is so wide that morphological variation may well exist, but no subspecies have been described.

© C. H. Chan

This photo shows the generally very flat, circular, and uniformly brown appearance of the species. A light line runs around the edge of the carapace. The head is minuscule, and the limbs are wide, facilitating quick burial in the mud.

Natural history. This species is found in estuaries, deltas, and sometimes in downstream sections of slow lowland rivers. It swims well at sea and may nest on beaches in company with marine turtles. It spends all its time in the water and uses pharyngeal and possibly cutaneous respiration. The diet is carnivorous—fish, crustaceans, and mollusks—and also includes certain vegetable species. This turtle is not an aggressive species, but the blows that it delivers with its head can be violent. Females nest just once per season, and the clutch consists of about 30 spherical eggs.

Protection. The great size and the sought-after meat of this turtle have made it a preferred food source for people within its range for a long time. Capture and consumption have both been conducted on a devastating scale and continue in all of the range states. A few decades ago this turtle could be seen in markets, but it has become rare in countries such as India, Myanmar, Thailand, China, and Papua New Guinea. It remains in reasonable numbers in the great estuaries of Bangladesh and Cambodia. In Indonesia, it may have better protection in reserves such as the Berbak National Park. But, all in all, its numbers are declining rapidly, and it is listed as one of the 48 most endangered turtle species by the Turtle Conservation Fund.

Pelodiscus sinensis (Wiegmann, 1835)

© F. Bonin

Distribution. This species occurs throughout central and eastern China (with the exception of the provinces of Xinjiang, Qinghai, Xizang [Tibet], and Ningxia) and also on the islands of Hainan and Taiwan and in northern Vietnam. It has been introduced to Thailand as well as Japan (Bonin Islands) and Hawaii. Reports from Korea and extreme southeastern Russia need to be confirmed.

Description. This is a small species, not exceeding 350 mm, and in the southern areas it is even smaller, less than 250 mm. The carapace is oval and flat, with a border defined by a slight bulge above the neck and limbs. Some individuals have an area of conjoined warty tubercles at this location. The general color is grayish brown, darker or lighter, and sometimes yellowish. The skin is very thin, and the bony parts poorly developed. The plastron is white or yellowish, with no special markings, but with seven reasonably visible plastral callosities. The head is of medium size and ends in a short snout, conical at the base and tubular at the extremity. The head and limbs are brown or olive, and there are a number of dark lines on the head. The throat and the underside of the limbs are light, with vermiculations under the neck. The lips are thick and fleshy and conceal the strong jaws. Males are smaller than females, and the latter may also have a slightly domed carapace. The hatchlings have an almost circular shell, olive in color, with large round yellow ocelli and longitudinal series of fine tubercles. A narrow light line defines the border of the carapace. The plastron may be white but also may be yellow, orange, or even red, with large black markings that disappear rapidly with growth, as do the white, dark-edged spots on the sides of the head and neck. A dark line runs between the orbits, and another descends from the anterior corner of the eye to the upper jaw. The snout is slightly upturned.

Subspecies. Some authors claim that there are morphological differences between the populations of southern China and the islands. In Vietnam, there is a light-colored form in which the carapace is dotted with black ocelli. As of now, no subspecies are recognized.

Natural history. A wide variety of aquatic environments are utilized: slow-running rivers, canals, rice paddies, lakes, swamps. This turtle is often observed basking on land but also spends much time buried in the mud or clay. This is a temperate zone form for the most part, and it sometimes hibernates when water temperature drops below 15°C. Time to maturity is variable: three years in Vietnam, four in China, five to six years in Japan. In commercial farms, it may mature very rapidly, reproducing at just one year of age and a weight of 1 kg. The diet is largely carnivorous: fish, crustaceans, bivalve mollusks, insects, worms, and various larvae. But this species may also eat leaves and seeds of various plants. It is a very active species and a good swimmer, but it has an evil nature and stretches its neck around to bite the hand that holds it. Certain populations are more aggressive than others, as in the Red River of Vietnam (Maran, 2003). It is capable of excreting a nauseating, smelly fluid from small pores opening on the anterior edge of the shell, but it does this only under great provocation. Males are always ready to copulate. Courtship is very brief. The male seizes the female by brute force, and they mate for several minutes while he continues to bite her with some violence; but it should be remembered that most of these observations take place under captive conditions. Multiple nesting occurs, usually from March to September, but sometimes year-round. The clutch size ranges from 10 to 35, and the eggs are spherical, white, and about 25 mm in diameter and are placed in a nest about 150 mm deep. Incubation runs for 40 to 80 days, according to latitude, but Mitsukuri has reported hatching after just 28 days, which may be a record for any turtle. Sex determination is independent of temperature of incubation. The young can grow very fast, doubtless in order to outgrow predators; in captivity the young grow from 5 g to 1 kg after just one year of good nutrition.

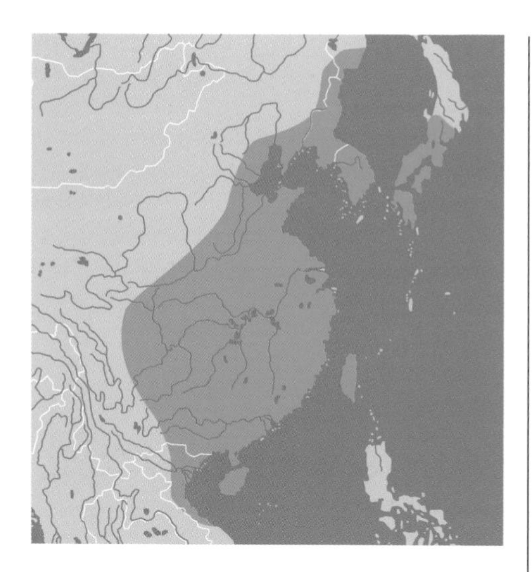

Protection. This species has always been valued for its meat in all countries of Asia, and millions are produced annually on farms in China, Thailand, and Vietnam. In addition, many Chinese farmers raise these turtles in small basins, as chickens or rabbits might be raised in the United States and Europe. These are the animals currently sold and consumed in China, just as chickens are the United States, and the brand name "Carrefour" is carried by 30 supermarkets in China, indicating that living turtles are available and can be decapitated before the eyes of the buyer. Certain jokers will also drink the blood of this species in order to gain sexual prowess. In nature, these turtles have almost disappeared, and their habitats are often completely cleared or altered. In certain parts of China, they exist nowadays only on farms, and the same seems too be the case in Vietnam. The status of the species in the wild needs to be evaluated so that the threats can be identified.

Common name

Chinese softshell turtle

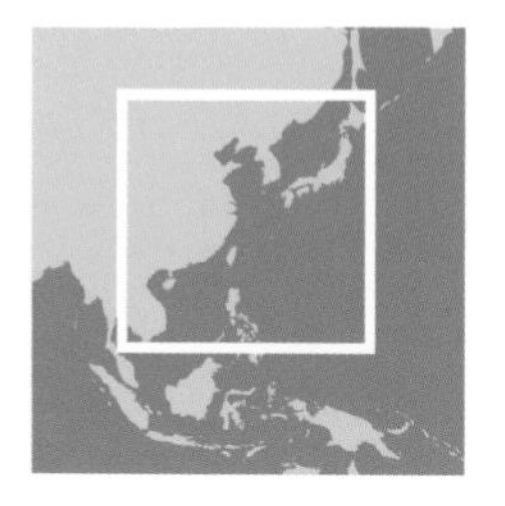

© A. Dupré

The snout is short, and the skin very fine and mostly yellowish.

Opposite: Individuals of this species are always small. The carapace is elongate, grayish in color, and completely plain. Raised in specialized farms, it is widely sold for food in China. It is much sought after for the delicacy of its flesh.

Pelodiscus spp.

Three species, recently described and found in Chinese territory, are not yet unanimously accepted by specialists, and further details need to be published. They may just be subspecies of *P. sinensis*. They are *P. axenarius* (Zhou, Zhang, and Fang, 1991), from the Hunan region; *P. maackii* (Brandt, 1858), the largest of the three, from northern China; and *P. parviformis* (Tang, 1997), the smallest of the three, found in the south of China.

Rafetus euphraticus (Daudin, 1802)

Distribution. This turtle is found in the Tigris and Euphrates Rivers and their tributaries, in southeastern Turkey, Iraq, Syria, and Iran (Khuzestan). Some authors also report it from northern Israel (Ernst and Barbour, 1989).

Description. The maximum length of this turtle is 500 mm, and the carapace is oval to circular, olive in color, and marked with irregular dark spots. There are tubercles on the anterior border of the shell, above the neck. The plastron is whitish to cream and has only two rather indistinct callosities. The head is of medium size and has a short tubular snout. The head, neck, and feet are greenish on the upper surfaces and lighter (whitish to cream) below. The limbs are strong and well webbed. The young have light spots on the head and a figure made up of yellow, cream, or white dots on the carapace, as well as several longitudinal series of tubercles. These details fade out within a few years.

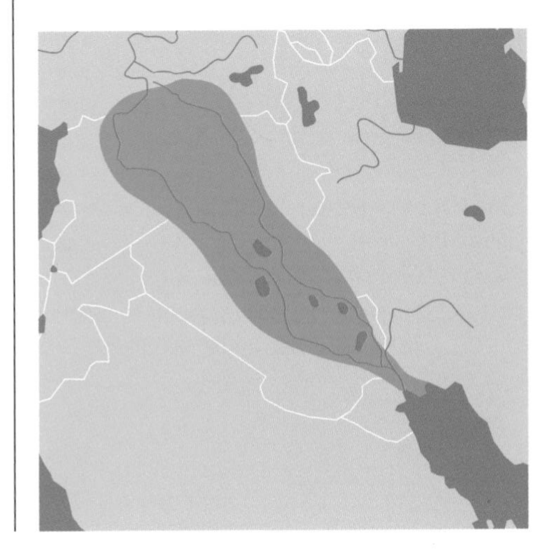

Euphrates softshell turtle

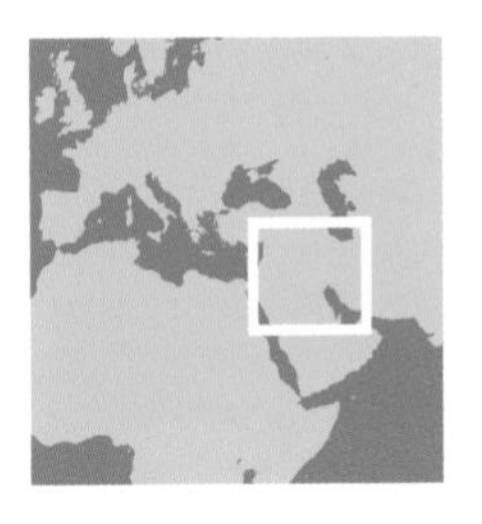

Natural history. This species prefers shallow, calm waters, dead branches of rivers, small lakes, ponds, and swamps. It is often seen basking on the banks. It tolerates surprisingly hot water and can also survive in saline conditions. It is diurnal for the most part but may also be active at night during the summer months. The diet is largely carnivorous, including fish, mollusks, crustaceans, amphibians, and insects and their larvae; but these turtles also eat carrion, and several individuals have been observed pulling apart a dead cow near the edge of the water. Stomach content analysis indicates that vegetable matter is also consumed, including algae, plants, tomato seeds, melons, and cucumbers. In winter, this turtle will hibernate if the temperature drops below 18°C. Irascible in nature, it can deliver extremely painful bites. In Turkey, the nesting season extends from the end of April into June. There may be several nestings per year. The eggs (about 10 on average) are white, almost spherical (23 mm in diameter), and brittle-shelled.

Protection. This species is sometimes eaten by local people, and it may also be sold as a pet in Turkey. But the main problem is the political situation in this region, along with environmental degradation. Wars, poverty, urban development, intensive agricultural practices—all these are harmful to *R. euphraticus*. This turtle is very dependent upon an intact ecosystem and is sensitive to any perturbation. The numerous hydroelectric dams constructed on the streams and rivers where it lives have profoundly upset the aquatic environment. According to E. Taskavak and M. K. Atatür (1998), the populations of this species are in sharp decline everywhere, and subadults and juveniles are hardly ever seen. The species is still not included in the *Red Data Book* of IUCN, but it is classified by the Turtle Conservation Fund as one of the 48 most endangered turtle species. In Turkey, programs for the evaluation of wild populations and for captive reproduction are under way, with the objective of releasing animals back to the wild.

The carapace is often more circular than shown in this photograph and is plain olive to blackish in color. The head is rather small, and the snout short. The turtle hides itself in the mud bottom of the Euphrates River, in the Middle East.

© F. Bonin

Trionyx triunguis (Forskal, 1775)

Distribution. The range of this turtle is truly enormous, encompassing much of the African continent as well as the Middle East, from Turkey, Syria, Lebanon, and Israel, down through the basins of the White Nile and the Blue Nile, to Lake Turkana in Kenya to Uganda, Ethiopia, and Somalia, and over to the coast of Tanzania. The species also occurs in Central and West Africa, reaching Angola to the south and the Democratic Republic of the Congo, Cabinda, Gabon, Equatorial Guinea, Cameroon, Chad, Niger, Central African Republic, Nigeria, Benin, Togo, Ghana, Ivory Coast, Liberia, Sierra Leone, Guinea-Bissau, the Gambia, and Senegal, and into Mauritania.

Description. This is a very large species, reaching 1,200 mm and a weight of 60 kg. The carapace is oval, rather flat, olive to dark brown in color, without any clear design. The anterior border of the carapace forms a swelling above the neck and the forelimbs, but there are no tubercles. The plastron is whitish or cream and is unmarked anteriorly, but dark spots may be present toward the rear. Adults have four clearly marked callosities on the plastron. The head is of medium size, and the eyes are small and located high on the head. The iris is black, with a gold ring. The snout is thin and pointed and is slightly expanded at the level of the nostrils. The head, neck, and limbs are rather dark olive-green, adorned with small yellow or whitish spots. The chin and throat are yellow or white, as are the undersides of the limbs. The limbs are strong, well webbed, and provided with three claws on each, recalling the scientific name *Trionyx* (*tri* = three, *onyx* = claws; *triunguis* has exactly the same meaning as the generic name but is derived from Latin instead of Greek). The young have a modest vertebral keel. The carapace bears dark spots, finely accentuated with yellow. There are several rows of small tubercles, visible only in the youngest specimens and disappearing with growth.

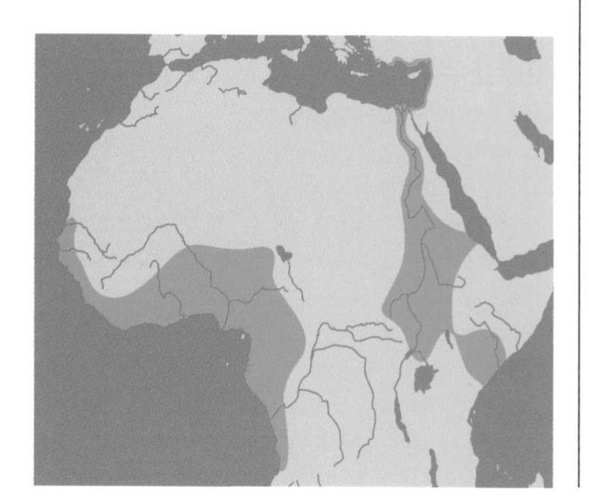

Common name

Nile softshell turtle

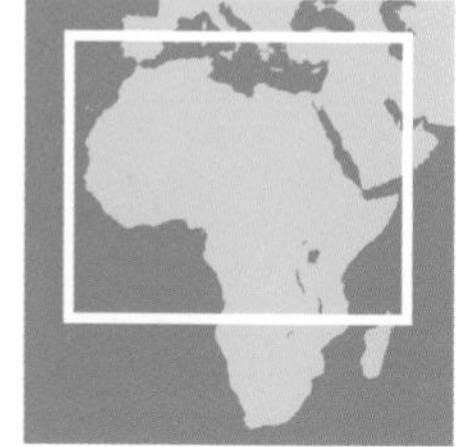

This young *Trionyx triunguis* is still brightly colored, with light markings on the entire body, on a brown background. Ultimately it may become uniformly brown or gray. This species has been represented in various ancient bas-reliefs of Egypt.

Natural history. This turtle is found in a wide variety of aquatic ecosystems—rivers and streams, lagoons, swamps, and lakes—with the key requirement being that the system includes sandbanks suitable for nesting. The species enters estuaries and even the open sea and utilizes nesting beaches that are also used by true marine turtles, as happens in Turkey. It can tolerate salinity as high as 38 parts per thousand for protracted periods. Omnivorous in diet, it consumes fish, amphibians, mollusks, insects, plants, and fallen fruit, including dates. It also consumes carrion and will readily seize upon carcasses of large mammals near the edge of the water. It is often seen basking on land, but it takes flight at any sign of disturbance and dives into the nearest water. It is unpopular with fishermen because it can damage their nets while trying to get at the fish therein and becomes very aggressive if one tries to pick it up. During the daytime it spends long periods of time down in bottom mud, using its cutaneous and pharyngeal respiration to satisfy oxygen demands. It is an excellent swimmer and also is able to race across land with considerable speed. Nesting occurs from March to July, according to latitude. Big females lay more than 100 eggs, which are hard-shelled, spherical, and white, measuring 25 to 30 mm in diameter. Incubation takes 75 to 80 days.

Protection. This large softshell has for a long time been an object of veneration by peoples of Egypt and Mesopotamia and has been represented in many archaeological bas-reliefs. But in addition to veneration, it is also heavily exploited. Nowadays it is caught in virtually all of the numerous range states. Its large size as well as its excellent meat make it a worthwhile target. Furthermore, while at sea it is often caught in fishing nets. And to make matters worse, many of its habitats have undone ecological degradation. The species has become rare in Syria, Lebanon, Israel, and West and Central Africa. On the other hand, it is still commonly seen in many of the widespread nations of Africa and even in the Middle East. It is listed as being in critical danger by IUCN in the *Red Data Book* (1996 and 2000), and the Mediterranean population does not include more than 1,000 adult specimens. It is also listed as one of the 24 species in critical danger by the Turtle Conservation Fund. Headstarting and reintroduction programs should be undertaken in those zones and areas where the environment is still suitable.

Carettochelyidae Boulenger, 1887

This family was quite widely represented in the Tertiary period of Europe, North America, and Asia. Today only a single genus and species survives, morphologically very distinctive: Carettochelys insculpta, *from northwestern Australia and the southern part of Papua New Guinea.*

Carettochelys insculpta Ramsay, 1886

Distribution. As of 1970 this species was thought to be confined to certain rivers in southern Papua New Guinea (the Vailala River, east of Purari), but in that year H. Cogger found it in a limited area of northern Australia, in the Darwin region and also in the Alligator River in the Kakadu National Park. Later it was found farther south in the Daly River, where large populations have been seen at Ooloo Crossing. In recent years, numerous observations have made it clear that this species can be found in a good part of northwestern Australia, although always with reduced populations.

Description. The pig-nosed turtle today constitutes an entire family and genus as well as a species. It is the only nonmarine cryptodire in Australia, and it is so distinctive that it could not be confused with any other turtle. It gives the impression of being halfway between a soft-shelled turtle and a sea turtle, and it is indeed well adapted for marine conditions; it is regularly seen at sea and will come to nest on the same beaches as marine turtles. Its being a cryptodire suggests the possibility that it is of more northern origin and that it arrived in Australia by accidental, or "waif," dispersal relatively recently. Its main distinctive features are as follows: it is the largest Australian turtle, with a maximum length of 700 mm and a weight of 30 kg; it is the only turtle in the world to have a thick, porcine snout, which gives it the vernacular name "pig-nosed turtle." Its limbs are more like flippers than like the webbed feet of a freshwater turtle (and it has been separated from the trionychid lineage for 70 million years). The shell is covered with a thin skin, in which the outlines of rudimentary vertebral scutes may be seen in hatchlings. The color ranges from silvery gray to brownish black. The skin of the carapace has series of light spots in the adult, but there is no sign of scutes or scute seams. The carapace is somewhat roof-shaped (tectiform), with a well-developed median keel, as in certain of the marine turtles. There may also be various light yellow spots on the sides of the first peripheral bones. The plastron is a uniform creamy yellow, sometimes touched with pink. The underside of the neck is yellowish or almost white. The neck is very short, and the head large and wide, with a perfectly smooth top and coloration identical to that of the shell. There is often a white streak behind the eye. In the males, there are two claws on the leading edges of the anterior flippers.

The young are quite different from the adults. Upon hatching, they are flat and circular, with a diameter of 50 mm. They have a very wide, skirt-like extension of the edge of the carapace, along

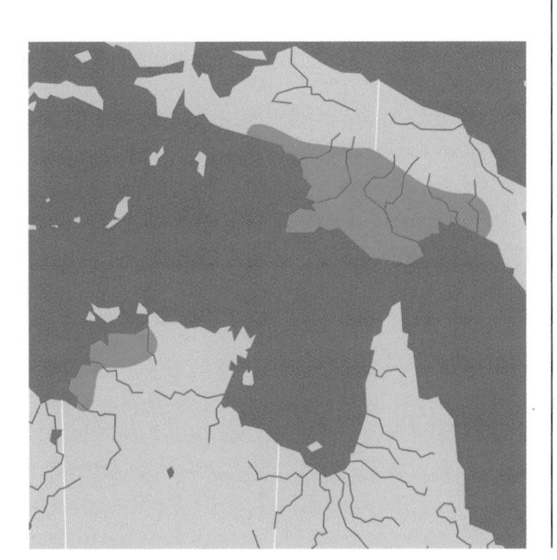

Common name

Pitted-shelled turtle, pig-nosed turtle

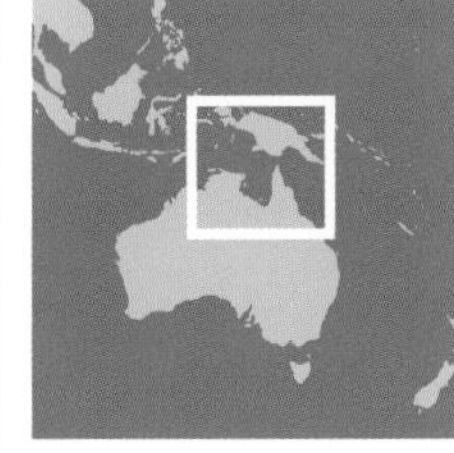

This turtle is out of its element, water, and thus seems awkward and inactive. In the photo, the very large forelimbs, the modestly domed shell, and the thick snout at the front of the round head, as well as the large, expressive eyes, are evident. Underwater, it recovers its vivacity and becomes a supple and elegant swimmer.

both the sides and the rear, which is deeply serrated and incised, giving the animal the appearance of a circular saw. The center of the carapace has vestiges of the neural bones that look like a string of pearls, and one can detect radiating ridges along the sides that correspond to the underlying bony ribs. The color is pearl gray, and the marginal flap is supple and soft. The young grow quickly, and in two or three years they take on the coloration and the form of the adults, conserving a yellow line around the carapace. The tectiform shape of the shell becomes exaggerated, and the limbs are very large. The carapace is slate gray, and the plastron is pink and without markings, later becoming cream or yellow. The neck is a clear yellow, the stripe behind the eye is well developed, and the snout takes on its characteristic porcine form. Out of the water, this species is ill at ease, remaining immobile with its head resting on the ground, like a sea turtle. It is only when it is in water that it comes into its element, and—once again—does as the sea turtles do.

Natural history. In captivity as in the wild, this species seems quite placid and spends much time immobile. It is never seen basking on sandbanks, tree trunks, or rocks. It maintains a low profile in calm rivers and isolated lagoons, and it even adjusts to saline waters and may enter the sea. In the Daly River, it will follow the advancing waters and move downstream when the river is low, and when the water rises, bands of these turtles will head for Ooloo Crossing, far in the interior. It is more nocturnal than diurnal and feeds upon snails, fish, and crustaceans, as well as fruits that happen to fall into the water. The snout seems to help it forage in the mud, looking for larvae or absorbing essential nutritive elements.

In the dry season, between July and October, the females make their nocturnal nesting emergences. Sometimes several females nest together, without reaching the point of being truly gregarious. The females have been observed spending some time on the beach before nesting. Usually, there are two nestings per year, at somewhat unpredictable intervals. The body pit is excavated with the hind limbs, and then the nest cavity itself is prepared; the cavity may reach a depth of 50 cm. The eggs look like tennis balls and are at least 40 mm in diameter, with a weight of 35 g. The shell is brittle and very smooth and is covered with mucus when laid. The critical temperature for sexual differentiation is rather high (31.5°C). The embryos develop normally for several weeks during the dry season and then enter a period of dormancy until the first touch of rising waters stimulates them to emerge. The hatchlings weigh about 30 g and feed upon insect larvae, small shrimp, and young snails. The adults are more omnivorous and enjoy aquatic plants, like *Vallisneria spiralis* and figs *(Ficus racemosa)*, as well as the fruits of *Pandanus aquaticus* and *Syzygium forte*. The main predators on the species are marine and freshwater crocodiles, but the large herds of feral buffalo are also a severe problem because they trample the banks and nesting beaches and destroy eggs and sometimes young in the nest.

Protection. Although this species is protected, capture by local tribal peoples for subsistence use is authorized. And because of its large size and succulent flesh, it is quite widely consumed by aborigines. To capture the turtles, the turtle fishermen lay out line baited with buffalo meat. A drain on the population is caused by the great herds of buffalo (see above). There seems to be no commercial use of this turtle in Australia, but in New Guinea the eggs are widely consumed, and each year more than 20,000 eggs are sold in the markets of Kikori. With the growth of human populations in these regions and the development of transportation options, it is inevitable that more and more pressure will be felt by the *Carettochelys* populations. In the last 20 years, the numbers on New Guinea appear to have diminished considerably. Frequently, these turtles are bought in the markets of New Guinea and exported to the United States or Europe. Many fail to survive their first few days in captivity or the transportation process. In Australia, Kakadu National Park has undertaken major efforts to protect this species. The feral buffalo have been controlled. The main remaining problem is the progressive extension of agricultural and pastoral areas, but one must worry just as much about gold and uranium mining operations, in the northwest of the country. Captive rearing, which has been attempted by several operations in Australia, does not yield very good results, this being a species that presents difficulties with both basic maintenance and with reproduction in captivity.

Ethnozoology. *Carettochelys insculpta* plays a very important role in the legends and in the art of the aboriginal people. It features frequently in rock art, with the animal presented in anatomical fashion, to show the muscles and viscera. These representations became evident to naturalists even before the species was "discovered" in Australia, because the aboriginal people had been capturing and consuming them all along. One story explains how the Warradjan (as this turtle is called) and Manbirri (the green sea turtle) were two sisters. Manbirri was the younger of the two. When the seawater retreated from the land, Manbirri decided to go off wandering, traveling great distances at sea, while the Warradjan remained in freshwater with her other sister, Nadwerrwo, the snapping turtle *(Elseya dentata)*. This story demonstrates how the aborigines are keen observers of biology and ecology of local turtles. Nevertheless, in their eyes, turtles are not sacred but simply represent an edible resource.

This species was discovered some time ago in New Guinea, but in recent years numbers of specimens have been found in northwestern Australia, as far as the Kimberleys.

DERMATEMYDIDAE Gray, 1870

During the Jurassic this primitive family was widely represented in Europe, and later, during the Cretaceous to the Miocene, in the American continents and in eastern Asia. Today it survives only as a single species in Central America, Dermatemys mawii. *This is a large turtle, with thin scutes and a smooth skinlike surface. Wide inframarginal scutes separate the marginal scutes from the plastron.*

Dermatemys mawii Gray, 1847

Common name

Central American river turtle

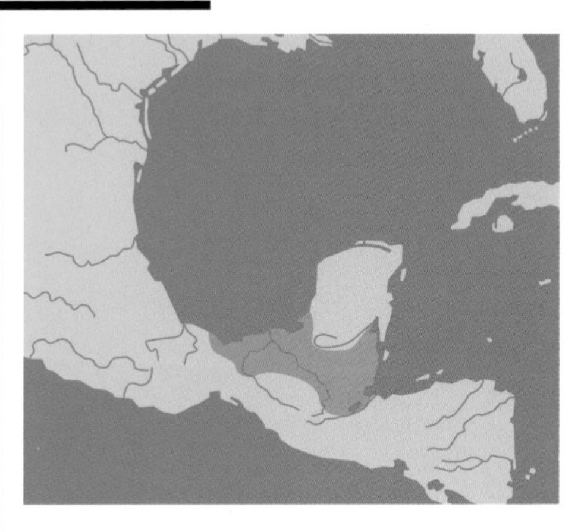

This large river turtle is particularly maladroit on land. It is even incapable of supporting the weight of its own head, which slumps passively on the ground.

Distribution. This turtle occurs in Mexico (Veracruz, Tabasco, northern Chiapas, Campeche, and southern Quintana Roo), in Belize, and in northern Guatemala.

Description. The species name makes reference to one Lieutenant Mawe. It can reach a length of 650 mm and a weight of 25 kg. The carapace is rather flattened, oval, unkeeled, very thick, and well ossified. With age, the scute seams tend to disappear, giving it the appearance of a small boulder polished by running water. The marginals are neither broadened nor serrated. The carapace is gray to olive or brown, and the scales, worn thin by the abrasions of life, may become so transparent that the structure and grayish color of the bones beneath may show through. The plastron is large, well developed, and unhinged and ends with a well-developed anal notch. The bridges are cream in color, like the plastron and the undersides of the marginals, and each includes a series of inframarginal scutes (three to six in number on each side). The inguinal is larger than the inframarginals, and the axillary is smaller. The head is rather small for a turtle of such size. The snout is somewhat projecting and rather long. In color, it is grayish on the sides and yellow to reddish on the top of the head. Sometimes a light line separates these two colored areas. This light background may include some dark, vermiculated patterns. The neck is rather wide and long. The limbs are strong, well webbed, and dark gray. A series of large scales is present along the edge of each limb. The head of males is usually brighter yellow (or red) on top, while that of the female is grayer. The hatchlings have an olive-brown carapace, a well-developed vertebral keel, and strongly serrated marginals. They have a yellow stripe extending from the nostrils, traversing the orbit, and ending at the base of the neck.

Natural history. This is a highly aquatic species, frequenting streams and rivers, lakes, lagoons, and indeed any deep-water habitat with abundant vegetation. *Dermatemys* even takes to brackish water in estuaries and on occasion may become encrusted with barnacles. These turtles are primarily crepuscular or nocturnal, but by day they may float on the surface, possibly for thermoregulatory purposes, or they may rest in small groups on the bottom. The pharyngeal respiration allows them to capture oxygen while underwater and to remain submerged for a long time. They

are never seen basking on land, and indeed when placed on hard ground, they seem maladroit and clumsy, hardly able to lift their head and neck from the ground. The diet is entirely vegetarian and includes aquatic plants, leaves, and fallen fruit. Nesting occurs during the rainy season, at which time the high water allows the female to reach a point well above normal water levels for nesting. Three clutches may be laid in a season, in nests that are quite shallow or even rudimentary, and very close to the water's edge. The eggs are white, elongate, and hard-shelled and measure 60 × 32 mm. A clutch consists of 6 to 20 eggs. The juveniles are more omnivorous than the adults.

Protection. The young are consumed by otters and other predators. Because of their large size and excellent taste, the adults are heavily harvested and fished and feature in the diet of numerous local peoples. In Mexican markets, the meat is especially valued, and a single adult turtle may fetch 30 to 40 euros. Legislation forbids the capture or sale of these turtles if they measure less than 400 mm, and there is a closed season from September 1 to December 31. Sale of the eggs is forbidden, but the enforcement effort is thin on the ground. There are captive rearing facilities in Mexico, whose objective is to release part of the production (25%) and to market the remainder. The species is classified in Appendix 2 of CITES and is one of the 48 species declared in the greatest danger by the Turtle Conservation Fund in 2002. Populations are steadily diminishing everywhere.

KINOSTERNIDAE Agassiz, 1857

This family includes three genera and extends most of the length of the American continents, from southern Canada to the center of South America. Fossils kinosternids have not been found outside this region. Kinosternids are mostly small to medium-sized turtles, with large heads and pebble-shaped shells, often flattened on top, and sometimes with one or three keels. The plastron may be narrow and cruciform or with anterior and posterior hinges permitting total closure of the bony box. The structure of the feet indicates that these turtles are not great swimmers, mostly preferring bottom-walking. They are largely carnivorous and opportunistic. Some species are confirmed burrowers and may spend several months of the year underground. Many secrete a foul-smelling musk from special glands close to the bridge on each side.

Kinosternon acutum Gray, 1831

Distribution. A rather limited range in western Yucatán, Mexico, to the coastal plains of Veracruz, Tabasco, and Campeche and into northern Guatemala and Belize.

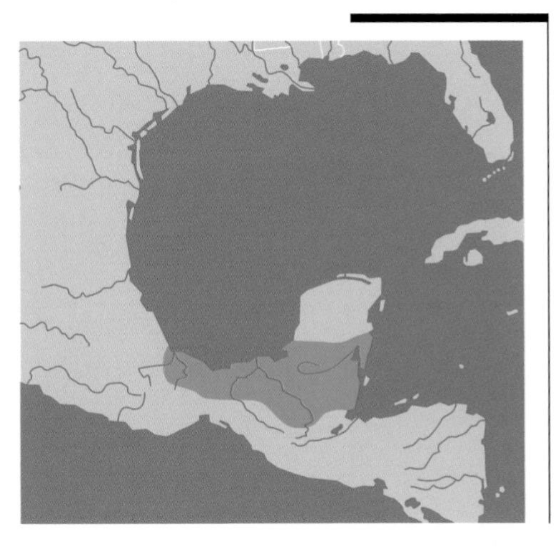

Common name

Tabasco Mud Turtle

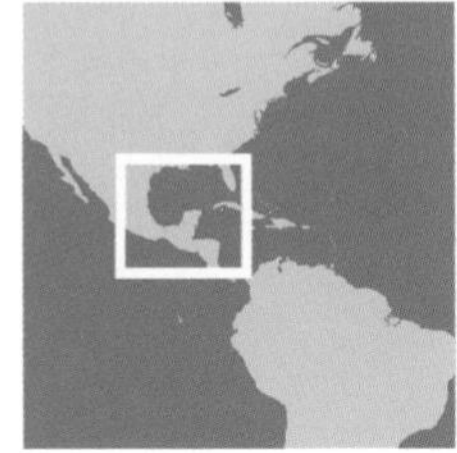

Description. Females may reach 120 mm, but males just 105 mm. The carapace is moderately domed and detectably higher behind than in front. The adults have a vertebral keel, and the subadults have two lateral keels that disappear with age. The carapace is dark brown. The scutes are separated by darkly pigmented seams. The plastron is large enough to completely close the shell and is double hinged; there is no anal notch. In

© F. Bonin

Identification of the species of *Kinosternon* is not easy, and photos do not always show the defining characteristics of the species. The keel of this individual has almost disappeared. The head is elongated and terminates in a point protected by a single large nasal scale.

color it is yellowish to beige, with no real pattern or markings. The head is relatively small and is yellow to reddish, or sometimes gray, ornamented with darker streaks ranging from dark brown to black, the same color as the neck and limbs. The nasal scale is large and covers the major part of the surface of the head, a characteristic that gives rise to the Spanish name *casquito*. The upper jaw has a hooked beak, more developed in males than in females. The chin, with a single pair of barbels, is light-colored, ranging from white to dirty yellow, with fine brown or black marbling. In females, the chin is lighter than in the males, and the tail is truncated and very short. Both sexes have a hornlike scale on the tip of the tail and lack patches of thigh tubercles.

Natural history. This turtle occupies temporary wetlands and small forest streams, at an altitude not exceeding 300 m. It lives in sympatry with *K. scorpioides* and *K. leucostomum* in the Peten region of Guatemala. The females nest in March and April.

Kinosternon alamosae Berry and Legler, 1980

Alamos mud turtle

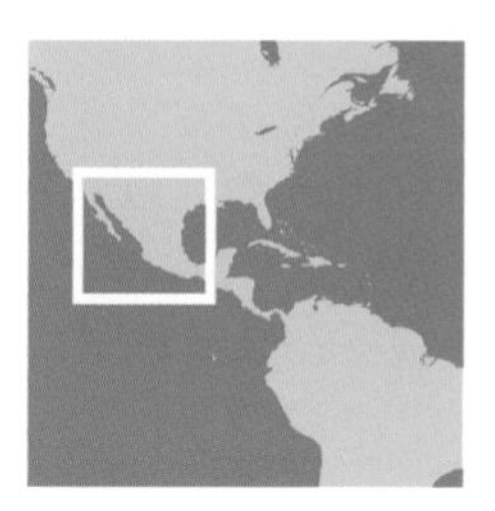

Distribution. Endemic to the Pacific coastal plains of Mexico, this turtle occupies a stretch of coastland up to 1,000 m in altitude in the basins of the Sonora, Sinaloa, and Yaquí Rivers of Mexico and possibly farther north into the basin of the Rio Magdalena.

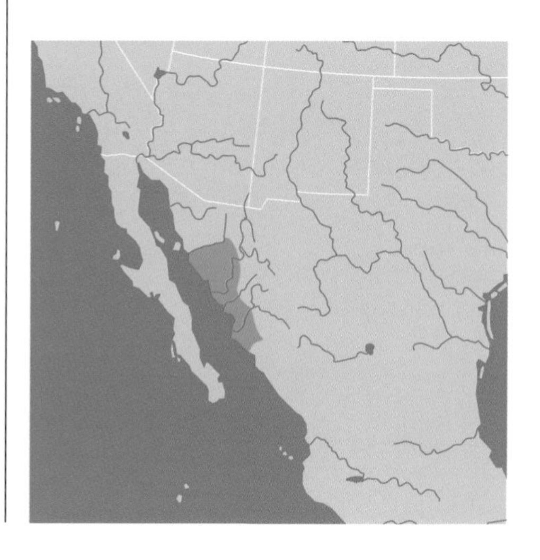

Description. The species name is derived from the town of Alamos, where the first specimens were collected. Males may reach 135 mm, and females 126 mm. The carapace is rather domed and sometimes slightly depressed on top. The adults are without keels. The first vertebral scute fails to contact the second marginal scutes, while the fourth costal scute does make contact with the tenth marginals. These marginals are the only ones that exceed the overall height of the marginal scute series, and they are notable for being vertical-sided. The shell is beige to brown, sometimes olive, sometimes with darker spots. The plastron is large and doubly hinged and closes the shell openings completely. It is pale yellow to orange-brown in color, sometimes with dark markings. The seams are very dark, and in old animals the growth rings may be very distinct. The inguinal and axillary scales on each side are separated. There is no anal notch. The plastron of the male is narrowed at the level of the femoral scutes. The head is of medium size and brown, with yellow

markings on the sides. The neck is brown on top, lightening to yellowish underneath. The chin, also light in color, has a pair of small barbels. The neck is rather short, with a large nasal scale. The feet are brown and have well-developed claws. The webs are clearly visible. Both sexes have a hornlike scale on the tip of the tail and lack patches of rough femoral scales. The tail of the female is short and truncated.

Natural history. This turtle lives in temporary ponds, small creeks, cattle tanks, and ditches beside paths and highways but is never found in large rivers or lakes. It seems to be somewhat terrestrial in habits despite its webbed feet. It can tolerate extremely high temperatures, and some individuals have been observed existing temporarily in waters as hot as 42°C. Activity is mostly confined to the months of July to September, one of the shortest known activity periods for any turtle species. Some individuals will estivate during the long, dry months in Sonora, at least in the coastal areas. This is primarily a diurnal species, although some individuals may be active during the night. The diet is strongly carnivorous, including arthropods, frog eggs, tadpoles, amd frogs, along with some vegetable matter. This is a very shy species that will close up tightly in its shell if disturbed. It never tries to bite. The female attains maturity in 5 to 7 years, and copulation has been observed at the end of July near Alamos. There may be two or three nestings per season, each with just two or three eggs, measuring about 26 × 15 mm.

Protection. Populations of this turtle appear to be small. Nevertheless, it is so cryptic that it does not seem to be particularly endangered. The places where it lives are not threatened by habitat degradation or urbanization. The greatest stress is the sheer aridity of its habitat, which may be extreme.

© F. Bonin

This small turtle, spotted on the road leading to Alamos, Sonora, has very light coloration, a well-developed carapace, and a short brown head. The *Kinosternon* species are generally shy and hide quickly under vegetation or in the natural crevices of the habitat, making haste on their long, agile limbs.

Kinosternon angustipons Legler, 1965

Distribution. The range of this species is quite limited. It occurs in the Atlantic coastal plain of Nicaragua (Rio San Juan), Costa Rica (Puerto Viejo de Sarapiqui, Los Diamantes, Rio Suerte, Tortuguero, Barra del Colorado) and Panama (Bocan del Toro).

Description. Discovered and described as recently as 1965, this species received its scientific name from the Latin *angustus* (narrow) and *pons* (bridge), or "narrow bridge." Females reach 120 mm and males 115 mm. The carapace is rather low, slightly flattened on top, and without a keel at any age except for some hatchlings in which traces of three keels may be visible. The color is a uniform dark brown. The first vertebral scute is in contact with the second marginals. The tenth marginals are higher than the rest of the series. The posterior marginals are slightly serrated. Sometimes in old animals there will be some imbrication of the carapace scutes, like the tiles of a roof. The plastron is narrow, cruciform, and doubly hinged. In old animals, the seams may be concealed beneath soft, whitish skin. The plastron

This very young *Kinosternon angustipons* is still brightly colored, with a light line around the shell (which is strongly keeled) and the head marked with light patches, notwithstanding a dark plastron.

© A. Hell-Kevorkian

and the bridges are pale yellow to orange-yellow, sometimes with dark spots. The inguinal is in contact with the axillary on each side. The posterior lobe is narrower and longer than the anterior lobe. There is a significant, rounded anal notch. The head is medium in size, with a bulbous snout, a defining feature of the species. The upper surface is dark brown, becoming lighter below. The chin and neck are cream below. The jaws are unserrated, light in color, and marked with darker lines. There are three to six small, light-colored barbels on the chin. The limbs are rather feeble, dark above and lighter below. They are well webbed and have five claws on the forelimbs and four on each hind limb. The males have patches of roughened scales on the rear of the thigh. In both sexes there is a horny spur on the tip of the tail.

Natural history. This turtle lives in shallow waters and slow-flowing creeks, often with high turbidity and abundant vegetation. It is very cryptic, and its populations seem to be low. It can tolerate remarkably high temperatures and is often encountered in the same places as other turtles, including *Chelydra serpentina, Rhinoclemmys funerea,* and *Trachemys scripta emolli.* The diet is largely opportunistic; this turtle eats insects as well as fallen leaves, and it may be caught in traps baited with sardines or banana skins. Reproduction occurs from May to August. A single nesting seems to be the rule, the clutch size averaging four eggs, which are whitish, elliptical, and about 40 × 22 mm. The shell has a slightly roughened texture, with small irregularities, although not actually granular.

Protection. The IUCN considers this species to be very rare and recommends that its status be studied without delay. Natural predators include jaguars, pumas, and coyotes. It often carries numbers of leeches on its soft parts. There are no real data on the impact that humankind may have on this turtle or on its habitats and ecosystem.

Common name

Narrow-bridged mud turtle

Kinosternon baurii Garman, 1891

Distribution. Known only from the southeast corner of the United States, from King and Queen County in southern Virginia to the Florida Keys, including North and South Carolina and Georgia.

Description. The name is a patronym for Georg Baur, a famous osteologist and turtle expert. This small species does not exceed a length of 120 mm. The carapace is often wider than high, being depressed along the vertebral area. The surface is smooth, and adults lose the three keels that are visible in young specimens. The carapace is very dark, almost black, with three longitudinal yellow lines. The plastron is double-hinged and yellow to olive in color, with dark spots and marks. The anal notch is very reduced. The head is of medium size, conical in shape, and blackish in color, as is the remainder of the integument. There are two light lines on each side that start at the point of the snout, passing above and below the orbit, respectively, and continuing along the length of the neck. These lines are sometimes supplemented by light spots on the minor protuberances of the skin. The limbs are also extremely dark, and the forelimbs each bear five claws and are well webbed. Males have a large tail, with a horny conical tip and patches of opposing rough scales on the thighs. At the time of emergence, the hatchlings have a very rugose carapace that is dark gray to black, with light yellow lines along the three well-marked keels. The edges of the marginals are also marked with yellow. The plastron is yellow, with a dark marking in the middle in the form of an arch, with the sides following the seams. The plastron shows kinesis only after about three months of growth.

Subspecies. Some authorities recognize two subspecies, *K. b. baurii* in the south, and *K. b. palmarum* in the north, but they do not appear to be valid.

Natural history. This mud turtle is active year-round and is certainly the most terrestrial member of its genus. It is often seen on land, crossing roads in its search for a flooded field or a nesting place. The diet is mainly carnivorous, including palm seeds, leaves, aquatic vegetation, algae, small snails, insects and insect larvae, and also carrion.

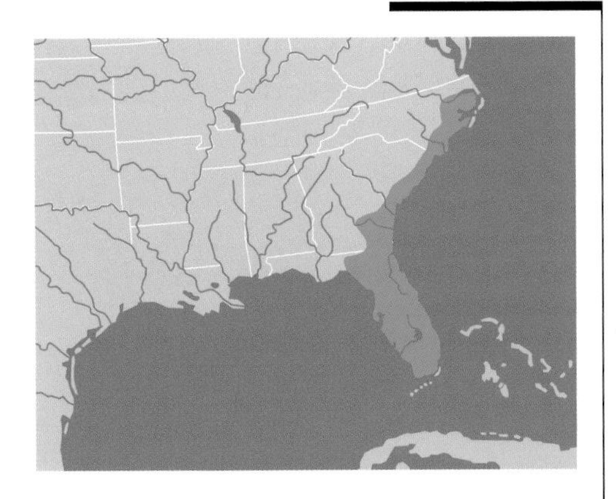

Common name

Striped mud turtle

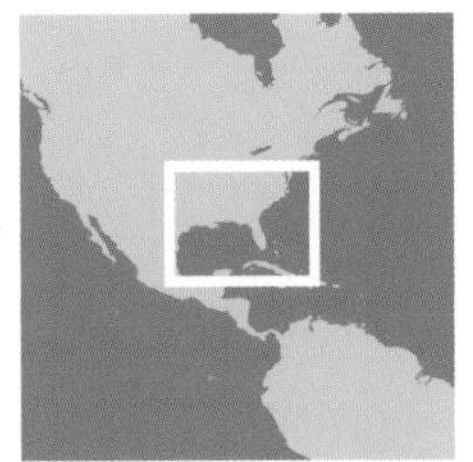

They often take the bait of fishermen, and they seem to be attracted to cattle feces, which they search minutely for concealed insects, a habit that gives them the name "cow-dung cooter." Females mature at five to six years of age. Copulation occurs mainly from March to October but may take place in any month of the year. The eggs are laid in the sand or in vegetative debris, close to the water. Maximum clutch size is seven eggs, which are white or slightly pinkish, elliptical, and about 30 × 17 mm in size. There may be as many as three nestings in a year, with incubation lasting for 80 to 145 days.

Protection. This species has been well studied by American biologists, but it still appears to be

High and rounded in front, the shell of this species forms a perfect oval, like a stone well polished by the passing years. The head, with two light lines on each side, is very strong and has a short snout.

endangered. This entire region of the USA is very developed, and the habitat has been profoundly altered by humans. The progressive drying of some areas of Florida is also a threat to the species.

Kinosternon carinatum (Gray, 1855)

Common name

Razor-backed musk turtle

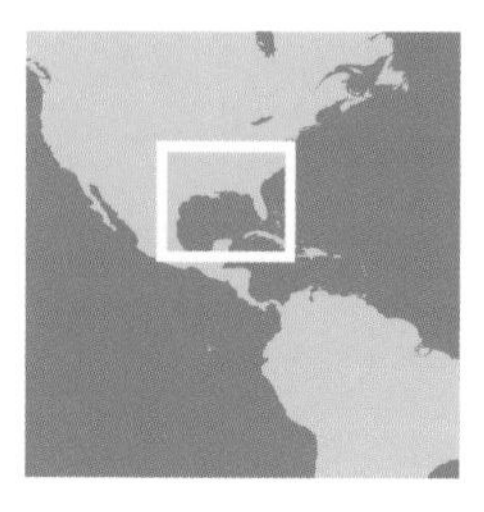

Distribution. This species has a wide distribution in the southern and central United States, from southeastern Oklahoma, Arkansas, and Texas to as far as southern Mississippi.

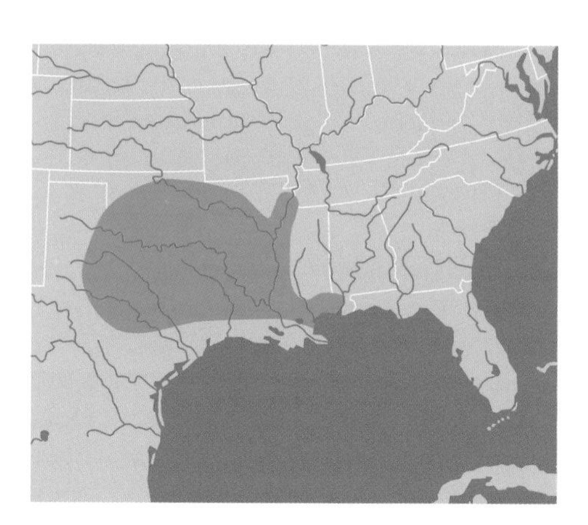

Description. The Latin *carinatum* means "keeled" and refers to the roof-shaped carapace of this species. This turtle does not exceed 160 mm in length, and the vertebral keel is extremely well pronounced, the sides of the shell meeting at an angle of about 100 degrees. Adults have no lateral keels. The scutes of the carapace are imbricate, the rear edge of each extending over the anterior edge of the scute behind. The marginals are somewhat serrated. The shell is light brown to orange, each scute being edged with a thin black line or adorned with series of dark streaks that disappear with age. The plastron is moderately developed and lacks the gular scute. There are thus just 10 plastral scutes, although other *Kinosternon* species have 11. A modestly developed hinge is detectable between the pectorals and the abdominals. The seams disappear beneath wide bands of soft tissue. The color of the plastron ranges from pink to orange in the young, becoming dark yellow to brown, sometimes with dark streaks, in old animals. The anal notch is rudimentary. The skin of the head is adorned with clouds of small black circular spots. The jaws are streaked with black, and the upper jaw is lightly hooked. The snout is pointed. The eyes appear black, with a light ring, and there is a pair of barbels under the chin. The limbs are light brown, with black dots. There are five claws on each forelimb and four on each hind limb; the limbs have well-developed webs. Only the males have a horny tip to the tail and rugose patches of scales on the thighs. The young have three dorsal keels, and the imbrication of the vertebral scutes is stronger than in the adults. The coloration of the young is also brighter and more developed. Upon hatching, they weigh about 4 g.

This young turtle has a light head, with white jaws and neck. The underside of the head and the neck is olive to brown. The plastron has light bare bands of clear skin on a background ranging from beige to quite dark.

Natural history. These turtles live in slow rivers, marshes, and mud-bottomed swamps with abundant aquatic vegetation. They will spend long periods of time basking on the banks of the water. Activity is mainly from March to November, and they hibernate in the mud at the bottom of the water or in cavities close to the water's edge. During the summer, they may estivate, remaining immobile in the coolest waters they can find. Thermoregulatory behavior is very developed, possibly because of the unusual form of the carapace, which may be especially effective at capturing the sun's rays. The diet is omnivorous, including insects, crustaceans, mollusks, amphibians, carrion, and aquatic vegetation. When searching for food, this turtle walks boldly along the bottom substrate of the water, holding its neck as if it were on land. This is a high-density species in some places, and at certain localities in Oklahoma, as many as 229 individuals per hectare have been documented. This turtle is shy by nature and gentle of disposition and does not try to bite when picked up, nor does it secrete the malodorous musk with which it is indeed equipped. The females mature in four to five years. Courtship takes place from April to June. The clutch consists of two to four white, elliptical eggs, with hard shells and measuring about 15 × 28 mm. Two nestings per season have been observed, with incubation lasting for 110 to 120 days.

Protection. The young are subject to predation by fish, snakes, and other turtles. This species, along with the sympatric *K. minor,* is caught in great quantities by fishermen.

Kinosternon chimalhuaca

Berry, Seidel, and Iverson, 1997

Distribution. This recently named species has a very small range along the Pacific lowlands of Mexico, in the states of Jalisco and Colima, from the basin of the Río San Nicolas in the north to the Río Cihuatlan in the south.

Description. The name *chimalhuaca* is taken from an Amerindian legend told by Gary Jennings in his 1980 book *Aztec.* The species is very close to *K. oaxacae* and *K. integrum.* It is of medium size, the males reaching 160 mm and the females 130 mm. The carapace is relatively broad and flattened and has three modest keels and imbricated scutes. The annual growth rings are visible in some old individuals. Marginals 8, 9, and 10 are widened. The overall color is dark brown or sometimes olive, often spotted with darker brown, or the whole shell may be black. The plastron is rather small and is smaller in males than in females; it is concave in males and is too small to close the carapace openings completely. Two plastral hinges are present. The anal notch is deeper in males than in females. The axillary and inguinal scutes on each side are in contact. The plastron is dark yellow to brown. The seams are often outlined in black, and various black spots may also be present, especially on the bridges. The head is rather wide, especially in males. It is dark green to dark brown, irregularly spotted with orange-yellow. There are between one and four pairs of pointed barbels on the chin, and there are four to eight rows of small tubercles on the upper and laterals surfaces of the neck. The limbs are brown above, cream to yellow below. They are equipped with well-webbed digits and with strong claws. Opposed patches of roughed scales on the thighs are not manifest in either sex, but there is a conical "tail horn" in both sexes, and the tail of the male is also provided with six rows of small tubercles.

Natural history. This species prefers calm swamps with a muddy bottom but is also found in clear waters, although rarely in fast-moving streams. It is primarily an opportunistic carnivore, and stomach contents have included insect remains, mollusks, and crustaceans, along with vegetable substances. The females apparently mature in seven to eight years and lay two to five eggs, probably several times per season. The eggs measure approximately 33 × 18 mm and are laid in July and August.

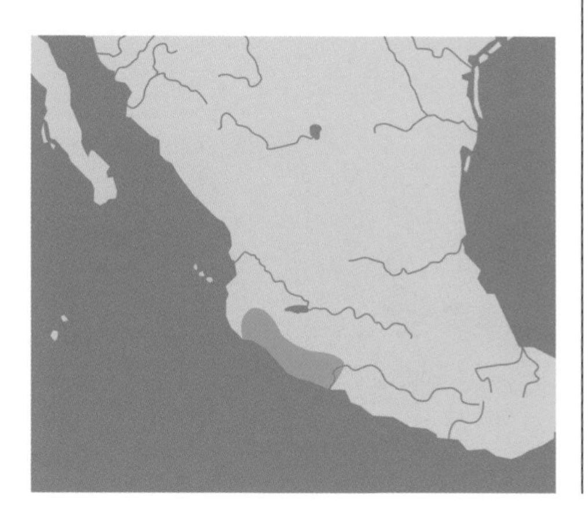

Common name

Jalisco mud turtle

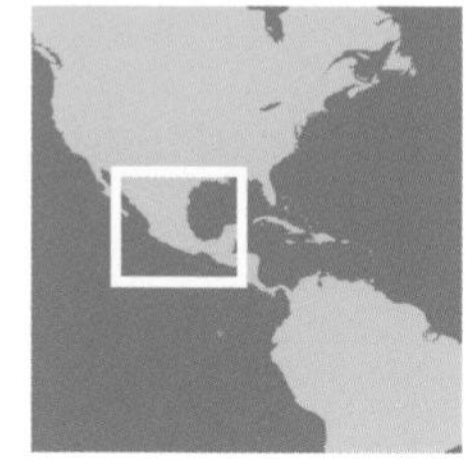

Kinosternon creaseri Hartweg, 1934

Common name

Creaser's mud turtle

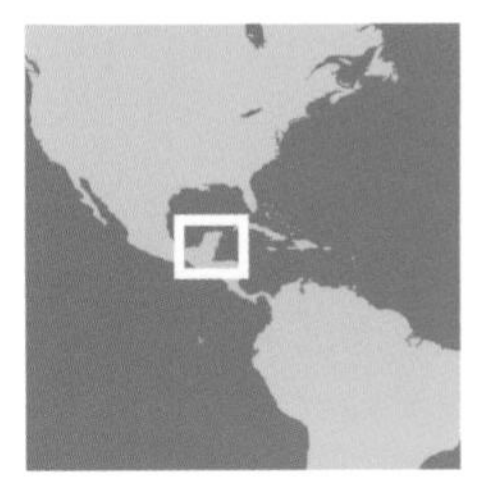

Distribution. This species is endemic to Mexico and is found only in the southeastern end of the country in the states of Yucatán, Campeche, and Quintana Roo.

Description. The name is a patronym for Edwin P. Creaser, a biologist at the University of Michigan, who collected the holotype on June 29, 1932, close to Chichén Itzá. It is a small species (males 125 mm; females 115 mm). The carapace, slightly elevated toward the rear, has a vertebral keel. The juveniles have an additional pair of lateral keels, but these disappear with growth. The shell is dark brown to dark gray, with no special pattern. The plastron is large and wide and generally yellow in color, with irregular dark spots also found on the bridges. The seams are black. The gular scute is large and triangular. There are two well-developed plastral hinges. The anal notch is very shallow and indistinct. The head is large, especially in the males. It is dark gray to black on top, with light marbling, and becomes light on the sides and pale gray below, with dark marbling. The upper jaw is provided with a strong hooked beak, accentuated by lateral cusps that become more developed with age. The females have a decidedly hooked beak, but the lateral cusps are less obvious. The nasal scale is triangular and covers the top of the head. There are no chin barbels. The limbs are rather feeble, gray-brown in color, dark above and lighter below; they have moderate claws and digital webs. The males have a hooked, conical spur at the end of the tail, a feature that is less developed in females. There are no roughened scale patches on the thighs.

Subspecies. No subspecies are currently admitted, but in 1965 Duellman observed two light-colored specimens in an underground cave, which may represent an adaptation to life in caves where daylight fails to penetrate.

Natural history. This turtle lives in water holes and temporary puddles in a karst region where there is little surface water. It estivates during the hot months and is active only during the rainy season, from June to October. It may be seen in the cenotes, natural wells that communicate with one another by subterranean channels and constitute the only aquatic retreats during dry periods. It is possible that the only surviving populations are the small groups in these cenotes. The reproductive habits are poorly known, but it seems that these turtles lay just one or two rather large eggs at a time and that two nestings per season are possible.

Protection. This shy species is poorly known, even by the local people. Fires in the dry forest have destroyed the habitat little by little. The rarity and unusual habitat of this turtle suggest that it is vulnerable. Field studies to determine its current status are essential if its progressive diminution is to be forestalled.

This *Kinosternon* has a rectangular, boxlike body form, with very enlarged costal scutes. The color shown here is light, but in general, specimens of this species are darker than this. The head is large, with various light markings.

Kinosternon depressum (Tinkle and Webb, 1955)

Distribution. The range of this species is very small. It is confined to the Black Warrior River system in central Alabama, USA, reaching upstream of the fall line.

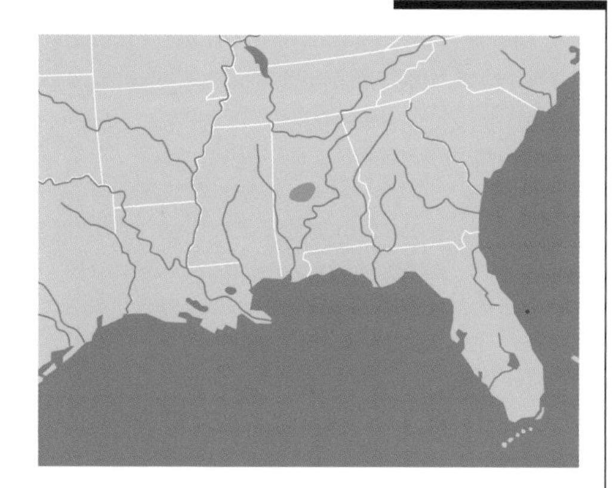

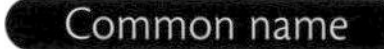

Common name

Flattened mud turtle

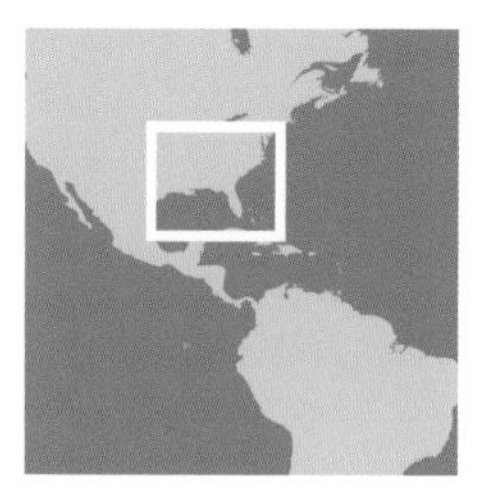

Description. Long considered a subspecies of *K. minor*, the flattened mud turtle has been elevated to full species status, and its name alludes to its most obvious characteristic: a very flattened body form. It is a small species, not exceeding 120 mm. The oval shell is quite wide, with a dorsally depressed central area and a modest keel in some individuals. The marginals are slightly serrated, and the carapace scutes slightly imbricate. The color ranges from brown to very dark green, spotted with yellow and olive areas. The edges of the marginals are marked with yellow, except in the oldest animals. Each scute of the shell carries a black line along the seams. The plastron is small and has a single gular scute. There is a barely distinguishable hinge between the pectorals and the abdominals. The plastron is pink in young animals, becoming yellow with age, without dark markings. A small anal notch is present. The axillary scute is in contact with the inguinal on each side. The head is rather large and has a distinctly pointed snout. The nasal scale is well developed and covers a large part of the head. The upper jaw may be slightly hooked. The skin of the head and the neck is dark green, with a vermiculation of yellow spots and lines and sometimes a bolder line between the tip of the snout and the orbit. There are two chin barbels. The skin of the limbs is dark above and becomes lighter below. The limbs are strong, well webbed, and of medium size, with powerful claws. The males have a horny conical spur at the tip of the tail and patches of rough scales on the opposing surfaces of the hind limbs. The young are more colorful than the adults, and their plastron is somewhat pink.

Natural history. This turtle is very aquatic, living in shallow, clear waters with rocky or sandy bottoms. It can stay a long time underwater, hiding in rocky crevices. The highly aquatic lifestyle explains the frequent incrustation of algae on the carapace and the equally frequent infestations of leeches on the soft parts. This turtle hibernates during the cold months, but the location of the

The vertebral scutes are slightly depressed, a feature that gives this species its name. Otherwise, the carapace is relatively small, and the animal as a whole rarely exceeds 110 mm. The head is wide and pointed, and the turtle seldom emerges from its built-in hiding place. This is a shy species that is difficult to photograph.

hibernation sites is still unknown. It is mainly diurnal but during the summer may become crepuscular or even nocturnal. It has the habit of taking the sun early in the morning. The diet is primarily carnivorous, including insects, crustaceans, and above all snails and bivalve mollusks. The jaws have well-developed masticating surfaces that bear witness to this durophagy, but it is also perfectly capable of eating seeds that have fallen into the water. The females lay twice per season, with one to three eggs per nest. The eggs measure 35 × 166 mm on average, and the incubation season lasts for 47 to 122 days.

Protection. This turtle is attacked by raccoons, skunks, and foxes and also by large fish and numerous birds. The area of the Black Warrior River where it lives is heavily industrialized, which implies numerous threats to the integrity of its aquatic habitats. Furthermore, the rarity of the species has resulted in its being highly favored by hobbyists. Additional governmental measures need to be taken quickly to safeguard the habitat and the populations of this animal.

Kinosternon dunni Schmidt, 1947

Distribution. This species is endemic to the department of Chocó, Colombia, in the Baudó, Decampa, San Juan, and Pepe Rivers and some of their tributaries.

Description. Named in honor of Emmett Reid Dunn, a celebrated U.S. herpetologist who resided in Colombia for many years, this species has a very small range. Males may measure 175 mm and females 150 mm. The carapace is elevated, slightly constricted around the "waist," and keelless in adults. The carapace is dark brown, marbled with reddish. The plastron is rather narrow, and has two well-developed transverse hinges. There is an anal notch. The plastron is generally light yellow with darker shadings and with dark seams and growth rings. Some individuals may have a very dark plastron. The axillary scute is in contact with the inguinal on each side. The head is large, with a bulbous and protruding snout, and bears a nasal scale so large as to envelop the whole of the skull roof. The beak is somewhat hooked. The head is brown, with several yellow spots on the sides and a light gray or yellowish zone under the neck. The limbs are webbed, with strong claws, and the tail is brown. There are several chin barbels. Both males and females possess a horny conical tail spur and opposed patches of roughened scales on the hind limbs. The young have three keels, which all disappear in adults.

Common name

Dunn's mud turtle

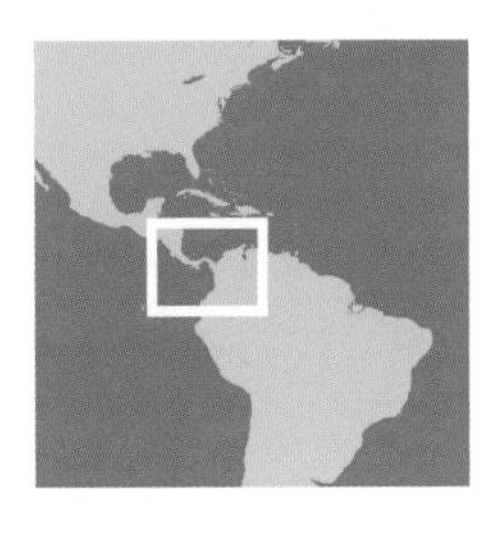

Natural history. This turtle lives in minor, tranquil watercourses with bottom-growing vegetation, which allows the animal both to bury itself in detritus and also to find food. The diet is composed mainly of aquatic mollusks. This species apparently nests year-round, and there may be several nests laid annually, each with about two shiny white elliptical eggs, measuring about 45 × 25 mm. These are the largest eggs laid by any members of the genus *Kinosternon.*

Protection. This species is listed by IUCN as rare, and studies on its current status are fundamental to any attempt to offer it better protection within its habitat.

Kinosternon flavescens (Agassiz, 1857)

Distribution. This turtle occupies a wide territory of the central and southern United States and northern Mexico, including southern Nebraska, western Kansas, western Oklahoma, western Texas, southern and eastern New Mexico, southern Arizona, and also the states of Sonora, Chihuahua, Coahuila, Nuevo León, Tamaulipas, Durango, and Veracruz.

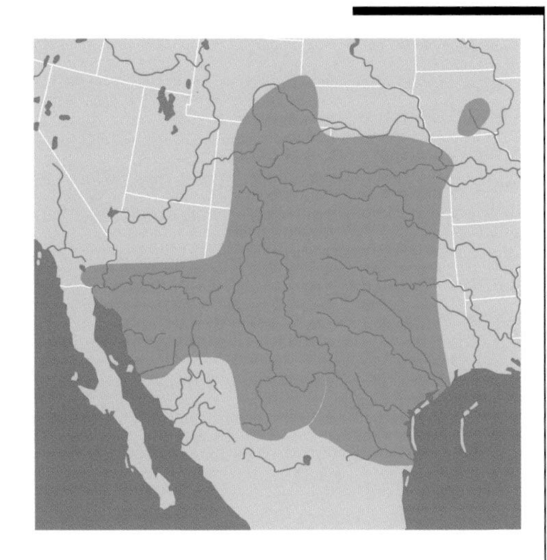

Common name

Yellow mud turtle

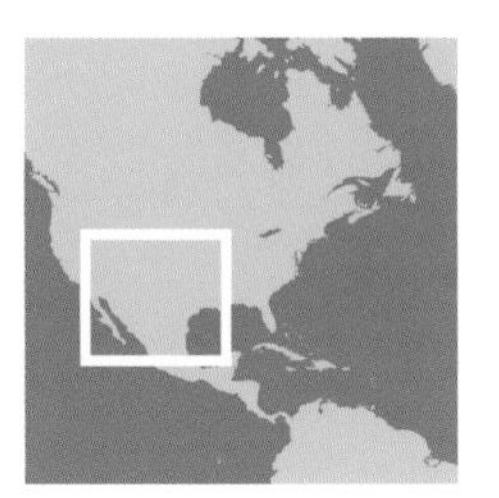

Description. The name derives from the Latin *flavesco,* meaning "tinted with yellow." This turtle is of medium size, reaching 165 mm, with a smooth, unkeeled shell, flattened on top. The first vertebral scute is very elongated and touches the second marginals. The ninth and tenth marginals exceed their neighbors in height and also are generally widened. The carapace is dark yellow, sometimes with an olive to dark brown tinge. Each scute has a fine dark border. The plastron is wide, with two well-developed transverse hinges at the front and the rear of the abdominal scutes. The gular reaches well forward and extends beyond the anterior border of the carapace. The plastron is yellow to light brown, often without spots. Only the seams and some of the growth rings may offer a somewhat darker color. There is a minor anal notch. Males have quite a strong midplastral concavity. The head is of medium size and has a slightly protruding snout. The upper jaw is slightly hooked, especially in males. The head is gray-yellow above and lighter below. A single yellow line is visible on the sides of the head and neck, accompanied by dark spots. One or two pairs of chin barbels are present. The limbs are strong, webbed, and equipped with powerful claws. Only the males have a conical spur on the end of the tail and roughened scales on the opposing sides of the hind limbs. At birth, the young have an almost circular carapace, with a slight keel. Marginals 9 to 10 are no different from their neighbors and become elevated only later on, when the carapace length passes 67 mm.

Subspecies. *K. f. flavescens* (described above): This subspecies occurs in southern New Mexico (USA) and has a shorter gular and a wider posterior plastral lobe, as well as reduced bridges. *K. f. arizonense* (Gilmore, 1922): This race is found in southern Arizona and in the state of Sonora in Mexico. It has a longer gular and a narrower posterior plastral lobe, and the bridges are wide. *K. f. durangoense* (Iverson, 1979): This subspecies occurs only in Mexico, in the states of Chihuahua, Coahuila, and Durango. The gular is long, the anterior lobe of the plastron is short, and the bridges are long.

Natural history. This turtle occurs in humid environments that vary in nature according to latitude, but it mainly frequents slow-running rivers, swamps, marshes, and cattle tanks, with muddy or sandy bottoms, often with abundant aquatic vegetation, and at altitudes of up to 2,000 m. The color of the animal often reflects that of its envi-

This turtle, photographed at Cuatro Ciénegas in Mexico, has a wide, flat shell, well rounded and yellowish to beige in color. The head is the same color as the shell, providing an excellent match with the coloration of the Mexican desert.

© F. Bonin

© J. Maran

This turtle is more colorful and darker. The snout is relatively pointed, as is true of most *Kinosternon.*

ronment. This turtle may undertake extensive overland travel. In the northern areas, it hibernates in old burrows or in heaps of dead leaves and is active from April 15 to October 15. In the south, it may estivate, buried 25 cm deep in dried mud. It has no fear of high temperatures and may be seen basking when the air temperature reaches 45°C. The diet is omnivorous and opportunistic, with a predominance of insects, crustaceans, mollusks, amphibians, carrion, and certain aquatic plants. It usually feeds underwater, but some individuals have been seen eating at the surface or even on land, a long way from water. Males mature in five to six years and females after four years. Copulation occurs primarily in the water, and males may become decidedly aggressive toward each other. Nesting occurs from May to August but in the south may occur at any time of the year. The females sometimes show a unique behavior pattern: they bury themselves with their eggs during times of extreme drought, for up to 40 days or so, possibly dampening the soil with their urine. From one to six eggs are laid in each nest, often just once per season. These eggs, which are hard-shelled and white, are elliptical and measure 26 × 15 mm. Fearful and timid, this species never seeks to bite the hand that holds it.

Protection. Some populations of this species are seriously threatened by local urban development and habitat alteration. In the United States it is listed on the endangered species inventories of Illinois, Iowa, and Missouri. It is often run over on highways and is subject to predation by water snakes, raccoons, skunks, and eagles.

Kinosternon herrerai Stejneger, 1925

Distribution. This species occupies a somewhat limited area in the Gulf lowlands of Mexico, in the states of Tamaulipas, Veracruz, San Luis Potosí, Hidalgo, and Puebla, from the basin of the Rio Tamesi in the north to the Rio Actopan in the south.

Description. This species bears the name of Alfonso L. Herrera, former director of the National Museum of Mexico. It is a rather large species, with the males reaching 170 mm and the females 150 mm. The carapace is somewhat domed and is posteriorly expanded. A weak median keel is present in young adults but disappears with age. The first vertebral scute is narrow and does not make contact with the second marginals. The tenth and eleventh marginals are higher than the ninth. The carapace is brown-olive, with darker seams. The plastron is reduced and is too small to close the shell openings completely. There is a small anal notch. The anterior lobe of the plastron is longer than the posterior lobe in males, but the reverse is the case with females. There are two plastral hinges, but only the anterior one is significantly flexible. The plastron and the bridges are dark yellow to a more or less light brown, with darker seams. The axillary always contacts the inguinal on each side. The head is wide and has a prominent snout, capped with a large nasal scale. The upper jaw is strongly hooked. The skin of the head is gray-brown, with dark reticulations that become lighter on the sides. The throat is unmarked light gray, and two pairs of chin barbels are present, the first being well developed and the second smaller. The jaws are very light cream, with dark streaks. The skin of the neck and limbs is rather dark gray-brown. The males have a horny

Herrera's mud turtle

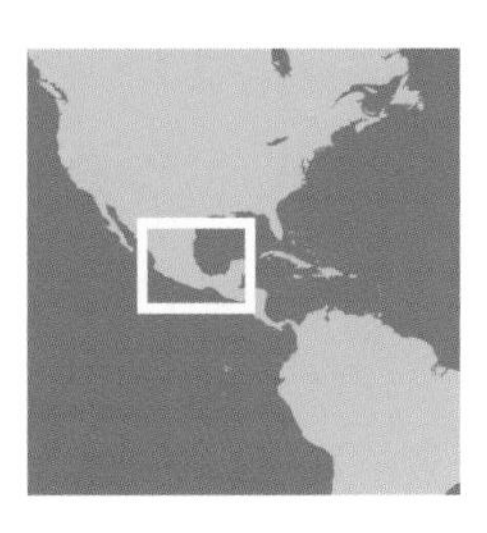

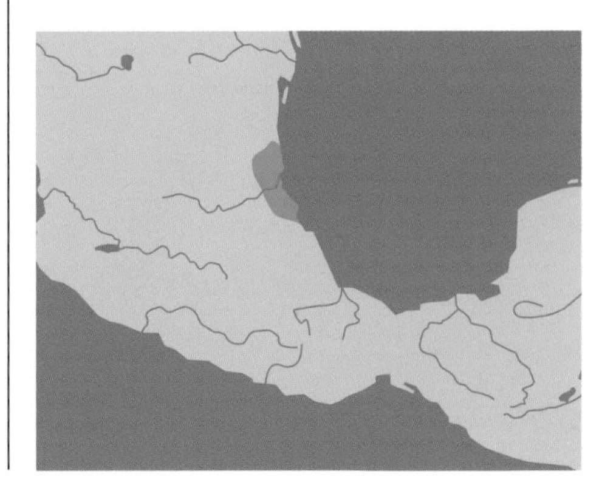

spur on the end of the tail and opposed patches of rough scales on the thighs. The young have three indistinct keels, the lateral pair disappearing quickly with growth.

Natural history. The species is still poorly known. It inhabits slow-flowing rivers and seasonal pools, at altitudes of up to 800 meters. These turtles are frequently seen on land and seem to be less aquatic than other *Kinosternon* species. The diet is mainly carnivorous.

Protection. Poorly studied, this species is at risk of disappearing through general apathy. In 1979, P. Pritchard reported that, even then, stuffed specimens were offered for sale in some markets in Tampico. It is probable that populations are diminishing. Studies of the status of the species are necessary.

© P. Pritchard

The protections provided by a strong head and a kinosternid shell are apparent in this photograph.

Kinosternon hirtipes Wagler, 1830

Distribution. In the USA this species occurs only in the border country of southwestern Texas (Presidio County) and in a narrow Pacific range in Mexico, some distance from the sea, in the basins of the Santamaría, Carmen y Conchos, Chapala, Zapotlan, San Juanico, and Patzcuaro Rivers and in the Valley of Mexico.

Description. The Latin name *hirtus pes* means "rough-footed," in reference to the rough scales on the hind limbs of this species. This turtle is of medium size, the males reaching 185 mm and the females 160 mm. The carapace has an obvious median keel, with lesser ones on each side. Aged individuals may become completely smooth. The tenth marginal is higher than the others, although in certain individuals the eleventh is equally high. The carapace is light brown to almost black, with dark seams on the lightest animals. The plastron has two transverse hinges and is narrow and short, not capable of closing the shell openings completely. The axillaries are always in contact with the inguinals. The plastron and the bridges are yellow, more or less dark, sometimes spotted with red, brown, or black. There is a minor anal notch. The head is dark brown with light reticulated markings, but certain individuals have a lighter coloration with dark patterns. The numbers of chin barbels varies according to subspecies. Both sexes have a horny spur on the end of the tail, but the patches of roughened scales on the thighs are present only in the males. The hatchlings have a brown carapace and a red or orange plastron, with a dark marking in the center.

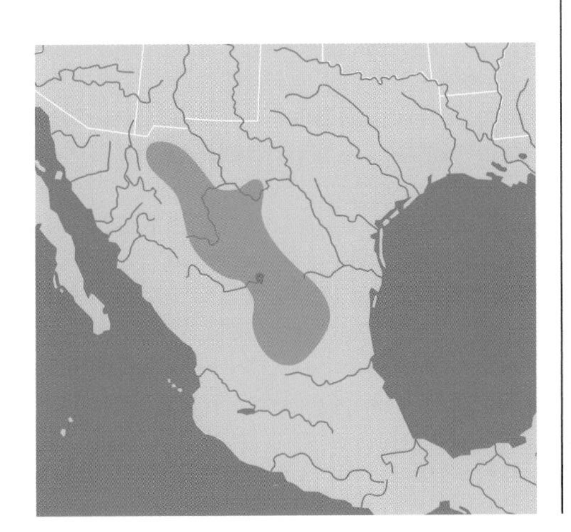

Common name

Mexican rough-footed mud turtle

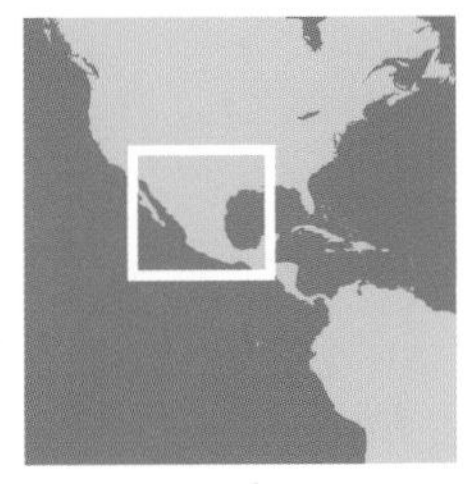

This species has a strong median keel, and its blackish head is marked with light areas. The various subspecies, often vicariant in nature, are mostly geographic varieties.

Subspecies. Six subspecies have been described, some with very limited distribution.

K. h. hirtipes (described above): The Mexican mud turtle occurs in the south, in the Valley of Mexico. It does not exceed 140 mm. The bridge is narrow, and the head is marked with a light line that extends backward from the corner of the mouth. There may be one or two pairs of chin barbels.

K. h. chapalaense (Iverson, 1981): The Chapala mud turtle is found only in the basins of the Chapala and Zapotlan Rivers (states of Jalisco and Michoacán). The gular scute is short, and the pigmentation of the head is light, with some dark markings. Two dark postorbital lines mark the sides of the head. There may be one to three pairs of chin barbels. Maximum size is 152 mm.

K. h. magdalense (Iverson, 1981): The San Juanico mud turtle is named after the valley of the Rio Magdalena in Michoacán. It does not exceed 94 mm. The gular is very short, and there are two pairs of chin barbels.

K. h. megacephalum (Iverson, 1981): The Viesca mud turtle is known only from two localities in southwestern Coahuila, near the village of Viesca. The race is almost extinct, because the aquatic habitats where it lived have been totally altered by human action. The females measure up to 117 mm, and the head is very wide, with strongly developed jaws. This turtle has a short gular scute, a small plastron, and three or four pairs of barbels.

K. h. murrayi (Glass and Hartweg, 1951): The Mexican Plateau mud turtle bears the name of Leo T. Murray, from Texas A&M, a famous college. It occurs on the frontier between Texas and the state of Chihuahua in Mexico and extends as far as the state of Mexico. The bridges are very long, and males reach 182 mm. A long gular scute is present, as well as two pairs of well-developed barbels.

K. h. tarascense (Iverson, 1981): The Patzcuaro mud turtle occurs only in the Río Patzcuaro (Michoacán). It does not exceed 136 mm. The bridges are narrow, and there are fine vermiculations on the head. It has two pairs of chin barbels.

Natural history. These turtles are generally nocturnal but may sometimes be seen during the day. They occupy water holes in primarily arid regions, as well as temporary swamps and small rivers draining into lakes, up to an altitude of 2,000 m. This species is very aquatic in relation to other *Kinosternon.* The diet is carnivorous and includes various insects, worms, and other invertebrates. There are two to four nestings per season, which (at least in Mexico) runs from May to August, and each includes up to seven eggs. The eggs are hard-shelled, brilliant white, and elliptical, measuring about 30 × 17 mm. The incubation time averages about 200 days.

Protection. The status of this species is poorly known. Certain populations seem to have become very rare, as is the case with *K. h. megacephalum.* Some of the habitats have been seriously disturbed by human action. Specific studies on each subspecies are necessary to evaluate the status of the species as a whole.

Kinosternon integrum (Le Conte, 1854)

Distribution. This species has a wide range along the Pacific side of Mexico, from the state of Sonora in the north, through Durango, Nuevo León, and Guerrero, to the state of Oaxaca. Iverson (1998) thinks that the species may have been introduced by humans to the Valley of Mexico. The same thing may have happened on the Tres Marías Islands, located in the Pacific 80 km from the port of Tepic, where the species also occurs.

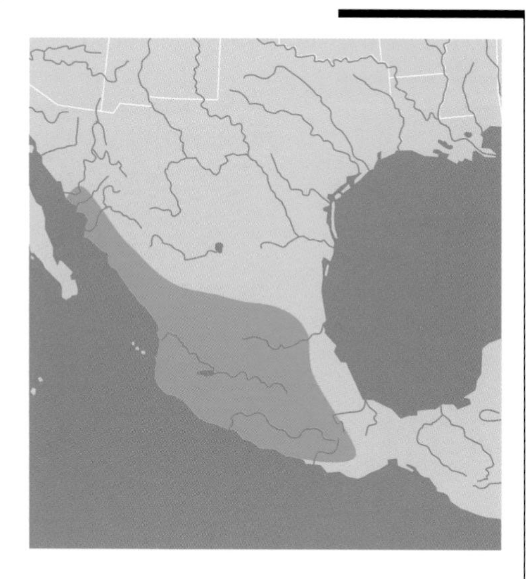

Mexican mud turtle

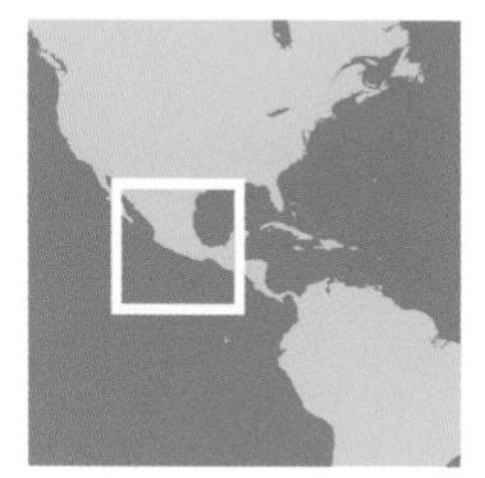

Description. The name comes from the Latin *integra,* in reference to the plastron, which closes the shell openings in a complete, or integral, fashion. This is a rather large species, which may reach 210 mm in males and 195 mm in females. The carapace is oval, a little flattened on top, with three parallel keels that become slightly more elevated toward the rear. The first vertebral scute is in contact with the second marginals. The color of the shell is variable and often depends upon the environment in which the animal has lived, but it ranges from olive to light beige to almost black. The whole shell shows dark marbling. The plastron, dark in color, closes the carapace completely. The two hinges are fully functional, and the color, like that of the bridges, is orange-yellow with brown seams. The inguinal scutes do not touch the axillaries. There is a limited anal notch. The head is wide, with a protruding snout and a hooked beak. There are three (or sometimes more) pairs of chin barbels, but only the first ones are conspicuous. The head is dark brown above and becomes lighter on the sides, with yellow-gray below. The whole head bears light reticulations. The jaws are yellow, streaked with black. The neck is smooth, without tubercles. This turtle lacks rough scales on the thighs, as are present in many other members of the genus. Both sexes have a horny spur on the end of the tail. The young are quite similar to the adults and have three distinct keels.

Natural history. This species leads a very aquatic existence and may sometimes be spotted in small groups in sluggish streams, deep ponds, and roadside ditches. It is mostly nocturnal but may walk about on highways during the daytime. The activity period extends from May to October. The diet is entirely carnivorous. Nesting occurs in late spring, and the elongated eggs are white and shiny and measure 30 × 16 mm. The young hatch from the end of July to September.

Protection. This species is abundant in certain rivers in Sonora, where the habitat has not been too disturbed by humans.

The Mexican mud turtle often has mottling on the head and a nose that appears to protrude. The brownish carapace is flattened above the vertebrae.

© F. Bonin

Kinosternon leucostomum

Duméril and Bibron, 1851

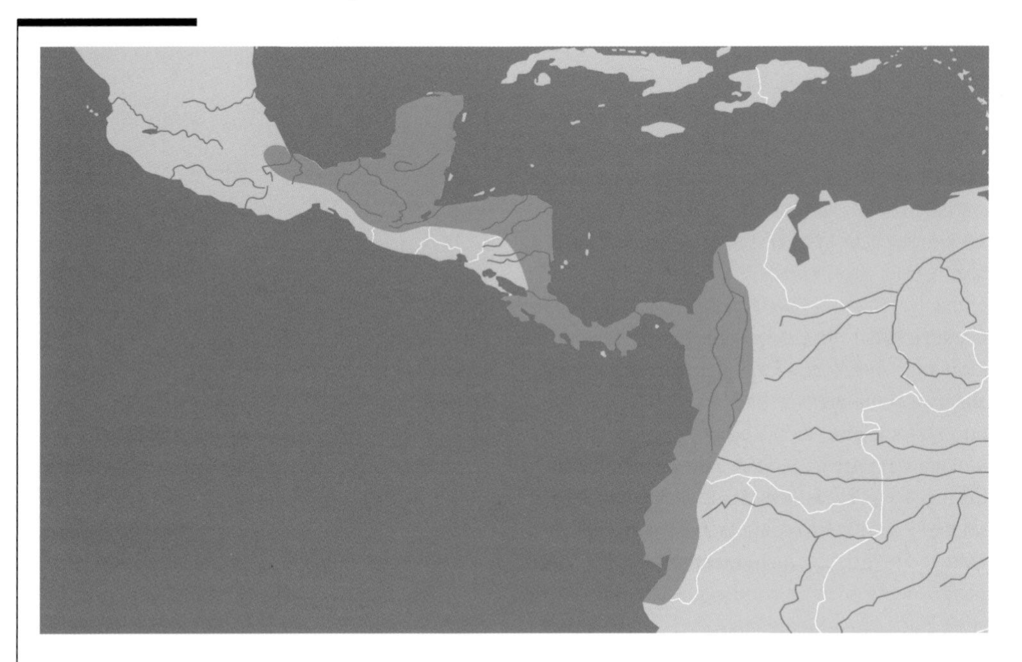

Common name

White-lipped mud turtle

Distribution. This species occupies a very wide range from Mexico to Ecuador. In the north, it occurs in the rivers that drain into the Gulf of Mexico, in all the watercourses along the Caribbean coast, and, beyond Panama, in the Pacific drainage basins as far as northwestern Ecuador. The presence of this species in northwestern Peru has been questioned (Carr and Almandariz, 1990).

Description. The name derives from the Greek *leukos* (white) and *stoma* (mouth), in reference to the white borders of the jaws. This is a medium-sized species, the males attaining 175 mm and the females 165 mm. The carapace is oval, with a single keel that may disappear with age. The first vertebral scute always touches the second marginals. The eleventh marginal is almost always higher than the tenth, which in turn is higher than the ninth. The posterior marginals are sometimes broadened. The carapace ranges from yellow to brown and sometimes even black. In the lighter individuals, the seams are rather dark. The plastron is large enough to close the shell openings completely and is concave, and slightly smaller, in the males. Two transverse hinges are present, the anterior one being more developed than the posterior one. The plastron is yellow, rather light but with darker seams. The anal notch is barely detectable. The head is of medium size and more or less dark brown, with a light cream-colored line from the orbit to the base of the neck. This line is clearly visible in the young but is less distinct in the adults. The edges of the jaws are cream to whitish, finely streaked with black vertical lines, especially in old males. Elsewhere the skin is gray-brown, often with scattered dark dots. Two or three pairs of chin barbels are present. Patches of rough scales are present on the opposed surfaces of the hind limbs, and the males have a strong horny claw at the tip on the tail, which is also present but less developed in females. The limbs are strong and have well-developed claws and digital webs.

Subspecies. Two subspecies are recognized. *K. l. leucostomum* (described above): The northern white-lipped mud turtle occurs from southern Mexico to northern Nicaragua. The shell is slightly flattened, and the gular scute is rather long. The plastron is large, and the inguinal is long but almost never touching the axillary. The thigh scales are feebly developed. The head coloration is variable, and the light postorbital line is often absent. *K. l. postinguinale* (Cope, 1887): The southern white-lipped mud turtle occupies the southern section of the range, from Nicaragua to Ecuador. The carapace is more flattened than in the nominal race, and the gular scute is smaller. The plastron is also somewhat reduced, and the inguinal scute small, almost never touching the axillary. Males may develop rugosity of the scutes. The light postorbital line is well developed.

Natural history. This turtle enjoys calm, muddy-bottomed waters, including lakes, swamps, and lagoons with dense aquatic vegetation, in forested regions at low and medium altitude. It often ventures onto land and is not as aquatic as some of the other *Kinosternon* species. It even tolerates saline water (Medem, 1962). It is primarily nocturnal, spending its days underwater, hidden among the rocks. The diet is omnivorous, including mollusks, worms, insects, and carrion, along with grass and aquatic vegetation. It reproduces almost year-round, except for the month of January. The greatest clutch size is only five, and the eggs are hard-shelled and elongated, measuring 37 × 20 mm on average. The nest is often quite rudimentary, sometimes at ground level with dead leaves covering the eggs. Incubation may last for 126 to 148 days.

Protection. This species is regularly consumed by local people, despite the presence of larger and fatter turtles of preferred species. Nevertheless, it still seems to be abundant.

© J. Maran

The head of this turtle has an angular shape that is quite characteristic, as well as a pronounced beak. The two white bands that border the mouth give the species its scientific name. The carapace ranges from yellowish to black, according to age and individual.

Kinosternon minor (Agassiz, 1857)

Distribution. Found only in the USA, this species occurs in southwestern Virginia, eastern Tennessee, eastern Georgia, and a far as central Florida, Alabama, Mississippi, and the extreme south of Louisiana. In in the Black Warrior River Basin the center of Alabama, it is replaced by *K. depressum.*

Description. The Latin word *minor* means "smaller," and this is indeed a rather small species, with large females reaching about 135 mm. The carapace angle is less than 100 degrees, and the shell forms a pronounced dome. The vertebral keel is strong, the lateral keels less so, and all keels may disappear with age. The first vertebral scute never touches the second marginals. Each vertebral scute overlaps its posterior neighbor, giving an imbricated appearance to the shell as a whole. The last four vertebral scutes are distinctly wider than long. The posterior marginals are lightly serrated, the tenth and eleventh being higher than their neighbors. The carapace is brown, with light to dark orange tones. The seams are dark, each scute having dark or even black streaks. The streaks are often radiating, but they fade with age. The plastron is reduced, pink to yellowish in color, unspotted, and equipped with a hinge between the

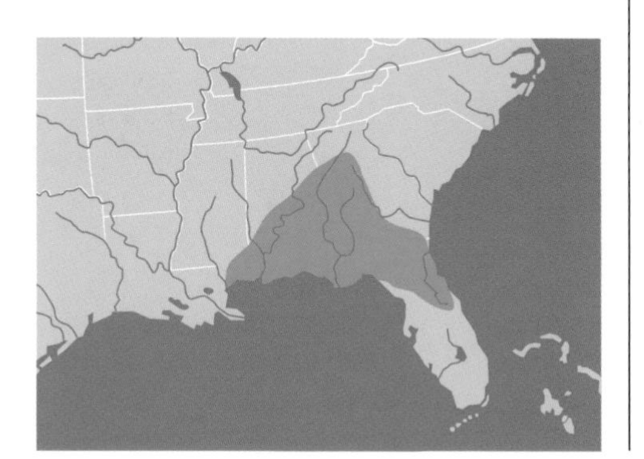

Common name

Loggerhead musk turtle

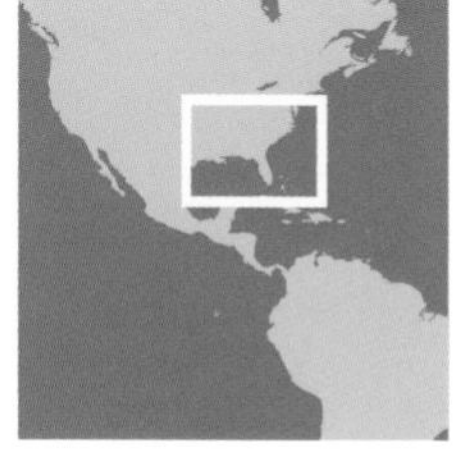

pectorals and abdominals that is sometimes poorly developed. There is a single gular scute and a distinct anal notch. The axillaries are in contact with the inguinals. The head is covered with gray to pinkish skin, dotted with dark spots, and the pattern continues on the neck and the upper surface of the limbs. Some individuals may be more orange, with the skin of the neck streaked with black. The head is rather wide, and the snout is protuberant and tubular or conical in shape. The nasal scale is large. There is a single pair of chin barbels. The males have a spur at the end of the tail and roughened scales on the thighs.

Subspecies. Two subspecies are recognized. *K. m. minor* (described above): This race lacks dark lines on the head and neck. The young have three keels on the carapace, and the plastron is pale pink. This subspecies occurs in Georgia to Alabama and southward to central Florida. *K. m. peltifer* (Smith and Glass, 1947): The stripe-necked musk turtle has a dark line on the neck, which is wide and distinct, and a vertebral keel that disappears with age. The young have an orange-yellow plastron. This subspecies occurs in southwestern Virginia, eastern Tennessee, and Alabama to as far as southern Mississippi. Some authorities consider this to be a full species.

Natural history. This turtle may be found in rivers, streams, swamps, cattle tanks, and shallow marshes with muddy or clay bottoms. It is highly aquatic, but it may often be seen on branches and other precarious basking platforms, and it can climb remarkably well to reach an ideal sunning spot. In the water, it walks about on the bottom in search of food. Buccopharyngeal respiration allows it to remain immersed for long periods of time, utilizing the oxygen dissolved in the water. It is active by night and by day but is most often found in the mornings. It hibernates in the northern parts of its range from December to February, but in the south it maintains activity year-round. In diet it is strongly carnivorous (mollusks, insects, crayfish, other invertebrates, carrion), but it also consumes algae and other aquatic plants. The subspecies *peltifer* is thought to be less insectivorous than the nominate race but consumes large amounts of mollusks and snails, as is demonstrated by its expanded jaw surfaces. This is a mean-spirited turtle that always tries to bite. It may also excrete a nauseating musk, and it has been reported that this material may be secreted even while the embryonic turtle is still in the egg. Males reach maturity after four years; the females, after eight. Copulation occurs from April to November. There may be two to five nestings per season, with a maximum clutch size of five. The nest is quite shallow or may even be at ground level, at the foot of a tree. The eggs are small and elongated, about 28 × 15 mm in size, white, and slightly translucent initially, later becoming opaque. Incubation lasts for 61 to 199 days.

Protection. The nests are raided by numerous mammals and also by birds. The adults may fall prey to alligators, but it is the destruction of habitats that is the worst threat to this species. Water pollution, which contaminates or kills mollusks, has brought about drastic reductions of this turtle's populations since 1942, when Marchand could find 500 specimens in one day in Florida. Today the species also suffers from drying of its habitats, especially in Florida.

© F. Bonin

This young *K. minor* will not exceed a length of 135 mm. It has a well-defined keel, a large strong head with extensive spotted decoration, and lively, curious white eyes, which become dull with age. The habitat includes lakes and marshes of the southeastern United States.

Kinosternon oaxacae Berry and Iverson, 1980

Distribution. So far, this species has been found only in certain rivers in the state of Oaxaca, on the Pacific coast of Mexico, but it could possibly also exist in certain streams in the neighboring state of Guerrero.

© C. Guerrero

Description. The name is a toponym, referring to the state of Oaxaca. This turtle may reach 175 mm in males and just 140 mm in females. The carapace is flattened on top and has three strong keels. The color ranges from brown to black, with dark marbling and seams outlined with black. The scutes are slightly imbricated. The first vertebral scute is in contact with the second marginals. Only the tenth marginals extend higher than the remainder of the marginal series. The plastron is rather small and cannot close the shell openings completely. It is double-hinged, both hinges being fully functional, and there is an anal notch. The axillary and inguinal scutes are in contact. The plastron is yellow, with brown seams, and the scutes sometimes have dark infuscations. The head is large, dark brown to black above and becoming cream or yellow on the underside, and has some dark vermiculations. The beak is strongly hooked. The nasal scale is large and has the shape of a V. There are three or four pairs of chin barbels. The skin of the limbs and of the tail is dark gray to black above and cream below. The males have darker coloration under the neck and have a spur on the end of the tail, but there are no roughened scales on the thighs.

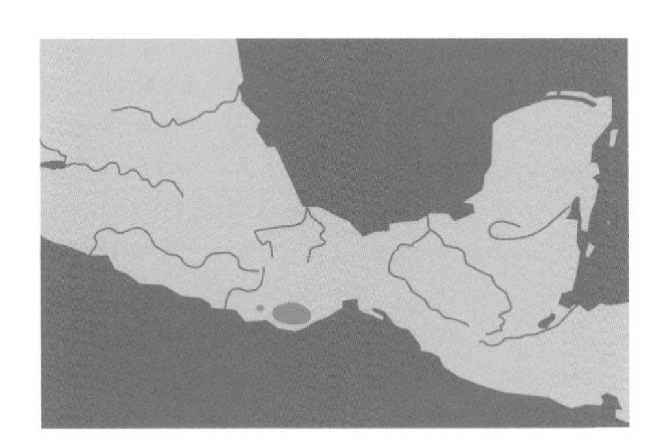

Natural history. This turtle occurs in permanent waters draining from the Sierra Madre del Sur but is also often found in the coastal plain, where it occupies a variety of aquatic habitats. Nothing of substance has been learned about its biology, and studies should be undertaken to determine its status and behavior, especially since it is a species with a very restricted range.

This *Kinosternon* is small and very flat, and the head is rather large. The plastron is often yellow, but it can be dark or almost black, according to the individual.

Common name

Oaxaca mud turtle

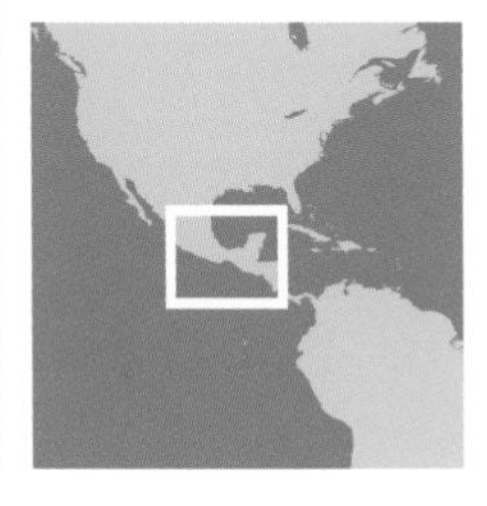

Kinosternon odoratum (Latreille, 1801)

Distribution. This turtle is one of the most widespread members of the family, with a range extending from southern Canada (Quebec, Ontario) to the eastern half of the USA, from New England to the tip of Florida. There may also be isolated populations in southern Kansas, west Texas, and even the state of Chihuahua, Mexico.

Description. The name refers to the foul-smelling glandular exudate that this species sometimes produces. The carapace does not exceed 136 mm, and it is rather steeply domed or tectiform. The shell is elongate and rather narrow, sometimes slightly flattened on top. Younger animals show traces of three keels, which disappear

© J. Maran

This turtle looks at us with curiosity and appears not too friendly. The head is pointed, with an obvious snout. The carapace is grayish to blackish and is flattened along the vertebral scutes.

Common name

Common musk turtle, stinkpot

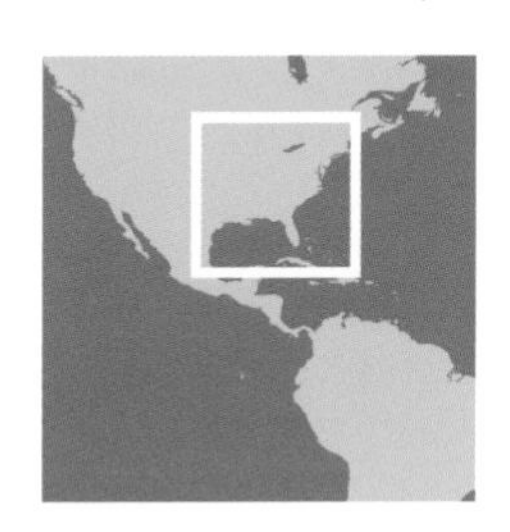

with age. The scutes are not imbricate, as they are in other musk turtles. The first vertebral is long and does not touch the second marginals. There are 23 marginals (including the nuchal scute). Only the tenth and eleventh on each side are higher than their neighbors, and the marginal edges are not serrated. The carapace ranges from olive-gray to black and is sometimes spotted or streaked in the younger and lighter individuals. The plastron is reduced and has a single, slightly functional hinge. Wide bands of soft skin are visible along the plastral seams. The bridges are extensive and are made up of both the axillary and inguinal scutes and extensions of the abdominals. A single, well-developed gular scute is present, and there is an anal notch. The plastron is more or less dark yellow, sometimes with black spots. The musk glands are located near the border of the carapace. The head is of medium size, and the snout is conical and somewhat projecting. The skin is a rather dark gray and may actually become black. There are two light lines, yellow or white in color, along each side of the head, extending from the point of the snout, passing above and below the orbit, and reaching to the base of the neck. In some individuals these lines may become broken or even disappear to some degree. The beak is lightly hooked, and the jaws are smooth. There are one or two pairs of chin barbels. The skin of the neck, the limbs, and the tail is darker above than below. The limbs are well webbed and clawed. Males have a spur on the end of the tail and two small areas of roughened scales on the opposing flanks of the thighs. The young have a blackish, rugose carapace with a high vertebral keel and less distinct lateral keels. Each marginal is decorated with a light spot, and the plastron is black, with the hinge essentially undetectable. The overall latitudinal range of the species is great, and there are some morphological differences between the populations.

Natural history. The species is found in a wide range of aquatic situations—rivers, lakes, bayous, swamps, marshes, cattle tanks, canals, and ditches—but they may also be found in fast-flowing creeks with rocky bottoms. The diet is omnivorous, the young eating small insects, algae, and carrion, while among the adults the diet is more varied and also includes vegetable debris, snails, fish eggs, and amphibians. The turtle may bask on a tree limb or on floating wood, but it also likes to float at the surface or rest on aquatic vegetation. To defend itself, it relies upon its nauseating musk, secreted by glands under the plastron; but it is also quite aggressive and may bite anyone who bothers it. In the north, hibernation occurs, and sometimes dozens of turtles may be found together in retreats and cavities. Mating takes place from mid-March to May, but this species can also reproduce in September, and as late as December, in the south. Courtship occurs in the water, but some coupling may also be seen occurring on land. Three successive nestings have been described, in May to June in the north but from February to September in the south. The eggs may be laid on the ground or perhaps hidden in vegetative litter. The clutches number about five eggs on average but may have as many as nine. The eggs are elliptical, thick-shelled, and white, measuring about 28 × 15 mm.

Protection. This small turtle is not rare, and as many as 194 individuals per hectare have been recorded. The eggs and the juveniles are eaten by mammals, snakes, birds, and even bullfrogs. But the species suffers mostly from human actions: boat propeller impacts and the pollution, drying, or modification of aquatic habitats.

Kinosternon scorpioides (Linnaeus, 1766)

Distribution. The six subspecies share a huge range, from southern Tamaulipas, Mexico, to northern Argentina, passing through Central America and a substantial part of Brazil, the Guianas, Ecuador, Peru, and Bolivia.

Description. For the name of this turtle, Linnaeus was inspired by the Latin *scorpio,* which refers to "the stinger on the tail"—actually just a conical spur on the tip of the tail. This is a turtle of medium size, males usually about 200 mm and

© J. Maran

This specimen is quite light-colored, with a yellowish head and an unkeeled shell. This appears to be a very old individual. The species is generally a brownish color, and the head is darker, with light reticulations. But the jaws are often cream or yellowish, with dark streaks.

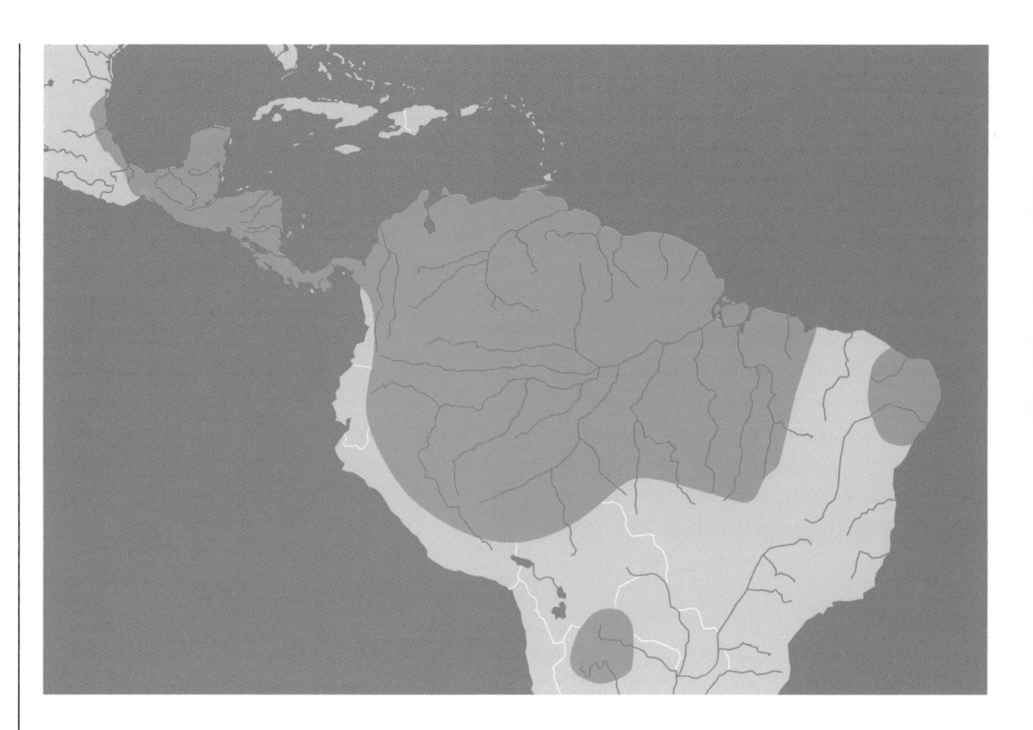

Common name

Scorpion mud turtle

females about 180 mm, although exceptional males of certain subspecies may reach 270 mm. The carapace bears three keels, developed to varying degrees and disappearing in old animals. The first four vertebral scutes each have a notch in the middle of their posterior edge, and they may also be lightly imbricate. The oval carapace is rather domed, and the posterior marginals may be widened. The tenth marginal is higher than the ninth and the eleventh (in most cases, but not always). The overall color ranges from beige to brown, or even olive or black, with midnight blue reflections and dark seams in the lighter-colored animals. The males often have a carapace that is most strikingly elevated anteriorly. The plastron has two wide, fully functional hinges, in front of and behind the abdominals, which allow for almost complete closure of the shell openings. The anal notch is minor or lacking. The plastron varies from gray to yellow but may also be orange, brown, or black. When the colors are vivid, one sometimes also sees irregular dark spots. The head is of medium size and provided with a triangular nasal scale and a hooked horny beak in the shape of a crescent. The coloration of the head is quite variable and depends upon the subspecies or population (as does the overall morphology of the species); its background may be brown, gray, or blackish, often marbled or spotted with reticular cream areas, which may in turn be yellow, orange, pink, or even blood red. The jaws are usually yellowish, with dark vertical streaks. The remainder of the skin (on neck, limbs, and tail) is brown, gray, or olive, with swirls of dark spots. The underside is lighter. There are three or four pairs of barbels on the chin. Both sexes have a spur on the tip of the tail, more developed in the males, but they are lacking the patches of roughened scales on the hind limbs.

Subspecies. Six subspecies are currently recognized.

K. s. scorpioides (described above): The Amazonian mud turtle occurs south of Panama, and the three keels are strongly developed. The nuchal scute is narrow, and the head rather small. The plastron is narrow in front, and the anal notch is of reasonable size.

K. s. abaxillare (Baur, in Stejneger, 1925): The Chiapas mud turtle, whose name refers to the absence of the axillary scutes, occupies the central plateau of Chiapas in Mexico and reaches the Pacific coast at Petraparada. In addition to the feature of the missing axillary scutes, the carapace is often almost black. The plastron is dark brown, lightened with reddish or orange, and it is large, completely closing the carapace openings.

K. s. albogulare (Bocourt, 1870): The white-throated mud turtle, named from the Latin *gulares* and *albus* (white-throated), is found from Panama to Honduras. The carapace is dorsally flattened but still has three keels. There is no anal notch in the plastron. The lower jaw and chin are very pale—cream or white.

K. s. carajasensis (Cunha, 1970): Carajás mud turtle) is only known from the Sierra dos Carajás—hence the name—located in the state of Para, Brazil. It is small (130 mm), and the shell is pointed and has a strong vertebral keel, while the lateral keels are indistinct or lacking. The head is large and wide.

K. s. cruentatum (Duméril and Bibron, 1851): The red-cheeked mud turtle gets it scientific name from the Latin for "blood-spotted," in reference to the bright orange or red color of the head spots, especially in the males. This subspecies ranges from northeastern Nicaragua and Honduras to southern Mexico (Veracruz and Tamaulipas). It is the most vividly colored subspecies, in that the sides of the head have bright red spots. The carapace is often light-colored, yellow or orangish, with darker seams. The plastron is orange and closes the shell openings completely. Three keels are visible, but they disappear with age.

K. s. seriei (Freiberg, 1936): The Argentine mud turtle is found in northern Argentina and in Bolivia. The head is quite small. The carapace has three keels, of which the vertebral one is the best developed. There is an area of exposed fibrous tissue between the abdominals and the femorals.

Natural history. This species is found in flooded forests, drainage canals, streams, and all kinds of humid environments with stagnant water and mud. When the wet areas dry up, the turtles bury themselves in the mud and await whatever happens next. Opportunistic in diet, they eat fish, snails, and plants, as well as the fallen fruit of various species of palms. They move mainly at dusk, but they may be seen by day as well as by night, in the course of their search for food. Bold and fearless, they may try to bite if molested. The males are particularly aggressive among themselves, dealing serious wounds to one another in the course of their combat. According to latitude, the mating season and the laying time vary greatly; in Mexico, for example, it extends from March to May, whereas in Guyana it has been observed in September. Each year there are several nestings, with three to six eggs laid each time. They are elongate and white, with fragile shells, and measure about 40 × 18 mm. Incubation lasts for 90 to 150 days, depending on the site and latitude.

The young have a strongly sloped shell. The beak is always well developed, and the eye incisive.

Protection. The species is still very widespread, but certain parts of the range have been extremely disturbed by humans. Deforestation is particularly harmful to this species. In French Guiana, these turtles may be encountered in the Saint-Laurent-du-Maroni region and along the coast to Cayenne, in the swamps of Kaw, and as far inland as Saint-Georges. Fretey (1987) also speaks of populations in the upper Oyapoque River. Because this territory is an overseas department of France, the species is completely protected there.

Kinosternon sonoriense (Le Conte, 1854)

Distribution. This turtle is found in the lower Colorado River in Arizona, California, and New Mexico in the USA and into northern Mexico as far as the basin of the Yaquí River and northwestern Chihuahua.

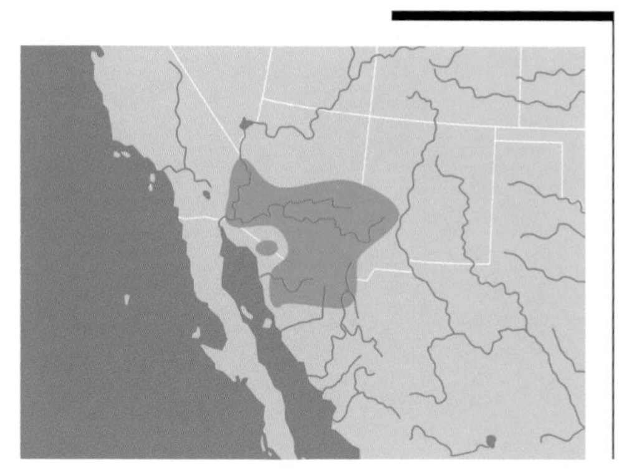

Common name

Sonora mud turtle

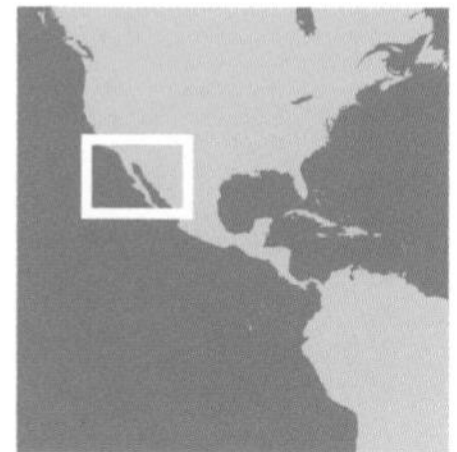

Description. The name makes reference to the Sonora region, near Tucson, in the USA, which is the type locality. This turtle may attain 175 mm in females and 155 mm in males. It has a rather elongated carapace with three keels, the vertebral one being the most developed. Some adults have just one keel, and others none. Usually, the first vertebral scute touches the second marginals. The ninth marginal is not elevated, although the tenth and eleventh ones are. The posterior marginals are somewhat widened. The inguinal is in contact with the eighth marginal, as well as with the axillary. The carapace is brown or sometimes olive, with dark sutures but with no particular pattern. The plastron has two fully functional hinges, one before and one behind the abdominal scutes, and it is almost immaculate, more or less light yellow in color, with darker sutures. By contrast, the bridges are dark brown. The skin of the head and neck is grayish, marbled with cream. There are two light lines on each side of the head. The first extends from the orbit to the tympanum, and the second commences at the corner of the mouth and extends to the neck. The head, with its prominent snout, is of medium size. Three or four pairs of chin barbels are present. The jaws are cream-colored, with black streaks. The beak forms a pronounced hook. The limbs are moderately webbed and equipped with strong claws. Only the males have the conical spur on the tip of the tail and roughed patches of scales on the thighs. The young have an almost circular carapace, with a strong median keel and two less conspicuous lateral keels. In color they are brown, with each marginal bearing a black ocellus. The plastron is cream, with a dark star-shaped figure running the length of the seams. The young also have two clearly visible light yellow lines on each side of the head.

The shell is elongated, keeled, and almost black. The head is fine, with the snout somewhat projecting and marbled with white. Very fearful, this turtle does not like to emerge from its shell even when molested.

© F. Bonin

Subspecies. Two subspecies exist. *K. s. sonoriense* (described above): The common Sonora mud turtle occurs in New Mexico, Arizona, California, and western Chihuahua. *K. s. longifemorale* (Iverson, 1981): The Sonoyta mud turtle) occurs in Arizona and in the Mexican state of Sonora. In this subspecies, the seam between the femoral scutes is longer than in the nominal subspecies, and the seam between the anal scutes is shorter.

Natural history. This turtle is active almost all year round, although in Arizona certain populations living at 2,000 m altitude will hibernate. The species is mostly diurnal during the winter. It is found in slow streams, especially in a forested milieu. Although thoroughly aquatic, it may wander far from its natural habitat in search of a new site. It bottom-walks in search of food, with its neck stretched out, seemingly using vision as well as olfaction. The diet is opportunistic, consisting of snails, insects, worms, amphibians, and fish, as well as algae and certain aquatic plant species. It rarely basks except in winter. Shy and fearful, it will retract its extremities for a long time when disturbed. Copulation occurs in spring, and nesting takes place from May to September. Two nests may be laid annually, with up to nine eggs in each. The eggs, which are elliptical and 32 × 16 mm on average, are deposited in shallow nests.. The eggshell is white, brittle, and slightly granular, giving the appearance of a multitude of small pores. Incubation may take as long as 345 days.

Protection. The introduced bullfrog preys upon the young turtles, as do mammals and birds. The adults of this species were consumed by the Indians in former times, but the current threats pertain to habitat destruction, especially the new methods of irrigation, which destroy the habitats of aquatic turtles.

Kinosternon subrubrum (Lacepède, 1788)

Distribution. This species has a wide range in the southeastern quadrant of the USA, from Connecticut to Florida and including central Texas, the Mississippi Valley, Missouri, southern Illinois, and southern Indiana. There are isolated populations in northwestern Indiana and central Missouri.

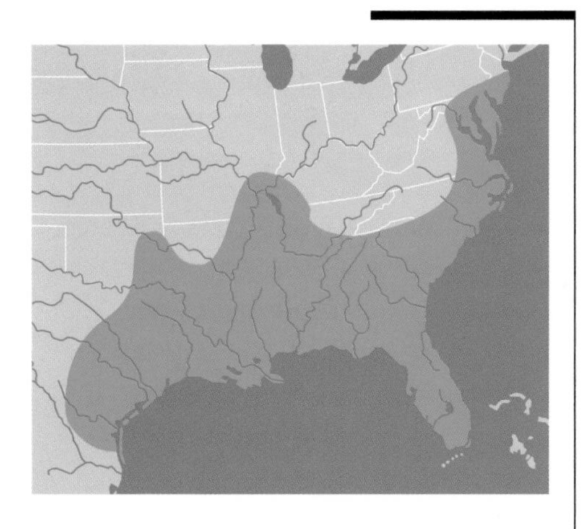

Common name

Eastern mud turtle

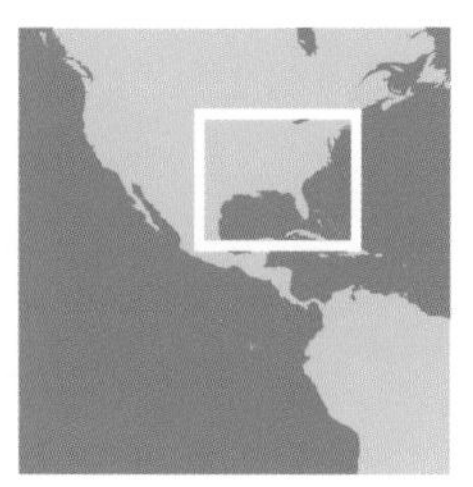

Description. The name derives from the Latin *sub* (beneath) and *ruber* (red), in reference to the orange-red plastron of the young. This is a small species, not exceeding 125 mm. The carapace is oval, smooth, unkeeled, and sometimes somewhat flattened middorsally. The sides of the shell are almost vertical, the hind part appearing truncated. The marginals, none of which is higher than the overall series, are not serrated. The general coloration ranges from dark yellow to olive or even black. The edge of the shell is sometimes somewhat lighter. The first vertebral scute is narrow and long and never makes contact with the second marginals. The plastron is rather wide, and the two hinges are quite mobile, although they may stiffen in old individuals. There is a slight anal notch. The plastron is yellow, with brown or dark brown seams. The bridges are strong, and the inguinal makes contact with the axillary. The head is of medium size and has a somewhat protuberant snout. The upper jaw has a slight hook. The skin of the head is grayish or brown, marbled with yellow. In some specimens, two light stripes (sometimes discontinuous) run from the orbit to the tympanic region. The skin of the limbs, which are well webbed and clawed, is brown, and darker above. Both sexes have a conical spur on the tip of the tail, and the males have patches of roughened scales on the thighs. The young have a strong vertebral keel and two less distinct lateral keels. The carapace is dark brown to black and has light spots on each marginal, giving it a rugose appearance. The plastron is reddish or orange, with a dark spot in the middle. The hinges are not yet apparent.

Subspecies. There are three subspecies. *K. s. subrubrum* (described above): The eastern mud turtle occurs from southern Connecticut and along the Atlantic lowlands and inland to southern Indiana and to Illinois. The bridges are rather wide, and the anterior lobe of the plastron is smaller than the posterior lobe. The head is marbled with light spots. *K. s. hippocrepis* (Gray, 1855): The scientific name of this subspecies derives from the Greek *hippo* (horse) and *krespis* (shoe), in reference to the form of the light marking on the head of the animal. Commonly known as the Mississippi mud turtle, it occurs in the Mississippi Valley of Louisiana, Texas, Missouri, and western Kentucky. The bridges of this race are wide, the anterior lobe is smaller than the posterior lobe, and the head is decorated with two light linear markings in the form of a horseshoe. *K. s. steindachneri* (Siebenrock, 1906): The scientific name derives from that of Franz Steindachner, a famous Viennese herpetologist. Commonly known as the Florida mud turtle, this subspecies occurs only in Florida. It has narrow bridges and a small plastron, with the anterior lobe often larger than the posterior one. The head is marbled but does not have light lines.

Natural history. The periods of activity for these turtles vary according to latitude. Those in the north hibernate, whereas those in Florida do not. A very aquatic species, it still may make lengthy terrestrial excursions especially after heavy rain. The habitats include slow, shallow watercourses, with muddy bottoms and abundant vegetation, as well as swamps, lakes, bayous, flooded cypress forest, and the like. Individuals may also live in brackish water. During the summer, they may bury themselves and estivate. Omnivorous in diet, they patrol the bottom to seize insects, crustaceans, amphibians, and so on, but they also consume aquatic vegetation. The species has a somewhat shy disposition but can also be aggressive and inclined to bite. It can also

© F. Bonin

The *Kinosternon* species are somewhat aquatic, but this species spends part of its life underwater.

secrete an extremely nauseating musk. Mating takes place from mid-March to the end of May and usually occurs in the water and only rarely on land. The eggs are laid from June to September, but in the south they may even be laid in February. The eggs are often simply laid on the ground, in vegetative litter. A single nesting per year is the norm, with two, three, or four eggs per clutch. The eggs are ellipsoidal, pinkish white or slightly bluish, and fragile. In size they measure about 25 × 15 mm. Incubation lasts for 90 to 100 days, and sometimes the young overwinter before emerging.

Protection. This species has many natural predators, including snakes (which may also consume the eggs), mammals, and raptors. By modifying wetlands, humans have caused much elimination of the habitats preferred by this species. Many individual turtles are also run over on highways. In the past, densities of up to 260 turtles per hectare were recorded (Mahmoud, 1957), but this is far from the case today.

Staurotypus triporcatus (Wiegmann, 1828)

Common name

Mexican giant musk turtle

Distribution. This turtle occurs in Mexico and Central America, from the central part of the state of Veracruz to the base of the Yucatán Peninsula, including Belize and northeastern Guatemala, and continuing into western Honduras.

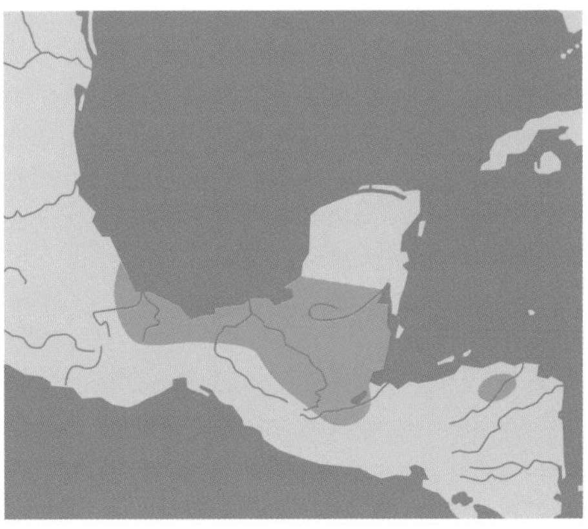

Description. The shell of this giant species may reach 380 mm. It is oval in form, with three extremely strong keels. The vertebral keel follows the five vertebral scutes; the two lateral keels lie along the four costal scutes on each side. The posterior border of the shell is smooth and is very slightly upturned, especially on the ninth to eleventh marginals. The plastron is very small, with a hinge between the pectorals and the abdominals. The posterior lobe is longer than the anterior lobe, in contrast to *S. salvinii.* The sides of the plastron are almost parallel. The posterior lobe is very narrow and pointed, and there is no anal notch. The humeral and gular scutes are absent. The bridge is rather wide and may be one-third the length of the plastron. The overall color is yellow, with darker seams in some individuals. The head is very wide in the temporal region and extends forward to the pointed, upturned snout. The upper jaw, quite blunt, is not hooked in front. The head is

© F. Bonin

This is a massive species, with a carapace bearing three strong keels. The head is broad and spotted with white on a black background. The animal has an aggressive disposition.

olive or yellowish, with fine dark reticulations that extend to the edge of the jaws. These decorations are well marked and numerous. Two chin barbels are present. The limbs and the tail are gray-brown. The limbs are webbed, and the tail bears two rows of conical tubercles. The males have small roughened scales on the thighs and lower limbs, but these scales are absent in the females.

Natural history. This turtle lives in slow watercourses, swamps, small lakes, and oxbows of large rivers, at an altitude not exceeding 300 m. It is carnivorous and opportunistic, devouring any animal matter that passes by. Its agility and great size make it a formidable predator, and it consumes crustaceans, amphibians, mollusks, large insects, fish, and occasionally other, smaller turtles, such as *Kinosternon leucostomum* and *K. acutum.* The chase technique includes forceful opening of the mouth, which brings about a powerful inrush of water, carrying the prey in with it. This turtle is capable of inflicting serious bites if not handled carefully. The clutch size is around six, and the eggs, about 40 × 24 mm in size, are usually laid in September.

Protection. This species is heavily fished and consumed by people, and its populations are dropping fast. In Mexico, there are ranches and farms established for the production of this species; some of the animals produced are sold, and the other rest are released into the natural environment.

The small, cruciform plastron hardly protects the underside of the animal. The energetic and aggressive nature of the species gives it sufficient defense against predators.

© B. Devaux

Staurotypus salvinii Gray, 1864

Distribution. The range is complementary to that of *S. triporcatus,* including Pacific lowlands of Mexico in Oaxaca and Chiapas and with an eastern extension into Pacific El Salvador and Guatemala.

Description. The elongate, oval carapace may reach a length of 250 mm. The three keels are well developed in the juveniles; they flatten somewhat with age but remain visible in the adults. The central keel starts at the rear of the first vertebral and extends to the level of the eleventh marginals. The two lateral keels run across the four costals on each side. The posterior border of the shell is smooth. Overall, this species presents a wider, flatter shell than that of S. triporcatus. The color is dark brown to olive gray, with perhaps some dark, indistinct

Common name

Pacific Coast giant musk turtle

© J. Maran

This species is very close to *S. triporcatus* but does not reach as large a size, and the head is smaller and more pointed.

shadings. The plastron is small and has a hinge that provides for the mobility of the anterior lobe. The posterior lobe, shorter than the anterior lobe and pointed, has no anal notch. The bridge is narrow, not exceeding one-fifth of the length of the plastron. The gulars and humerals are lacking. The general plastral color is uniform yellow or grayish. The head is quite wide and has a pointed snout and a very blunt upper jaw. The beak is not well developed. The lower jaw is more delicate and is more cornified at the tip. The head is gray, with fine yellow and orange patterns, and the jaws are entirely yellow. When the animal reaches great age, the head becomes uniformly dark. There are two chin barbels. The tail and limbs are brown to grayish. The limbs are well webbed, and the tail has two rows of small conical tubercles. The males have patches of rough scales on the thighs and lower legs. The hatchlings have a yellow plastron, covered with irregular black spots.

Natural history. This turtle prefers small, mud-bottomed watercourses with abundant aquatic vegetation. It also may be encountered in certain bodies of calm water, as well as in swamps. It is carnivorous, consuming mollusks, insects, larvae, other invertebrates, and small fish. It is an aggressive, fast animal. Courtship occurs in January, in the water. The male climbs onto the carapace of the female, the rugose patches on the hind limbs securing his position, while the female makes attempts to bite him. Two or three nestings per season are possible, each including about 10 eggs, 38 × 20 mm in size. Incubation takes 90 days.

Protection. This species is collected and eaten, and its aquatic habitats are disturbed or destroyed by human activities. The status of populations needs to be better known.

Claudius angustatus Cope, 1865

Common name

Narrow-bridged mud turtle

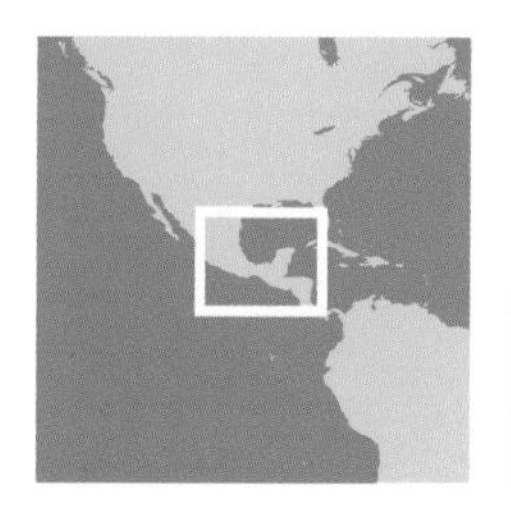

Distribution. The range of this species includes southern Mexico and northern central America, from northern Oaxaca and central Veracruz to Belize and northern Guatemala.

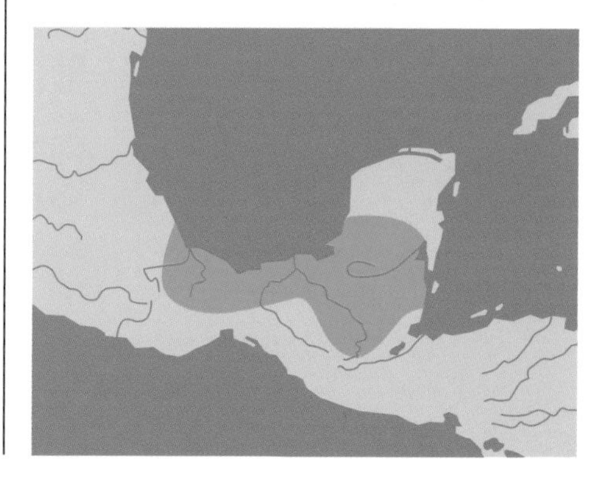

Description. The carapace does not exceed 165 mm. The oval shell has three strong keels, but they become obliterated with age. The posterior border of the carapace is always smooth. The first vertebral scute is very wide and touches the four anterior marginals. The scales of the carapace are rugose in the young, with well-developed annual growth rings. The overall color is yellowish brown to dark brown, with black seams. In the juveniles, the scutes are decorated with dark radiations. The plastron is extremely small and is cruciform in shape. The bridge that attaches the plastron to the

© F. Bonin

carapace on each side is minuscule. It is held in place by a solid ligament, and usually the axillary and inguinal scutes are lacking. The anterior and posterior lobes are triangular. The plastron is uniform yellow in the adults, although the young have a dark figure in the middle, with extensions following the seams. The head is very large and has powerful jaws, with a horny beak on the upper one. There is a sharp point on each side of the upper jaw behind the anterior hook, giving the animal a menacing appearance. The lower jaw is long and also has a powerful hook. There is only one pair of chin barbels. The head varies from brown-yellow to gray. Dark areas or various degrees of development are seen on the sides and top of the head. The jaws are yellow, with thin vertical black rays. The neck is grayish, with dark shading, and bears several series of small tubercles. The limbs are gray to brown and have strong webs. On their upper surface, the forelimbs have three well-developed transverse lamellae. The males are larger than the females and have a horny spine at the tip of the tail. Upon hatching, the young measure 35 mm and are blackish, with a bright yellow plastron.

Natural history. This species lives in shallow waters with muddy bottoms. It may be found in small watercourses, swamps, and small lakes. It leaves the water with some frequency and estivates during the dry season in underground cavities not far from the water. It is carnivorous and eats all kinds of accessible prey types, including fish, frogs, newts, snails, earthworms, insects, and larvae. These are very active, agile animals that can extend and turn the neck adroitly when picked up, with extremely painful results for the person holding one of them. Nesting takes place at the start of the dry season, in November. Several nestings may occur in a season, with two to eight elongate, hard-shelled eggs being laid per nest, each egg measuring 32 × 18 mm. Incubation lasts for 100 to 150 days.

Protection. This animal has been collected for consumption for a long time. It is also collected for sale to hobbyists, and in some regions its populations are very reduced. In Mexico, several farms have been established to breed and raise this species. Some of the animals produced are sold, and the remainder are released into the natural environment.

This species has a solid shell, with three keels. The head is strong, with cutting jaws. The lower jaw has a strong hook.

The very small plastron is connected to the carapace only by a ligamentous band on each side. Usually yellow, the plastron may also be brown to black in old animals.

© B. Devaux

CHELONIIDAE Oppel, 1811

*The marine turtles appeared relatively recently in the overall history of the chelonians. There were four families of them during the Cretaceous, evolved from different branches that led later to the development of the ancestors of both the land and the marine turtles. Today the marine turtles comprise two families (Cheloniidae and Dermochelyidae) and eight species. These turtles are all well adapted for the marine environment and have a streamlined morphology and very specialized limbs, which permit rapid swimming and also effective excavation of nest cavities. Advanced physiological adaptations allow marine turtles to remain underwater for long periods without surfacing and to tolerate cold waters despite their characteristic poikilothermy. The Dermochelyidae, with a single living species (*Dermochelys coriacea*), shows extreme adaptations to marine life, including a carapace covered with a sort of supple, thick leather. This species also has truly pelagic habits and is able to navigate cold, subpolar waters and to dive to great depths and for long periods of time. This is also the heaviest turtle in the world, with a maximum weight of 950 kg.*

Caretta caretta (Linnaeus, 1758)

Common name

Loggerhead turtle

Distribution. After *Chelonia mydas,* this is the most widely distributed species in tropical and subtropical seas. It may also be found in cold waters, including those of southern Norway, Nova Scotia (Canada), Vladivostok (Russia), and Anchorage (Alaska, USA). The distribution extends to latitude 70° north near Murmansk, and 35.5° south near the Rio de la Plata in Argentina. On the

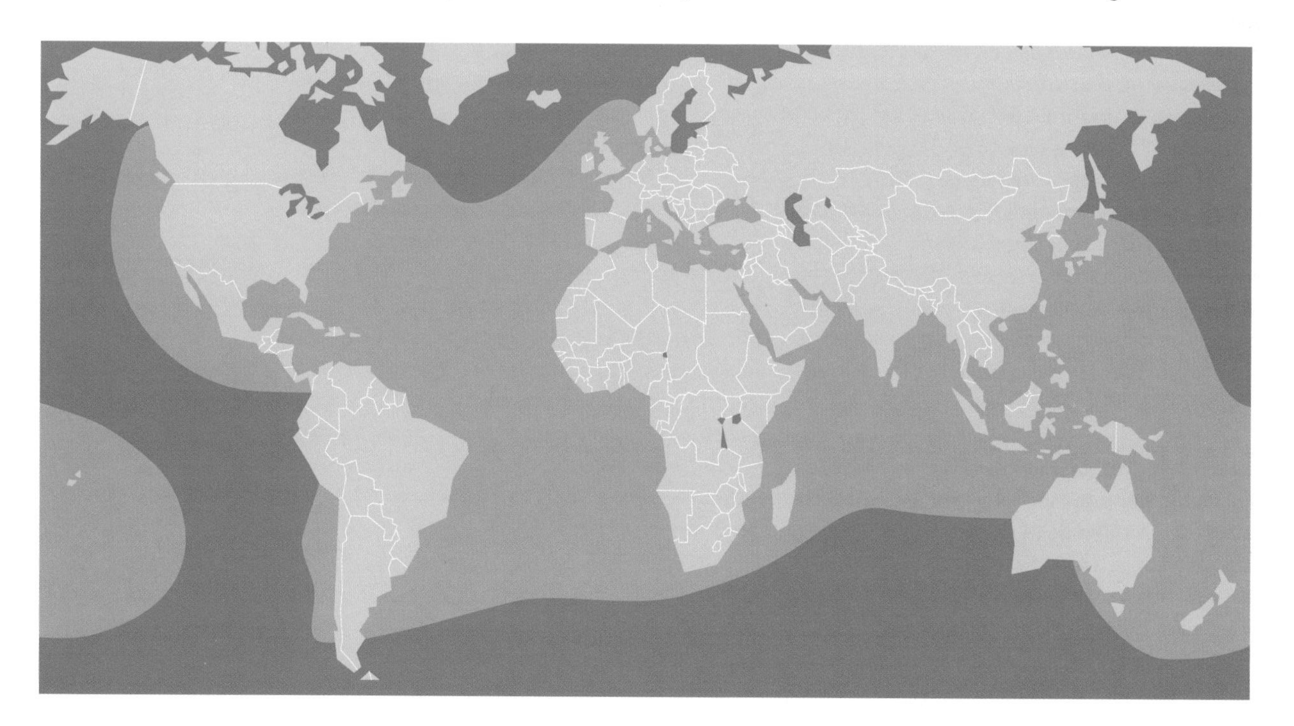

other hand, it seems to be absent in parts of the Pacific Ocean and in northern South America (Colombia, Ecuador, Galápagos Islands, Peru).

Description. This is a large species, reaching up to 1,200 mm in length and weighing 150 to 250 kg. Some weights of up to 450 kg have been reported but are questionable. The carapace is cordiform, with a vertebral keel that is very pronounced in younger animals, forming a series of distinct elevated points along the midline. This keel is reduced in adults but may remain visible on the first two vertebral scutes. The carapace includes five pairs of costal scutes, of which the first pair is in contact with the nuchal. Twelve to 13 marginals are present on each side, but in extreme cases there may be 11 to 15, a rare circumstance among turtles in general. The rim of the marginal scutes is serrated in the young, but the serrations flatten out with growth, and in old animals the edges of the shell simply show modest undulations at the sides. The color is dark or chocolate brown to orange-yellow, and the marginal scutes have a yellow edge. The plastron is yellow to light brown and sometimes actually orange; the plastral bridges are orange, and each is covered by three inframarginal scutes that lack pores. The plastron is smooth in adults but has a pair of longitudinal keels in the juveniles. The head is very large, and this characteristic is the source of the vernacular name "loggerhead." The front of the head is somewhat pointed, with a strong horny beak, and a large frontoparietal scale is present on the top of the head, along with three large postorbital scales behind the eye on each side. The snout is slightly protruding. The head is brown above and lightens up ventrally, becoming orange-brown or even yellow, as do the neck and the limbs. The limbs are especially well adapted for swimming, and each bears two claws. The anterior limbs are edged with large scales, enclosing areas of smaller scales toward the center of the surface of each flipper. The males have much thicker, longer tails than the females.

Natural history. This species is not very pelagic but rather is found along the coasts of the continents; it is also quite migratory, leaving the feeding areas in order to swim to the nesting beaches. It can maintain a body temperature higher than that of the ambient water, thanks to muscular activity at the micro level and to good natural insulation, which allows the animal to frequent cold waters. During the winter, this species may spend long periods of time in a lethargic state at the bottom of the water. It is occasionally found on the Atlantic coasts of France. It is mainly carnivorous, eating crustaceans, mollusks, echinoderms, and fish, as well as aquatic plants and algae. The swimming action is slow and majestic, but as divers well know, it can instantly depart at high speed when it needs to. It is somewhat aggressive in nature and will bite if molested. There is a film that shows a loggerhead being attacked by a tiger shark; the turtle turned and bit the fins of the shark, which decided to abandon the attack. The jaws are extremely powerful and can easily smash a crab, a lobster, or the shell of a large mollusk.

Mating occurs at the surface, the male hooking onto the shell of the female with his anterior claws. Sexual maturity may be precocious; some reports speak of just four years of age and a shell 600 mm in length at first reproduction, but these are unlikely. Nesting usually takes place at dusk or by night, on a rising tide, in the fine sand of temperate and subtropical beaches. Nesting occurs in spring or early summer, in opposing months in the Northern and Southern Hemispheres. There may be three to seven nests per season, at intervals of about 15 days, but females usually have nesting seasons separated by two or even three years. About 100 eggs are usually laid (maximum about 150), and they are spherical in shape, with parchmentlike shells, and 35 to 49 mm in diameter. The

Usually, sea turtles are photographed at the time when they are laying eggs. This photo shows the lachrymal secretion, which in marine turtles serves to purge the excess salt ingested from the marine medium in which they live.

females are nervous while digging and may interrupt the process and return to the sea if disturbed. Oviposition itself is quick, taking just 10 to 20 minutes. The whole process, from emergence from the sea to return, is often completed in less than 60 minutes. Incubation takes 46 to 71 days, according to temperature, and the temperature of incubation affects or controls the sex of the resulting hatchlings. When the eggs are kept above 32°C during the first weeks of incubation, only females will be produced, whereas temperatures below 28°C produce only males.

Upon hatching, the young have a grayish or brownish, well-keeled carapace. They are about 55 mm long. If they are fortunate enough to escape from the numerous predators (crabs, birds, mammals, and then fish and sharks), they swim in a frenzied fashion for several days, until they reach a zone of floating seaweed, such as *Sargassum,* in which their cryptic coloration makes them difficult to see. They then have a phase of rapid growth, nourished by zooplankton and small medusans. The major nesting sites for this species are in Oman (Masirah Island), Florida, the Antilles, northern Australia, and Africa. These turtles also nest in the Mediterranean, Greece (Zakynthos), Turkey, Cyprus, and Libya. They are also rare visitors to French Guiana.

Protection. Many of the nesting sites for this species have been affected, degraded, or even destroyed by urban development and tourism. The species has been exploited for centuries and is still pursued for its flesh and its shell. As recently as 1980–1990, one could see mounted heads of loggerheads for sale in the French Antilles. In the sea, these turtle are adversely affected by drift nets and shrimp trawls. It is vital that turtle excluder devices (TEDs) be used in trawls throughout the wide range of the species, but unfortunately, few countries actually implement this system. Nevertheless, numerous conservation programs for this species are in place, including that at Zakynthos. Also, loggerheads have been equipped with satellite transmitters in order to follow their migratory routes. Similar programs have been undertaken in the Azores. This turtle, classified as Appendix 1 by CITES, is also protected by the Bonn Convention on Migratory Species.

Chelonia agassizii Bocourt, 1868

© J. Fretey

The carapace is more domed in *C. mydas,* and the scalation of the head is finer and darker than in other species.

Distribution. This species occupies a wide stretch of coastal waters (at least 500 km wide) in the northern part of the North American continent, from southern British Columbia to Chile, and including the Galápagos Islands at the equatorial level.

Description. For a long time considered a subspecies of *Chelonia mydas,* this species differs in its generally darker coloration, often appearing almost black. The carapace is streaked or spotted with light brown or olive. The scutes have light markings. The carapace is often high-vaulted,

much more than that of its congener, and this is especially true of the large females. The scalation of the head is slightly reduced and darker than that of *C. mydas.* The eye is often smaller also. This species measures 650 to 1,200 mm in length, with a weight rarely exceeding 150 kg. The young are dark brown to dark bluish black, with flippers bordered by white, as are the undersurfaces of the limbs and the plastron.

Natural history. The life history is similar to that of *C. mydas.* This is a migratory species, but the Galápagos population appears to complete its life cycle within the archipelago—the capture of tagged Galápagos turtles in Costa Rica or Peru is exceptional. The main nesting grounds are in Mexico, El Salvador, and Peru and on the islands of Bartolomé, Santa Fe, and Isabela in the Galápagos. Aquatic plants and algae compose 90% of the diet, which also includes sponges, echinoderms, jellyfish, and sometimes fish and crustaceans. Nesting season depends upon the latitude but occurs in August to January in Mexico and from December to June in the Galápagos. Courtship is very similar to that of *C. mydas,* and groups of competing males surrounding individual females have been observed. Several (three to eight) clutches are laid each season, but some females do not return to nest for two or three years. Seventy to 80 eggs are deposited in each nest, with a maximum of about 140. They are spherical, white, and about 42 mm in diameter. Other species of sea turtle may share the mainland nesting beaches of *C. agassizii,* including *Dermochelys coriacea* and *Lepidochelys olivacea.* But the nesting seasons are staggered for the various species—with the smaller species nesting later—to reduce the destruction of eggs by subsequent nesting females. Incubation lasts for 50 to 55 days.

Protection. The status of this species is poorly known, but it is certainly becoming rare in Mexico and has been subject to heavy exploitation. Poaching continues in the nations of Central and South America for the meat, oil, and leather. Many turtles are also caught in shark nets. It is important that the populations of this species be evaluated to determine the current size of stocks and the threats that they face—all the more important in view of the restricted overall range. It is urgent also that the species be listed in Appendix 1 of CITES, like *C. mydas,* since *C. agassizii* is not protected by law in either Mexico or the Galápagos Islands.

Common name

Pacific green turtle

The close-up of the front of the turtle shows the light borders of the scales, the short snout, and the very large, dark eyes.

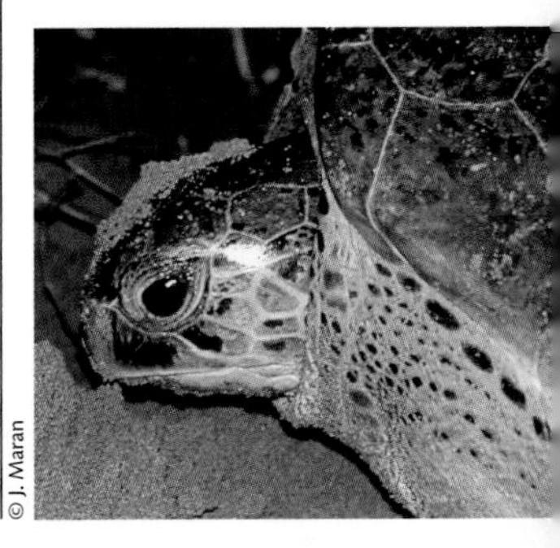

© J. Maran

Chelonia mydas (Linnaeus, 1758)

Distribution. This species is widely distributed in the seas of the world where the water temperature exceeds 20°C. This turtle may be found in waters of Nova Scotia, northern Ireland, and southern Japan, as well as in Argentina, South Africa, and northern New Zealand. It occurs in the Indian and Atlantic Oceans, in the Pacific (Micronesia, Polynesia, Melanesia), as well as in the Japan Sea, the China Sea, and the Mediterranean. On the other hand, it is absent from part of the Pacific, along the coasts of the Americas (North, Central, and South), where *Chelonia agassizii* is present and was long considered to be just a subspecies of *C. mydas.*

Common name

Common green turtle

Description. This is the largest of the Cheloniidae, reaching a carapace length of 800 to 1,300 mm and a weight of 140 to 160 kg. Occasional large individuals may exceed 230 kg. The flatted, cordiform shell has juxtaposed (nonoverlapping) scutes. The nuchal scute does not make contact with the first costals, and the latter number four on each side. The shell is very rounded, often steeper and more domed toward the front and sloping off toward the rear. The marginals are not serrated but curl upward in some individuals, similar to the configuration in the flatback turtle *(Natator depressus).* The carapace is brownish, grayish, or greenish, light or dark, and decorated with irregular streaks and spots. The word "green" in the common name refers not to the turtle's external color (which is rarely green) but to the color of the fat. Subadults usually have a light brown carapace with radial streaks, but some individuals are reddish, and the scutes are marked with yellow along the seams, as are the scales of the head and limbs. The plastron is rather wide, with rigid bridges and with four inframarginal scales on each side. In color it is white or yellowish. The head is rather small, with a blunt, rounded snout and a strongly developed beak. The cutting surfaces of the jaws are strongly serrated and denticulate. The eyes are large and almond-shaped. The head has two pairs of prefrontal scales, separated from the large frontoparietal by the small frontal. The limbs and the head are grayish and greenish in adults, with dark scales bordered by light cream or yellow. The limbs are large and well adapted for swimming; they are wide and covered with large, elongate scales surrounding a series of smaller scales. The undersides of the limbs, the neck, and the shoulders are cream to light yellow.

This species shows many color variations. Some, in the Pacific, may be reddish or yellowish. But in general the carapace is very wide, heart-shaped, somewhat domed, and mainly greenish to grayish.

© F. Bonin

Subspecies. No subspecies are currently recognized, but *Chelonia mydas* in the Indo-Pacific region differs morphologically from that of the Atlantic and could be considered as a subspecies, according to J. Fretey.

Natural history. Different life stages of this species inhabit very different environments. Where the young reside for the first few years is still unknown. During the subadult period, the species may take refuge in mangrove areas, as on Europa Island in the Mozambique Canal. Later they migrate to nest in distant locales, perhaps thousands of kilometers away from their feeding areas. The diet changes with development; at first carnivorous (eating jellyfish, zooplankton, and other invertebrates), they gradually become primarily herbivorous. The adults browse upon submarine sea-grass prairies, where they show tendencies toward aggregation—for example, in waters south of the island of Mayotte. The digestion of the cellulose fibers of aquatic plants is comparable to that in ruminant mammals on land. The low nutritive value of the algae explains the slow growth of these turtles, as well as their late achievement of sexual maturity. It may also be the reason why *Chelonia mydas* is the only sea turtle that crawls out onto beaches in the daytime to bask, possibly to benefit from ultraviolet rays necessary for the synthesis of vitamin D. Even the males may be seen sunning themselves. Although considered to be thoroughly herbivorous, the species will on occasion take jellyfish, crustaceans, mollusks, and other small animals encountered in the sea grass and marine algae feeding grounds. The shell is often covered with algae and parasites, and some individuals benefit from the attention of small "cleaner" fish.

Courtship takes place not far from the nesting areas. Mating itself seems to be rejected by the females except at certain times of the year; at other times they decline the male's attentions and protect their cloacal regions with their posterior flippers. Intromission itself—when it is accepted—may last as long as six hours. Sometimes one sees females courted by cohorts of males numbering a dozen individuals or more. The super-

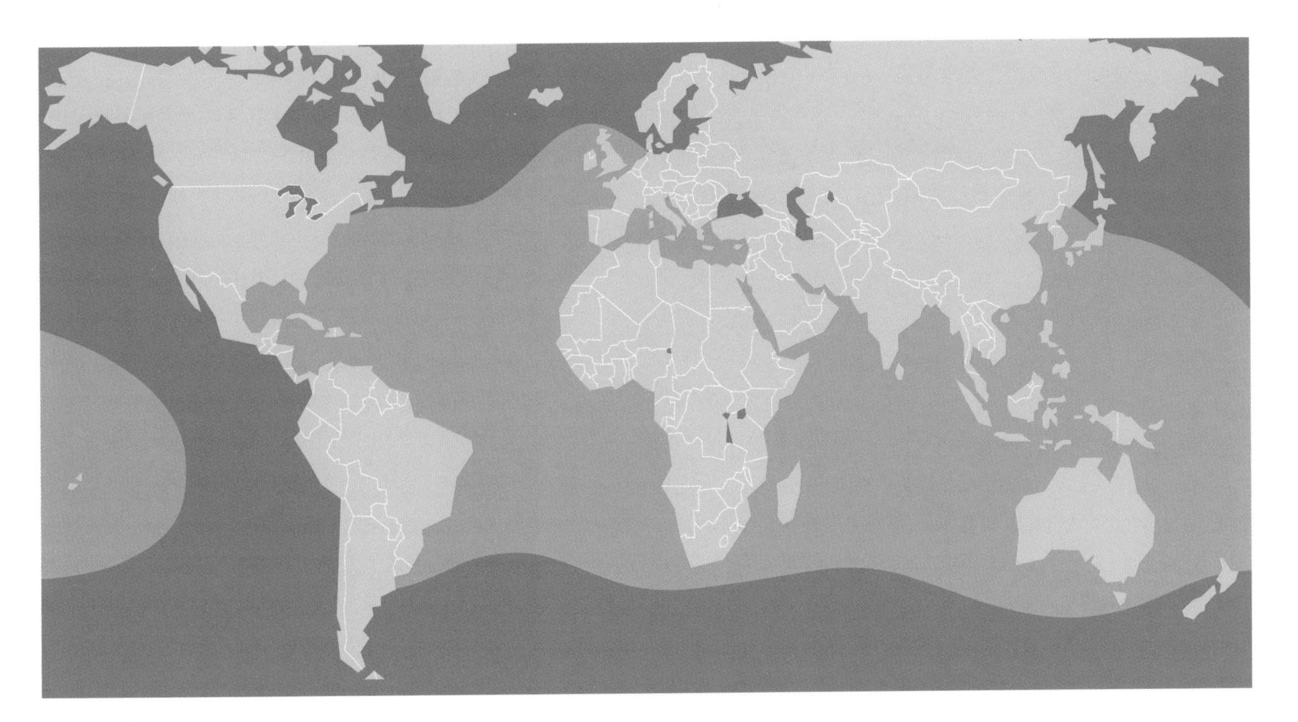

numerary males will push and bite, trying to throw off the single male that is fully mounted on the female and in the process of intromission. Sometimes it may happen that the female is so weighed down by males that she drowns. Several clutches —up to about six—may be laid in a season, at intervals of about 15 days, and are usually deposited by night and at high tide. For actual nesting, the female turtle will select a place close to the vegetation line. She makes a preliminary body pit by throwing sand back with her fore flippers; then she digs the nest itself, which is a hole 300 to 600 mm deep, widened at the base, and corresponding to the size and shape of the hind flippers. The eggs are then dropped in groups of 2 to 4, accompanied by abundant thick clear mucus. Usually, about 100 eggs are laid, but clutches of 200 and more have been described. The female fills in the cavity, using an automatic sequence of alternating movements of the hind flippers; then she switches over to powerful thrusts of the fore flippers, projecting sand backwards with great force in order to obliterate clues to the location of the nest itself. The whole process takes between an hour and an hour and a half. The eggs measure about 45 mm in diameter and are white and round, with a supple, parchmentlike shell. Incubation takes 48 to 74 days. The critical temperature for sexual differentiation of the embryo is 28.75°C.

The newly hatched young remain buried for several days, after which they all emerge together under cover of night, which reduces the impact of predation. The young have a dark gray, almost black carapace, and the marginals and the edges of the flippers are edged with white. The young suffer assaults from numerous predators (crabs, mammals, frigate birds, fish, sharks). The survivors swim vigorously for several days, after which they move toward feeding areas. The nesting beaches are numerous for this species. At certain preferred sites—including Costa Rica, French Guiana, Suriname, Ascension Island, Indonesia, and tiny Raine Island, Australia—thousands of females may nest in one night, forming a veritable *arribada,* such as the ridleys may offer at other times and places. The French possessions include many important nesting sites, such as Tromelin, Europa, Mayotte, Guadeloupe, and Martinique, not to mention New Caledonia and Polynesia.

Protection. This turtle without doubt has the history of the heaviest exploitation of any turtle species, inspiring the name "buffalo of the sea" in certain places. The species is listed in Appendix 1 of CITES, which should protect it globally, but numerous ethnic and subsistence people and communities continue to catch the turtles and collect the eggs. This species is caught for its carapace, its cartilage (for the famous green turtle soup), its skin, its meat, and its green fat. Farming operations in some parts of the world, such as the

© G. Hontebeyrie

During mating, the male grips the carapace of the female, using specially curved claws on the forelimbs. The female may have difficulty reaching the surface, and this can sometimes lead to her drowning.

Cayman Islands, were initially designed to produce animals for consumption but have had to convert to other purposes following protective legislation (e.g., on the island of Réunion). Numerous field conservation programs exist in various countries (Costa Rica, Sri Lanka, Malaysia), with the general goal of putting enough young turtles into the sea to lessen the diminution of the wild populations. Comprehensive turtle conservation requires that the nesting sites be totally protected, marine pollution be reduced, and substitute meat (chicken, pig, sheep) and other products be found for local ethnic groups to utilize instead of turtle. At sea, large numbers of *C. mydas* fall victim to various fisheries. Utilization of turtle exclusion devices is essential for controlling incidental mortality of green turtles (see text for *Caretta caretta*). The public must be educated not to buy marine turtle products, including soup, carapaces, and trophies, and ecotourism must be developed in ways that permit only a few sites to receive tourists, with the remainder being protected to avoid any stresses or disturbances to turtles nesting there.

Eretmochelys imbricata (Linnaeus, 1766)

This old turtle has an unusual, very flattened shell, with many barnacles and marine algae and mosses. The head is black, with white seams clearly seen between the scales.

Distribution. This species is widely distributed in all tropical seas, as well as some subtropical and even temperate areas of the Atlantic, Indian, and Pacific Oceans. It does not enter the Mediterranean except on very rare occasions, when it presumably arrives through the Suez Canal.

© J. Fretey

Description. The cordiform carapace of this turtle is somewhat elongated and is remarkable for the imbrication of the scutes of the shell, each of which overlaps those posterior to it like the tiles of a roof. This feature lessens with age, and old animals may have smooth shells similar to those of other hard-shelled marine turtle species. Four pairs of costals are present, which fail to make contact with the nuchal scute. The vertebral scutes may be lightly keeled, a feature more noticeable in the hatchlings. The marginals form a straight line laterally, but posteriorly they are often strongly serrated, and expanded and upcurved to form a shallow gutter. The two posteriormost marginals, above the tail, are strongly pointed. The carapace is yellow to light brown or reddish orange with dark (sometimes black) streaks, often also marked with cream or yellow. These features disappear gradually with age and are replaced by a less contrasting appearance. The plastron is wide and pale, doubly keeled, and sometimes it has a few black markings. The forelimbs are powerful but relatively narrow and

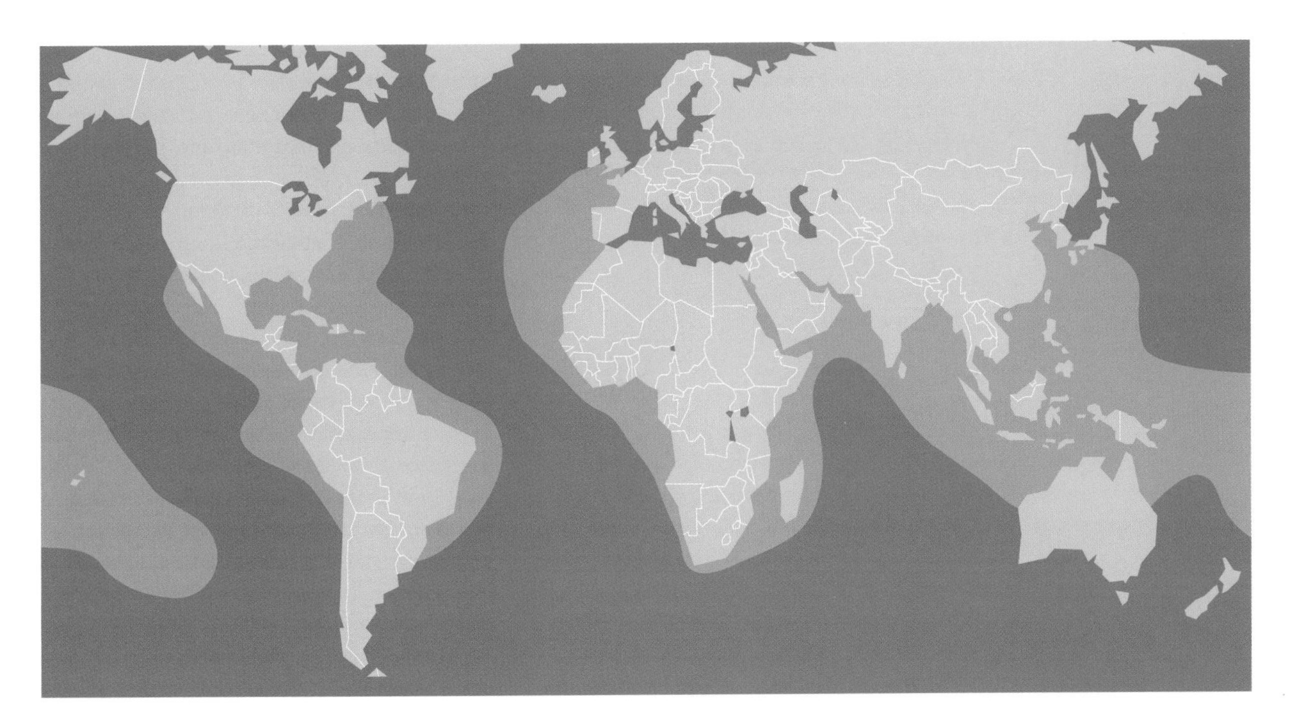

are rounded at the tips and equipped with two claws. The edges of the flippers are protected by a series of large, somewhat elongate scales, while the central area of the face of the flipper is covered with smaller scales. The head scales, dark brown or black in color, are outlined with yellow. The underside of the limbs and tail are very pale yellow, darkening toward the distal ends. The head is small and narrow and has a strong beak, somewhat similar to that of a bird (although not a hawk). Two pairs of prefrontals and a single frontal scale form a shield on the top of the head, which is red-brown above and light yellow below. The head scales are outlined with yellow. The carapace rarely exceeds 1,000 mm in length, and the maximum weight is about 60 kg, although exceptionally a length of 1,500 mm and a weight of 140 kg may be attained. The young have a brown or slightly reddish carapace, with a well-developed keel, and the plastron is dark, lightening up along the seams. The marginal scutes are strongly serrated, and the fore flippers are quite long.

Subspecies. *E. i. imbricata* (described above): The Atlantic hawksbill occurs in the western Atlantic from Massachusetts (USA) to Brazil and in the eastern Atlantic from Scotland to the coast of West Africa. *E. i. bissa* (Rüppell, 1835): The Indo-Pacific hawksbill is a vicariant form that occupies the remainder of the range, living in tropical and subtropical waters of the Indian and Pacific Oceans, from Madagascar to the Red Sea in East Africa; in the Philippine Sea, the Sea of Japan, and northern Australia; and along the coasts of the Americas from California to Peru.

Natural history. This species rarely occurs far from land and is not highly migratory. Usually, the movements from feeding to nesting areas are less than 500 km, but some exceptional journeys have been reported, including one from Yucatán to the Dominican Republic (1,622 km). This is the most omnivorous of the marine turtles; it is herbivorous during its first years, then later feeds upon invertebrates such as cephalopods, crustaceans, sponges, oysters, and corals, the latter being broken up by means of the fore flippers and more finely pulverized by the horny beak. This turtle has also been seen to feed upon mangrove shoots. The species nests on tropical and subtropical beaches, especially in the Antilles, Yucatán (in Mexico), the Red Sea, Oman, the Mascarenes, Madagascar, North Africa, Australia, Indonesia, and Malaysia, as well as the French Antilles. Mating takes place close to the nesting beaches, in the shallow water of lagoons. Copulation may last for several hours, the male pursuing the female, sometimes almost up to the beach itself, before intromission occurs. For actual nesting, the female chooses a well-vegetated, isolated locale, and she excavates behind the first tier of vegetation. Sometimes, nests are excavated in very hard soil or in very course

Common name

Hawksbill turtle

sand, and nesting may occur by day or by night. The whole process is usually completed within an hour. Three or four nestings may occur within a season, with 50 to 200 eggs laid each time, according to the size of the female. The eggs are white and round and have a parchmentlike, flexible shell. They are about 355 mm in diameter. Incubation takes 58 to 75 days.

This turtle, dozens of years old, shows strongly overlapping scutes that are clean and beautiful. This explains why the scutes have been intensively sought after by artisans for centuries. When hawksbills become very old, the carapace scutes become thinner and lose their quality and their commercial value, perhaps thus saving the turtle's life.

© B. Devaux

Protection. The hawksbill has borne the brunt of the folly and greed of humankind more than any other sea turtle and perhaps more than any other turtle of any kind. For a long time, techniques of "welding" the plates of its beautiful shell by heat treatment have been known, making possible the manufacture not only of combs and jewelry but also of large items like boxes and tables. Nowadays, even though the species has been placed in Appendix 1 of CITES and is banned from international commerce, one still finds hawksbill products offered for sale in many countries. The vendors usually argue that they are using up old stock, but it is clear that traffic continues and that the commerce in the shell is no longer always legal. Furthermore, the sale of these products is poorly understood by the public, even though one may insist that the turtles are protected and that it is forbidden to sell the shell. It remains essential that the public be made more aware of the problem, particularly in places where the sale of the artifacts persists. It is also important to realize that both Japan and Cuba make regular attempts to downlist the species with CITES so that they can continue to trade in the shell. In the Pacific Ocean, this turtle is also exploited for its meat, even though this practice is dangerous because the species can accumulate poisons known as cheloniotoxins, related to ciguatera in fish, and consumption of hawksbills is sometimes even fatal. The species is also readily disturbed by tourists during its nesting processes, and this occurs at least in Malaysia, Sri Lanka, and the Antilles. Artificial hatcheries have been created in Mexico and Malaysia, with the goal of protecting and incubating the eggs under controlled conditions and thus augmenting natural populations. Effective conservation requires not only rigorous protection of the nesting sites but also the confiscation of products offered for sale and the dissemination of information to local ethnic communities and the general public.

Lepidochelys kempii (Garman, 1880)

Distribution. This rarest of marine turtle species occurs only in waters of the Atlantic, from the Gulf of Mexico in the west (Louisiana, Texas, Yucatán) to western Europe and North Africa in the east. Rarely it may be found in Ireland and the Netherlands, along the French coast, and in Spain, the Canary Islands, Madeira, and the Azores. Occasional individuals have been found in the Mediterranean.

Description. This is the most retiring of the sea turtles and the smallest, measuring between 600 and 700 mm as an adult, with a weight of 45 kg. The cordiform shell is wider than long and is elevated in the front, giving the animal a bulbous appearance. The first pair of costals are in contact with the nuchal. The vertebral scutes are small, although the costals (five, sometimes six, in number) are very large. In young individuals, there may be pointed projections from the vertebral scutes, which flatten out with growth. The marginal scutes number 12 to 14 and are quite narrowed toward

© D. Grolin

This turtle, digging its nest, is working in broad daylight, although in general other species of marine turtle lay their eggs at the end of the day or by night. The head is broad and large and is provided with a strong beak.

the front of the shell. The overall color is gray to olive, lightening to yellowish at the edge of the marginals. The plastron bears two rows of tubercles that form keels separated by a modest central depression. The bridges have large inframarginal scutes, each perforated by a pore corresponding to the outlet of a Rathke's gland, which has several functions. These glands are particularly active in young individuals and may assist in both sexual recognition and avoidance of predators. They may also function in bringing large numbers of females together into the nesting aggregations called *arribadas* (see also under *L. olivacea*). The plastron is whitish to yellowish, darkening toward the front. The head is wide, is slightly pointed at the level of the nostrils, and has an upper beak that forms an anterior hook. There are two pairs of prefrontal scales. The forelimbs are relatively short and do not exceed half the length of the carapace. Each one bears two claws, but the distal one is very reduced. The forelimbs are bordered by a series of large scales, and the scales of the central part are larger than those seen in other sea turtles. Upon hatching, the young are dark gray to black. The carapace is oval and somewhat elongated and bears three longitudinal keels. The limbs and the shell are finely edged with white. The plastron is whitish and has four ridges.

Natural history. This turtle is a coastal dweller and is not found in deep water. It may be seen in the Florida Keys near the mangroves, out on the flats, and in the lagoons. It is almost entirely carnivorous, eating crustaceans, mollusks, sea urchins, cephalopods (and their eggs), jellyfish, and sometimes fish. It may also ingest some aquatic vegetation. The species does not appear to tolerate cold waters below 10°C. Below 8°C, it is in mortal danger and needs to get back into warmer water as quickly as possible. Possibly because it is usually near the surface and exposed to solar rays, this species rarely carries barnacles, whereas loggerheads and hawksbills may have many. When handled, this turtle is aggressive and may bite.

Sexual maturity is reached at six years of age. Mating takes place close to the nesting beaches. The male may be quite brutal, biting the female on the neck and the limbs before gripping her shell with the claws of his forelimbs. In extreme cases, a male may actually be brought onto the nesting beach by a female and may continue to attempt copulation while on dry land (A. Carr). The nesting is diurnal and takes place from mid-April to mid-July. The principal nesting sites are along the Gulf of Mexico, from Mustang Island (Texas) to Isla Aguada (Campeche, Mexico). High concentrations may be seen at Rancho Nuevo (Tamaulipas), where thousands of females come to nest at the same time. One of these spectacular *arribadas,* filmed on June 18, 1947, led to an estimate of 40,000 turtles at this one site. Nowadays only a few hundred turtles nest in each *arribada.* This aggregating tendency is still mysterious and may be a strategy to coordinate the emergence of the young turtles and thus limit the impact of predators, but now that the *arribadas* are reduced to a few hundred turtles, one may ask if this effect is still being achieved. Two to three nestings occur per season, at intervals of about 15 days. The clutch size ranges from 50 to 185. The eggs are white and spherical and have soft shells; they measure about 40 mm in diameter. Incubation lasts for 45 to 60 days. If the young turtles (weighing about 20 g) escape their multitudinous pred-

Common name

Kemp's ridley turtle

ators, they will spend their first years in the *Sargassum* flotillas in the Gulf of Mexico, where they grow rapidly.

Protection. This is the most endangered sea turtle species, at least in the Gulf of Mexico. Populations have dropped precipitously in recent years. The *arribadas* of 60 years ago are now distant memories only, and the scattered nestings of today bear witness to a rapid decline. This turtle is especially liable to capture in shrimp trawls, because it swims near the surface and remains in coastal areas. In the year 1980, it was estimated that 20,000 turtles of this species were perishing in trawls each year. Since the introduction of turtle exclusion devices (TEDs), these figures have come down somewhat, but there are still lots of trawls that do not use TEDs. Furthermore, the Gulf of Mexico is heavily polluted with effluents that come down the Mississippi, and impacts with motorized vessels may kill turtles directly. A reintroduction experiment has been undertaken with some success since the 1980s: eggs laid in Mexico have been translocated to beaches in Texas. After seven years, some females have come back to nest at the sites of their hatching. The species is included in Appendix 1 of CITES. Only effective conservation measures will prevent the complete disappearance of this species. Recent news has been encouraging; the species has been increasing in numbers at Rancho Nuevo for 15 years or so, and in 2006 more than 10,000 nests were reported for the year.

Lepidochelys olivacea (Eschscholtz, 1829)

Distribution. This species has a circumglobal distribution in the Atlantic, Pacific, and Indian Oceans. It is absent from Polynesia and Hawaii but may be encountered along the Pacific coast of the Americas from Oregon in the north to Chile in the south, and in the western Atlantic from the Gulf of Mexico to Brazil. It also occurs along the West African coast, from Senegal to the Cape of Good Hope, as well as shores of the Indian Ocean. It may be found around the Indian and Arabian peninsulas as well as in the China Sea and the Philippine Sea and into Melanesia and northern Australia.

At Ostional Beach in Costa Rica, olive ridley turtles make their aggregated and rapid nesting excursion, often on moonless nights. After nesting, the turtles rock from side to side to cover up their nests and compact the sand.

© B. Devaux

Description. This is one of the smallest species of marine turtle. The largest individuals measure 750 mm in length and weigh about 45 kg. The carapace is cordiform, almost as wide as long, and triangular in front, with a significant emargination above each of the forelimbs. The shell is rather flat, except anteriorly, where the nuchal and costal area is elevated into a rooflike peak. The scalation is quite variable. The nuchal is always in contact with the first costals (but situated significantly far back, exposing the long neck), but the costal scutes range from five to nine or sometimes even more on each side, and there is often a different number on each side. The vertebrals are sometimes irregular in shape and vary between four and six or seven. The marginals range from 12 to 14 on each side and are unserrated and rounded at the edges. The carapace is gray, gray-green, or brown-ocher, with the scutes sometimes bordered with yellow. Old animals assume a very dark olive-brown, dark gray, or almost black color, with the seams finely outlined by light lines. The plastral scutes are more stable that those of the carapace, though the gular may be absent, single,

or double. There are two strong longitudinal keels on the plastron. The bridges are composed of four inframarginal scutes, each of which is perforated by a pore toward its hind margin—the openings of the Rathke's glands. The plastron and bridges are whitish, greenish, or yellowish, but always light. The top of the head, which bears two pairs of prefrontals, and the top of the limbs are greenish gray; the undersides are colored similarly to the plastron. The anterior limbs are medium in size but are wide and covered with a mosaic of scales of different sizes, surrounded by a series of enlarged scales. The fore flippers each have two claws, although one of them is often hardly visible. The head is of medium size, rather broad, and triangular from above, with a short snout but very well-developed mandibles. The beak of the upper jaw is also well developed. The carapace of the hatchlings is more elongate than that of the adults, has three clearly visible keels, and is dark gray to dark green. There are four raised crests on the plastron, which is also dark gray or even brown. There is a white spot on the supralabial scale.

Natural history. The diet of the olive ridley appears to be opportunistic (J. Fretey), with a carnivorous preference; echinoderms, crustaceans, mollusks, and jellyfish are all eaten, but algae and aquatic vegetation may also be taken. The usual feeding grounds are in shallow waters and in lagoons. These turtles may sometimes be found in small groups, sunning themselves at the surface. They may dive deeply to locate favored prey—one was captured incidentally in a net at a depth of 110 m.

Sexual maturity seems to be precocious, occurring at an age of seven to nine years. Copulation has been seen close to the nesting beaches and also occurs in the open sea. It may last for one to three hours. The male holds tight to the shell of the female with his recurved claws. As a result, mature females often exhibit a characteristic scarred area on the marginals. Actual nesting does not take longer than about an hour. After nesting, the species shows a characteristic behavior pattern in which it rocks from side to side to compress the sand over the nesting site.

This turtle has numerous nesting beaches, many of which are isolated, but it is remarkable for the habit of nesting in enormous aggregations of thousands of females in events known as *arribadas,* or *flotas.* These occur on the beaches of Nancite and Ostional in Costa Rica, as well as certain beaches in western Mexico (states of Jalisco, Guerrero, and Oaxaca) and on the beaches of Rushikulya and Gahirmatha in the state of Orissa in India. One assumes that the Rathke's glands play a role in the formation of these massive groups, perhaps by producing a secretion that facilitates both the grouping of the animals offshore and also the formation and timing of the *arribadas*

Common name

Olive ridley turtle

© S. Bonneau

In the early morning, on the black beach of Ostional, just a few dozen turtles remain, finishing off their nests before returning to the sea. Soon the last one will be back in the watery medium.

themselves. D. Robinson gives an example of a completely blind female that participated in an *arribada* in Costa Rica on October 14, 1982 and that may have been able to join the group of its fellow female turtles by using the olfactory sense. Females also have an interesting behavior pattern when they come ashore to nest: they lower their snouts and push them through the sand, as if using keen olfactory sense to detect that this is the right beach. In addition to the *arribada* beaches mentioned above, the species nests on numerous beaches in Micronesia, northern Australia, Indonesia, Malaysia, Sarawak, Vietnam, Pakistan, the Seychelles, Sri Lanka, French Guiana, Venezuela, and West Africa, including Ghana, São Tomé, and Bioko. Females may nest three or four times in a season, at intervals of 17 to 29 days. The nest contains 30 to 170 eggs, which are white, usually spherical or occasionally ovoid, have a soft parchmentlike shell, and are 32 to 45 mm in diameter. Incubation lasts for 45 to 62 days.

Protection. In nature, these turtles fall victim to numerous predators and are eaten by coyotes, coatis, frigate birds, and vultures. Adults may be killed by sharks, and on land, females may be attacked by jaguars or crocodiles, as happens at Playa Nancite. They also suffer from pollution, and some individuals may be infested by fibropapillomas, as is more typical of *Chelonia mydas.* But the worst problem is exploitation of the adults, a serious drain on populations that has persisted for centuries and includes take of turtles for meat, for skins, and (mainly) for eggs. All the beaches where *arribadas* occur have seen their ridley populations collapse drastically during recent decades, as, for example, in Suriname. Only Costa Rica attempts to operate rational exploitation of eggs; the local people are permitted to take eggs only during the first day of an *arribada,* and local guards are posted until the end of the *arribada* to make sure that no more eggs are taken. At sea, shrimp trawls continue to be a serious threat. The species was legally exploited for its leather in Mexico, Ecuador, and Pakistan until recently; the product was exported to Japan and Italy, where the work of trained artisans generated major added value. The species is listed in Appendix 1 of CITES.

Natator depressus (Garman, 1880)

Common name

Flatback turtle

Distribution. This turtle is endemic to Australian waters. It is confined to the northern part of the continent, from the Great Barrier Reef in the east, through Torres Strait to Papua New Guinea and into the Gulf of Carpentaria, the Arafura Sea, and the Timor Sea to the west, and in the Indian Ocean to just north of Perth.

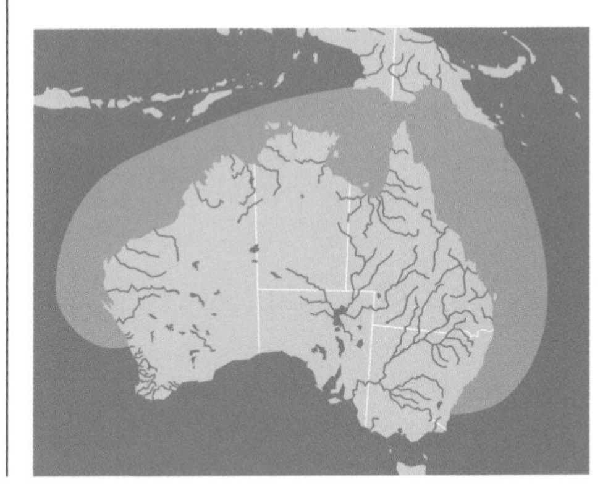

Description. This species is rather similar to *Chelonia mydas* but is recognized as a completely separate species and is characterized by several important differences. The largest animals measure about 1,000 mm in carapace length and weigh about 70 kg. The shell is notably flat and is depressed along the third and fourth vertebral scutes. The lateral marginal scutes are often rolled upward at the sides; the front marginals may be serrated. The entire shell is covered with a sort of keratinized skin that can be damaged, with blood drawn, simply by the passage of a fingernail. The skin seems slightly oily to the touch. The general color is gray

or gray-green, always of a very pale shade. The plastron, which lacks an anal notch, is wide at the level of the bridges but narrowed anteriorly and posteriorly. It is cream or yellow in color, like the undersides of the limbs and neck. The head is of medium size and has an indistinct beak. The ridges of the lower jaw are denticulated. A single pair of elongate prefrontal scutes is present. There are three large postocular scales and numerous small scales above the eye in the flatback, a feature that distinguishes it from the green turtle *(Chelonia mydas)*. The fore flippers are of medium size and covered with a network of very small scales, surrounded by a series of larger scales that are wide and distinct on the leading edge but even more so on the trailing edge. The tips of the flippers are rounded, and the skin of the limbs is extremely fragile. The limbs are gray to olive on top, as are the neck and the head, and they lighten up to cream or yellow on the sides and below. The young are much larger than those of *Chelonia mydas,* and this may help them survive in the face of certain beach predators, especially if the small green turtle hatchlings are more numerous or accessible.

Natural history. This species seems to prefer shallow, somewhat turbid waters, rather than the waters of the lagoons of the Great Barrier Reef, where it may be found on occasion. These turtles frequent coastal bays, where they feed upon sea cucumbers, jellyfish, bivalve mollusks, and various other invertebrates. They rarely enter deep water, such as may be found in Torres Strait, and they are captured by accident by prawn trawlers operating out of Papua New Guinea. They are often seen immobile at the surface of the water, warming themselves in the sun.

Copulation occurs not far from the nesting sites. Nesting takes place primarily in November and December but can occur in any month of the year. The females dig their nests at the foot of sandy dunes in early morning or late evening. They do not disdain nesting even in full midday sun when the day is not too hot. Up to four nests may be laid per season, with an average interval of about 15 days between clutches. The female digs a shallow cavity, in which she shapes the actual nest. Each clutch includes just 50 to 65 eggs (maximum 70) of great size, 51 to 55 mm in diameter; they are white, with soft shells. Incubation is quick—just 42 to 50 days. Emergences are triggered by a significant lowering of the temperature of the sand and occur under cover of night. Those hatchlings that escape predation take refuge in the mangroves. The strategy of the species is to lay fewer eggs, but unusually big ones, so that the hatchlings may be able to escape the depredations of some of the smaller predators, such as ghost crabs.

© P. Pritchard

This young turtle has an especially round, flat shell, somewhat serrated at the rear. The strategy of this species is to produce rather large hatchlings, and even though this necessarily reduces the number of eggs per clutch, their size will allow them to survive attacks by silver gulls, ghost crabs, and other beach predators, whereas the smaller hatchlings of other species would probably be killed.

Protection. Predators upon the young include sea eagles, monitor lizards, gulls, cephalopods, and certain carnivorous fish. The flatback is surely the least exploited sea turtle in the world as far as humankind is concerned, and its meat is reputed to be much less tasty than that of the sympatric green turtle. Nevertheless, in past centuries factories have been dedicated to exploitation of the oil of this species. Today the take of this species is illegal (except by aboriginal people who have the right to catch and consume them). At the nesting beaches, such as Mon Repos, near Bundaberg (in Queensland), this turtle has been studied and protected for many years. Nevertheless, its nearshore lifestyle renders it vulnerable to degradation of the marine habitat, and some animals have been found to be infested with fibropapillomas, an index of pollution in the ambient waters. The species is listed in CITES Appendix 1.

© P. Pritchard

This female has excavated her body pit at the base of a bush and then carefully fashioned the nest itself, about 60 cm in depth. This species usually nests by daylight, often early in the morning and late afternoon to avoid the intense midday heat.

DERMOCHELYIDAE Fitzinger, 1843

Dermochelys coriacea (Vandelli, 1761)

Distribution. This is certainly the turtle species with the widest global range. It occurs in all the oceans and seas of the world: the Atlantic, Indian, and Pacific Oceans and the Red, Mediterranean, and North Seas. It may venture even into the cold waters of Labrador, Iceland, Norway, Alaska, and the Bering Sea. At the southern extreme of its range, it reaches Argentina, Chile, South Africa, Australia, and New Zealand.

Description. The leatherback is the only living representative of the family Dermochelyidae. It is also the largest turtle in the world. The heaviest known specimen weighed 950 kg, but it is possible that some individuals may exceed a metric ton. The leatherback also is unique in the nature of its shell, formed from a mosaic of thousands of small bones that overlie a thick layer of oily connective tissue and are covered externally by a thin, smooth skin resembling fine leather (a feature that gives the turtle its common name). The leatherback is perfectly hydrodynamic and has extreme adaptation to the marine environment. For example, it has seven longitudinal, tuberculate keels running the length of the carapace (one along the vertebral region and three on each side), of which the lowermost pair define the flanks of the carapace, and the fusiform structure terminates in a long supracaudal spike. The shape of the "pseudocarapace" gives the animal some similarity to ancient Greek stringed musical instruments, inspiring the vernacular name "luth," meaning "lute." The carapace length may reach more than 1,500 mm in length. The skin is dark blue, dark gray, or almost black but is dotted with numerous whitish spots whose abundance and size vary considerably among individual turtles. The same patterning continues on the head and the limbs. The crown of the head, in the parietal region, has a depigmented irregular white or pink area known as the *chanfrein*—the French word for the white "blaze" on the forehead of some horses. For a long time it was thought that only females had this feature, but various pictures of males demonstrate that some of them also display the *chanfrein*. The head is huge, wedge-shaped, and flattened on top, with a short snout and upper jaw; two deep notches are situated on each side of a strong median hook on the upper jaw. The neck is short and thick and connects smoothly with the body at the level of the shoulders, which themselves are massive, and this structure permits little independent movement of the head. The flippers are immense and of great power. They are indicative of the pelagic life and the migratory behavior of this species. The tail is conical, with a very thick base, and is outlined by a keel that may bear white tubercles. The males have an extremely well-developed caudal appendage. The plastron has three easily visible keels. In color it is generally white or pink, with some degree of dark shading, as are the underside of the head, the throat, the limbs, and even the tail. The buccal cavity is quite unusual: the surface is entirely covered with long, pale, conical tubercles, which are used for the separation of the jellyfish prey from ingested water and possibly also have a respiratory function, retaining oxygen dissolved in the water.

In this photo taken in French Guiana, a large female uses her powerful anterior flippers to cover her nest before returning to the sea. Everything about this species evokes the hydrodynamic streamlining of pelagic species: dorsal keels and enormous flippers.

© A. Dupré

Natural history. This is the most pelagic of all the marine turtles. It frequents the great ocean basins as well as shallower waters in coastal regions. The life history of this species is still poorly known, and it spends much of its life in places vir-

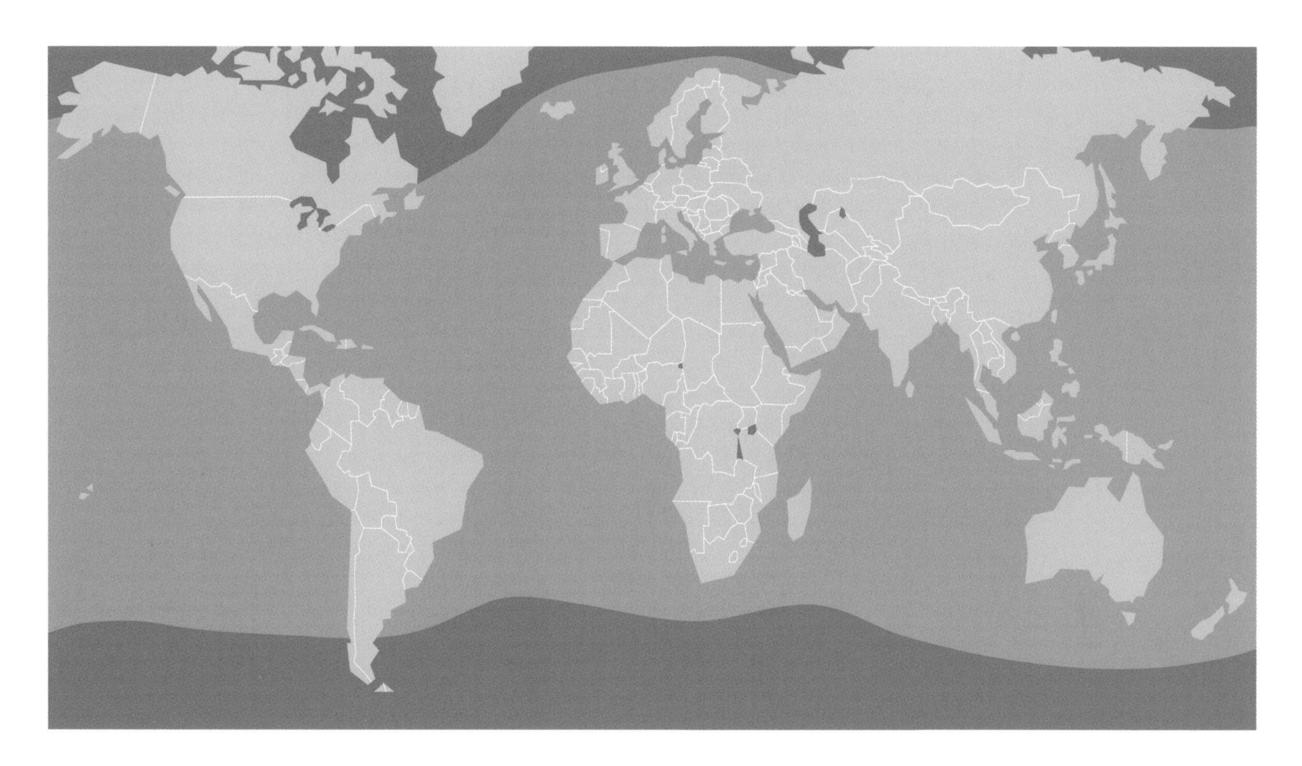

tually inaccessible to researchers. Even so, this species is the most intensively followed and thoroughly studied of all turtles. Some specimens have been equipped with satellite transmitters and depth gauges, which have revealed migratory routes and pelagic behavior, especially the duration and depth of dives. For example, turtles in French Guiana often travel north or east to reach areas of abundant jellyfish (their preferred food) in the middle of the Atlantic. The diet, on occasion, will also include cephalopods, crustaceans, mollusks, echinoderms, fish, and even young marine turtles. Algae and vegetation may also be ingested. But the special appetite for jellyfish is the cause of a significant level of mortality nowadays, as the turtle mistakes floating plastic bags for jellyfish and eats them. Individuals washed up dead on the coast of France have had as much as 5 kg of plastic in their stomachs. Other aspects of the life of these turtles are interesting, including the capacity to dive suddenly to depths of as much as 1,000 m and to regulate their body temperature, even in cold waters, thanks to the thick oily insulating layer and the internal muscular activity. The species also has perhaps the best of all chelonian orientation instincts, allowing it to roam the oceans of the world and then return to its natal beach from the other side of the planet. The age of first sexual maturity is still unknown. Furthermore, the difficulty of finding subadults in the wild prevents researchers from getting any accurate idea about the first years of life. It is basically impossible to study or to raise this creature in a tank, because it blunders repeatedly into the tank walls and soon dies. It was only as recently as March 1999 that three juvenile specimens, 170 to 210 mm in length, were accidentally captured near Príncipe, in the Gulf of Guinea, and gave a possible hint that this area might be an important habitat for young leatherbacks (J. Fretey et al., 1999). In the Atlantic, animals observed at large in French Guiana have been located a few months later in the Pertuis Charentais, on the west coast of France, where they were eating jellyfish.

Mating has rarely been observed, and it is not clear whether it occurs exclusively near the nesting areas, as it does with other marine turtles. During copulation, the male embraces the leathery shell of the female with his long, prehensile fore flippers. A second fertilization may provide for 10 successive nestings (a record number of 17 nestings for one female has been observed in French Guiana). These nesting are spaced by intervals of 10 to 15 days. The best nesting sites in the world change with time and the natural evolutionary shifts of the beaches, but currently Suriname, Guyana, Costa Rica, Colombia, Australia, West Africa, and various islands of the Lesser Antilles are included on the list. One of the most remarkable sites is in western French Guiana, on the beach of Ya-lima-po (Les Hattes), where a record number of 957 females was observed on the beach on June 27, 1988. Nevertheless, the nest-

Common name

Leatherback turtle

The huge head of the leatherback turtle, with the characteristic tricuspid jaw.

© B. Devaux

ing cycles are complex, and certain years have fewer nestings than others, the "good" years sometimes having five times the nesting of "bad" years. In recent years, important nesting areas have been discovered in West Africa—Pongara Point in Gabon, near the Conkouati region of Congo (J. Fretey). By contrast, the former nesting sites in Malaysia, especially at Terengganu, have been almost abandoned by this species.

Nesting almost always occurs by night, at high tide. It includes seven stages: ascent of the beach (about 10 minutes); preparation of the nesting site (15 minutes); digging of the flask-shaped nest cavity, 700 to 800 mm in depth (25 minutes); deposition of the eggs, several at a time, and accompanied by abundant mucus (20 minutes); filling of the nest cavity (10 minutes); flinging sand with its long fore flippers; and the return to the sea, perhaps with several feints and circlings to hide the traces as well as possible. The eggs laid are of two kinds. The most numerous are the large ones (50 mm in diameter), which are white, have a supple parchmentlike shell, and are normally fertile. Others (up to 40% of the number of "good" eggs) are smaller, variable in size and shape, yolkless, and nonviable and apparently serve only as an aid to the packing of the fertile eggs in the nest. At each laying, 100 to 150 eggs are laid, which may correspond to up to 1,000 eggs per season for a single female. During the excavation of the nest, the females are sensitive to bright lights and to disturbance and may cease their efforts and return to the sea. But when they are actually laying, they seem immune to external disturbances.

The rear flippers function like human hands as they carefully excavate the nest cavity.

© B. Devaux

Incubation of the eggs takes 60 to 70 days, and the sex of the embryo depends on the temperature of incubation during the early weeks. Below 28°C, only males are produced, whereas higher temperatures produce females. The hatchlings, 70 to 80 mm in length, have relatively enormous fore flippers. Midnight blue or gray black in color, the hatchlings are identical to the adults, with the keels well defined, but they have a complete covering of small, beadlike scales instead of smooth skin. They are consumed by numerous predators, and even the eggs are attacked by insects, including mole crickets. Once the hatchlings have passed through the gauntlet of predators, they evade the eyes of all observers, and we do not know where they spend their early years. The adult turtles fall victim to marine pollution and various kinds of fishing gear, including trawls and long-liners. They suffer also from collisions with boats and with propellers and are attacked by killer whales and sharks. When a leatherback is attacked by a shark, the turtle turns its carapace toward the attacker, presenting it with the broad expanse of the back, a shield that offers some degree of protection.

Protection. The nesting sites of this species are often disturbed by humans, causing the disappearance of some populations, including that at Terengganu in Malaysia. The eggs are gathered almost everywhere for food. This is also the case with the flesh, even though in some countries it is rumored to be inedible or even poisonous. Several nesting sites have benefited from studies and monitoring by conservation teams, as in French Guiana and on the west Atlantic coast. The development and deployment of TED (turtle excluder device) technology is equally vital for the survival of marine turtles, and problems with longline fishing may exist. But the specialists disagree among themselves over the future of this species. Some consider it to be one of the most endangered turtles in the world, in danger of disappearing completely almost everywhere. Others maintain that the populations are already quite large and, in many cases, expanding, and that it is far from being the most endangered sea turtle. The species is listed in CITES Appendix 1, but it remains one of the best-studied and, in some ways, best-protected turtles in the world. Nevertheless, to correct the overall diminution of the worldwide populations, it is still urgent to monitor nesting colonies and populations. Without doubt, the leatherback is a true flagship species for the conservation of turtles everywhere.

TESTUDINIDAE Batsch, 1788

The Testudinidae, which figure among the very first chelonians to be studied and named, are entirely terrestrial. Their key features are as follows: solid, often domed carapace; annual growth rings on the scutes often very clear and well defined; herbivorous diet; wide stumplike, or "elephantine," hind feet; and ability to withdraw the neck and limbs entirely within the bony shell. Certain species have a plastron (or even a carapace) with articulations or hinges, allowing for better protection of the soft parts. Some of the testudinids attain enormous size (up to 300 kg) and are s urpassed in this respect only by the Dermochelyidae.

Agrionemys (Testudo) horsfieldii

(Gray, 1844)

Distribution. The range of this tortoise is huge, encompassing the area from Iran in the west to China in the east, Russia in the north, and the Gulf of Oman and Pakistan in the south. It is well represented in the nations around the Black Sea. It has also been reported from Nepal, but this needs confirmation.

Description. This species, formerly included in the genus *Testudo,* is phylogenetically very close to Hermann's tortoise, although the latter is still included within the genus *Testudo.* The new genus *Agrionemys,* proposed by Khozatsky and Mlynarski in 1966, includes tortoises specialized for burrowing and adapted for extremely severe climates (F. Lagarde, 2004). This tortoise is easily recognized by its pebble-shaped, rounded form, a feature that it shares with other burrowing tortoises and that allows it to slide easily in and out of the underground burrows and chambers it excavates. This is the only tortoise over a huge range of eastern Europe and southern Asia that excavates deep burrows in which to hide and avoid extremes of climate, and it is also the smallest tortoise that does so. The shell is rather flat and is oval in shape, and the marginals are rigidly sutured to the plastron. The limbs are long and delicate, giving great mobility to the elbows to facilitate the work of digging. When this tortoise of the steppes ventures forth, it moves rapidly, appearing to "scamper away," with a characteristic gait imparted by the specialized form of the limbs. In color it is uniformly brown to reddish, some individuals also showing a greenish tinge. The areolae of the scutes may be somewhat darker, especially in the young, but in general the adults lack black spots and have a uniformly brownish tinge. The growth annuli are often very well marked during the early years, but they are gradually abraded smooth as a result of movement in and out of burrows. This species has only four claws on the forelimbs, in contrast to members of the genus *Testudo,* but they are powerful and scoop-shaped, to assist the

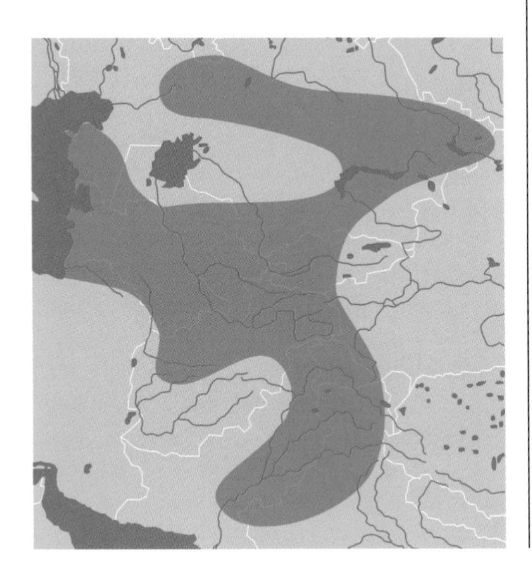

Common name

Horsfield's tortoise

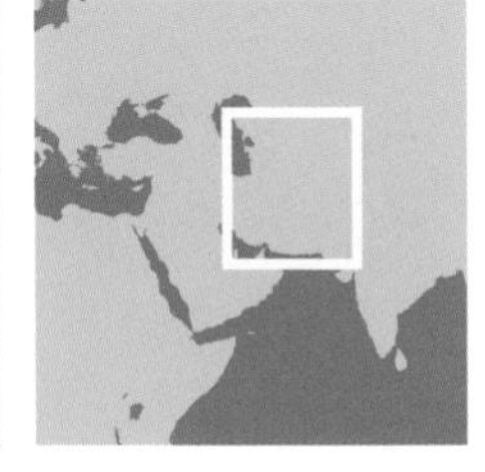

© F. Lagarde

In this field study in Kazakhstan, the tortoises were marked with bright colors to facilitate following them and observing their behavior.

digging process. On the plastron, which is oval, there are large black blotches delimited by light seam lines, but old animals often have a uniformly brown plastron. The head is strong, and the muzzle is well developed and squared off. The color of the head and limbs is the same as that of the carapace, but with the addition in some specimens of lighter spots under the chin and the throat.

Subspecies. The geographic range is so vast that one would expect to find some differentiation into subspecies, but these are not yet accepted by all parties. Listing by country, F. Lagarde (2004) recognized three subspecies: *A. h. horsfieldii* in Afghanistan, with a relatively high, vaulted, globular carapace, a rounded profile, a maximum size of 280 mm, a very upcurved bridge, brown coloration, and wide, irregular dark spots; *A. h. kazakhstanica,* from Kazakhstan, with a low, depressed, even flattened shell, a maximum length of 200 mm, a very upcurved bridge, and yellow-brown to greenish yellow coloration marked with a dark, contrasting spot on each scute; and *A. h. rustamovi* in Turkmenistan, with a low, convex carapace, overall brown color, and nearly flat bridges.

Natural history. Little was known of the ecology of this species until about 1995. This is a species specialized for life in the dry steppes and deserts, with altitudes up to 2,500 m and with very hot summers and severe winters, as in Russia and the Black Sea area. It has evolved a unique way of life, digging deep burrows that may be several meters long, and following a very restricted rhythm of life. A team of French biologists (F. Lagarde, J. Corbin, X. Bonnet) have observed this species in Kazakhstan and in Uzbekistan. The activity period is no longer than four to five months per year; the tortoises emerge from hibernation in mid-March and start courtship and mating immediately. Males have a territory of about 10 hectares, and females have more like 30 hectares. The males patrol their territory constantly and mate with all females encountered, before the tortoises have even started to feed. The courtship is animated. The male follows the female wherever she goes, harassing her and biting her violently on the legs and bobbing his head tirelessly in front of her, desperately trying to win about three minutes of her time for copulation. By late April, temperatures are rising, and the males abruptly switch from courtship activities to feeding and to retreating to shade during the heat of the day. They come out mainly in the early morning and the evening. The females start to look for nesting sites and may move great distances. By late May, the males have gone underground in their burrows. The females nest once or several times and then also go underground. The nesting season lasts for a month and a half, and some females may lay as many as four times, with clutches ranging from one to five eggs per nest. Emergence is observed about 100 days later, after the onset of scarce late summer rains. The young tortoises look very like the young of *Testudo graeca* but have flatter shells. It is only after five to six years that they take on the characteristic pebble shape of the adults.

During the season of activity, the tortoises eat copiously and grow fast. Some males may gain 100 g in five months, or 20% of their initial weight. According to the researchers, these male *A. horsfieldii* have an energy output higher than that of any other reptile, to take maximum advantage of the four months of the potential activity cycle. Later, the tortoises will spend eight to nine months in their burrows, in a kind of estivation state in June, July, and August, followed by hibernation in the subsequent months.

It is often said that this species is very resistant to cold and that it is still active at temperatures around 10°C. Outside the reproductive period, it seems to be timid and retiring, with each individual spending most of its time hidden in its burrow, which it digs in the sandy substrate of its environment. When it is imported into European coun-

tries, it does not tolerate the climate and the humidity and often dies quickly. This is a primarily herbivorous species, but in the last analysis it is quite opportunistic. Perhaps to stoke its high metabolism, it does not disdain the carbohydrates provided by a large insect, a snail, or a dead animal, given that the dry grass of the steppes are not a great source of energy. It drinks more rarely than the *Testudo* species but may abruptly take long drafts of water, to the degree that some observers thought that they had heard a frog croaking (Annandale). Like all burrowing tortoises, this species finds sufficient humidity within its burrow to stay there for long periods of time without feeding, but doubtless it would prey upon small mammals, reptiles, or insects that it encounters, to help fuel its underground sojourn.

The species hybridizes readily with *Testudo hermanni* (Kirsche), at least in captivity, a phenomenon presumably facilitated by the taxonomic closeness of the two species. Probably their common ancestor occupied a wide band of territory from western Europe right through to eastern Asia.

Protection. Up to the year 1970, this tortoise was not exported to Western countries. It was consumed so some degree by local people, but populations remained high. Twenty years later, by contrast, it was the subject of intensive collecting, especially in Russia and the Black Sea countries. For example, exports from Kazakhstan attained 126,000 in 1975 and 150,000 in 1982. And these are just the data for a single country. Certainly, along with American *Trachemys,* this is the most heavily trafficked chelonian in the world. Populations have started to decline rapidly in most of the range states, and capture for export is augmented by habitat destruction (e.g., intensive irrigation of semidesert areas for culture of cotton or rice) and direct consumption, as in China. In Pakistan, it is threatened by demographic expansion. But local wars, as in Afghanistan and Pakistan, are also destructive. For a long time considered to be the most widespread of all of the *Testudo* group, it is today subject to such heavy collecting that observers are worried about a major population collapse. Furthermore, mortality in the Western importing countries is high, and the widespread sale feeds the demand for "garden tortoises," which puts pressure on a host of other species. It is vital that the export of this tortoise be controlled within quotas and limitations on sale and that the exporting nations understand its status, preserve it, and oppose the pet-shop concept of "garden tortoises," because of course this is a highly specialized species adapted for a completely different climate and ecosystem.

Astrochelys radiata (Shaw, 1802)

Distribution. This species is found in southwestern Madagascar, in a band of coastal lowlands about 100 km wide, from Tuléar to the Mandrare River. It occurs principally in Antandroy territory, from Cape Sainte-Marie to Itampolo.

Description. Various species of "star" tortoises exist, from South Africa to India, but the species endemic to Madagascar is easily identified. This is the largest of the star tortoises, with a maximum weight of 22 kg. In contrast to *G. elegans,* which lives in India and which it closely resembles, *A. radiata* has a nuchal scute, and the carapace is evenly rounded, with vertical marginals on the sides, whereas the Indian species has somewhat conical scutes and, when very young, denticulate marginal scutes. Furthermore, the forelimbs of the radiated tortoise are quite distinctive, adapted for the sandy terrain where it lives. The forelimbs are large and covered with fine skin and small rounded scales, with a rounded tip, like a slipper. The elbow is very mobile, and it makes possible wide sweeps with the forelimbs similar to the strokes made by sea turtles. The head is bicolored—brown-black on top and yellow below a line that originates at the back of the eye. The overall coloration is blackish, with numerous brilliant yellow rays that emerge, fanlike, from the areola on one-third or

This tortoise, still young, shows strongly sculptured carapacial scutes, but the plain coloration is atypical.

© B. Devaux

one-half of the surface of the scute. Old animals may lose their markings and become almost unicolored, with tones ranging from brown to beige. Males are a little smaller than females (15 kg maximum) and often have a very concave plastron and a thickened xiphiplastral area. The penis is half the length of the carapace and is often extended after rain, when the animal is excited and gets up on its legs to seek a partner. The penis has an arrowhead-shaped extremity, designed to locate and penetrate the cloaca of the female.

The young are like living jewels, with no two having identical designs. It has been said that Queen Victoria offered a major reward to anyone who was able to find two identical juveniles, something that nobody has ever done. The decor may range from yellow or cream marked with black to black marked with yellow, with a great variety of markings, dark or light rays, and an overall color ranging from whitish to orange.

Natural history. Growth is slow, and sexual maturity is achieved only after about 20 years. Courtship is quite gentle, and the male never bites the female. He turns toward her and bobs his head up and down once or twice in search of sensory inputs, then climbs onto her carapace without further ado. Mating is seen especially after rain, and nesting occurs from September to March, with intervals of a month and a half to two months between nestings. The eggs are large and spherical, with a diameter of 45 to 50 mm. They may number between 8 and 22. Incubation is fast, taking 52 to 70 days. The young weigh about 30 g.

This tortoise is mainly herbivorous but will also bite at dead animals—insects, rodents—and organic detritus. The favorite food appears to be cactus pads, which they gulp down without any attempt to remove the spines. They walk about and feed mainly in the morning during the hot season, hiding the rest of the time under bushes and in herbaceous litter, where the shell markings allow the animals to virtually disappear among the narrow, yellow leaf remnants. This species lives in sympatry with the little spider tortoise, *Pyxis arachnoides,* at least in the southwest of Madagascar, but the latter prefers treeless, gravel areas, whereas the radiated tortoise lives in sandy areas with spiny plants, baobabs, and acacias. The radiated tortoise has two main styles of walking. In one mode, it drags itself over the ground, like other reptiles, with its plastron touching the ground, hauling itself forward with the forelimbs in sandy soil. In the other gait, the legs are extended and the body held as much as 8 cm above the ground. This mode is utilized in the rutting season and after the rains and is a dominance posture identical to that of the other *Astrochelys* species, *A. yniphora.*

Ethnozoology. For the Antandroy and Mahafaly people of southern Madagascar, this species is *fady,* or taboo. It may not be consumed nor even touched. For a long time this status protected the radiated tortoise from any collection or consumption. But this *fady* is not shared by the Vezo and the other ethnic groups of Madagascar, who increasingly have migrated to the southern parts of Madagascar, and they collect radiated tortoises without restraint.

Protection. This tortoise is included in Appendix I of CITES, but it has the bad luck to be considered the most beautiful tortoise in the world and in thus very sought after by hobbyists all round the world. Its gentle nature, the brilliance of its shell, and the beauty of the juveniles, coupled with the fact that it does very well in captivity and rarely poses sanitary problems, have made this animal a great prize for collectors. Furthermore, it is still eaten in certain villages of the south in which *fady* is no longer operative.

Since the year 1990, there has been heavier and heavier exploitation of this species. Toward Cape Sainte-Marie, the density of specimens is still strong and may reach 50 individuals per hectare, but in recent years, heavy collecting seems to have eliminated many such areas of high density. Wholesalers in the capital and overseas (Comores, Réunion) offer 1.30 euros for each tortoise collected, which may permit an impoverished villager to earn 40 euros per month. The tortoises are brought to the coast and taken by pirogue to La Réunion or directly to South Africa or the

Common name

Radiated tortoise

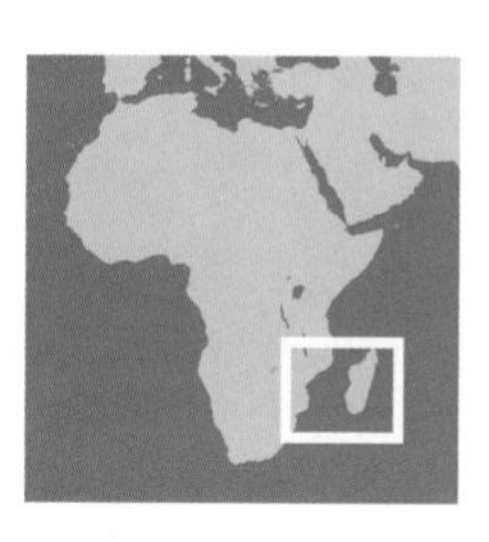

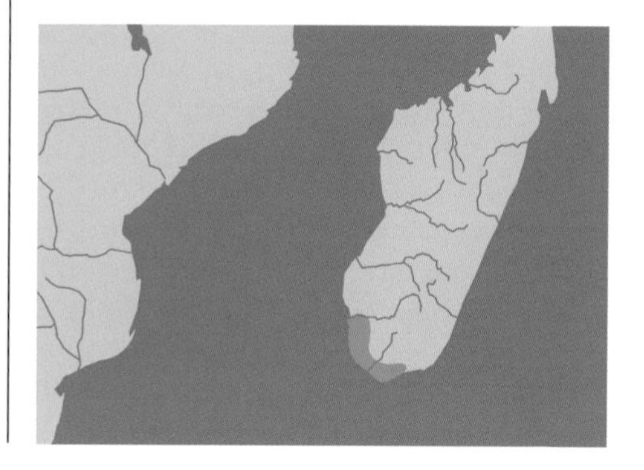

© B. Devaux

Old specimens of *A. radiata* sometimes become very pale, and the light radiations become wider. Note the raised position of the limbs, allowing for rapid locomotion.

Comores. After passing through the hands of a series of middlemen, they are resold in Western countries for up to 2,000 euros each. Near Tuléar, whole villages are dedicated to this practice, and the numbers of tortoises handled must be huge, involving both the big *A. radiata* (the most easily sold), and the little *P. arachnoides* (sold for about 0.80 euros). The various nature reserves in the area, such as Tsimanampetsotsa, are too small and too poorly patrolled to assure protection of the tortoises that live there. The inspectors of the Water and Forest Service also lack any real means of surveillance or intercepting the traffic.

In the face of this serious situation, which imperils both species of tortoise found locally, an ambitious project has been developed since 1999—the Sokake program, initiated by SOPTOM (Station d'Observation et de Protection des Tortues des Maures, the organization behind the Tortoise Village in Gonfaron, France) and the Association for the Protection of Nature (ASE) of Tuléar, with the help of the Malagasy authorities (Water and Forest Service, ANGAP). This program is based at Ifaty, north of Tuléar, where a "tortoise village" has been built, destined to receive confiscated tortoises, as well as tortoises brought in wounded or seized from traffickers. The first part of the plan is to transform the commercial collectors into guides and sources of information and to pay them a salary as naturalists. They become resident at the Ifaty center and have vehicles available to patrol the region, and everywhere they go, they increase awareness of the necessity of protecting the local fauna. The objective is to augment the local economy by means of ecotourism and, furthermore, to replace the role of tortoises in the local diet by substituting other products like fish, fowl, pork, and so on, along the principles of sustainable development. The tortoises gathered at Ifaty are raised in captivity and are destined for reintroduction programs.

© B. Devaux

This tortoise poacher fixes his animals to a pole, to carry them to market in Tuliar. Happily, we can now report that he has changed his profession and become a fisherman.

Astrochelys yniphora (Vaillant, 1885)

Distribution. This is one of the rarest tortoises in the world. Its range is very restricted and is situated south of Mahajanga, on the west coast of Madagascar near Baly Bay and Cape Sada.

Description. This turtle measures up to 400 mm in length and reaches a maximum weight of 8 kg. The general form is similar to that of *A. radiata,* but it differs in the following points: the gular is very extended, in the form of a plowshare, and its coloration is not radiating but tends toward plain yellow. The elongated gular is rather thin and upcurved, is unique among chelonians, and permits the males to challenge each other vigorously during the breeding season. In the female, the gular is much smaller. During the early years of life, this tortoise is yellow and black, with wide, very dark seams, well marked growth rings, and sometimes quadrangular markings on the scutes. Later this pattern evolves into a uniform cream or whitish. The plastron is yellowish, and the sutures are brown. There is very little plastral concavity in the males. The head is beige-yellow to brown and has well-demarcated scales and a more "smiling" physiognomy than its southern cousin, *A. radiata.* The limbs are covered with wide, well-developed scales. One often sees these tortoises raised up high on their limbs, immobile and with neck extended, as if hypnotized. The hatchlings weigh about 25 g. They are yellowish with large brown markings on each scute.

© B. Devaux

The extraordinary gular "ploughshare" of this tortoise is not yet fully developed, but it may ultimately reach 100 mm in length. The neck is often carried fully erect in a posture of dominance.

Natural history. Rather inactive during the daytime, plowshare tortoises live in dry forest with abundant bamboo. They emerge early in the morning, at the end of the day, and after rain. The rest of the time they hide under bushes and move very little. This is essentially an herbivorous species that seeks out wild fruits, local grasses, and succulents. In captivity, these tortoises are happy to eat fowl eggs, dead chicken embryos, and various animal foods. In nature, they are active at the beginning of summer (October–November) and again when the rains return, in February and March, which generally produces two growth rings on the scutes for each year of life. The eggs are rather large (47 × 42 mm) and almost spherical. The nest cavity is excavated in friable soil, almost always in the morning, before the real heat starts. Each female may nest four or five times per season, from January to May, with four to eight eggs each time. Incubation lasts for about four months, but may last as long as eight months, according to the climate; a period of high humidity is necessary for the emergence of the hatchlings. The young grow rapidly during their first years, increasing in size by 20% in a single year.

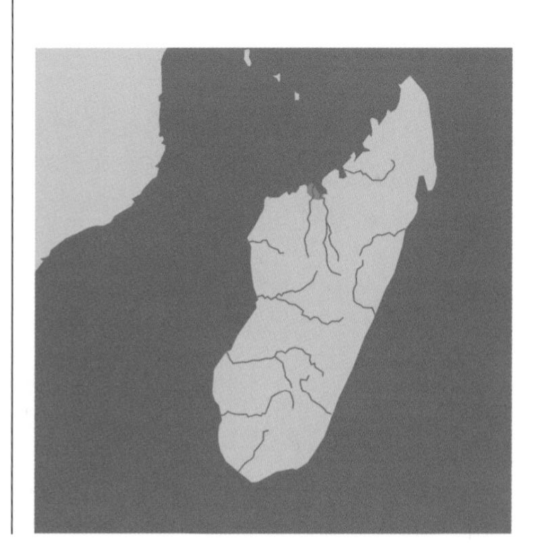

Common name

Ploughshare tortoise, plowshare tortoise

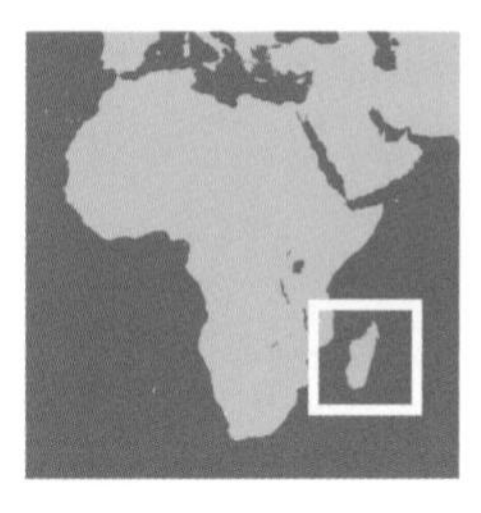

Protection. During the 1980s, the very reduced populations of this species led to the impression that it was "the rarest turtle on earth," virtually extinct. But recent observations have modified these somber conclusions somewhat; populations are not really as low as had been thought. The reduction of the populations had various causes: the numerous raids by Arabs from the Comores during past centuries, and in our own times wildfires, deforestation, and the impact of introduced predators, including rats and wild pigs. But the local villagers do not consume this tortoise. The animal is even thought to bring protection against certain maladies, such as cholera, which may explain the presence of captive tortoises in certain local villages with the role of "protector of health." According to some herpetologists, the low reproductive rate, the reduced habitat, and the dependence upon a very restricted habitat type, as well as the narrow dietary regime, have all been contributors to the slow but progressive disappearance of the species.

Since 1985 there has been a conservation program for the species at Ampijoroa, south of Mahajanga, organized and run by the Durrell Wildlife Conservation Trust in England, on the island of Jersey. A rearing center has been open since the 1990s, and several animals in the hands of private individuals were obtained to provide initial breeding stock. Don Reid, an Englishman, was in charge of the program for many years. Now he concentrates on captive rearing of the freshwater turtle *Erymnochelys madagascariensis* in neighboring enclosures. The reproduction of *A. yniphora* was difficult in the early years, until the detailed life cycle of the species became known, and husbandry took account of the wet seasons and the need for appropriate food. Following these discoveries, several dozen tortoises were successfully hatched each season, but unfortunately, as a result of a theft in the year 1995, the program was severely set back. Today the colony has been built up to about 100 subadults, and successful experimental releases have been conducted by Miguel Pedrono. In the years to come, releases of subadult tortoises of about 2 kg in weight will be made regularly, in appropriate habitat. The species is classified as "in danger" by the Turtle Conservation Fund.

Young ploughshare tortoises resemble *Astrochelys radiata,* but the pattern consists of rectangles, and the central part of the scute is clear, without streaks or spots.

Centrochelys (Geochelone) sulcata

(Miller, 1779)

Common name

African spurred tortoise

Distribution. This tortoise formerly occupied a vast range, about 500 km wide, from Mauritania to Ethiopia. Recent investigations (B. Devaux, 2000) have revealed relictual colonies only, isolated groups in northeastern Senegal (Ferlo), southern Mali, northeastern Chad, western and northeastern Sudan, northern Eritrea, and northeastern Ethiopia.

Description. This species is remarkable for more than just its name. It is the largest continental tortoise in the world, the males reaching 98 kg; it survives in the most arid habitats on earth (the sub-Saharan Sahel); it digs the longest burrows (15 m), with its entire morphology adapted for digging; it has the greatest gular development; and it has a uniform color that corresponds to the substrates of its habitat (the orange-yellow of the sands of the Sahara). The females attain 60 kg, and the sexual dimorphism is very pronounced, the plastral concavity of males sometimes being as much as 12 cm deep. The xiphiplastra are quite thickened, and the forked gulars are very developed. The general body shape is oval to quad-

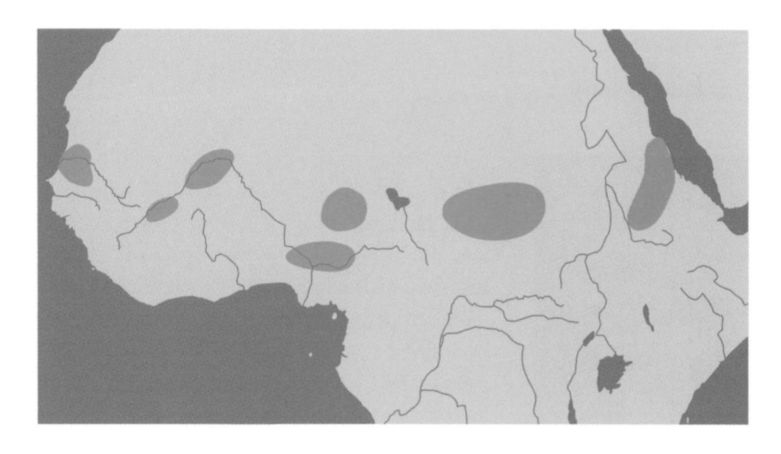

rangular, and the shell forms an extremely flattened dome. Only the females remain somewhat rounded, and although the males get more and more flattened, they may develop a significant depression on the second, third, and fourth vertebral scutes. Very old males have the fifth vertebral almost vertical, giving them the appearance of a chair or seat, on which ancient Dogon sages may sit to pass out decrees and sentences. The rest of the animal is totally dedicated to the burrowing role: powerful scaly plaques on the forelimbs, deep grooves on each scute, multiple long spurs on the thighs, head wide and covered with heavy scales, and enormous muscular strength. This species can hit a person with the front of its carapace so violently as to break a limb. In color it is uniformly beige to cream, or pale brown-orange, according to the individuals and to geographic variation. The plastron is the same color as the carapace and also has deep growth annuli. In this species, the bony parts correspond to about half the total weight of the animal, which may be a record for any chelonian. Those from the western regions seem to be somewhat orange in color, whereas the eastern ones are whitish, which corresponds to the color of the sands in the regions they frequent. With age, the marginals become strongly recurved in the males, especially above the forelimbs, forming veritable "gutters." The head is triangular, with a beak in the upper jaw and toothed serrations in the lower jaw. The eyes are placed high on the sides of the head, and the snout is square. In some males, one may see an inflation or swelling in the temporal region, forming two domes at the back of the skull. In certain very old tortoises, the head and face appear to be so emaciated, with sunken orbits and a projecting snout, that they seem to form a sort of eerie "death's head," which gives such animals a totemic role. Young individuals are yellow, with whorls of scales and a brown areola. They have quite slender marginal scutes, serrated and toothed at the back like the wings of a butterfly.

The spurred tortoise spends long hours in the entrance of its burrow, thermoregulating. From this position it can also retreat rapidly in case of danger.

© B. Devaux

Subspecies. The range across Africa is so huge that one might suppose that different subspecies could be identified in the various geographic sectors. M. Lambert (1994) demonstrated that this tortoise presented a remarkable physical homogeneity despite the distance of more than 5,000 km from one side of the range to the other. But genetic studies in 2001 (B. Devaux) have established that different genotypes do exist, distinguishing the Western populations (Mauritania, Senegal) from those of the east (Ethiopia, Sudan). Subspecies will probably be described in the future.

Natural history. The spurred tortoise is adapted to the Sahel-Sudanian ecosystem between the isohyets of 150 and 700 mm, south of the Sahara. This environment is characterized by a sandy substrate, little rain, and a xeric vegetative community of spiny plants, acacias, and baobabs. The activity cycle of *C. sulcata* is constrained by the challenging climate of its environment, with a rainy season in July and August, a warm, moist climate in January and February, and extreme drought and heat until June. In order to protect itself from the heat and aridity, this tortoise digs very long burrows, where it spends the hot hours of the day and also the entire dry season. These galleries have a cross section identical to that of the tortoise itself, with various wide chambers for resting and turning around, where it finds a little humidity during the dry season and perhaps even some dead animals (snakes, insects, fennecs, rodents) upon which it might feed. Tortoises have even been seen to push clumps of grass into their chambers, using the forked gulars as a bulldozer, to provide for the long months when no fresh food is available. In autumn, after the rainy season, they dig their burrows, mate, and eat great quantities of dry vegetation. They nest from early November to the end of May, and their nests may be as deep as 40 cm. Two nestings normally occur, with a maximum of four per female per season. The eggs weigh between 45 and 60 g and number 13 to 31, with an average of 19 per nest. The egg is almost spherical, like a golf ball. Incubation takes 120 days approximately, but the emergence of the young is controlled by the approach of the wet season and occurs in June and July. Ninety percent of the eggs produce viable offspring. The young weigh about 50 g, and about 5% of them show aberrations of the carapacial scute pattern. Growth is rapid: one kilogram after one year, and five to six kilograms after three years. There are numerous predators upon the hatchlings, but the most significant is the monitor lizard. The subadult tortoises also have much to fear from hyenas, warthogs, sand foxes, ratels, zorillas, various raptors, and African crows.

This tortoise eats what it can find in its harsh environment. Largely herbivorous (in captivity it mainly eats spiders' webs, but it also enjoys mangoes and watermelons), it also eats dead animals

© B. Devaux

Mating of these large tortoises is always spectacular. The males are larger than the females. They place their forelimbs on the anterior marginals of their partner and give vent to raucous sounds that may be heard 100 m away.

and detritus, along with spiny vegetation, tree bark, and in certain swampy areas algae and aquatic plants. Geophagy is common in this species, perhaps in search of calcium to help settle its stomach. It rarely drinks, and the integument is so thick that the animal surely loses little hydration through the skin, but after a drought period it has the typical giant tortoise capability of taking in 15% of its weight in water at one sitting. It has been said to aspire water through its nostrils, in the manner of *Dipsochelys elephantina,* but this may be merely an anecdote, because the spurred tortoise lacks the cartilage in the nasal chambers to separate air from water.

The juveniles share the burrows of the adults during the early years, or they retreat to natural cavities in the sand. They do not start to dig a personal burrow until they have reached an age of five to six years. Each tortoise digs one or several burrows each year, according to the size of its home range. Some burrows become flooded during the rainy season, and others may collapse. But a tortoise is able to excavate several meters of burrow in a single day, in a substrate of compact sand and often under trees and bushes that keep the soil coherent and well drained. The tortoises live to be very old, and certain captive animals have been known to surpass a century. In Dogon communities one may see emaciated, flattened specimens, with heavily ossified shells, that probably exceed 100 years of age.

Ethnozoology. Among the various ethnic groups of this region, numerous animist legends and stories include the spurred tortoise as a central character. In Senegal it is called Ngom, and tradition says that the placement of one in the garden will ensure good luck and fecundity. "Tortoise water" is used as a cure for certain infirmities (it may be either tortoise urine or water in which tortoises have spent a long time). The stories are especially prevalent in Mali, among the Dogon people, and at one time each village had an old tortoise that represented the spirit of the ancestors and that, naturally, lived for a longer time than the people did. The flat carapace allowed a judge to sit on the animal as the tribunal of the village. Sentences were passed down according to how the tortoise advised the chief, by turning either to the right or to the left.

Protection. During colonial times, this tortoise was considered a common species that was consumed by lots of people, collected by the colonists, and exported to Western countries. In this region of the Sahel, tribes were extremely nomadic and carried these tortoises on their camels for food or as an offering to the chiefs of other clans. Up to the middle of the twentieth century, the tortoise populations were still quite strong, but, with a brutal clash, various factors came together to provoke a rapid collapse. These factors included collection of animals for Western zoos and private collections; capture for food or local commercial exchange; ever-growing desertification; campfires and bushfires; and growing settlements around the edges of large towns and cities like Dakar and Khartoum. To these may be added the worst problem of all: constant increase in domestic hoofstock (such as zebus and goats), which overgraze the pasture. The herds destroy the natural environment, sterilize the Sahel, and increase desertification, and the tortoises cannot survive in these impoverished areas. It became imperative to stop all commercial collection. CITES has placed this animal in Appendix 2, but with a quota of zero. Furthermore, it is now forbidden to collect this species in its natural environment. On the other hand, it is so abundantly represented in zoos and Western collections, and it reproduces so freely in captivity, that exchanges and sales between Western countries can now be done entirely with captive-bred stock.

In nature, it is still necessary to protect the relict populations in order to strengthen certain depleted populations and to create conservation plans. In Senegal the program SOS Sulcata was started in 1994 by the French organization SOPTOM and Senegalese field teams. A center was constructed in Noflaye, near Dakar, with the goal of breeding tortoises that had been seized by customs officials, repatriated from Europe, or brought in wounded. The captive breeding has resulted in the production of about 200 young annually, and controlled releases should by now be in effect, with the goal of stopping the rapid decline of wild populations. SOS Sulcata serves also as an information center for local people and for tourists, urging them to refrain from collecting the animals and to increase awareness of chelonian conservation in general. The Noflaye Tortoise Village also is on the path to financial autonomy, thanks to ecotourism and the economic opportunities it brings to the region. This center, unique on the continent, is an example of indigenous, sustainable development, with the full participation of the villagers themselves to safeguard their ancestral fauna. The tortoises produced at Noflaye are then liberated in natural environments in the western arc of the range, from Mali to Mauritania, following the best principles of conservation.

Chelonoidis carbonaria (Spix, 1824)

The four tortoise species of South America, formerly included in the genus *Geochelone* but now considered to be species of *Chelonoidis,* have been separated since the Oligocene, according to Auffenberg. One of these groups includes the closely related *C. carbonaria* and *C. denticulata,* and the other includes the small Argentine land tortoise *(C. chilensis)* and the Galápagos giant tortoise *(C. nigra).*

Common name

Red-footed tortoise

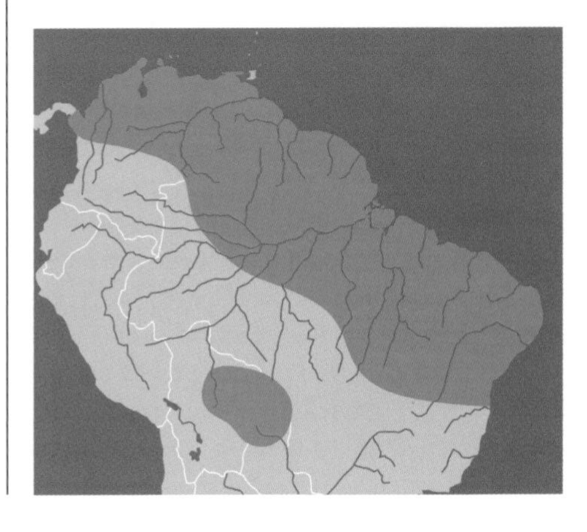

Distribution. The ranges of *C. carbonaria* and *C. denticulata* overlap and interdigitate. The territory of *C. carbonaria* includes savannas and dry forests as well as humid forests to some degree, while *C. denticulata* lives exclusively in humid (rain forest) areas. *Chelonoidis carbonaria* is found in much of northern South America, from western Brazil to Bolivia and into northern Venezuela and parts of the Guianas. It also occurs on some of the Caribbean islands, where it may have been introduced by humans.

Description. The two species are sympatric in places and have many points in common. We here present both their points of similarity and their points of difference. They both have shells of elongate, cylindrical form. The shell is narrow and rather low. There is no nuchal scute. They are me-

dium to large in size (300–700 mm). The limbs are very long, giving them a characteristic gait: they walk with long steps, their legs stretched out fully, and this permits them to cross muddy areas or small creeks. The overall color is dark, almost black, with a light spot on each areola and bright points on the forelimbs that may be yellow or red, according to the species. The head is rather large, and the snout pointed; the head is blackish, with bright yellow or orange scales. The growth annuli on the shell are very clear.

The special characteristics of *C. carbonaria* are as follows: the male has an hourglass shape, with a narrow waist, and has bright red scales on the limbs and sometimes yellow markings on the sides of the head. The shell is much darker than in *C. denticulata*. On the head there are small isolated scales, and in the plastron the inguinal scutes are rather larger and more triangular than in *C. denticulata*. Finally, the front margin of the shell, above the head, is not denticulate, although in *C. denticulata* it is often saw-edged (hence the scientific name of the species), especially in young specimens. Finally, the length rarely exceeds about 350 mm, although *C. denticulata* may reach 800 mm.

The reddish color of the scales on the limbs is not always pronounced and may tend toward orange or yellowish, as in *C. denticulata*. The plastron is orange-yellow, with dark markings along the seams in juveniles. Sexual dimorphism is pronounced; some males, seen in profile, have the fifth vertebral scute almost vertical.

Natural history. This species is adapted for open environments, drier than those preferred by *C. denticulata*. It is more herbivorous than the latter and consumes fallen fruit and numerous vegetative species, but it may also be coprophagic or cannibalistic, as the opportunity may occur. In French Guiana it seems to like larvae, dead animals, termites, and large galley-worms. It is most active in the morning and the evening. During the warmest months, it hides in vegetation or beneath a layer of plant material, where it may remain for days at a time without feeding. When it does move, it may cover great distances, making circuits of known localities where different types of food—for example, certain kinds of fallen fruit—may be available at certain seasons, as has been revealed by a study by Bruno Josseaume. It also swims very well and is not afraid to cross ponds, lakes, and other bodies of water.

© F. Bonin

Note the red color of the forelimbs and the very light areolae, on a blackish shell. This very young animal does not yet show much closure of the bones of the shell.

Courtship is quite animated. The male initiates matters by means of violent head movements, bobbing the head up and down, to facilitate olfaction. The female may respond with an abbreviated version of these movements. This is followed by sniffing at the cloaca of the female, nibbling her limbs, and finally actual copulation. There are multiple nests each season, as many as four per female, with a dozen or so slightly elongate eggs being laid, about 45 × 35 mm in size, and weighing about 40 g. In this humid, detritic environment, the actual nesting place may not be easy to find; the nest may be located against the buttress of a tree, sometimes beside a pathway, or in a dry cleared area. Incubation lasts three to six months, depending on the humidity and the amount of sun reaching the site. The young weigh about 30 g upon hatching.

Ethnozoology. This tortoise has had involvement with humankind for a very long time. It is certainly well known to indigenous peoples, who frequently consume it, and European colonists also export significant numbers. It has been known in Europe since the seventeenth century, in small zoos and in "cabinets of curiosities." Indigenous people, who consider this tortoise to be a cherished dietary item, roast it over charcoal. One method of immobilizing animals after capture is to pass sticks across the shell openings, front and back, and to tie them together on each side with lianas or cord. During Holy Week each year, great numbers may be caught for consumption in some cities of Brazil, as a result of an ancient papal declaration that tortoises are fish, not "meat."

Protection. One would assume that populations of this species must have been seriously reduced in recent decades. Happily, reproduction continues abundantly, and one can still find good numbers throughout the range of distribution. This tortoise does not seem to be directly threatened. The most worrying concern is the destruction of ecosystems, especially in northern Brazil and in Venezuela.

Chelonoidis chilensis (Gray, 1870)

Distribution. The vernacular name is accurate; this species mainly occurs in Argentina, but it has also been found in Bolivia and Paraguay, south of the distribution of the other *Chelonoidis* species, such as *C. denticulata* and *C. carbonaria.* The distribution coincides with the driest provinces of the Chaco and the province of Monte. It also occurs farther south in the Calden District and in the province of Espinal.

Description. This tortoise looks remarkably like *C. sulcata* of Africa, and it also lives in dry, xeric places with meager vegetation. The general appearance (oval and flattened), the thick scutes with deeply inscribed growth rings, the uniformly yellow color, the grayish skin, the square shape of the head—all approach the condition of *C. sulcata.* The main differences are as follows: size not exceeding 400 to 450 mm; no nuchal scute; scutes circled with brown in juveniles; upper jaw more protruding than the lower, forming a strong beak; no major gular extension; and vertical, not recurved, marginal scutes. The limbs do not have the heavy spurs of *C. sulcata,* and this species just digs modest retreats underground, not long burrows. There is a claw at the end of the tail. Old specimens sometimes have a uniformly grayish tint. The plastron is yellowish, with brown seams, and becomes uniformly yellow-brown with age. The males have a modest plastral concavity, but in general there is little sexual dimorphism. The young resemble juvenile *C. sulcata* in all ways, except for the contrasting yellow-brown coloration.

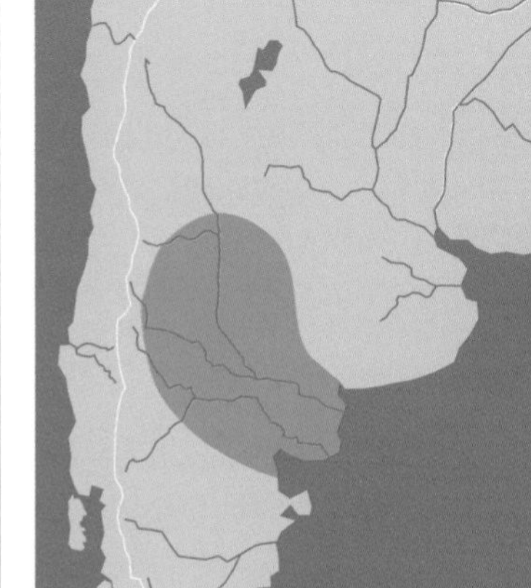

Common name

Argentine land tortoise

Subspecies. Two subspecies have been proposed but have not been accepted by all specialists. *C. c. donosobarrosi* Freiberg, 1973, in southern La Pampa Province and as far as Rawson, in Argentina, is larger and darker than the other subspecies. *C. c. petersi* Freiberg, 1973is from the Chaco of Paraguay and Argentina.

Natural history. This tortoise lives in areas with little vegetation, where it is found among cacti and spiny bushes. These plants also serve to keep it fed as well as to protect it from the sun. It frequents saline basins, lowlands scoured out by rivers in the semidesert, and sometimes the pebble-strewn flanks of ancient volcanic tumuli. It uses natural holes and crevices to protect itself from the sun. These tortoises are often encountered walking about in the early hours of the night, especially during the rainy season. During the summer, they emerge early in the morning and late in the evening. The annual cycle is as follows: they are active in November and December and breed at that

© F. Bonin

This species strongly resembles the African spurred tortoise, but the scutes are marked with brown grooves and there is no nuchal scute. The head is also darker than that of *C. sulcata.*

time; then they take refuge from the end of March in their winter retreats. They appear to be exclusively herbivorous and enjoy vegetables, cactus pads, and fallen leaves. They also eat grasses and ears of wheat. During the mating season, the males engage in violent combat. Later comes the time of courtship, accompanied by loud groaning sounds. Nesting occurs in January and February: three to seven eggs are laid, almost spherical and 40 mm in diameter. Hatching occurs about three months later or longer, after the rains have come. The rains are rare and unpredictable, and the wait may be for a year or more. The young measure 35 mm, and they grow slowly. Sexual maturity is rather precocious, with fertile couplings occurring at an age of 10 years, and the longevity may be quite modest for a tortoise—about 40 years. The eggs are eaten by foxes, polecats, and armadillos; the subadults fall prey to pumas, foxes, and black-chested buzzard-eagles, or *águilas moras.*

Protection. In nature, this species does not seem to be very common, and it suffers from overgrazing of its pastures and from the spread of cultivation and urbanization of lands. In recent decades, great numbers have been collected for sale as pets (up to 75,000 per year), destined for customers in the large cities, and this is causing a rapid drop in populations of the species. A program of study and conservation was launched in 1986, coordinated by various specialists, including Honegger, Waller, Cranwell, and the Argentinean Wildlife Foundation (FVSA). The main effort is to bring information to the public and to curtail commerce as drastically as possible. The species is included in Appendix 2 of CITES.

Chelonoidis denticulata (Linnaeus, 1766)

Distribution. The range includes the humid areas of South America, from southern Colombia and Venezuela, through the Guianas, and throughout the Amazonian lowlands. It does not occur south of the Amazon, in the state of Bahia, or in the Tocantins River, but this could be due to deforestation or to overcollecting. It also occurs in Trinidad, where it is indigenous, as well as in Guadeloupe, where it was presumably introduced by people.

Description. Morphologically very similar to *C. carbonaria,* this tortoise differs in the following points: oval shape without a narrow "waist"; maximum length 700 mm (some records of up to 800 mm); bright yellow scales on the limbs and head; denticulated edges to the anterior marginals in juveniles; overall coloration light brown, much lighter than *C. carbonaria;* large, connected scales on the top of the head, forming a sort of shield or helmet; and small inguinal scutes. The head is rather light in color, with orange-yellow spots. The plastron is yellow with brown edges, later becoming beige. In the males, the posterior plastral lobe is very recurved, and the xiphiplastron extremely thick. The young are yellow-brown, with small

The cylindrical form is more or less typical, as well as the "tiptoe" gait, permitting this species to cross wet or marshy areas.

yellow spots on the areolae and on the edge of the marginals. The head is large and brownish black, with bright yellow spots and yellow prefrontal scales.

Natural history. This species lives in the shaded, humid understory of tropical rain forests, always with high rainfall. It often hides under the substrate and is rarely seen in open situations. It moves around with some trepidation, like its congener, but in the dry season one can hear the sound of its moving among dry leaves. It loves water and swims well. Daily activity is bimodal: early morning, and end of the day. According to J. Fretey, it is often parasitized by ticks and nematodes. Its morphology makes it easy for it to move about in humid areas and small creeks. With feet held high, head extended and immobile, it wanders freely through its subforest habitat, looking for insects, larvae, and fallen fruit. It can crack open a grasshopper or a galley-worm with a single bite.

A study by B. Josseaume has demonstrated that this tortoise is a major disseminator of seeds ("zoochorie"), thus participating in the regeneration of tropical forests. The tortoises gather around certain vegetative species, like *Jacaratia spinosa,* the moment they come into fruit. The tortoises eat the fruit and thus disperse the seeds in their feces, depositing them especially in the open areas generated by tree falls. It has been calculated that 87.7% of the tortoise droppings contain these seeds. In the area of French Guiana where the study took place (Les Nouragues), they certainly prefer a frugivorous diet, but they also enjoy cadavers of rodents and other animals, as well as invertebrates. Their overall dimensions are reduced in this area, however, and none exceed a length of 400 mm.

Copulation is brief, and fights between males are violent. During courtship, the male rapidly swings his head from left to right, and he makes loud sounds at this time. The eggs are spherical (up to 50 mm diameter), 10 or so in number, and with a maximum of 20 (but only between 4 and 8 at Les Nouragues). Nesting occurs during the dry season (in French Guiana, July to October); the eggs are sometimes laid directly on the ground, with no attempt at excavating a nest. Incubation lasts for four to five months, and the hatchlings weigh about 40 g and grow rapidly.

Ethnozoology. Lots of stories and legends surround the *morrocoy* throughout its range. It is thought to have medicinal value, and the shells are sometimes made into musical instruments. Some tribes make small portable boxes out of the shells—with wax used to seal the opening and leather thongs used for carrying them—such as the Bushmen make out of the small *Homopus* tortoises in South Africa.

Protection. The large size of this species makes it a choice food item for the people who live in the rain forest. The degradation of the habitat is also a factor in the retreat of the species. But it can conceal itself perfectly in its environment, and we know little about the current status of populations. There is no major traffic in this species. Several organizations have conservation programs for this species, including FUDENA (Fundación para la Defensa de la Naturaleza) in Venezuela.

Common name

Yellowfoot tortoise

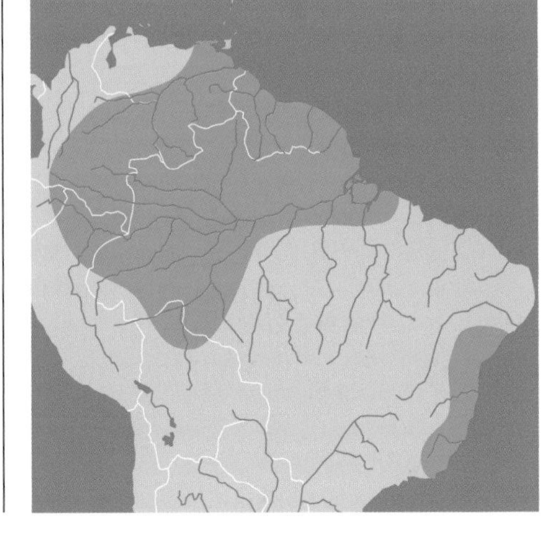

Chelonoidis nigra (Quoy and Gaimard, 1824)

Distribution. Formerly known as *Geochelone elephantopus,* this tortoise is found only in the Galápagos Archipelago, 1,000 km west of the coast of mainland Ecuador. The islands take their name from *galapago,* an old Spanish word for chelonians. The origin of this species of *Chelonoidis* on this remote Pacific archipelago must have been from the mainland across the ocean, because the islands were never connected to the continent. Most probably, a tortoise that was already large and possibly already bearing eggs arrived hundreds of thousands of years ago by passive flotation.

Description. Today there are only two species of surviving insular giant tortoises, following the extirpation by humankind of the tortoises of the Mascarene Islands (Indian Ocean). These are the Aldabra tortoise, *Dipsochelys elephantina,* and the Galápagos tortoise, *Chelonoidis nigra.* The ranges of the two could hardly be farther apart, but the species do have a rather similar appearance. RC. *nigra* has no nuchal scute, and the different subspecies and island populations offer a great variety of carapace shapes. The neck is very long and rather thin, and the shell is blackish, sometimes with fine white lines along the seams. Its classical bearing, noted by Charles Darwin, is characteristic: shell held high in front, head and neck stretched forth, limbs fully extended, a proud stance indeed. The head is more emaciated and triangular in this species than in the Seychelles tortoise, especially in old males. Certain ancient individuals acquire a strong hooked beak, sunken orbits, and withered muscles on the rear of the head. In captivity, a Galápagos tortoise may attain a weight of as much as 400 kg, and specimens of up to 300 kg do exist in nature on Santa Cruz and Isabela. Sexual dimorphism is well developed, males achieving three times the size of the females and having an impressive plastral concavity, as much as 120 mm deep. The hatchlings weigh about 50 g and are rather flat, oval, and either uniformly dark or with modestly contrasting black rings on each of the costal scutes. The longevity of this species is often overestimated, but it seems that some individuals (e.g., on Pinzón Island) may achieve an age of 120 years.

Subspecies. This species varies from island to island because each population is reproductively isolated. A monograph by Peter Pritchard (1996) allows us to understand these tortoises, which inspired Charles Darwin and a host of scientists after him. We offer here some details of the current status of these tortoises as well as their distinguishing features, although in the terrain of the Galápagos Islands it is extremely difficult to count them with precision. Eight extant subspecies have been recognized (several others are extinct).

—*C. n. abingdoni* (Günther, 1878): Shell reaching highest point at second vertebral scute. Pinta; a single living individual, resident at the Charles Darwin Station on Santa Cruz since 1973.

—*C. n. becki* (Rothschild, 1901): Shell extremely variable (domed, elongate, flattened, or saddle-backed); relatively small. Volcan Wolf, northern Isabela; population about 8,000.

—*C. n. chathamensis* (Van Denburgh, 1907): Rather small, somewhat saddle-backed. Northeastern San Cristóbal; population about 3,500.

—*C. n. darwini* (Van Denburgh, 1907): Very large, rounded, with some elevation of anterior margin of carapace. Santiago; population about 1,000.

—*C. n. duncanensis* (Garman, 1917): Small, strongly saddle-backed, anterior shell opening high and broad. Pinzón Island; population about 250.

—*C. n. hoodensis* (Van Denburgh, 1907): Small, strongly saddle-backed, shell narrow above the head and neck; neck thin. Española; population about 1,300.

—*C. n. phantastica* (Van Denburgh, 1907): Quite large, very high in front. Only one specimen ever found, on Fernandina; probably extinct.

—*C. n. porteri* (Rothschild, 1903): Very large, highly domed. Southwestern Santa Cruz; population about 3,500.

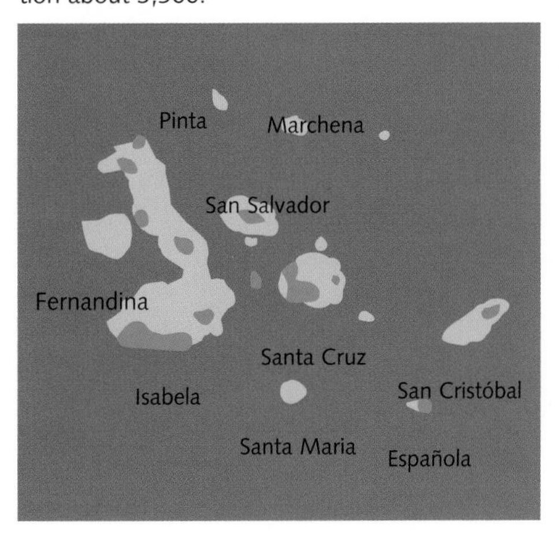

© T. de Roy

Certain subspecies have a very long neck and a carapace that is raised and opened up in front, allowing them to reach both fallen leaves and high-growing bushes.

Common name

Galápagos giant tortoise

—*C. n. vicina* (Günther, 1875): Very large, domed; somewhat variable but rarely if ever saddlebacked. One population in southern Isabela with very flattened shells ("guntheri") probably should be recognized. Isabela (Volcan Darwin, Volcan Alcedo, Sierra Negra, Cerro Azul). Population estimated at 10,000, mostly on Alcedo.

Natural history. Each subspecies follows its own ecology and way of life. Some of the races live in dry, grassy places, perhaps close to the ocean, while others live halfway up the great volcanoes or in moist caldera and dry, rocky places. The ocean moderates extremes of temperature, which does not change greatly throughout the year, and there is a rather short wet season in August and September. Mating take place toward the end of the wet season. The conduct of the males is unforgettable. They select and approach or chase one of the small females, trying to bite her limbs. Then they place their huge forelimbs upon the front part of the female's shell, making the observer worry quite seriously that the latter will be completely and catastrophically crushed. The bellows uttered by the male can be heard from hundreds of meters away. Happily for the female, actual copulation is relatively brief. The eggs are laid in certain habitual lowland sites, where the sun shines brightly and the digging is easy, even though this brings the risk of nests being placed on top of each other, with drastic consequences for both. The eggs are spherical and measure about 50 mm. The average per nest is about 17 eggs, and incubation lasts for 80 to 100 days. The young weigh about 60 g and grow rapidly.

The adults have the capacity to go without food or water for weeks, which allows them to survive during prolonged droughts. In the 1830s, Darwin had already taken note of their resistance to dehydration. They rest quietly during such times and, from time to time, make long hikes to certain rare springs or water sources. These journeys are conducted over ancient pathways, probably thousands of years old, which are maintained by the very passage of the tortoises themselves. For two or three days, they remain in the vicinity of the springs and then return to their usual feeding spots. They are basically solitary animals, but they may form sizable groups as they assemble for their spa visits or when numerous animals simply inhabit the interior of a grassy crater. They also take dust baths to cool themselves off, turning their heavy body around in friable soil, and such baths may also help relieve them of parasites. One also often sees the male tortoises being cleaned by finches; with shell raised high off the ground and head and neck fully extended, the tortoise appears in a trance as the finches hop over and around it, extracting seeds and ticks from the wrinkles in the skin. Some have also reported a singular mode of feeding, in which the animal raises itself high until small birds move in to investigate the area under the plastron, at which point the turtle allows its heavy body to fall suddenly, killing the birds beneath and stepping back so it can eat them. (This may not work with adult males, in which the plastron is highly concave.) They prefer vegetable food, however, eating cactus *(Opuntia)* when they can reach the pads, but they also have been known to eat dead goats, fecal matter, and organic detritus. They also have a behavioral detail that causes them to drop to the ground like a massive hydraulic hammer. As they walk casually upon their habitual routes, raised high on their four limbs, shell raised in front and head held high, they may suddenly become aware of an alien presence and collapse abruptly to the ground, head and limbs retracted, at which time the compressed air in the lungs is released as a forceful, audible blast through the nostrils.

Sexual dimorphism is considerable in this species, and the males may have three times the weight of the females, as here on Volcan Alcedo (Isabela, Galápagos).

Protection. These tortoises were so heavily exploited by whalers, buccaneers, and settlers for two to three centuries that some island populations were extirpated. Others were massively reduced, subjected to the continuing ravages of rats, dogs,

pigs, and goats. The goats form huge herds that completely destroy the natural vegetation, eventually leading to the death of both tortoises and goats. In recent years, there has been an additional stress, particularly on Santa Cruz and Isabela, which is the conflict between wild tortoises and the management of cultivated areas for crops, cattle, and horses. Tens of tortoises have also been killed by settlers and fishermen as a form of protest against the national park restrictions. There have, in addition, been major fires on southern Isabela, which have destroyed habitat and killed tortoises; in some cases it was possible to launch a helicopter-based rescue operation to save the unique "flattened" tortoises ("guntheri," of southern Isabela). The Charles Darwin Station on Santa Cruz has been the focus of many tortoise conservation programs, each of which is tailored to a particular subspecies and its special requirements and priorities. "Lonesome George," the only survivor of *C. n. abingdonii,* has failed to father any offspring with females of neighboring subspecies during his more than thirty years in captivity. On Pinzón, by contrast, there has been an excellent program for the reintroduction of subadult tortoises hatched and headstarted at the Darwin Station. On other islands, such as Isabela, the priority has been to eliminate the introduced domestic mammals. Access to certain islands is now regulated or even banned, and on the islands open to tourism, it is forbidden to touch, move, or (most certainly) take away tortoises. The principal contemporary problem is the constant immigration of continental Ecuadorians seeking a "better life" in this alleged tropical paradise. Galápagos tortoises are well represented in the zoos of the world and now reproduce quite well, at least in some collections. The offspring are not needed for reintroduction to the wild at this point, because they mostly result from "mixed marriages," but their existence may help eliminate illegal collection from the archipelago itself. But still, passing vessels have been known to collect tortoises illegally, for profitable sale to unscrupulous buyers.

© T. de Roy

In the Alcedo caldera, the tortoises become quite gregarious and spend long hours in shallow pools to protect themselves from heat and parasites.

Chelonoidis petersi (Freiberg, 1973)

Distribution. This species is found in the Argentine provinces of Santiago del Estero and La Rioja and also in the southern Chaco of Paraguay.

Description. Very close to *C. chilensis,* this vicariant species is of small size (maximum 250 mm) and has relatively smooth scutes, especially on the plastron.

Natural history. The tortoise lives in an arid environment dominated by cacti and succulents, and it hides in bushy areas and rocky crevices; it does not dig earthen retreats as other tortoises of arid lands (such as *Gopherus*) may do. Males emit sonorous yelps when copulating. The adult females lay about 10 oval eggs, around 42 mm in diameter.

Protection. Unfortunately, there was heavy collection of this species during the 1980s and 1990s, mainly for export. It is listed in Appendix 2 of CITES and classed as vulnerable in Argentina.

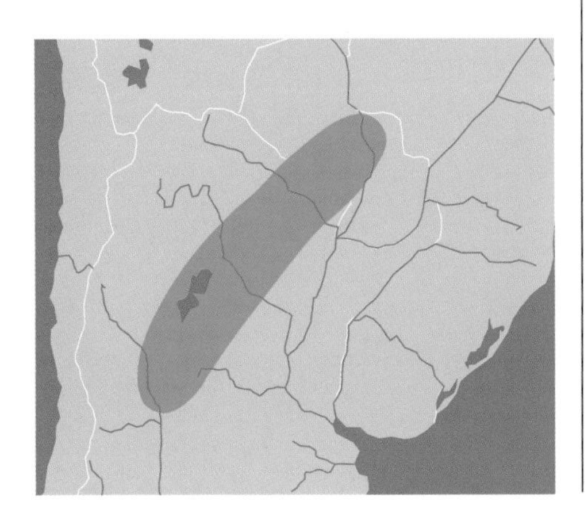

Common name

Peters' tortoise

This tortoise is reminiscent of *Testudo ibera* of the Middle East, but its length may reach 215 mm and the gular is strongly developed. This specimen has perfectly formed annual growth annuli.

© T. Vinke

Chersina angulata (Schweigger, 1812)

Common name

Angulated tortoise

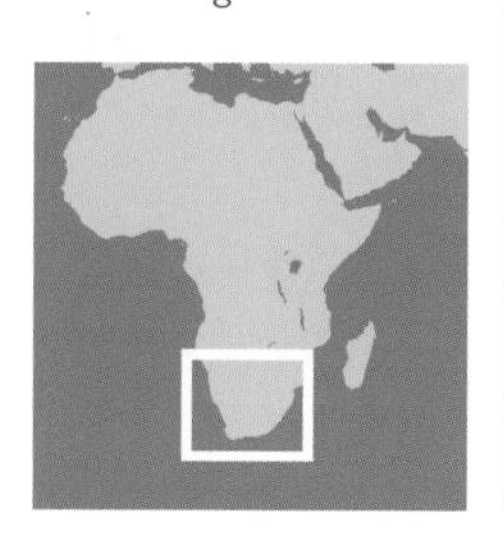

Distribution. This species is found in the southwest of South Africa (with the exception of the southern extreme of the continent) and in a small part of the southern coastal region of Namibia.

Description. The name derives from the characteristic black triangles on the marginal scutes. This tortoise somewhat resembles *Testudo graeca,* from the opposite extreme of the African continent, but differs in the following points: carapace flattened, less rounded, and taking the form of an elongate cylinder; plastron very long, anteriorly triangular, with wide humeral scutes and a very developed, single gular spur, which extends forward under the head. The anterior shell opening is closed in and narrowed, and the anterior marginal scutes are quite long. Males are larger than females—the males averaging 272 mm and the females 216 mm—which is unusual among small tortoise species. The nuchal is narrow and elongate, and the supracaudal scute is undivided. The coloration is extremely variable, even within a single population. The two basic colors, yellow and brown-black, are differently arranged in different individuals. Sometimes the animals are very light, with dark seam lines; others are primarily dark, with light seams. But the carapace scutes always have a dark center around the areola and a large, dark surrounding area, while the marginal scutes offer a series of regular black triangles, each one bisecting the scute on which it lies. The young show a comparable variety of chromatic contrasts. The soft parts are dark gray to brown, with a light spot on top of the head and on the lower jaw. The plastron is very

light, with variable designs in brown, salmon, or even pink. Extremely old animals tend toward uniformity of coloration and range from beige to brown.

Natural history. Within this tortoise's extensive coastal range, the climate varies from a Mediterranean model in southern Western Cape Province to an arid, almost desert climate in southern Namibia. These tortoises like sandy areas with a dry vegetation community like the Languedocienne *garigue.* This diversity of environments is the key feature shaping the different ways of life. *C. angulata* may be active year-round, but in the south it may hibernate. It hides is little natural tunnels but is not a burrower. In nature, it is lively and quick. Males are aggressive during the breeding season, utilizing their gular prongs in violent duels. Populations may be remarkably dense—for example, in the West Coast National Park, where the density may attain several hundred per hectare, or close to Port Elizabeth, where 400 tortoises were found in a 17 ha plot (B. Branch). The diet includes dry grasses and maquis plants, as well as snails and the feces of rabbits and goats. The species is thus somewhat omnivorous in diet. It may remain without drinking for long periods of time, but when the rains finally arrive, it may take on a large amount of water. It may adopt the posture of certain *Homopus* species during rainstorms, with the rear of the body raised, so that the water runs forward along the shell and is available for ingestion at the mouth. Mating may occur at any time of the year in the hottest parts of the range, but in the south it occurs mostly from September to April. This species, despite it medium size, usually lays only a single egg or at most two to four. The eggs are oval, about 35 mm × 28 mm on average. Nesting may be observed several weeks after mating; the eggs are laid in small sandy banks well exposed to full sun. One female may nest five or six times per year, a frequency that greatly enhances its annual reproductive potential. Incubation is lengthy, and emergence is dependent on rainfall; according to B. Branch, the maximum may be 200 days, and the minimum 95 days. The hatchlings weigh about 12 g, and growth is slow, maturity not being reached for 10 to 12 years. There are numerous predators upon the young, including black-backed gulls, certain raptors, baboons, and numerous carnivores, as well as monitor lizards. Fires and highway accidents also cause heavy mortality. Furthermore, a respiratory infection has been reported in this species, giving rise to numerous casualties.

The fact that the female usually only lays one or two eggs has led to a specific strategy. There are advantages for a given male to maintain a "harem" of females. The males fight among themselves, utilizing their gular extension to conquer their adversary. Then the winner becomes the "proprietor" of an additional small group of females. This drive to dominate and to fight explains the greater size of the males in comparison with the females of this species.

Protection. This tortoise has long been collected and sold to pet shops in South Africa and also exported to the United States. It has even been consumed, and may still be, in the northern part of the range, in the course of traditional barbecues *(skilpadbraii).* Presently it is no longer collected, but the danger comes from the extension of land cultivation and urbanization, because in South Africa it shares its territory with the human species. Natural predation is also considerable, especially by the black-backed gulls, locally known as kelp gulls, which hang out around trash heaps and garbage dumps. The species is well represented in various national parks of South Africa and does not seem to be too endangered.

© B. Devaux

Even within a population, color may vary considerably, from blackish to beige, all more or less decorated with angular markings for greater effect.

The "prescribed" coloration is as follows: a balance between the white of the marginal scutes and the black rectangles on the costals and vertebrals. The species loves to conceal itself in holes in the ground or in old burrows.

© F. Bonin

Dipsochelys elephantina

(Duméril and Bibron, 1835)

Distribution. This species is only found in the Seychelles, principally on the coral atoll of Aldabra. The tortoises present on the granitic islands or the other coral islands of the Seychelles are of uncertain origin and may have been brought there by humankind in past centuries—or more recently.

Description. *D. elephantina,* often compared to the Galápagos tortoise *Chelonoidis nigra,* is easily recognizable by the presence of a nuchal scute and by the domed, regular form of the shell, which is anthracite gray. In contrast to the Galápagos animals, it does not show color or morphological variation, except as a result of captivity. An enormous living specimen named Esmeralda on Bird Island, Seychelles, is the largest known at 310 kg. The color ranges from bronze to black, with the seams of growing animals finely outlined in white. The growth rings are easily visible in younger animals (two rings per year), but once maturity is reached, the carapace becomes smooth and even, like polished metal. The costals and the vertebrals are much less spreading than the marginals. The marginals, although vertical along the sides, are somewhat turned up anteriorly, above the forelimbs. The head of *D. elephantina* is more rounded and less emaciated than that of the Galápagos tortoises, with a pointed snout and a more "smiling" mouth. On the top of the skull there are two long, well-defined scales, which form a sort of helmet. The eyes are large, with a well-developed upper lid. A unique feature is the cartilaginous nasal chamber, which allows this tortoise to drink by sucking water in through the nostrils. This feature gives the tortoise its generic name, *Dipsochelys* ("drinking tortoise"). This detail explains how the animal is able to obtain drinking water on the dry atoll of Aldabra, where rainwater may gather only in hollows a few millimeters deep. The tortoise can ingest the water only by inclining its neck vertically so that the nostrils, at the anterior tip of the head, can break the surface.

The color of the skin ranges from uniform gray to bronze-brown. The similarity between the color the animal and that of its habitat is perfect: the blackish and metallic brown tones are those of the ancient coral of Aldabra. The plastron is also uniform gray. The limbs are long and armed with big, flat nails; the feet are elephantine. This tortoise often lies flat on the ground—limbs, head, and neck fully extended—to take the sun. The males are somewhat larger than the females (about 30% bigger), and sexual dimorphism is not especially pronounced. Only the oldest males have a deep concavity in the plastron, with the xiphiplastra greatly thickened.

History and geography. This species is the last survivor of a range of giant tortoise species that inhabited certain islands of the Indian Ocean until a couple of centuries ago. They probably originated from the African continent, reaching La Réunion, Mauritius, Madagascar, the Seychelles, and Rodrigues, where they differentiated into various species, all of medium to large size. Heavy collection during the seventeenth century by sailors and colonists brought about rapid extermination of most of these species. In Madagascar, extinction of the giant tortoises may have taken place before the first arrival of humans. The isolation of some of the Seychelles Islands, especially Aldabra, fortuitously allowed the survival of a single species, *Dipsochelys elephantina.* At the beginning of the last century, even on Aldabra the remaining population was estimated at just a few hundred individuals. But following protection of the atoll and prohibition of collection of tortoises, the population grew impressively, reaching a maximum of 152,000 individuals during the 1970s. Today this population has dropped back somewhat, the 2001 figure being 85,000 tortoises (D. Bourn). Scattered over other islands in the Seychelles are about 2,000 more tortoises.

Aldabra, a small atoll about 35 km long and

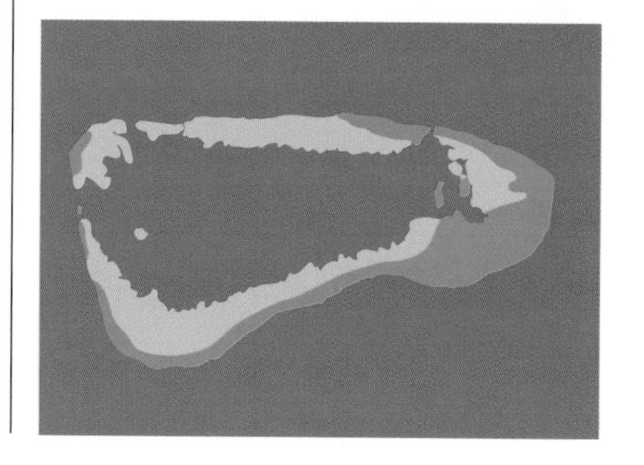

Common name

Aldabra giant tortoise

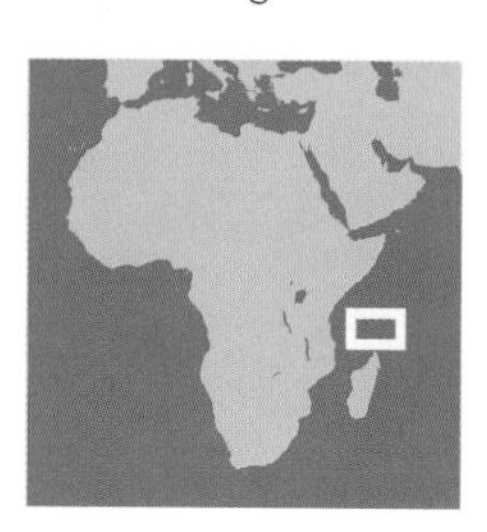

12 km high, has been designated as a biosphere reserve, and it constitutes an exceptional natural laboratory and refuge for a single species of tortoise. The biomass of tortoises on this island is the highest of any grazing animal on the planet, with 21 tons of tortoise per square kilometer, much higher than that of the large mammals of the Serengeti Plain. Aldabra is made up of four islets, of which three are inhabited by tortoises. Ninety percent of the total population of this tortoise lives on Grande-Terre, 7% on Malabar, and 3% on Picard, where there is a laboratory and dormitories for scientists. A few tortoises also live in small islets (Michel, Ile aux Cédres) in the middle of the central lagoon. A dozen or so people reside on Picard and act as guides for researchers visiting different parts of the atoll. The environment of this coralline atoll is very unusual and includes several distinct ecosystems. The densest tortoise habitat is a mixture of coral and flat rock, covered with *Sporobolus virginicus* and "tortoise turf," a low-growing plant community adapted for heavy grazing by tortoises. "Higher" plants include *Pandanus, Pemphis, Cyperus,* and *Ochna,* but around the lagoon the dominant vegetation is mangroves, which do not offer tortoise habitat. In Malabar a *Casuarina* forest offers good habitat, and on Picard the coconut palms planted by people are interspersed with *Casuarina* trees, offering a mixed habitat, highly appreciated by the tortoises despite its anthropogenic nature.

© F. Bonin

In this lagoonside paradise, under the shade of the coconut palms, the tortoises reach an excellent size and sit in state at the edge of the sea, looking for tortoise turf to graze upon.

Natural history. On each of the islands of Aldabra, the tortoises have developed unique strategies that shape both their morphology and their longevity. Courtship and copulation in this species involve neither biting nor lengthy preliminaries. The male emits at most a gentle cry, much quieter than that of the Galápagos tortoises. Most nesting occurs from July to November, before the rainy season. The eggs are the size of a tennis ball and weigh about 80 g. The studies of Swingland and Coe have yielded the following data on clutches for the three islets: on Grande-Terre, the average clutch has 5.3 eggs, each weighing 81.6 g; on Malabar, the mean is 14 eggs, averaging 90 g; and on Picard, the mean is 19.2 eggs, averaging 75 g. Incubation lasts for 85 to 157 days, and emergence depends on the arrival of the rains. About 80% of the eggs produce viable hatchlings. The young tortoises weigh 32 g on average and are entirely black and rather flat-shelled. Mortality of the young is considerable, resulting from the harsh coral terrain, the overall aridity, and predators such as land and robber crabs and introduced rats. The young grow rather quickly, and growth rings soon appear on the carapace scutes, together with a whitish ring around each scute corresponding to the most recent growth. The carapace does not take on the even black color of the Grande-Terre animals, or the bronze color of those from Malabar, until maturity is reached.

On Picard and Malabar the tortoises follow a classic lifestyle, with a morning activity period when they feed and mate, taking a siesta at the end of the morning until midafternoon, and with a short activity period at the end of the day. This applies to the strongest and largest animals, that is, adult males with weights of up to 150 kg on Picard and 80 kg on Malabar. In the Casuarina forest one may see small groups composed of females and some subadults and young, with every appearance of being happy families, although in reality they are simply miscellaneous animals that have selected the same resting site.

On Grande-Terre a lifestyle truly unique among tortoises has evolved, which was studied during the 1970s by I. Swingland. The harshness of the environment and the limited food supply have not led to the expected result of tortoises becoming quite scarce along with the survival of certain individuals of great size, but rather a remarkable degree of peaceful aggregation has developed, with a proliferation of tortoises of subadult size. Over several decades, the tortoises became extremely numerous, reaching densities of 250 animals per hectare in the Cinq Cases region. But this excessive population density has had the corollary of early mortality and the development of a sort of neoteny in this population.

This species, as its name implies, can "drink through its nose," by placing the snout vertically into shallow puddles and aspirating vigorously.

© B. Devaux

The tortoises rarely even reach maturity, and many remain as nonbreeding subadults. The ossification of their shells bears witness to this; the shell fontanels remain open throughout life, and the carapace is very poorly ossified and remarkably thin. The tortoises usually get no heavier than 30 kg, with a maximum of 60 kg among some males, and they often die even before their twentieth year by entrapment in pits in the coral, dehydration, or wasting away on an inadequate diet.

In order to survive at all, the tortoises have initiated a sort of ritual that is unique among chelonians. They pursue an extremely rigid and defined lifestyle designed to optimize their chances of survival. From 11 a.m. to 4.30 p.m., during the hot hours of the day, they remain hidden under shading vegetation, especially certain large-leaved trees of the species *Guettarda speciosa.* By 5 p.m. they all come out at the same time and walk en masse over to the tortoise turf plains, where they graze in close formation in vast herds. When night arrives, they stop moving at whatever location they have reached and spend the night sleeping on the spot. In the morning, at 6 a.m., they start to move about and eat, defecate, and mate. Then at about 10 a.m., they return to the high dunes and seek out the large *Pandanus* and *Guettarda* trees that offer shade. By 11 a.m. they are concealed once again for their daily siesta. Not a single tortoise varies from this routine. Any tortoise that ventures forth during the hot hours and gets onto the open coral is condemned to rapid death through dehydration and overheating. In this same Grande-Terre population, Swingland has discerned two other contrasting strategies: some tortoises remain inland and travel only a short distance, while others make the longer migration to the seaside tortoise turf fields. Each strategy offers advantages and disadvantages, measured in terms of longevity and average weight of individuals. This extremely demanding response to a harsh habitat is evidence of adaptive success, despite the short life span, as measured by the numbers and biomass of this population, which, as we have mentioned, exceeds that of any other vertebrates.

A few tortoises, confined to the islands of Malabar and Picard, do not play the game as regards to group activities and have been observed to travel exceptional distances. Some have traveled 25 km over several months, and others arrive by sea (having presumably tumbled in from a coral promontory), to arrive on one of the few beaches on Picard. They swim quite well, and this is the mode by which their ancestors presumably arrived from Africa. The adaptability of the species is exceptional, as its strategies for survival on the atoll demonstrate. This adaptability has also allowed them to resist the terrible raids of pirates and sailors, who kept them in rock corrals for months without food or water and slaughtered and ate them at their convenience. And they also adapt to captivity in completely unsatisfactory situations, including small cages, roadside zoos, hotel gardens, and parks in cold countries. Under such situations they often develop deformed carapaces, with high pyramiding of the scutes or grotesquely flattened shells, which demonstrates the great plasticity of the growth process. It is doubtless thanks to this plasticity and this adaptability that they have survived to our times, in contrast to all the other giant tortoises of the Indian Ocean islands. Their longevity has often intrigued humanity. Some have credited them with ages of 150 years, and others even several centuries. The truth is that they can live to 100 years without too much difficulty, but their astonishing capacity for survival, even under the worst conditions, stops at that point.

Mating is somewhat discreet in this species and less boisterous than that of the African spurred tortoise or the Galápagos tortoises. This male seems to be "attentive."

Protection. Although listed in Appendix 2 of CITES, this tortoise is still exported each year from the Seychelles and shipped to Western zoos and private collectors. On Aldabra the Seychelles Island Foundation maintains permanent vigilance and guarantees enlightened conservation. Nevertheless, one fears that this single island population could be eliminated by rising sea levels, a freak hurricane, a tsunami, or some other disaster. So it becomes necessary to develop other natural populations on islands closer to Mahé—like Curieuse, for example, where a free-ranging outlier population was established in the 1990s, although it has had recent setbacks. Certain more distant islands, like Mauritius, or sanctuaries like the Vanilla Crocodile and Tortoise Park have developed captive breeding programs, and there has been introduction of a population to Rodrigues also. In the Seychelles themselves, several tortoise breeding facilities exist, the original stock having been imported by Arab sailors, and tortoise populations have survived in Zanzibar for several centuries. This species is also displayed in almost all the zoos of the world, although this may not constitute an actual contribution to its conservation.

Possible *Dipsochelys* Species

Two species thought to have been extinct have potentially been rediscovered by Justin Gerlach (1998) in enclosures in hotel gardens in Mahé. About 10 of these adult tortoises are currently being maintained and are being bred on Silhouette Island, near Mahé, by the Nature Protection Trust of Seychelles (J. and L. Gerlach), and young have been produced in recent years (2003–2004). The parents have been attributed to the species *Dipsochelys hololissa* (Günther, 1877). This is a large form, reaching 1,140 mm in carapace length, and the males may reach 200 kg in weight. The second, third, and fourth vertebral scutes form a horizontal plane, often with very deeply incised seams. The midmarginal scutes form a right angle with the plastron, which is often enlarged and quite thick. The carapace has a very smooth, finely polished appearance, without rough areas or growth annuli, and often is an intense black color. *Dipsochelys arnoldi* Bour, 1982, reaches 875 mm. The carapace is lower and flatter than that of *D. hololissa.* The shell is somewhat domed at the level of the first vertebral scute, and there is a lateral keel between the first and second costals. Specialists are still divided on the subject of the validity of these two species.

© B. Devaux

© B. Devaux

Two photos taken on Silhouette Island, in the Gerlach Center. At the top is *Dipsochelys arnoldi,* with a furrow between the first and second costal scutes. The lower photo is of *D. hololissa,* with a flat area involving the second, third, and fourth vertebral scutes.

Geochelone elegans (Schoepff, 1794)

Distribution. The range of this species has two distinct parts, one in the northwest from Gujarat into Pakistan (Sind), and the other from the south of the Ganges Delta to the southern tip of the peninsula and into Sri Lanka. Some of the alleged records from Pakistan and Bangladesh are from humid localities and are probably in error, this being a species specialized for arid zones.

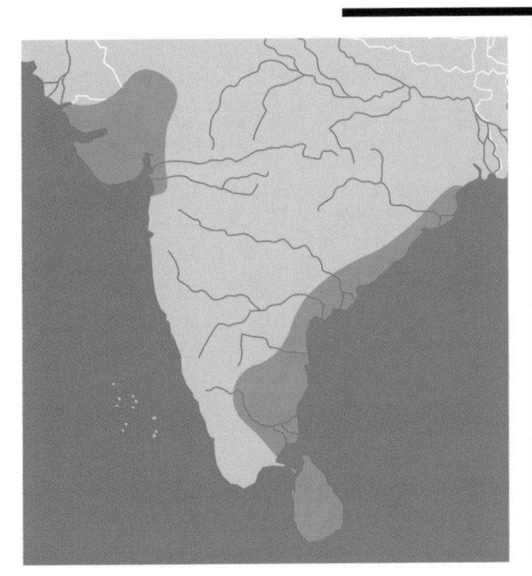

Common name

Indian star tortoise

Description. Numerous tortoise species in the world, from South Africa and Madagascar to India, may be described as "starred." Although these species may seem to be so similar as to be closely related, the patterns represent convergence of evolution in response to factors of climate and ecosystem, rather than relationship (R. Bour). To

© I. Das

In certain individuals the light lines are fewer in number but are always present, as is the pyramidal form of the vertebral and costal scutes.

The general pattern recalls the star tortoises of Madagascar or South Africa. However, the scutes often show pyramidal formations, and the radiating streaks are less fine than those of *A. radiata.*

distinguish the Indian star tortoise from the others, the following characters should be used. The Indian star does not exceed 380 mm and 7 kg; the larger radiating scutes of the carapace often have a conical form (especially in Sri Lankan females); the posterior marginals have lightly serrated edges, especially in juveniles; the middle scutes of the carapace are somewhat narrowed; and there is no nuchal scute. Furthermore, the black rays forming the stars on the scutes extend in all directions, while in *Astrochelys radiata* they form a fanlike design on only about one quarter of the scute. The areolae are often brown to orangish and form slightly raised platforms. The limbs are rather short and have large round scales, encircled with black. The head is also rather small, with small yellow scales on a black background. The plastron is yellow with black rays. The females are larger than the males, and the latter have a somewhat concave plastron, with thickened xiphiplastra. The young are almost entirely yellow or orange-yellow, with dark markings along the seams.

Natural history. This tortoise prefers the dry areas of Tamil Nadu, with a flora of herbaceous plants, acacias, and euphorbias, but it is also found in some green prairies farther to the north and the west of India. During the dry season, these animals are active in the morning and hide during the day. They become very active when the rains come, during which time they mate and feed heavily. This is an herbivorous species, showing a preference for fruits and vegetables, as well as succulent plants, while during the dry season it subsists upon dead leaves and spiny vegetation. In Sri Lanka it is known to consume the fruits of the pawpaw tree, although doing so may lead to death through intestinal impaction (Hunt). It has also been seen feeding on lizards, dead rats, and insects. In the western part of the range, it spends several weeks each year in hibernation when the temperatures become cool, especially at night. Sexual maturity arrives quite early: some tortoises have been seen copulating at just five or six years of age. The mating season coincides with the monsoon season. The first rains cause the males to sport impressive erections, as happens also with *A. radiata,* but their courtship efforts are brief and rather gentle; they do not injure the females nor do they bite them. The males emit a soft groan during actual intromission. Most nesting occurs from May to June and again in October. The eggs are elliptical or nearly spherical and are rather large for this somewhat petite species, measuring about 40 × 32 mm. The eggs number three to six, with a maximum of four nestings per season. Incubation lasts for 110 to 130 days. In the course of an entire season, a female may lay about 24 eggs.

© I. Das

Protection. The numbers of this species are dropping fast, and it faces three major threats: collection for international commerce, especially to the Far East and to Western counties; collection for human consumption (more than 100,000 are sold in a single Calcutta market each year); and growing pressure from humans, who transform habitat to agricultural land and urbanization. According to I. Das, formerly of the Madras Crocodile Bank Trust, India is being deforested rapidly, and high-density pastureland as well as the spread of cultivated zones have precipitated a rapid diminution in tortoise populations, which at one time occupied several tenths of the entire area of the nation. More-rigorous protection measures are needed, including the placement of this species in Appendix 1 of CITES (it presently is only in Appendix 2), the establishment of truly effective and well-patrolled reserves, and broad public education. Export must cease altogether, and studies of the status of populations need to be undertaken.

Geochelone platynota (Blyth, 1863)

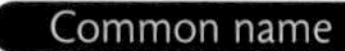

Burmese starred tortoise

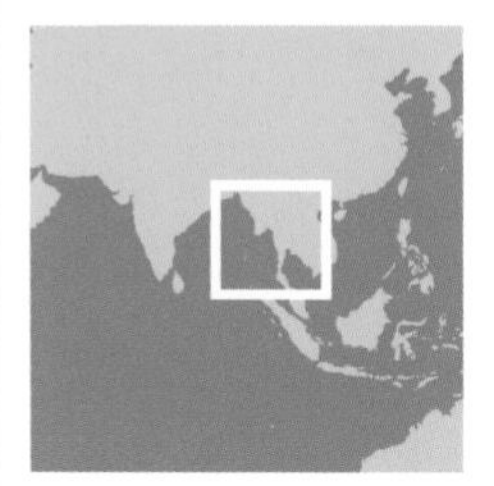

Distribution. Found only in Burma, this species is restricted to dry forest habitats in the central and western areas.

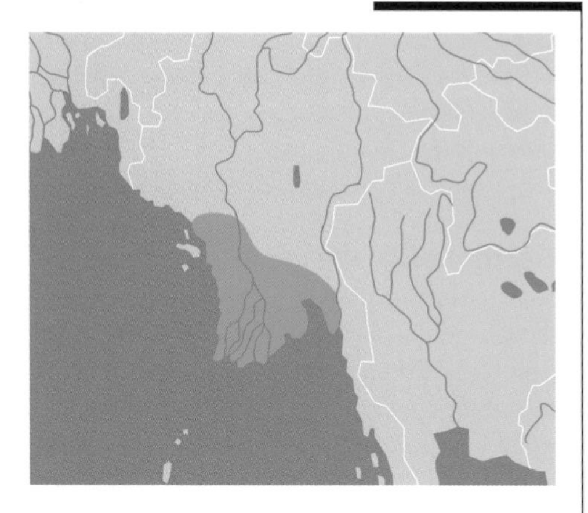

Description. This species is very close to the Indian *G. elegans,* of which it may be thought of as an eastern representative. It reaches a length of about 30 cm, and the carapace scutes are always flat, as its name indicates—that is, the scutes do not have the conical form of those of the Indian star tortoise. The shell is perfectly oval and lacks serrations along the posterior edge. The coloration is somewhat more restrained, with the dark rays fewer in number and the background creamy white rather than yellow. It strongly resembles *Psammobates geometricus* but is distinguished by the absence of the nuchal scute. The limbs are covered with small tubercles, yellow in color, on a dark background. There is a strong "claw" on the end of the tail, and spurs on the thighs. The plastron is generally decorated with a black triangle on each scute, with a light yellow or orange background. The head is small and uniformly brown-cream.

Natural history. This species shares its dry forest habitats in Burma with the more widespread *Indotestudo elongata.* It is rarely observed in the wild. Most activity takes place toward the end of the day, and the species is omnivorous in diet. The eggs are large and elongate (55 × 38 mm). They are deposited in February in beds of humus, which assure effective incubation.

Protection. The rarity of this species throughout its small range is a source for concern. The Burmese people have caught the tortoises for food for a long time, using trained dogs to find them. In former times at least, the empty shell was often used as a vessel. Deforestation has reduced habitat seriously, and in recent years, after being unknown to the outside world for a century, a disturbing number of specimens have shown up in the specialized pet trade. The low populations are a warning that better studies and better protection are required. This species is classified as endangered by the Turtle Conservation Fund.

The pattern is generally well proportioned. Wide light rays, relatively few in number, boldly mark the blackish background.

Homopus areolatus (Thunberg, 1787)

Distribution. Endemic to South Africa, this species occupies the southernmost extreme of the African continent, in the Cape Provinces, along the coast from East London to the east and Cederberg to the west, and reaching altitudes of 1,300 m.

Description of the genus *Homopus*. The name *Homopus* means "feet all the same," although in truth only two of the five species in this genus have the signature feature of four claws on both forelimbs and hind limbs. The five species, which are very similar, are small (maximum 160 mm) and flat and have a sort of dwarfed aspect, due no doubt to the aridity of the environments that they occupy and to their own special ecology. They live almost entirely in South Africa, within small isolated ranges, but one of them, now known as *Homopus bergeri (solos),* is endemic to Namibia. They are restricted to dry, rocky areas, where their low body profile as well as their long limbs and strong claws allow them to slip away into rocky retreats and to ascend vertical cracks like an expert rock climber. One of them *(H. signatus cafer)* even has a soft shell, similar to that of the famous pancake tortoise *(Malacochersus tornieri).* The bony structure retains some primitive aspects. The maxillary surface is not serrate or dentate, and the gular region of the plastron is very thickened. One of the *Homopus* is the smallest tortoise in the world, reaching a maximum of 100 mm in adult females. The head is relatively large in comparison with the body, and the scutes are dense and thick, with annuli tightly bunched together and sometimes deeply incised.

This species remains small, and the scutes become shriveled up and dwarfed to an extreme degree. Sometimes, the scales form bony "coffee beans," unique among tortoises.

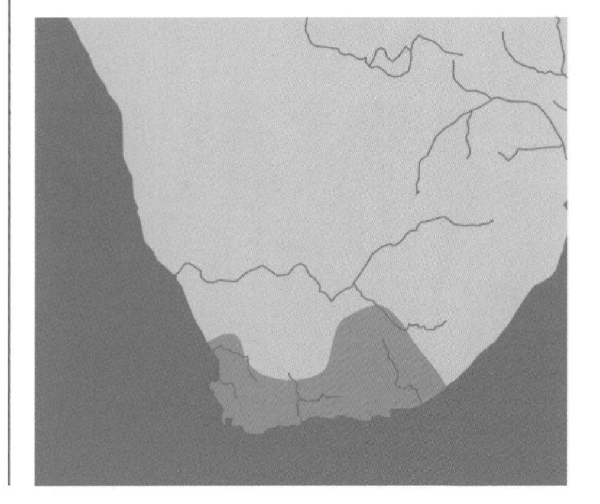

Common name

Beaked Cape tortoise

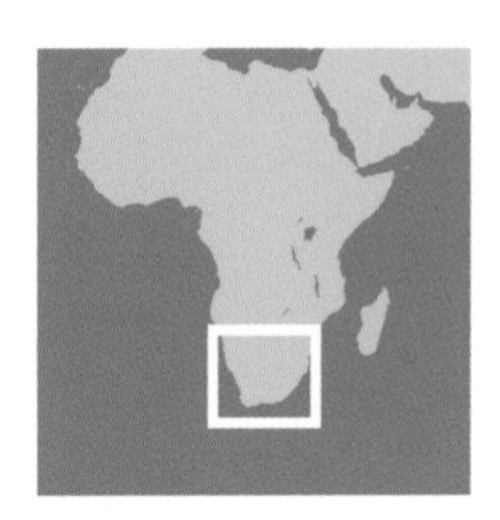

Description of *H. areolatus*. This species does not exceed 120 mm (in females). The scutes have deep annuli, each growing outward from a recessed, quadrangular, fully pigmented areola. The carapace is the least flattened in the genus, and it gives the impression, with its wide, flattened areolae, of a short skirt. It is wider behind than in front. The marginals are curved upward above the posterior limbs. The coloration resembles that of a Grecian frieze, beige and black, with brown areolae and sometimes a grayish base. The plastron is yellowish with several brownish areas, or with black sections around the middle. The head and limbs are yellowish, beige, or brown. There are just four claws on each of the four limbs. There is no plastral concavity in the adult males, but they have a stronger head and a more pronounced snout than the females. Adult males are often more orange and have an overall lighter color scheme than the females. The young are darker than the adults, almost brown-black. Their areolae are distinguished by being lighter and more flattened.

Natural history. This tortoise is almost invisible in its native habitat, and it occupies well-shaded areas and small hiding places under bushes, from the edge of the sea to the tops of the hills. When disturbed, it emits a loud groaning noise. During the mating season (June–January), the males are quite combative, squaring off and fighting vigorously. During such times the prefrontal scales become bright orange-red. Nesting occurs from August to November and sometimes in April also. The eggs average just two in number, with a maximum of four, and have an elongated form, measuring 35 × 22 mm. Incubation can be very long (5–10 months), but actual emergence occurs following rain. Hatching takes place in March and April, which is the wet season.

Protection. This species is placed in Appendix 2 of CITES. It is completely protected by national law, as are the other *Homopus,* and seems to be rarely collected. It is not found in Western collections. The principal stresses result from habitat disturbance, because it seems to be very specialized for its intact habitat, and human inroads are becoming more severe. There are also many predators, including ostriches, domestic

dogs, baboons, and numerous raptor species. Unfortunately, it seems that this species is slowly and progressively becoming rarer and rarer. Status surveys are necessary.

This specimen, with its contrasting colors, is more representative of the species, but it still has deeply sculptured scutes and sunken areolae.

© B. Branch

Homopus bergeri (solos) (Lindholm, 1906)

This species was not included by P. Pritchard in his *Encyclopedia of Turtles,* nor by IUCN in its classification. But it was cited and described by W. Branch, the South African tortoise specialist, and it was he who introduced the name *H. solos* (because of the species' isolation in Namibia).

Distribution. This species is endemic to Namibia, south of the desert, in the granitic mountains near Aus.

Description. This is the rarest of the five species of the genus, and it occupies the most arid habitat. It is very small, with a maximum length of 110 mm (in females). The carapace is very flat, with depressed areolae and deep seams between the scutes. The nuchal scute is small and narrow. Each of the forelimbs has five claws. The marginals are not raised posteriorly, in contrast to those of the other *Homopus.* The coloration is reddish brown, with light rings around the areolae and sometimes a reddish tinge on the seams between the scutes. The plastron is similarly colored, with dark borders. The head is uniformly brown, and the beak is somewhat pointed. The tail ends in a horny spur. Males have a slight plastral concavity.

Natural history. This tortoise occupies a very arid environment, with many rocks and stones, not far from the sea. In this particular habitat, the species is mainly active in the mornings and especially during rainy episodes. But it spends most of its time hidden under flat rocks, and it estivates during the warm months. Farmers near Aus sometimes have small populations of these tortoises on their lands, and these areas constitute informal mini-reserves where small groups of tortoises can live in peace. This is a very lively species, capable of climbing up steep rocks, and in many ways it resembles the pancake tortoise, *Malacochersus tornieri,* of Tanzania, although the former is smaller. It is mostly vegetarian but may consume the occasional dead insect or scrap of organic debris. The worst predators are crows and hyenas; the minute size of even the adult tortoises makes them extremely vulnerable to predators.

Protection. This species has only small, scattered populations. It is critical that status surveys and conservation efforts be instigated as soon as possible.

© B. Devaux

© F. Bonin

This rare, flat, and minuscule tortoise is able to hide under rocks and in crevices. The male (at left in the lower photo) is almost always lighter in color, livelier, and more aggressive than the females.

Common name

Berger's Cape tortoise

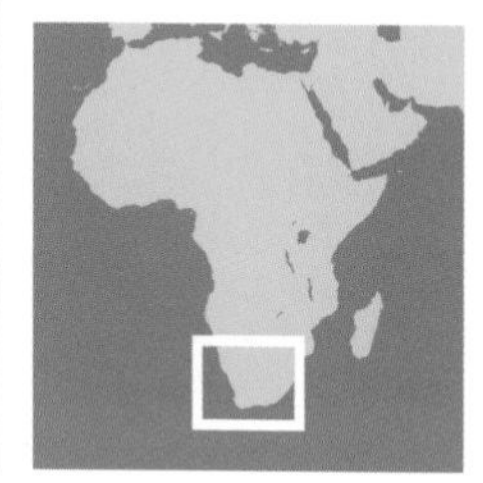

This specimen, observed in the arid country of Namibia, is just 80 mm in length. The head is large, the scutes already show deep growth annuli, and the limbs are covered with pointed scales.

Homopus boulengeri Duerden, 1906

Distribution. This species is endemic to South Africa and occupies a more restricted area than *H. areolatus,* in the far interior and not in the coastal zone. It occupies rocky escarpment country in the Great Karoo.

Description. The species name derives from G. A. Boulenger, a curator and herpetologist at the Natural History Museum in London. This tortoise is rather small, the females reaching 110 mm and the males 90 mm. In shape it is oval, without posterior widening, and it is the flattest member of the genus. The carapace is often depressed, and each scute is flattened, with large, pigmented areolae. The first vertebral scute is square and very large. Each of the forelimbs bears five claws, and there are four on each hind limb. There is no cornified spur on the end of the tail, which distinguishes it immediately from *H. bergeri (solos).* The males have a moderate plastral concavity. The beak is feebly developed and is slightly tricuspid. The coloration varies, but in general reddish brown and beige tints predominate. Young specimens sometimes have black lines following the carapace seams. The plastron is yellowish to reddish, with darker lines around the scutes. The skin is yellowish to orange, and the limbs bear large, imbricated scales.

Natural history. These tortoises live in rocky cracks, and their coloration matches their substrate. They emerge mostly on stormy days and also when the weather is cool. It has been observed that crows will carry these turtles away in their claws and smash them against rocks before eating them. This species is remarkable in that it lays only a single egg, but the egg is so large that successful oviposition seems to be an impossible task. F. Siebenrock noted that the elongate egg may be 55 mm long and 25 mm wide, whereas R. C. Boycott and O. Bourquin reported average dimensions of 35 × 24 mm. The actual laying of this egg necessitates some degree of mobility of the hind

Common name

Boulenger's padloper

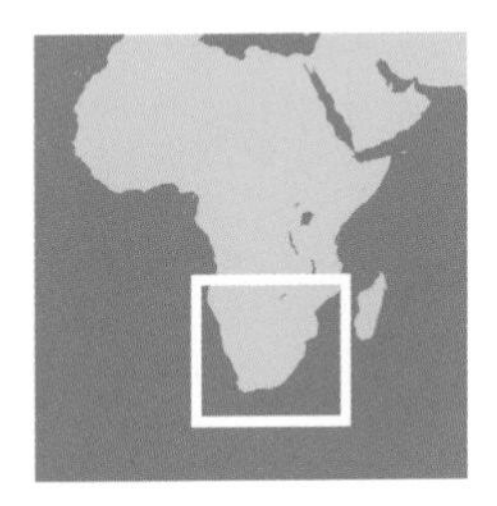

lobe of the plastron, but it is also possible that the egg itself is a little soft up to the time of laying.

Protection. This species is placed in Appendix 2 of CITES. The very low reproductive potential and the limited natural range suggest that the overall population is small, but the actual status needs to be determined. Happily, the ecosystem is not significantly damaged by agriculture or urbanization. The species occurs in parks and reserves in Western Cape Province, including the Karoo National Park. Special permits are needed to keep specimens in captivity. Furthermore, the species does not adjust well to captivity, and captives usually die within a short time.

This species in unicolored, often beige or reddish. The scutes are always deeply incised.

© B. Devaux

Homopus femoralis Boulenger, 1888

Distribution. This species is endemic to South Africa, but its range is the widest of any *Homopus.* The range extends farther east than that of either of the preceding species and also nearer to the ocean, from the Karoo Plain in the west to the Free State in the east. It also occurs in southeastern Northern Province, but probably only in relictual colonies.

Description. This is the largest member of the genus, with females reaching 153 mm and males 111 mm. They have paired spurs on the thighs, giving them their name of *femoralis.* The shell is oval and flattened, and the scutes and the areolae are perfectly flat, which gives the animal a "short-skirted" look. The supracaudal scute is single, and the rear shell margins are not elevated. All four limbs have just four claws each, and they bear numerous strong tubercles, typical of burrowing tortoises. The beak is tricuspid but is not projecting. The overall color is a uniform olive or brown, and it is the only *Homopus* that shows no contrasting markings on the scutes, although there may be fine white margins to the scutes. The plastron is yellowish or brown. In the juveniles, the plastral scutes are outlined with black.

Natural history. This tortoise is specialized for life in rocky areas and may also occupy abandoned termite nests; it hibernates in rocky crevices from June to September. Among its various predators, monitor lizards are notable for ripping open the nests and eating the young tortoises, but dogs, baboons, and raptors also consume their share. The clutches average about three eggs, each measuring about 30 × 26 mm.

Protection. This species is listed in Appendix 2 of CITES. In general it seems to be rather scarce.

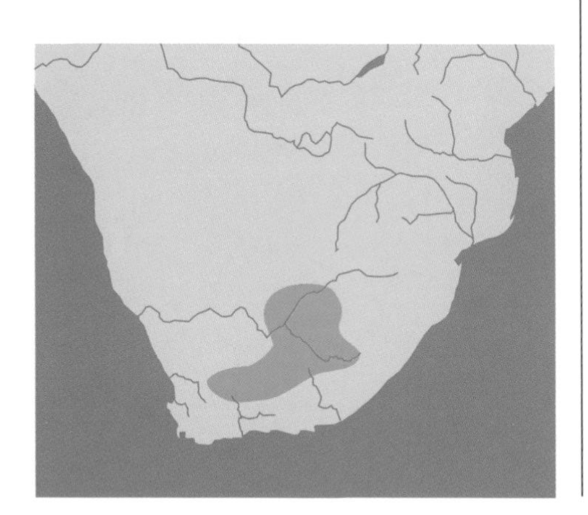

Common name

Karoo tortoise, greater padloper

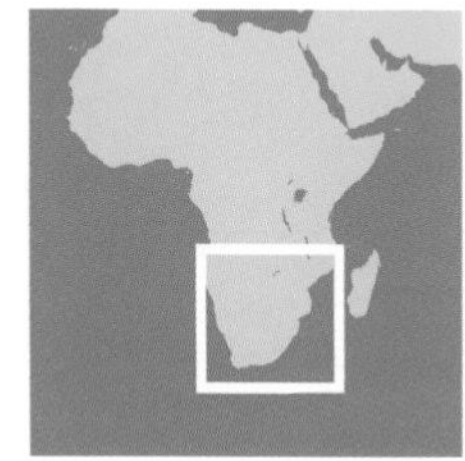

© B. Branch

In the rocky areas where it has the habit of concealing itself, it is presumably not seriously threatened, but wildfires are frequent in these dry areas and probably kill numbers of tortoises. The species occurs in certain national parks, including the Karoo, Mountain Zebra, and Addo Elephant parks. As with all *Homopus,* better understanding of the survival status of this species is necessary.

The best way to recognize this species is by the enlarged spurs on the thighs. The size is also useful, in that this is the largest of the *Homopus* species.

Homopus signatus (Gmelin, 1789)

Distribution. This species has a narrow range, extending from southern Namibia and along the western and coastal areas of Western Cape Province in South Africa.

Description. This is the smallest of all the Testudinidae, and indeed of chelonians in general, with the females of the subspecies *H. s. cafer* reaching a maximum of 100 mm and the males reaching just 80 mm. The appearance is quite attractive, with numerous fine regularly spaced spots and reddish or black rays on a light orange or beige background. This tortoise is very flat, with the scutes also correspondingly flat, or even actually depressed, and with deeply incised seams. There may also be a slight "waist," or narrowing around the middle of the animal. The posterior marginals are sometimes serrated, and there are five claws on each forelimb and four on each hind limb. The first vertebral scute is square and quite small, in contrast to that of *H. boulengeri*. The skin is orange, beige, or brown to grayish. The forelimbs are covered with pointed tubercles on their exposed sides. Large and strong, they allow the animal to climb steep rocks with ease. The plastron is ivory, beige, or brown, with diffuse black markings.

Common name

Speckled tortoise

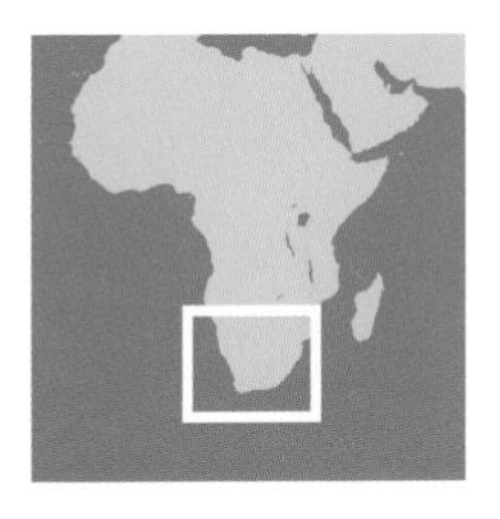

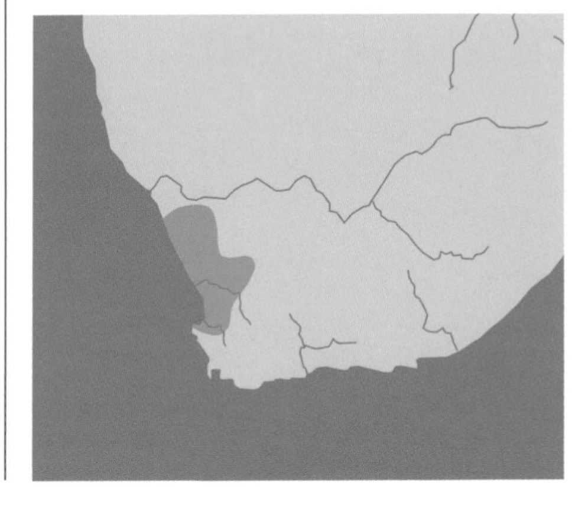

Subspecies. The nominal subspecies *H. s. signatus* (described above) is found in southern Namibia to northwestern South Africa. It occurs in some abundance near Springbok. With a maximum length of 106 mm (in females), it has little black dots on an ivory background (hence the name *signatus*), depressed areolae, recurved and serrated posterior marginals, and a nuchal scute that is wider than long. *H. s. cafer* (Daudin, 1801) occupies a very small range in northwestern South Africa, living in isolated rocky outcrops where its coloration offers almost perfect crypsis. This subspecies has an orange to reddish—or even salmon—tint, dotted with fine black spots. The areolae are large and are not depressed, and the marginals are less serrated than in the nominal subspecies. The nuchal is narrower than it is long. This is thus the smallest species in the world (100 mm for females and 80 mm for males). The subspecies name derives from the old geographic name "Caffraria."

Natural history. This tortoise lives only in rocky places, and in Afrikaans it is called *klipskilpadjie,* or "little rock tortoise." It lives on dry hillsides in arid places up to an altitude of 1,000 m, often not far from the sea. It climbs with ease along the edges of rocky outcrops and hides itself under scree piles; it can hold on to precarious sites with both its feet and its beak. This is a very agile tortoise, and the limbs are correspondingly elongated, as well as being quite limber. A true climber, it is adapted for its rocky habitat, and the coloration is evocative of the pebbles and gravel of its environment, as is the case with *Pyxis arachnoides* in Madagascar. It seems to tolerate the extreme aridity of its environment and has never been reported to drink. It may dig small tunnels, about 30 cm deep, to escape the extreme heat. These animals are most often seen in winter, resting immobile on a rock, busy with the task of thermoregulating, and at these times they show maximal activity. They eat insects, grass, herbs, earthworms, and dead animals, as well as vegetative debris. During the time when wildflowers are in great profusion, just after the September rains, they consume great quantities of both flowers and succulents. Copulation may be observed from September to October and from June to July. In the course of courtship, the males brush and stroke the face of the female in almost frenetic fashion. The eggs are huge for such a small animal (34 × 24 mm), but the female usually lays only one at a time (maximum is two). The plastron is somewhat flexible toward the rear, which facilitates oviposition. The growth of the young is rapid. At the age of two years, they may already have attained 60 mm, and sexual maturity is reached quickly.

© F. Bonin

This youngster measures just 40 mm.

Protection. This species is placed in Appendix 2 of CITES. These tortoises are highly sought after by collectors, because of their attractive markings and colors (especially in *H. s. cafer*). The nominal subspecies is still quite widespread, and field tracking studies have been undertaken by the Western Cape Nature Conservation Department. On the other hand, the subspecies *H. s. cafer* survives—and even then only in small numbers—in isolated places that urgently need to be protected to avoid its widespread disappearance, especially through wildfires and overcollecting. The nominal race is protected in the Hester Malan Nature Reserve (Goegap Nature Reserve) and in the Richtersveld National Park.

© F. Bonin

The adults are very flat and do not exceed 80 mm. Adults of this species are the smallest mature tortoises in the world.

Indotestudo elongata (Blyth, 1853)

Distribution. It is probable that, in former times, all three species of *Indotestudo* had a continuous range from the Western Ghats of southern India to central Indonesia. Today this range is fragmented, and the discontinuity probably results from the Pleistocene, according to Auffenberg, and was brought about by the progressive reduction of forest cover. *Indotestudo elongata* occupies the central part of this composite range and also has by far the widest range of the extant taxa. Its range includes southern Nepal, eastern India, southern Malaysia, and southern China (Guangxi).

Description. Until recently there was considerable confusion surrounding the three *Indotestudo* species. A publication by Pritchard (2000) clarified the situation and differentiated the three taxa (two of which had customarily been synonymized) by means of three key characters, which we use

During the breeding season, the males have a yellow or whitish head, which is easy to see in the humid understory vegetation where this species lives.

here. Generally, *Indotestudo* species are somewhat elongate in body form and rather flattened in shape. Males may reach 330 mm, and females about 290 mm. The carapace may be depressed along the third and fourth vertebral scutes. In *I. elongata* the highest point of the shell is at vertebral 3 (I. Das). The marginals are somewhat serrated at the sides, and the carapace as a whole is heavy and massive. *I. elongata* is distinguished by the following: a nuchal scute is present, always long and narrow, with parallel sides; the ratio between the interpectoral and the interhumeral seams is about 0.7 on average (i.e., the interpectoral is unusually long); and the color of the carapace and the plastron is cream to greenish yellow, with relatively small and scattered dark markings, so that some old animals appear "washed out" or almost albino. On the plastron, the black markings are few and often are limited to the abdominal scutes, in contrast to the other species. Some old specimens have a brown or reddish coloration, with dark spots.

The skin is yellow to gray-brown, and the scales on the forelimbs are well developed. This lends justification to the vernacular names and the astonishing color of the head during the mating season. Initially yellowish to grayish, the head becomes bright yellow or pink, or even completely white, from the snout to the rear of the skull and sometimes including the neck also, doubtless due to a change in the vascularization at these times.

Natural history. This is a tortoise of humid forests, but it also lives in rocky and moderately dry habitats, as long as there is some heavy rainfall from time to time. The animals hollow out retreats, or "beds," in the moist humus or in friable ground, taking advantage of their robust limbs and heavy scales, as well as their cylindrical, flattened body form. Gairdner (1931) remarked that activity was maintained even at extremely high ambient temperatures (48°C). According to Swindells and Brown, this species may use it own saliva to moisten and cool the neck and limbs. Primarily frugivorous, it also enjoys slugs, dead animals, mushrooms, and general detritus found in the moist substrate. Its diet and mode of life are very similar to those of the South American yellowfoot tortoise, *Geochelone denticulata.* During the dry season, these tortoises hide in leaf piles in a state of semi-estivation. They come out at night to feed, after several days of complete immobility. J. Anderson observed that they give forth a loud groan at the point of intromission and when they are disturbed. The eggs are large (50 × 40 mm), somewhat elliptical, and few in number (four or five) and are laid from July to October, with two or three nestings per female per year. Incubation is quite rapid in this warm, humid environment and is completed within 100 days.

Protection. This species is widespread in Asia and is consumed by various ethnic groups in China, Burma, Thailand, and Cambodia. In Vietnam it is not eaten but is collected in large numbers for resale in China. Some of the tortoises are destined for export to industrialized countries, especially Japan and the USA. They are used as a source of traditional medicines, the most famous being *gui-ban,* a concoction used for enhancement of virility in Hong Kong. The natural range of the species has dense human populations and is heavily exploited by them. Forests are being cut and burnt, and it is probable that the populations of tortoises are in fast decline. P. Pritchard (2000) considers that the natural retreat of the genus and species, which may have commenced in the Pleistocene, is now proceeding more rapidly than ever. I. Das (Madras Crocodile Bank Trust) recommends that the actual status be evaluated with all dispatch, and a brake put on the unsustainable commerce in the species.

Common name

Red-nosed tortoise

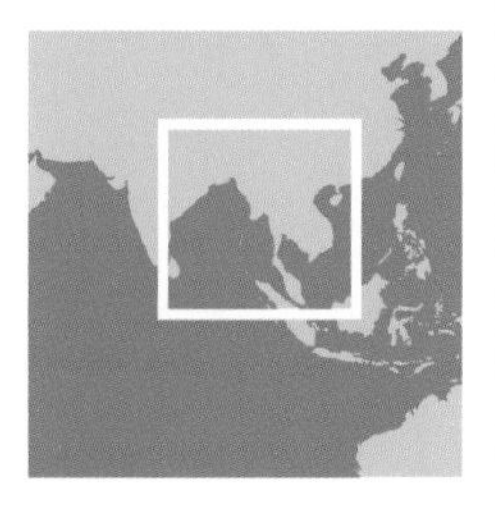

© F. Bonin

This old male takes the sun in a clearing in Vietnam. The white head suggests that the photo was taken during the breeding season.

Indotestudo forstenii (Schlegel and Müller, 1840)

Distribution. This species is the easternmost representative of the genus, inhabiting part of the island of Sulawesi (formerly Celebes) in Indonesia. The reported presence of the species on the island of Halmahera appears to be in error (P. Pritchard). Some have claimed that its presence in Sulawesi is the result of human introductions from India. But the paper by Pritchard (2000) demonstrates that the distribution of the genus is fragmented and naturally relictual and has led to speciation events such that the southwestern Indian tortoises *(I. travancorica)* are quite distinct from those of eastern Asia and of Sulawesi. Even though there may indeed have been much commerce and human relocation of specimens, which probably led to much consumption of tortoises, no new wild populations resulted, as far as we know.

Description. This species is smaller (males to 290 mm) than *I. elongata* and is distinguished by the following details: the nuchal may be present or absent (present in five out of nine, according to Pritchard); when a nuchal is present, it is short, is wide at the back, and has curved (concave or convex) sides; the ratio between the interhumeral and interpectoral seams is 1.93 (i.e., considerably greater than in the other species); and the coloration is beige-yellow to brown or slightly reddish, with strong black markings with square sides on all the scutes. Each of the costals has a black spot somewhat separated into a "high" and a "low" part, with little pigment near the center. In the plastron, which is yellowish or beige, there is a large black marking on each of the abdominal scutes, the anterior border of which is straight and parallels the pectoral-abdominal seam, and there are small marks on the other scutes. The head is pinkish to whitish during the mating season.

Natural history. Most known specimens of this tortoise are (unfortunately) in captivity, but many aspects of its life are probably somewhat similar to those of *I. elongata.* According to Frank Yuwono (cited by Pritchard), the habitat is composed mainly

© F. Bonin

Examination of the plastron allows one to distinguish between the species of *Indotestudo* (see text).

Common name

Sulawesi tortoise

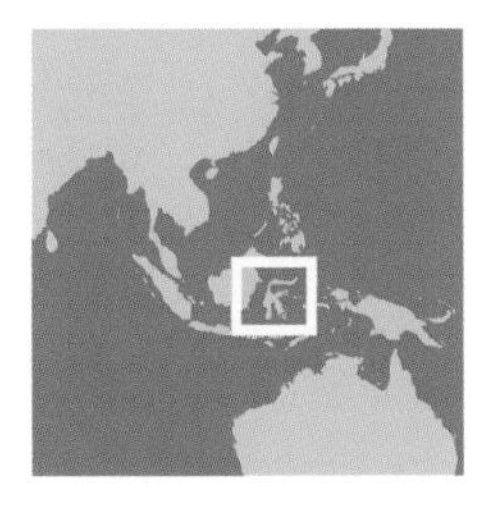

Much less dark and often pale yellow in color, this species is more retiring than *I. elongata,* but it is sometimes found in captivity, mixed in with its congener. It is exported to China for human food.

of rocky hills with low rainfall in a very limited area of north-central Sulawesi, between Palu and Poso.

Protection. The status of this species in Sulawesi is poorly known, but it is fair to assume that populations are extremely reduced. Large numbers of the tortoises are shipped to the West as pets; this anomaly derives from liberal export laws in Indonesia, whereas almost all of the range states for the other two species of *Indotestudo* do not permit commercial export. We recommend promotion of this species to Appendix 1 of CITES as an initial step. Conservation action for this severely endangered tortoise has been delayed by the assumption that the population was introduced from elsewhere and was thus unworthy of conservation efforts. However, it is classified as endangered by the Turtle Conservation Fund.

Indotestudo travancorica (Boulenger, 1907)

Distribution. This species occurs in the western part of the range of *Indotestudo,* in the southwest of India, particularly in the hill country of the state of Kerala.

Description. This species is slightly smaller than *I. elongata,* with a maximum length of 300 mm in the males. It often has a depression at the rear of the shell, of the fourth vertebral scute or between the fourth and the fifth. A conical spur is present at the end of the tail. The species is differentiated from the other two *Indotestudo* species by the following features: there is never a nuchal scute; the ratio between the interhumeral and the interpectoral seams averages 1.22 (i.e., it lies between the values for *I. elongata* and *I. forstenii*). The color ranges from chocolate to red-brown, with fairly large but diffuse and poorly contrasting black markings. The plastron may be immaculate (i.e., overall red-brown), or there may be diffuse black markings on each scute. The color of the head is similar to that of its congeners—creamy white, becoming reddish during the mating season.

Natural history. This species lives in hill forests and also in rocky areas. Apparently it will undergo considerable treks in order to find good habitat or to find food or mates. During the hot season, it remains immobile under the heavy leaf litter in the forest or in rocky retreats. It does not burrow, but it may occupy pangolin burrows. The diet is frugivorous at certain times of the year, and small groups of tortoises may be found under favored fruit trees in season, a habit also observed with *Geochelene denticulata* in French Guiana (B. Josseaume). However, it also shows enthusiasm for mushrooms, bamboo shoots, dead animals, and, from time to time, frogs and invertebrates. Courtship seems to be very protracted in this species, and so is the immobilization of the female, possibly so that the hormone levels in both sexes can be brought to optimal values (P. Pritchard). Nesting occurs from November to January (two to four clutches per season), and the eggs are quite large (55 × 35 mm). Incubation may last for about 150 days.

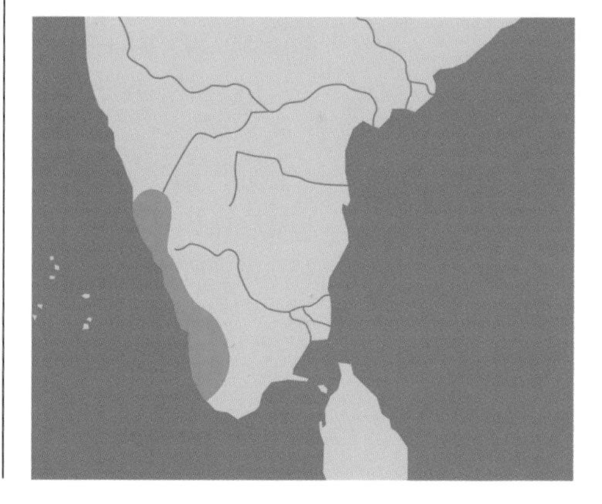

Common name

Travancore tortoise

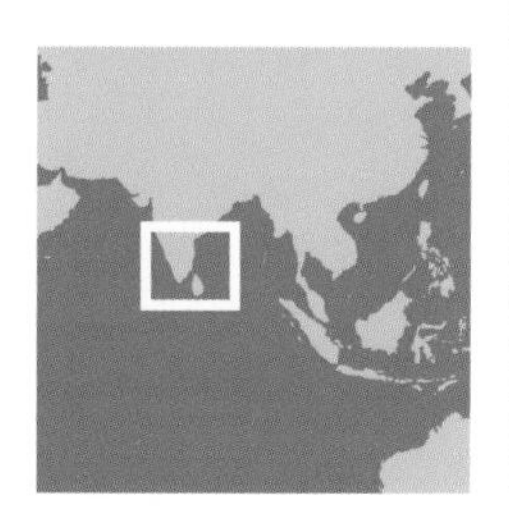

© P. Pritchard

This tortoise is uniformly dark, reddish, or perhaps almost black. The males always have a rather light head, especially in the breeding season.

Kinixys erosa (Schweigger, 1812)

Distribution. This tortoise occupies a wide range in west-central Africa, from the Gambia to the Democratic Republic of the Congo. It also occurs in Madagascar, presumably introduced.

Description of the genus *Kinixys.* These tortoises are distributed throughout much of central and southern Africa. The habitat is for the most part warm and humid, and the tortoises usually live in the forest understory and have nocturnal and somewhat carnivorous tendencies. They are of medium size, with a maximum length of about 400 mm, and have elongated, narrow shells, usually with a flat top. The unique feature of the genus is the hinge across the rear of the carapace. This hinge is arranged diagonally, from left to right, according to the species. The dorsal hinge does not develop until an age of five or six years has been reached. The marginals become deformed, their shape becoming more trapezoidal, and the lower/lateral ends of the hinge line become invaded by supple cartilage to replace the lost bony tissue. In some species, the hinge eventually reaches as far as the vertebral column, and in others the shell reaches a sharp downward angle between the fourth and the fifth vertebral scutes. This dorsal hinge is such a peculiar feature that it is reported that some veterinarians, when brought a *Kinixys* with an apparent wound across the carapace, have recommended pomades and unguents so that the healing might be facilitated—without success, one might add.

Description of *K. erosa.* This is the largest member of the genus, reaching 400 mm in length. It is also the most serrated, with very pointed marginal scutes forming a sort of toothed skirt toward the back. The anterior marginals are sharply turned upward above the head. The paired gulars are bifid and eventually develop a forked configuration like that of *Centrochelys sulcata.* All of these features serve to protect the animal from predators. The carapacial hinge goes from the back (lower down) toward the front (higher up) as in *K. homeana,* and the opposite of the condition in *K. belliana.* In contrast to both of these species, *K. erosa* lacks the nuchal scute. The head is remarkably elongated, with a projecting snout, and the eyes are placed very high and are separated by a slight depression. The plastron is quite wide, completely covering the underside of the animal, so that when the posterior section of the carapace is lowered, the tortoise is well protected within its bony box. The plastron is dark, with somewhat lighter seam

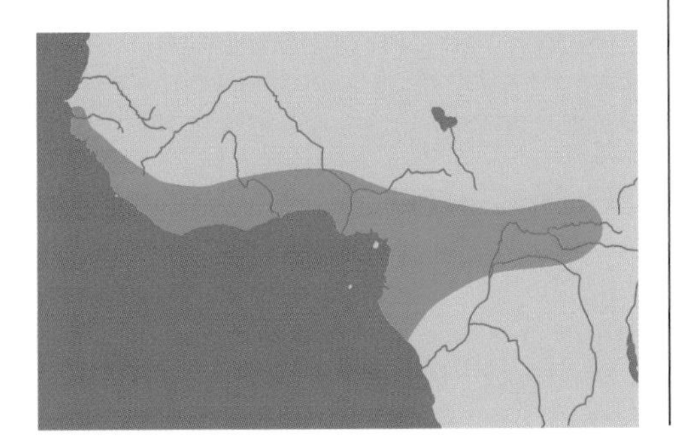

© J. Maran

In the middle of the night, this flash photo surprises two tortoises copulating. This species is generally shy and diurnal, and it is unusual to see them mating.

Common name

Serrated hinge-back tortoise

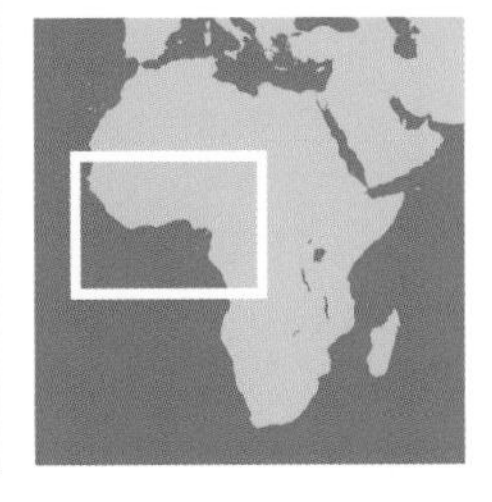

© J. Maran

Here one can see the hinge in the carapace and the very raised posterior of the shell, peculiar to the genus *Kinixys.* The head is small, with an upturned snout and a whitish color most often considered a characteristic of aged specimens.

lines. The carapace is orange-yellow to gray-brown, sometimes with large contrasting markings on the costal scutes. The head is almost uniform yellow, with darker areas above and at the base of the neck. The limbs are yellow, with heavy scales on the front. The young have a red-brown tint to the carapace, each scale being circled with light yellow. The plastron is already black in the young.

Natural history. All of the *Kinixys* species live on the edges of swamps, in humid areas and heavily vegetated substrates, and their limbs are big enough to allow them to progress easily across spongy substrates. *K. erosa* swims very well and feeds within the swamps and creeks that it frequents. It is omnivorous in diet, eating dead fish, amphibians, and the bulbs of aquatic plants. Nesting takes place on earthy banks, not far from the water, and the nest may be simply a small heap of dead vegetation and detritus in which the thick-shelled, oval eggs are deposited. The eggs measure about 45 × 35 mm. Each clutch consists of just three or four eggs, but oviposition takes place several times in the course of the season. The time for hatching is dependent upon the amount of sunshine that reaches the nest site, but incubation seems to be rather slow among all the *Kinixys,* lasting from 110 to 300 days. The young measure 40 mm and weigh about 30 g.

Protection. These tortoises are heavily consumed, and local people use trained dogs to find them. We have seen villagers use petrol to cook them and then eat them "grilled." The young are heavily predated by numerous species. The actual status of these *Kinixys* in their own habitat is poorly known, in that they hide very effectively and are hard to see. Several radio-tracking experiments have demonstrated that these tortoises will walk long distances at night, following a very straight course, presumably looking for food or mates. Intensive deforestation in some of the natural habitats of this species and the constant use of tortoises as "bush meat" make one fear that the numbers are dropping fast.

Kinixys belliana Gray, 1831

Distribution. This species has the widest range of any in the genus. In occupies all of central and southern Africa, from southern Senegal to Mozambique, and the eastern corner of South Africa. It is even found in the northern one-third of Madagascar, presumably brought over by humans.

Description. The name of this tortoise derives from Thomas Bell, a famous English zoologist. The overall shape of the carapace is oval, elongate, and lacking in denticulation or projections. The maximum length is 230 mm. The hinge of the shell extends forward at the lower edge and backward at the vertebral side, the opposite of the condition in the other species, but some individuals have an almost vertical hinge, which may reach to the edge of the vertebral scutes. A nuchal scute is present, and the gulars are very small. The profile of the shell is somewhat round, and the fourth vertebral (the highest) is curved, quite the opposite of the "square back" condition of *K. homeana.* The marginals are neither widened nor denticulate, even in the young. The color ranges from beige to brown, with darker areas or circles, and with the seams cream or whitish. According to the subspecies, the patterns and markings of the carapace

are extremely varied. The old adults often lack all markings, becoming uniformly beige or brown. The plastron is cream, sometimes marked with darker areas. The head is beige to grayish, with the lips somewhat lighter. The limbs are dark, with yellow scales, especially in young animals.

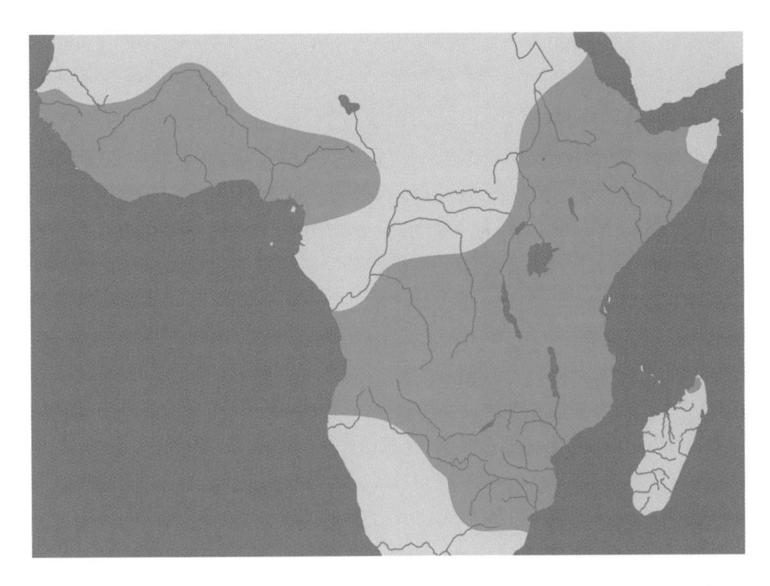

Subspecies. Five subspecies are recognized. *K. belliana belliana* is described above. *K. b. domerguei* (Vuillemin, 1972) occurs only in Madagascar and was certainly imported by humans. *K. b. mertensi* (Laurent, 1956) is found in the Democratic Republic of the Congo. The anterior lobe of the plastron is longer and narrower than in other subspecies. *K. b. nogueyi* (Lataste, 1886) lives in humid areas, in the western part of the range (southern Senegal to northern Cameroon). It only has four claws on the forelimbs, and the plastron is usually completely plain. *K. b. zombensis* (Hewitt, 1931) is found in the southeastern part of the range (Tanzania and southern Zululand). It has five claws on the forelimbs, and the plastron has symmetrical markings. The carapace is always marked with black.

Natural history. This range of this species is so wide that the details of the ecology and behavior are far from uniform. In the western areas, the subspecies *nogueyi* inhabits the humid forest understory. In the east, the species occupies drier areas and is most active after rains. In general, this tortoise is most active by evening light or even at night. It has been seen eating false scorpions and fungi, as well as snails of the genus *Achatina* and dead animals. When alarmed, it retracts abruptly, and the shell hinge flexes so that the animal is flat against the ground. When handled roughly, it defecates and emits a nauseating odor. Males often fight at the start of the rainy season. The objective is to overturn the opponent, but the flipped male gets back over quite quickly, being very agile and having long, supple limbs. During mating, the male grips the shell of the female with his claws and stands vertically, like a mating male *Terrapene.* It also emits frequent sounds during copulation. Nesting usually occurs in November to April in the southern part of the range, with six to eight elongate eggs (measuring 36 × 28 mm) per nest. Incubation can be very slow, according to the specifics of the nest site. The young weigh 18 to 20 g.

Common name

Bell's eastern hinged tortoise

This tortoise is abnormally light, and this condition draws attention to the sculpturing of the carapace. The posterior is upturned, in the shape of a helmet.

Protection. This tortoise is consumed almost everywhere it is found. In the past, from 1970 to 1985, it was also collected on a massive scale for export to Europe, but it fares badly in the European climate. Fortunately the fashion passed, and it is no longer exported. The great ground hornbill is a dangerous predator upon this species as well as others like the leopard tortoise; the formidable bill is used to extract the soft parts of the tortoise even when it is fully retracted, and young tortoises may be smashed before consumption. This tortoise is still fairly common. In certain parks and reserves it is considered to be a protected species, and capture is forbidden. In Mozambique the numbers have become low, and surveillance and protective action are needed.

Kinixys homeana Bell, 1827

Distribution. The range is long and narrow, extending around the Gulf of Guinea from Liberia to Cameroon, and with an enclave in the Democratic Republic of the Congo.

Description. This species is rather similar to *K. erosa,* but without the strong serration of the latter. The length does not exceed 220 mm. The scutes are very flat, and the vertebrals are horizontal, giving the animal a decidedly "angled" look, especially toward the back. The hinge goes from front (low down) to back (high up). The nuchal is very narrow and long, and the gulars are short. The rear marginals extend like a skirt and curve out horizontally, but they are smaller than in *K. erosa,* and the anterior marginals are extended forward and horizontally, as a sort of protection for the head. The overall coloration is rather dark, mixed with shades of brown, yellow, and red, with darker seam lines. The head and limbs are pale yellow. The plastron is yellow, with blackish spots in the centers of the scutes.

Here the posterior carapacial hinge is quite evident; it serves to protect the rear of the tortoise.

© B. Devaux

Natural history. This tortoise seems to prefer very humid areas, along watercourses and creeks. It undertakes a special posture in the rain, tipping the shell forward and raising the hind limbs as high as possible, so that the rain that falls on the carapace runs toward the mouth. This tortoise is omnivorous and feeds on whatever it can find in the water—amphibians, larvae, other invertebrates, and so on. The eggs are large and few, measuring 45 × 35 mm.

Protection. The habitat of this species is being degraded at a rapid rate, and since the population density is low, it is probable that the numbers of these tortoises are decreasing. Nevertheless, it is not collected for export and is perhaps somewhat less heavily hunted for food than the other *Kinixys* species. It is protected in certain parks in Liberia (Sapo), the Ivory Coast, Cameroon, and Gabon.

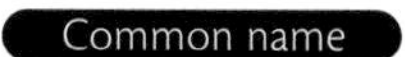

Common name

Home's hinged tortoise

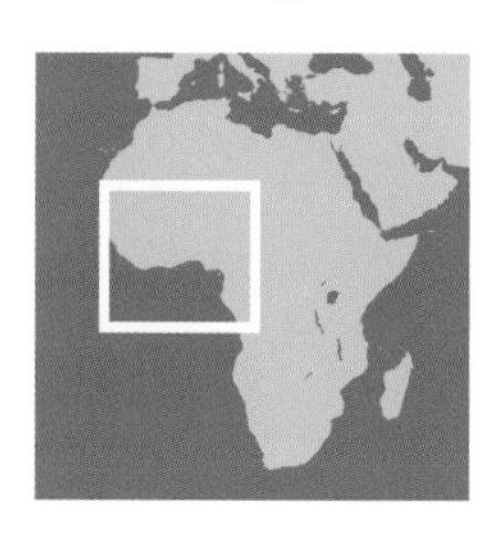

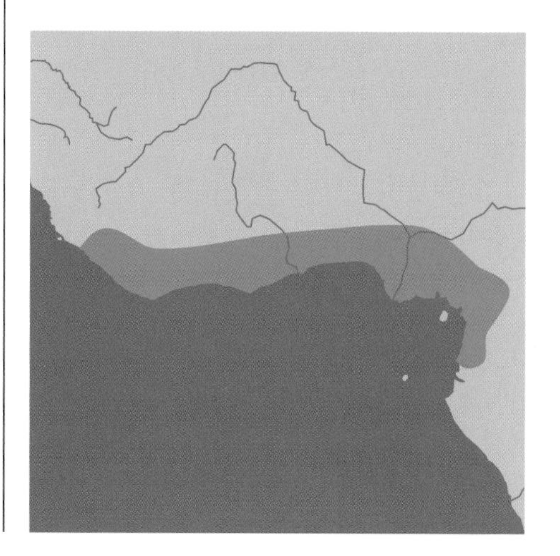

© J. Maran

The last vertebral is very elevated, but the hinge is less evident than in the photos of the preceding species. Serrated edges are evident on the marginal scutes.

Kinixys lobatsiana Power, 1927

Distribution. The most restricted of the *Kinixys* species, this tortoise is found only in northwestern South Africa and southeastern Botswana.

Description. The name derives from Lobatse, a locality for this tortoise in Botswana. It is similar to *K. belliana,* but with an elongated and rather flat configuration. The carapace is narrowed between the second and third vertebral scutes, and the carapacial hinge is poorly developed. Small in size, this species does not exceed 200 mm. The gulars are rather short, although still more developed than in *K. belliana.* There are five claws on the forelimbs and four on the hind limbs. The general color is rather contrasting—bright orange or brown with arcs of black, and wide, light areolar rings that are beige to yellowish. The plastron is pale yellow, with scattered black markings. In the males, the plastron is somewhat concave.

Natural history. This is an omnivorous species, feeding on insects, fungi, and snails. It is most active after rains, during the summer months. It hides in rocky retreats during the dry, cold season and may use the burrows of other animals. It hibernates from May to September. Reproduction occurs from November to April. One female laid six eggs, of impressive size (40 × 30 mm), and the incubation took 313 days, although this extreme duration is surely atypical.

Protection. The range is small and is subject to overgrazing and wildfires. The species occurs in three nature reserves (Ohrigstad Dam, Loskop Dam, and Nylsvley). The status of this tortoise is poorly known, as is its natural history. It is important that studies of both be undertaken.

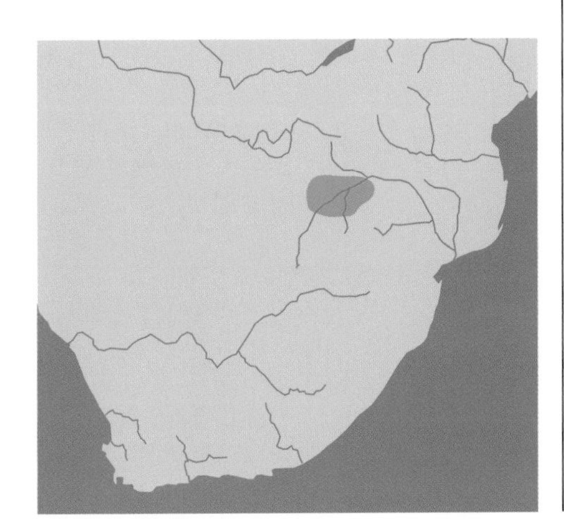

Common name

Lobatse hinged tortoise

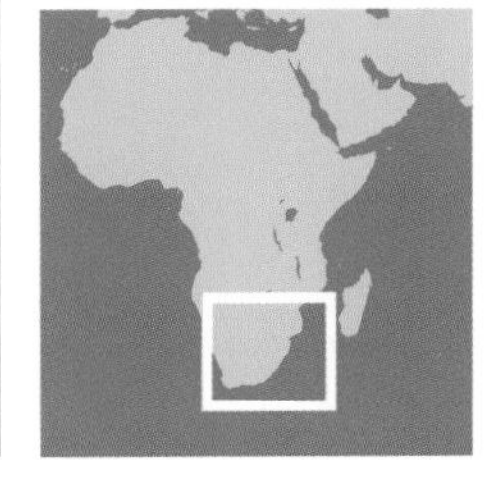

This species never grows very big, and the carapace is less sculptured and the hinge is less obvious than in the other *Kinixys*.

Kinixys natalensis Hewitt, 1935

Distribution. The range of this species is limited to southeastern Africa, from Swaziland to southern Mozambique.

Description. Absent from the lush plains of the veld, this tortoise is found in dry, rocky areas up to an altitude of 1,000 m. It is surely this harsh environment that makes this the smallest species of the genus, the greatest length being 160 mm (for females). The general form is elongate, oval, and rather flat, the greatest height being at the third vertebral, and the carapacial hinge is rudimentary, not extending higher than the marginals. The beak is tricuspid. The gulars are short and often scoop-shaped. The supracaudal is often divided. Sexual dimorphism is minor, and the plastra of the males and the females are equally flat. The males are smaller, however, not exceeding 130 mm. This tortoise has five claws on the forelimbs and four on each hind limb. The coloration is brown, with lighter central markings on the scutes that may be yellow or orange-yellow in color, but as the animals age, they become a plain brownish red or beige. There is great variation in the color details. The plastron is yellow with symmetrical dark markings and a pair of black rings on the abdominal scutes, a good feature for species recognition. The young are often uniformly light or dark brown, the contrasting furrows not developing until a year or two later.

Natural history. This *Kinixys* is the least specialized for humid habitats. During the daytime, it hides in rocks, and it does not become active until nightfall. It ranges the bushveld, looking for small insects, reptiles, and rodents, or eats dead animals, feces, and, in truth, almost anything even remotely edible. Eggs are laid in April, and hibernation occurs from May to September, with copulation observed in February. Two eggs are usually laid, each about 32 mm long, and incubation may take five to six months. Emergence is synchronized with the arrival of the rains.

Common name

Natal hinged tortoise

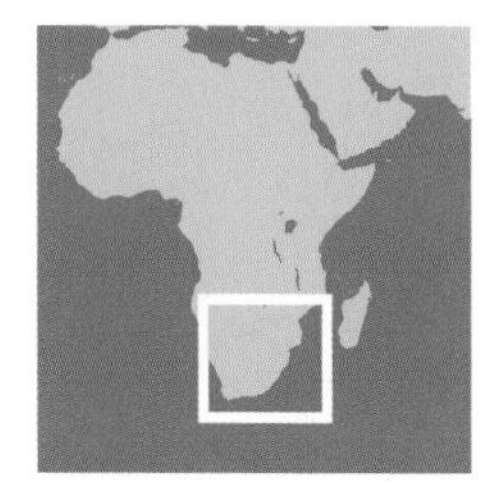

Protection. This species is endangered primarily by habitat destruction. Great inroads are made by agriculture, silviculture, and the numerous bushfires that rage through the Limpopo. Happily, the species is not caught in any significant numbers nor is it often eaten, and it has rigorous protection in certain parks in Kwazulu-Natal, the former Transvaal, and Swaziland. Furthermore, the populations are isolated, which may ultimately constitute a problem for genetic diversity, although on the other hand it may retard the spread of disease and provide a setting for further evolutionary change in small, genetically isolated demes.

© B. Branch

This tortoise is also small and has a modest hinge. Usually a bold beige or brown pattern is evident. The head is dark and strong.

Kinixys spekii Gray, 1863

Distribution. The range is wide and tropical, extending from northeastern South Africa to a small part of western Mozambique and almost all of Zimbabwe. All known localities are far inland.

Description. This tortoise is named after J. H. Speke, codiscoverer of the sources of the Nile. It is of medium size (females to 200 mm), elongate, oval, and flat, lacking serrations in the marginals, and the second and third vertebrals are slightly depressed. The carapacial hinge is not very developed, extending only to the middle of the third costal scute. The gulars are short but wide. The plastron is slightly concave in the males. This tortoise is often uniformly brown or dull yellow, without striking markings. It has five claws on the forelimbs and four on the hind limbs. The overall coloration is not very contrasting and is quite variable. In general, it ranges from beige to brownish, with light annuli in the young. The adult animals eventually become monochromatic, with a reddish brown tint. In the females, the carapace sometimes bears very dark rays, the decor being basically brown on a pale yellow background. The head is rather wide, and the beak has a single point.

Natural history. This tortoise lives in a variety of habitats, extending from rocky plains to tropical savannas, at low altitudes. It often hides in rocky crevices but survives very well in herbaceous savannas and seems less dependent on water than other *Kinixys,* with the possible exception of *K. lobatsiana.* This species is primarily herbivorous but also appreciates millipedes, snails, and insects. It shows intense activity once the rains start, and it hibernates during the cold season, which is also the time when wildfires strike. Nesting has been recorded from December to April. Oviposition occurs several times per season, with two to six eggs each time, each measuring about 42 × 32 mm. Hatching has been observed in September and

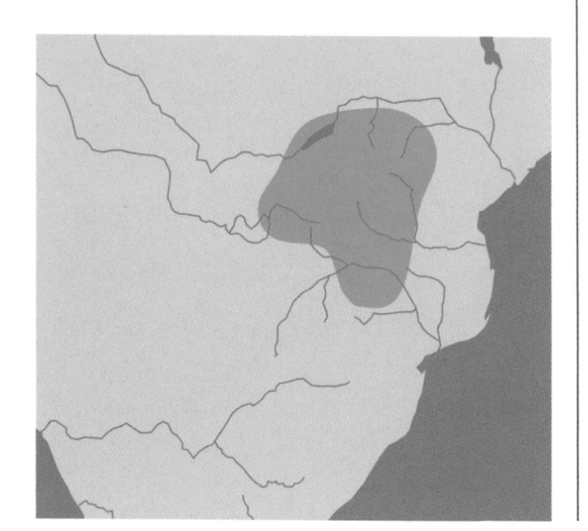

Common name

Speke's hingeback tortoise, savanna hingeback

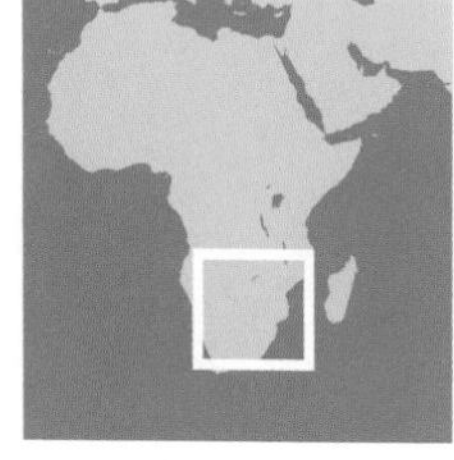

© B. Branch

October but also occurs in March and April. Incubation may take a full year, although this may be well above average. The young measure about 40 mm and weigh about 20 g.

Protection. The tortoise is consumed in Zimbabwe, but its habitat is not yet heavily disturbed by humans. Protected in the Loskop Dam Nature Reserve in Transvaal, it appears to be plentiful in some areas and rare in others, with the overall distribution quite fragmented. It is mainly threatened by major bushfires, which occur frequently in this region. Its status needs to be investigated.

This species closely resembles *Kinixys natalensis,* but there is a profusion of small circular markings on the carapace scutes. This is one of the most colorful members of the genus.

Malacochersus tornieri (Siebenrock, 1903)

Distribution. This species occurs in arid, rocky areas, at intermediate altitudes (average about 1,000 m), and in kopjes—isolated, rocky outcrops—between southern Kenya (Samburu District) and northern Tanzania (Lindi), with a preference for the Tarangire.

Description. The vernacular names describe it well: a soft-shelled tortoise, shaped like a pancake. The rocky habitat and the specialization for life in narrow horizontal cracks make for a morphology unique among chelonians. With a maximum length of about 180 mm (for females), the shell is squarish from above and is extremely flat. The bony structure of the shell is unique, with the bone eroded away except in the areas that lie directly beneath the external seams in the carapace scutes. Some of the costal bones ultimately disappear completely. The scutes are very thin, and the marginals almost translucent. The edges of the carapace and plastron are relatively stiff, but the middle sections of both are totally soft to the touch, an adaptation that allows it to accommodate to the irregularities of the rocky cracks in which it resides. Even the biggest specimens are only about 60 mm thick. Early observers misinterpreted these peculiarities; Siebenrock wrote that "this tortoise appears to have some kind of bone disease, because the carapace is flat and soft." This morphology is crucial for the lifestyle of the animal, which is described below. The head is quite large and strong and is well ossified, with a rounded snout and well-developed temporal scales. The limbs are long and flexible, facilitating rapid walking. Eglis (1967) described it as the fastest tortoise in the world, capable of running 18 m per minute.

The morphological details are also explained by its ecology. The claws of the hind limbs are especially strong, allowing the animal to hold fast to rock and to climb to the summits of kopjes. The coloration of the species is equally remarkable, in that it varies between regions and sometimes even within a colony, but it always serves a well-

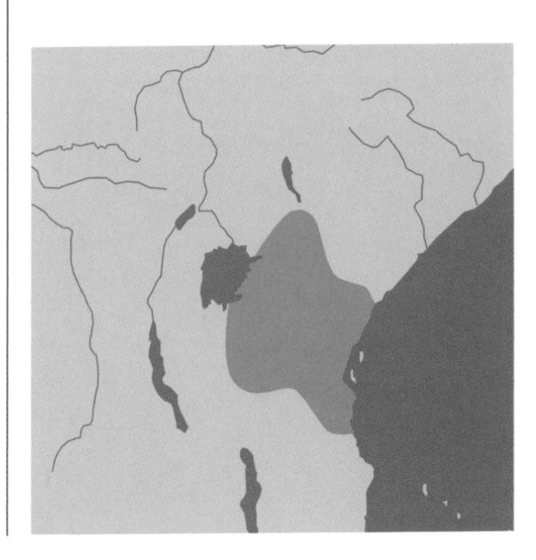

Common name

African pancake tortoise

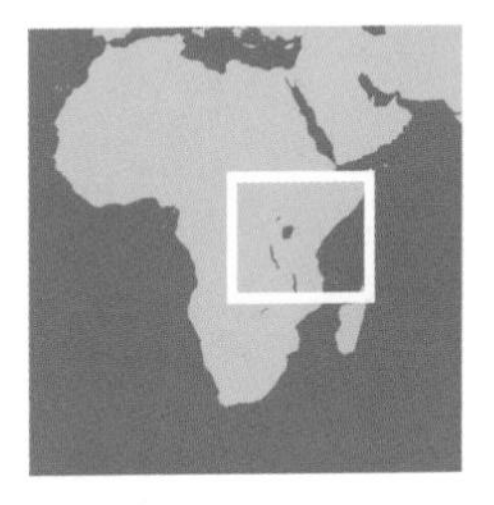

developed cryptic function. The background is reddish to beige, and the markings may be quite distinct, with a high density of dark reddish or blackish radiating streaks. Within a group of these tortoises, the pattern and the color of each are likely to be different. The oldest animals are perhaps the most monochromatic, but the species does not actually seem to be very long-lived, even in captivity. The young are rather rounded but offer quite variable, richly ornamental radiating patterns, in tones that range from mahogany to beige. It is only after six or seven years that they come to resemble their elders. Sexual dimorphism is modest, the males not needing nor having a concavity in the plastron. They are slightly smaller than the females.

Natural history. The flat, supple carapace, the long limbs with powerful claws, and the energy and speed that allow this species to walk rapidly correlate with the habit of living in narrow rocky cracks and of having an aversion to both sunlight and heat. Thus it can run rapidly from one kopje to another, climb up the rocky surface, and slip quickly into a crack in huge boulders. It used to be thought that this tortoise could forcibly expand its lungs so as to jam itself into these crevices, but studies have shown that this is not true; the compression of the body that occurs when the animal reaches the narrow end of a rocky fault or crevice is entirely passive. The pancake tortoise is omnivorous; it consumes vegetables and fruits but does not disdain dead prey or invertebrates. According to Ingo Pauler, it moves away from high temperatures, which would explain its aversion, in nature, to sun and warmth. F. J. Obst has observed that it maintains activity at quite low temperatures (12°–15°C). Its flat form and square shape allow for very rapid thermoregulation when necessary.

The love affairs are ungentle. Males fight violently among themselves, and they make frequent attempts at copulation. In captivity the males and the females may have to be separated to prevent permanent damage. We have even seen this species pursue the local African *Agama* lizards and bite them. Maturity in *Malacochersus* seems to be somewhat precocious and corresponds to a length of about 120 mm in males and 145 mm in females (i.e., about eight years of age). Because of the reduced space in the abdominal cavity, the female only lays one egg at a time, of medium size (40 × 35 mm) and oval form. Nestings are spaced about 25 days apart, and there may be as many as six per year. In spite of the minimal individual clutch size, the annual productivity may reach four or more eggs. Nest construction is not well studied. It seems that the female tests the temperature of the ground with her chin. When she has located a site with an appropriate temperature, she fixes her position with her forelimbs while scattering the substrate with vigorous kicks of the hind limbs. Oviposition is rapid, and the single egg is quickly concealed. Incubation seems to be quite long: 178 to 237 days. The egg opens along the longer axis, not near the end, as most turtle eggs do.

© I. Pavler

This fine tortoise is flat and posteriorly enlarged, and the scutes are strongly marked with radiating streaks, although these are not evident in all individuals.

Protection. *M. tornieri* started to appear in international trade in 1980. Its peculiar morphology, rarity, and lively coloration quickly made it a favorite among hobbyists. Even though placed in Appendix 2 of CITES, with legal export subject to special permission from the exporting state, illegal collection was frequent because of the rarity and high price of the animal. Some collectors did not hesitate to break open the rocky retreats with heavy machinery, and shipments of 100 animals or more were frequently intercepted by customs agents. An investigation by Don Moll and M. Klemens (1992) demonstrated that the habitats of the species were being damaged and permanently degraded; damaged rock does not heal. Tortoise

© B. Devaux

Two contrasting examples of carapace decor in the same species.

sites that are too close to tourist circuits may be adversely affected by too many visitors. No real captive breeding programs exist—annual productivity is too low, and the species not very long-lived. There remains a major demand for the species among hobbyists and collectors, which results in continuing collection activity and illegal traffic. A moratorium on exports was established by the government of Tanzania in 1994, pending a study of the exact status of the species and its precise distribution. It would seem to be urgent that the species be placed in Appendix 1 of CITES as rapidly as possible; this is a case where international trade is truly the problem.

Psammobates geometricus (Linnaeus, 1758)

Distribution. Endemic to South Africa, this species is found only in a minute area of extreme southwestern Western Cape Province, between the Piketberg in the north and Gordon's Bay in the south.

Description of the genus *Psammobates*. The name *Psammobates* defines the ecology of the genus: *psammos* and *bates* mean "inhabitant of the sands." There are three species, rather similar to the star tortoises of Asia and Madagascar but differing by their somewhat miniature size and by certain bony variations. One could be led to think that these various star tortoises had a common ancestor, but the pattern is even shown by some of the completely unrelated box turtles *(Terrapene),* and it appears to be highly adaptive for crypsis in certain plant communities.

Description of *P. geometricus*. This tortoise has a truly geometric appearance. It could be confused with *Geochelone elegans* of India, and Linnaeus himself was confused by them, identifying this species as "Asiatic." But it can be distinguished by the following details: size does not exceed 150 mm (females); there is a nuchal scute (absent in the Indian species); the marginals are not serrated; and the shell scutes are gently rounded, without the "pyramided" effect of some *P. tentorius* (and some *G. elegans*). The general shape is rounded and high-domed, with a center of gravity shifted toward the rear, so that the fifth vertebral scute is almost vertical. The ground color is saffron yellow to orange, with thick, intensely black radiations (four to seven per scute), regularly spaced around the areola, and giving the appearance of a bar code between the costals (but this species is never legally for sale). The head is rather small, is orange-yellow in color, and has small lighter scales—completely different from the head of *A. radiata* or *G. elegans.* The plastron is often black, with a few diffuse yellow streaks or with lighter seam lines. The young are rather similar to the adults or may, by contrast, have a very light background, with black lines or simply a black X on each scute.

Natural history. This species lives in the South African renosterveld, an environment of bushes and herbs that incorporate several sandy ecosystems. The climate is temperate; hot and dry in summer, it is cool and wet in winter. This tortoise hides during the winter, in a state of semihibernation. It lives not far from urban areas (Darling, Malmesbury), and sometimes it may be found cut off from its habitat by farms and agricultural areas. In some areas, the females are much more numerous than the males (J. Juvik, R. Rau). Basically herbivorous, this species is active during the warm hours of spring and autumn and in the morning and evening in summer. The females nest just once per season, from September to November, and produce two to four eggs, each measuring 30 × 24 mm. Incubation may last for six to eight months; hatchling emergences have been observed in April to May, along with the first rains. The growth of the young is quite rapid, and two growth rings may be added each year. Sexual maturity seems to be reached at a somewhat early age and may be attained within eight years.

Common name

Geometric tortoise

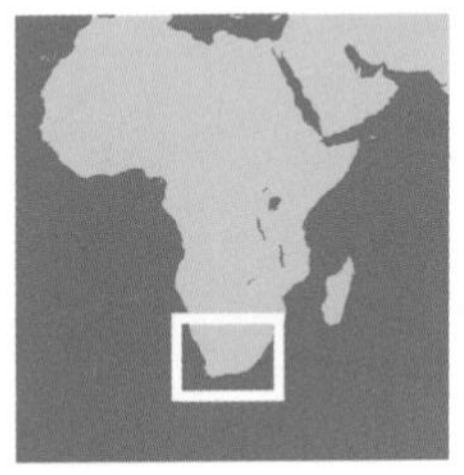

Protection. This is the only species of South African tortoise to be considered "endangered" in the *Red Data Book,* and it is also classified in Appendix 1 of CITES. It is thought to be one of the rarest tortoises in the world. Populations are estimated at about 4,000 individuals. In the past, it has been collected for local markets and also for export. Today the challenges lie in the extension of agricultural areas and changes in local vegetation, especially the acacias. Five reserves exist to protect the species: Eenzaamheid, Romans River, Hartebeest River, Harmony Flats, and Elandsberg. But altogether these protect only 1,200 ha of habitat. One could foresee the creation of a much bigger park, perhaps 6,000 hectares, that would also protect *Homopus areolatus,* under the aegis of Cape Nature Conservation. Several studies have been launched on the ecology of this tortoise, with the goal of focusing conservation efforts. One of these steps is the photographing of each individual as a sort of "identity card" (J. Greig). No concrete measures have yet been taken to enhance the population of this species, and one may fear that the decline will continue. The species is classified as endangered by the Turtle Conservation Fund.

© B. Branch

This tortoise closely resembles *Geochelone elegans,* but the length does not exceed 150 mm and the scutes are never pyramidal. Furthermore, there is a nuchal scute.

Psammobates oculiferus (Kuhl, 1820)

Distribution. This tortoise is the most widely distributed of the species of *Psammobates.* It occurs in the desert areas of southern Botswana, southern Namibia, and a large part of the Cape Provinces and the Free State.

Description. Females do not exceed a length of 143 mm, and they are less highly domed than *P. geometricus* and may be constricted around the middle of the marginal scutes. There is a small nuchal scute, narrow and long in shape, and the carapace margins, back and front, are strongly serrated. There are small horny spurs on the thighs. The key distinguishing feature is the coloration. The background is yellow-beige, and the dark radiations are less numerous, less wide, less contrasting, and less regular than in *P. geometricus.* It appears that the rays are not created by the lifetime growth pattern, but they may "stop and start" as the scute enlarges with growth, if one examines the growth annuli carefully. The result is a somewhat pointillist decor, which explains the species name *oculiferus.* The head is often dark, almost black, and has none of the light markings of *P. geometricus.* The snout is significantly projecting, and the beak is tricuspid. The anterior limbs carry large scales, yellowish to beige in color, but there are no small scales among these enlarged ones, as are visible in *P. geometricus.* The plastron is dark, with diffuse, radiating light streaks.

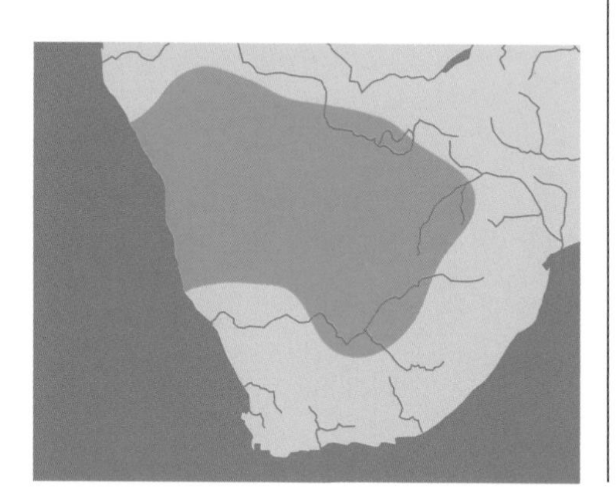

Common name

Serrated tortoise

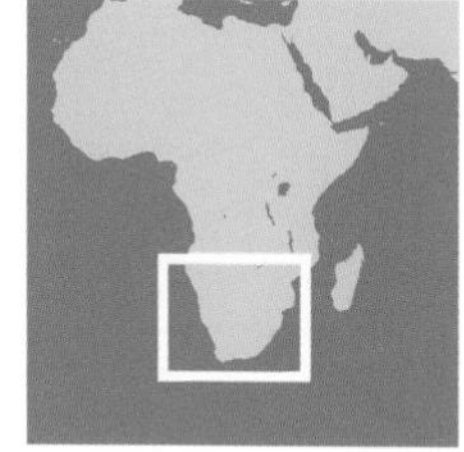

This small star tortoise with recurved and serrated marginals is more white than black. The limbs are equipped with heavy spurs.

Natural history. This species occupies the arid savannas and somewhat more humid prairies in the south of its range. The distribution includes the sands of the Kalahari, which actually include several zones of humid forest, as well as the scattered forests of Western Cape Province. It is adapted for various climates, estivating in summer and hiding in winter, according to the latitude. It is not often seen in its habitat, the absolute numbers being small and the coloration extremely cryptic. It is primarily an herbivore, but it has been seen to feed upon grasshoppers, hyena feces, and organic detritus. It is subject to numerous predators, against which its small size offers little defense. Eagles, hyenas, and various carnivores eat the young and even the adults. This is doubtless why it spends so much time in hiding, venturing forth a little at midday and hiding itself in sandy crevices or under accumulations of vegetation. There is little information available on its reproductive habits. Mating has been observed in November. The males charge and crash against the females with great energy and put forth loud groans. The eggs, a maximum of six per female, are small: 34 × 28 mm. Emergence may occur in March and April, about 100 days after oviposition.

Protection. This species is listed in Appendix 2 of CITES. For a very long time, the Bushmen of the Kalahari have used the shells as storage boxes for drugs or small personal effects. The attractive coloration makes them a magnet for collectors. Some level of illegal take, for resale, certainly occurs. However, this species is not yet collected to excess, and the huge range of distribution leaves the hope that many tortoises remain. It is protected in parks and reserves, in the north of the Northern Province and in the Cape Provinces. Rather few collections include this species, and it does not do well in captivity. Studies on the conservation and status are necessary.

Psammobates tentorius (Bell, 1828)

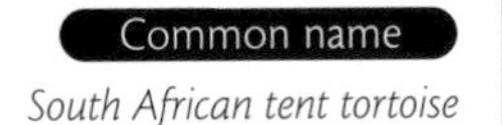

South African tent tortoise

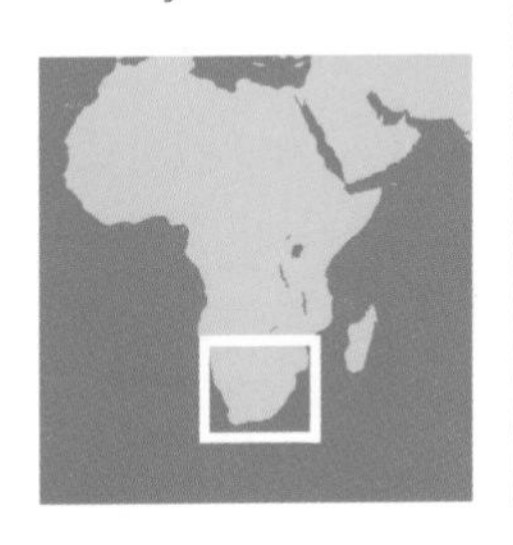

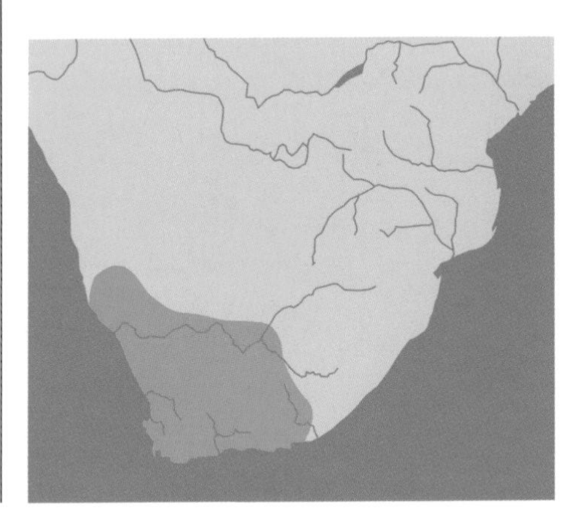

Distribution. The range of this tortoise overlaps that of *P. oculiferus,* but it extends farther south in Namibia and Western Cape Province. Occasionally it may be found in coastal areas, and indeed one of the subspecies, *P. tentorius trimeni,* occurs in Little Namaqualand, in northern Western Cape, near the sea.

Description. The length of this tortoise does not exceed 140 mm, and the general form is flatter than that of the other *Psammobates* species. What makes the real difference is the tentlike pyramiding of the scutes, which may be more extreme than in any other tortoise, apart from certain pet animals kept indoors on grossly unnatural diets.

The pyramiding is most evident in the adult females of *P. t. tentorius,* and indeed tenting is not characteristic of either sex of the subspecies *P. t. verroxii.* The coloration is also distinctive, because the dark radiations are so wide that they leave little space for the yellow background to show through, and they often form a prominent X shape, emphasized by the angular form of the scutes. The background is reddish beige, with the dark lines a very dark brown. In adult animals the coloration may fade, which again represents a difference from the Indian star tortoise. The plastron is also different in that it has a single all-encompassing black figure, with no star pattern. In summary, this species is less vividly marked than the other species of *Psammobates,* leaning toward a bronze color and with reduction of the starred effect. There are many small scales on the rear of the thighs, as in *P. oculiferus.* The head is of rather uniform color, with several light scales on top. It also has significant development of the upper jaw, which has a parrot-beak appearance. The limbs are light in color, with very well-developed and enlarged scales on the anterior face.

Subspecies. *P. t. tentorius* (described above) is found in the southern part of the range, in the Cape Provinces. It does not surpass 131 mm in length. The plastron bears a wide central black band, without any radiations. The carapace of the adult females is highly pyramided. *P. t. trimeni* (Boulenger, 1886) occurs in the western and central part of the range, from northwestern South Africa to southwestern Namibia. It has well-defined stars on the plastron and often has pyramided carapace scutes. As with some other arid-zone tortoises, this one has the ability to raise the rear of its shell high off the ground so that rain falling on the carapace will be directed toward the mouth. It is named after Roland Trimen, curator at the South African Museum. *P. t. verroxii* (Smith, 1939) probably derives its name from Jules Verreaux, a director of the South African Museum. It occurs over a wide north-south stretch of terrain from central Namibia to the Great Karoo in South Africa but is always well inland. It has a very rounded shell, without pyramidal scutes, and has rather diffuse radiating markings. This is the largest of the three subspecies, and sometimes the pattern is almost completely faded out. It has the same habit as *P. t. trimeni* in drinking from water rivulets running forward along the carapace.

Natural history. These are the least-known tortoises of the region, doubtless because they hide very well and their population density is quite low. In captivity they do not adapt well and rarely survive for long. They have the habit of hiding themselves in burrows or under vegetation, coming forth after rains. In Namibia they estivate, although farther south they may be active throughout the year. They like succulent plants but are generally omnivores. Their predators are not just raptors and typical carnivores, but they are also killed by ostriches, which strike with their powerful beaks. In nature, reproductive activities of the species are rarely observed. The tail of the male is longer and stronger than that of the female and has large scales on each side. Copulation is observed from September to January. The nests have one to two eggs, of medium size (35 × 24 mm). Incubation is lengthy, and emergence does not occur until February to April but is heavily dependent on the start of the rains.

Protection. There is no evidence that this species is endangered, but it has certainly diminished in the zones where huge ranches have been installed, as in the Great Karoo. It is protected in numerous parks, including the Hester Malan Nature Reserve (Goegap Nature Reserve) and the Karoo National Park, and provincial legislation prohibits any collecting. Permits may be obtained from South Africa's Department of Environmental Affairs and Tourism. Studies on this tortoise are necessary.

The pyramidal carapace scutes recall the Indian star tortoise, but the marginals are strongly widened and the light lines form a darker X mark than in *Geochelone elegans.*

© B. Branch

Pyxis arachnoides Bell, 1827

Distribution. This species occupies a coastal band, 50 to 100 km wide, in southern Madagascar, in extremely arid country. The range extends from Morombé in the north, to Cape Sainte-Marie in the south.. Major populations occur southwest of Tuléar. It is also known from as far as Békily, but at present it is rare there, at best.

Description. This is the only tortoise that has ever been compared to a spider, of all things. And, in truth, perhaps its small size and the yellow carapacial scutes with graphic, weblike markings do justify such a comparison. It is also somewhat evocative of *Psammobates* of South Africa, but one detail immediately differentiates the two (at least in two of the three subspecies): the plastron of *Pyxis* has an anterior hinge. The carapace length does not exceed 150 mm, and the shell is moderately elongate and rather rounded. The marginals are somewhat flattened toward the lower edge, and the scutes have somewhat raised centers. There is a spur on the tip of the tail. The general color is cream to orange-yellow, with black lines and spots radiating from each areola. In adult specimens the contrast lessens, and the shell may become almost plain. The plastron is rather light, with irregular spots. The hinge between the humerals and the pectorals is very supple and quite "floppy" and cannot be raised with great force, as it can in the box turtles. The head is small, with a brownish black ground color above, punctuated by light scales, and the jaws and the underside of the neck are yellow.

Common name

Common spider tortoise

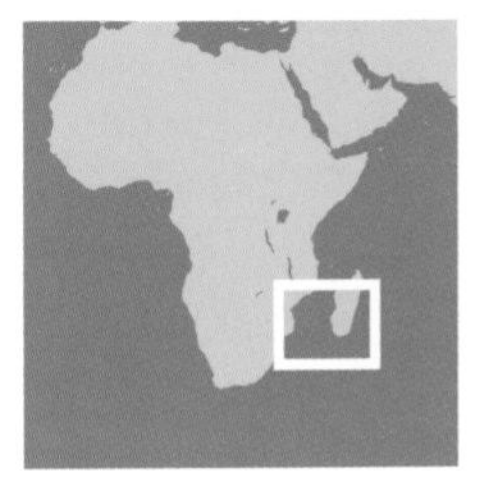

Subspecies. The three subspecies occur in separate areas and show differences in the plastron. From Morombé to north of Tuléar, one finds *P. a. brygooi* (Vuillemin and Domergue, 1972); it has a rigid plastron. *P. a. arachnoides* (described above) is found from Tuléar to south of Anakao and has a reasonably flexible plastron. *P. a. oblonga* Gray, 1869, occurs in the far southeast and has a flexible hinge and some black markings on the bridge and sides of the plastron.

Natural history. This species prefers a substrate of sand, pebbles, or organic detritus. It emerges during the wet season, particularly after heavy rains. During the dry season, it hides under the surface, whether it be leaf litter or roots, or conceals itself under sand or gravel. It is active toward evening. To find specimens in the natural environment, one needs to feel with one's feet under bushes and to probe the topsoil. These tortoises are able to estivate during the long dry season. But when heavy rains come, they emerge in droves and can be seen walking about in numbers that would not have imagined. In certain areas south of Anakao, there may be as many as 50 tortoises per hectare, at least in areas where the collectors have not done their work. This species is omnivorous, and in its arid habitat it has to be content with what it can find: a few insects here and there, some dried leaves, fallen fruit, dead animals. In captivity it is frugal in demeanor and somewhat difficult to feed, especially if the atmospheric pressure is too low. When it is active, however, it shows astonishing vivacity, runs around rapidly, and copulates noisily. The egg clutches are very small; generally only a single egg is laid. Incuba-

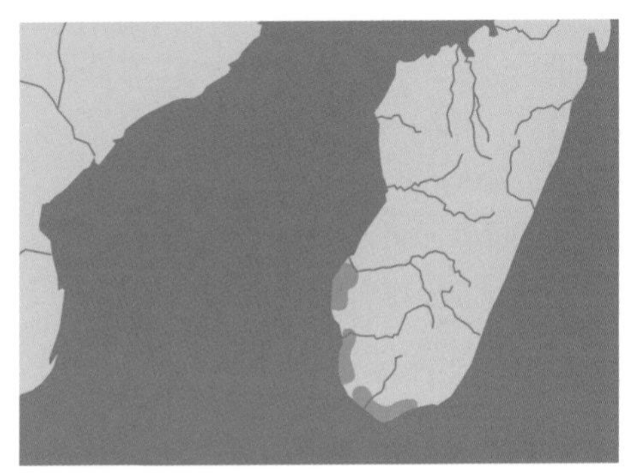

This golden-yellow tortoise shows the characteristic yellow-and-black pattern of the species.

© B. Devaux

tion is slow, and emergence depends on the rains. The young weigh about 8 g and look like tiny spiders, being very rounded, and with a strong black and beige or cream design on the shell.

Protection. The species is placed in Appendix 2 of CITES. In the year 2000 the capture quotas were set at a level of 2,000 tortoises per year. But the Madagascan government decided not to renew permission for any further collection of wild tortoises. The problem was that this species lives in a region where the large and much sought-after radiated tortoise is found. In every confiscated batch of tortoises coming out of Madagascar, the proportion of *Pyxis* became higher and higher, because they were more and more in demand among hobbyists, even though their poor prognosis in captivity was becoming increasingly evident. We have followed the traffickers in the field (Devaux, 2001); during the dry season, they can find about 10 *Pyxis* per hour, south of Anakao. The tortoises are then sold for about 20 euro cents each, although by the time they get to Europe they are each worth 200 euros. A portion of the catch is reserved for the Antananarivo market, and these end up in the gardens of wealthier Madagascan citizens. In order to control or stop the commercial collection, which especially affects *Astrochelys radiata,* the plan must have several components: set up information booths and a program of traveling instructors so that local people will become aware of the problem; replace the tortoise-based economy by one utilizing chickens, pigs, fish, or artisanal work; and convert the traffickers into naturalist-guides working for conservation organizations. This is what we have tried to do with the Association for the Protection of Nature–SOPTOM center, created at Ifaty, north of Tuléar. The center receives and rehabilitates tortoises seized by the authorities, including the largest specimens of *A. radiata,* and provides for the education of local people and the development of ecotourism opportunities for visitors, to the benefit of the local and the regional economy.

© B. Devaux

This subspecies has a moderately functional plastral hinge and lacks dark plastral markings. It is referable to *P. a. arachnoides,* here observed south of Tuliar. The animal on the right is aged, its coloration faded by the passing years.

Pyxis planicauda (Grandidier, 1867)

Distribution. This species is thought to be restricted to the Andranomena Forest and the region from Tsiribihina to Morondava, on the west coast of Madagascar, but it may well be that the distribution extends as far as south as Analabe, and it is possible that it reaches even farther into the west-central and southwestern parts of the big island.

Description. The scientific name refers to a special character: the wide tail of the male, which is rather flattened in a horizontal plane. This feature is not apparent in the females. On the other hand, the common name refers to the "flat back." This is the more elongated of the two species of *Pyxis,* and the shell is often flattened at the level of the second to fourth vertebral scutes. In general the coloration resembles that of *P. arachnoides,* and the similarity is enhanced by the spiderlike designs; the background is cream to beige, with radiating, geometric, dark brown to black designs on each of the scutes. The contrast, very marked in the hatchlings, is reduced with age, and very old animals may be quite plain. This species differs from *P. arachnoides*

Common name

Flat-shelled spider tortoise

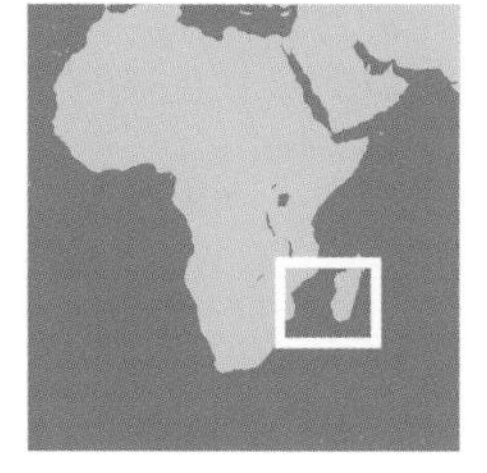

In the image at right, the flat tail of this species, and the reduced space between supracaudal scute and plastron, are evident.

This species is similar in size and general aspect to *P. arachnoides,* but the design is darker and more uniform, and the tail is flatter in the males.

in the following characters: no plastral hinge; flat tail in males; carapace more elongate, flatter on top, with individual scutes also rather flat, giving the animal the shape of a loaf of bread; gular area extended and bilobed; marginals serrated toward the middle of the series and forming a rounded surface where they connect to the bridges; front and rear marginals strongly serrated; head dark, usually greenish brown, with less contrast than in *P. arachnoides.* The carapace length reaches 160 mm. In this species, the space between the plastron and the carapace, at the level of the tail, is extremely narrow. In order for the eggs to pass through, some degree of breakdown of the area between the abdominal and femoral scutes occurs before actual oviposition, and the hind lobe of the plastron itself also become more flexible, thus providing space for the egg to pass.

© P. Pritchard

Natural history. Most of the observations of behavior in this species have been made in captivity. There are two main centers for study and captive breeding: St. Catherine's Island, Georgia (USA), in a facility overseen by John Behler until his recent sad demise, and on the Island of Jersey, at Gerald Durrell's famous zoo. In these facilities, about 40% of the diet of the tortoise is mushrooms, and in addition they eat selected fruit and small insects. If they are sprinkled with water, they tilt themselves forwards so as to let the water run from the shell towards the mouth, as do certain of the South African *Psammobates.* In nature, they are often coprophagous, but may also be almost omnivorous. In the Jersey and St. Catherine's facilities, copulation is seen from time to time, although rarely, and is triggered by certain climatological conditions: initially, a hot, humid period, with the thermometer at around 37 degrees and the hygrometer showing more than 80%. Then the humidity is diminished for four months, mimicking a temperate, dryer season, with a temperature of 27 degrees by day, 12–15 degrees by night, and a humidity of about 40%. The conditions are similar to those found in nature. Nevertheless, actual nesting is quite rare. One or two eggs are laid per clutch, and they are elongated, averaging 35 × 22 mm; the incubation (in an incubator) lasts for about 100 days. In nature the tortoises show great activity after rains. The rest of the time, they hide under the substrate of dead leaves, coming forth only at night. They keep away from damp areas. In old age the adults become covered with moss and take on a greenish appearance, as a result of their subterranean existence. This species does not like direct sunlight; they run away from it rapidly.

© B. Devaux

Protection. The rarity of the species, made clear by field investigations from 1980 to 1990, has stimulated the development of several raising centers, although the jury is still out as to whether these have been successful. In the field this tortoise is not hunted, eaten, or collected. Nevertheless, the habit of burning fields and woodlands and the expansion of agricultural areas, as well as highway development and mining and petroleum exploration, do nothing to promote a happy future for this tortoise. The distribution does appear to be considerably wider than previously thought, but a thorough scientific evaluation of this, as well as the survival status, remains a high priority. The species is listed as endangered by the Turtle Conservation Fund.

Stigmochelys (Geochelone) pardalis
(Bell, 1828)

Distribution. The leopard tortoise occurs widely through eastern and southern Africa, from central Sudan to Namibia and southern Angola, but is specialized for savanna and grassy areas. It does not occur in countries with uniformly high humidity, such as the Democratic Republic of the Congo.

Description. All the common names refer to the feline type of leopard and are a reference to the tortoise's usual pattern of black spots and markings on a light cream background. The markings are variable and irregular, some tortoises being covered with them and others having very few, depending on the region and the particular population. When the tortoises become aged, the design starts to fade, and some very old animals are uniformly beige or whitish. The maximum weight is normally about 20 kg, but W. Archer describes a population of leopard tortoises in the Graaff-Reinet area of South Africa, where they may be found at 2,000 m altitude and where some individuals may weigh 50 kg. Some really huge carapaces discovered by B. Branch must have corresponded to tortoises weighing more than 50 kg. This species has conical spurs on the thighs. The limbs and head are similar to those of *Centrochelys sulcata* and are also uniformly yellowish, but the snout is less squared off, and has two isolated frontal scales. The eyes are very small. The shell is well rounded and sometimes the vertebral scutes show traces of conical form. There is no nuchal scute. In the young the "leopard" pattern is well marked, with details that differ between any two individuals, and sometimes one may detect red-brown markings on the seams.

Subspecies. Formerly two subspecies were recognized: *S. p. pardalis* (described above), from southern Namibia to Western Cape Province; and *S. p. babcocki* (Loveridge, 1935), occupying the entire remainder of the huge range and characterized by a more pronounced dome and more intensely black patterning. But in reality the species shows extensive local variation, and according to B. Branch and other South African herpetologists, there are no recognizable or definable subspecies.

Natural history. This tortoise is adapted for different environments, which may explain its wide distribution on the African continent. It is at home in the dry savannas and semideserts of Namibia, as well as in the mountainous zones of eastern Cape Province, but lush grassy plains are generally devoid of this species. It is diurnal in habits, but when the day gets too hot, it hides in the vegetation. During the cold period or at high altitudes, it will utilize existing burrows and tunnels made by other animals, as well as abandoned termite mounds and niches in boulder-strewn areas. It swims very well, and individuals have been seen crossing Lake Kariba with consummate ease. The diet is essentially vegetarian: watermelons, succulents, herbs, and almost any wild vegetation, as well as mushrooms and other fruit. But it will also dine on the excrement of other animals and on bones of dead animals.

Copulation is preceded by a classic ritual with the male battering the female and emitting a sound described as "asthmatic" by Bennefield. Nesting takes place during the warm months, which may be May to June or October to November, according to latitude. The female expels cloacal water before digging her nest in a dry or stony area. The

© A. Hell-Kevorkian

This tortoise is walking across one of the plains of the Serengeti, and the Masai warrior has paused to watch it. The Masai do not eat nor otherwise disturb the tortoises, and for this reason the populations of *S. pardalis* are still strong.

Common name

Leopard tortoise

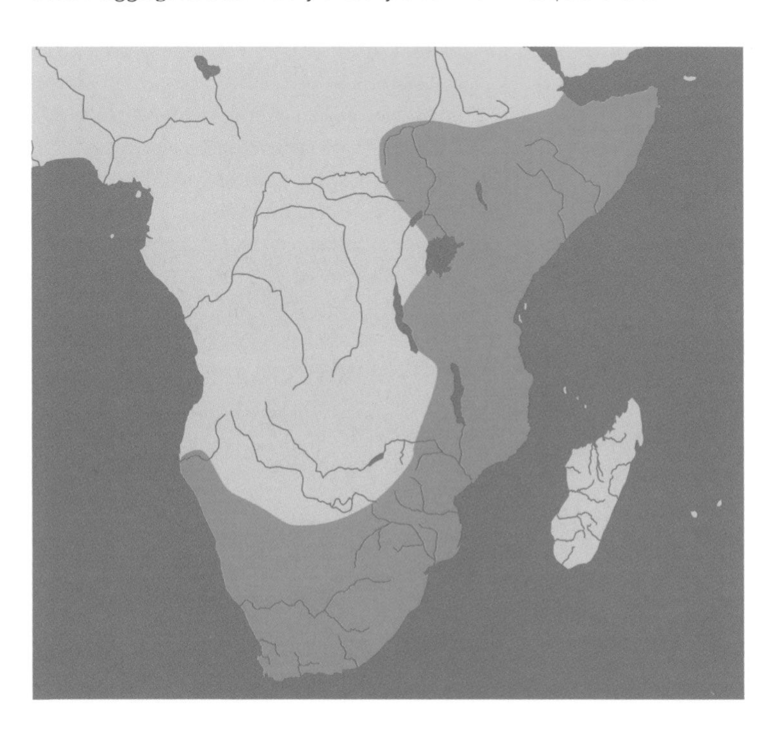

© B. Devaux

This large tortoise is emerging from a dry creek bed in the Great Karoo of South Africa. The plain shell with just a few, discrete black spots indicates that this is an old animal.

eggs are spherical and about 42 mm in diameter, and there may be four to eight per nest. There may be six nestings per year, with an annual output that may equal or exceed 30. Incubation is very slow, the precise duration being dependent on rainfall. It may take as long as 380 days, although this is presumably not typical. The young weigh about 30 g. A rather high degree of predation is imputed to bats, monitor lizards, and snakes, and in the Kruger National Park we have seen lions eat them, burrowing into the shell with their claws. Archer indicated that this species has a certain homing ability, being able to return to a known spot when displaced. The animals can cover 12 km in a month and a half and then return to the point where they started.

Protection. Since the export of *C. sulcata* was controlled, commerce has shifted to this species, whose populations are still numerous, and it is an equally spectacular animal from the point of view of hobbyists. Certain nations (Tanzania, Kenya) are exposed to heavy collecting pressure, and Western pet shops have started to offer large numbers of leopard tortoises. Furthermore, it is eaten by some ethnic groups within its range, although this does not happen very frequently. Fires, which are common in East Africa, destroy habitats and kill lots of animals of all kinds. This species is also subject to a skin disease that upsets the growth of keratin. In South Africa it is protected in numerous reserves, including Kruger, Bontebok, Karoo, and Mountain Zebra.

Testudo graeca Linnaeus, 1758

Distribution. This species occurs in North Africa, from Morocco into Libya, as well as southern Spain, the Balearic Islands, Sardinia, and Sicily.

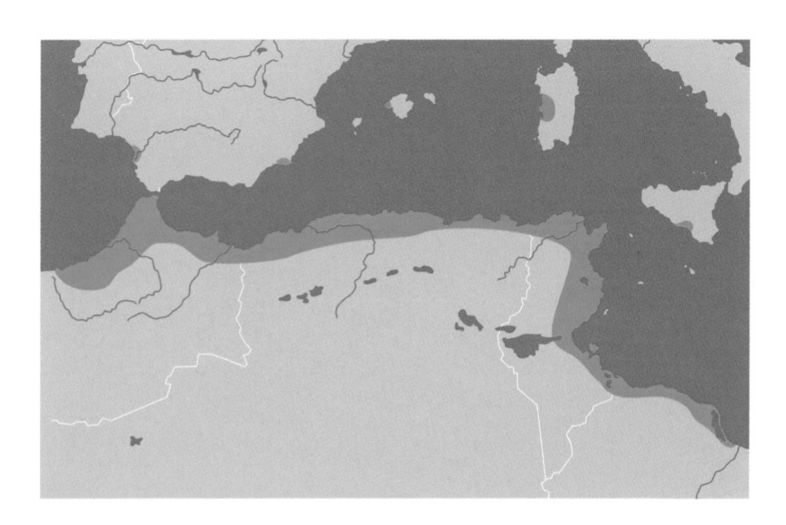

Description of the genus *Testudo*. The genus has recently been reviewed by R. Bour, J. Perälä, A. Pieh, E. Taskavak, et al. (2004), but a more complete general revision is expected in the years to come. One of the *Testudo* species was split off into a new genus *(Agrionemys horsfieldii),* while a sister species, *Testudo hermanni,* even though closely related, remains in the genus *Testudo.* New species and subspecies have been described, and we take note of them in the present work. Nevertheless, this taxonomy is a work in progress and should be treated with some caution. We recognize the greater part of the new species, but we retain four forms as subspecies within the species *T. graeca.* Tortoises of this genus are mostly rather small (*T. marginata,* reaching 450 mm, is the biggest). The carapace is elongate and generally domed, usually with a contrasting pattern of yellow and black. The diet is omnivo-

s© B. Devaux

The coloration of this species is quite variable, but this specimen shows the "classic" pattern of irregular black spots and markings on a cream or beige background. The carapace is very domed.

rous, with a strong emphasis on herbivory. The ecology of all of the species is very similar, and their conservation needs and challenges are generally the same: all have suffered from heavy collection for centuries, and they are all becoming scarcer and need a break.

Description of *T. graeca.* Linnaeus named this tortoise *graeca* in reference to the patterning of the carapace, which reminded him of a Grecian frieze, but the species is not found within the geographic limits of Greece. This species, the flagship of its genus, is distinguished by the following characters: it has horny spurs on its thighs (hence the English vernacular name), and there is no spur or horn on the tip of the tail; the supracaudal scute is not split; the plastron is yellowish to cream and is marked with irregular black markings that sometimes fade almost completely and never form heavy lateral bands; the carapace is domed and is yellowish to greenish, with regular black figures that form the "Grecian frieze" but often start to fade with age; the maximum length is about 300 mm; and the head is spotted with black and yellow. The plastron of the females often has a significant degree of mobility at the base of the posterior lobe, which facilitates oviposition.

Subspecies. Six subspecies are retained, but a further revision of the genus may change all of this.

T. g. graeca is described above.

T. g. cyrenaica Pieh and Perälä, 2002, is found from the Cyrenaica Peninsula as far as El Adem, in northeastern Libya. It does not exceed 200 mm in length and has very smooth forelimbs, a narrow carapace, an irregular, interdigitating line of contact between the marginals and the costals, and posterior marginals strongly incurved toward their upper borders, with vague and variable markings on an orange or beige background.

T. g. marokkensis Pieh and Perälä, 2004, occurs in Tarmilet, in central Morocco. The carapace is lower than in the typical form, with more numerous dark lines.

T. g. lamberti Perälä, 2004, occurs north of Tetuan, in northwestern Morocco. It has very strongly developed thigh spurs, and the posterior marginals are recurved. The coloration is rather plain and dark, with extremely fine, diffuse lines.

T. g. nabeulensis Highfield, 1990, occurs in Tunisia and western Libya. It has a rather "elegant" appearance (A. Pieh, 2004), a strongly domed carapace, and a maximum size of only 180 mm. The fifth vertebral scute often has a tarantula-like design. The gulars are rather wide and elevated.

T. g. soussensis Pieh, 2001, occurs in the valley of the Souss, in southwestern Morocco, and probably around Ouarzazate. The interpectoral seam is relatively long in comparison with the interfemoral, the third vertebral scute is short in males, and the coloration is dark and diffuse, on a yellow background. The juveniles may be distinguished from the nominal subspecies by an absence of dots on the areolar areas of the vertebrals and the marginals. Some specimens lack the thigh spurs. Females may reach a length of 249 mm.

Natural history. This species lives in arid places, and its diet is correspondingly reduced to that which is available: thistles, vegetative detritus, and so on. Some populations at higher altitudes (the Middle Atlas Mountains) may hibernate, but more often these tortoises will estivate during the hot months, hidden under bushes or in crevices. Males may be quite aggressive and confront and challenge each other during the breeding season.

Common name

Mediterranean spur-thighed tortoise

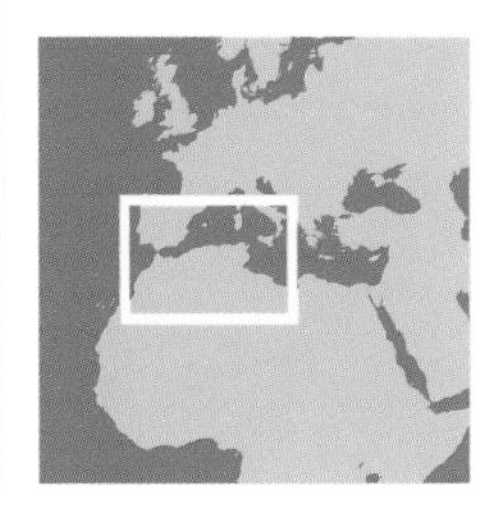

Courtship occurs in early spring and also after rare autumnal rains. The male chases the female, bites her neck and limbs with some violence, walks around her to keep her from moving away, and proceeds to climb onto her shell. During actual copulation, the male utters groans that may be heard from some distance. The male is smaller (150–180 mm) than the female (300 mm), and sometimes one sees very small males attempting to copulate with huge females. Nesting occurs from April to June. Often there may be three nestings, with a maximum of five for a single female. The maximum clutch size is about one dozen, but more usually 6 or 7 eggs are laid, each weighing 10 to 12 g. This species is credited with exceptional longevity, and in England one example lived for 120 years. At hatching, the young are often grayish and later achieve a more contrasting black and yellow pattern. Already, tiny spurs may be seen on the thighs, a useful identification feature. Very old animals may undergo loss of some of the scutes, with exposure of the underlying bone, but this does not seem to present any disadvantage. This species is specialized for a hot, dry climate, and shipping the animals to Europe as pets generally condemns them to a series of chronic pulmonary maladies leading to rapid death. Moreover, these diseases, of the herpesvirus family, may be transmitted to other, native species, including *T. hermanni.*

Ethnozoology. For millennia, tortoises have been part of the imagination and culture of the Mediterranean people. At Roc Saint-Cirq in Dordogne, a small sculpture, just 7 cm long and 13,500 years old, is the earliest known representation of a small *Testudo,* and it appears to be a *T. graeca,* judging by the constriction of the middle of the shell and the vertical posterior margins. Four hundred years before Christ, the Greeks used representations of both *T. graeca* and *T. marginata* on their coins, called staters. *Testudo* went on to be a theme for designs, statues, and figures, and (especially in Italy) representations were to be seen on palace walls, at the base of fountains, and in paintings. Later the tortoise theme spread throughout Europe, especially into France and the Netherlands. Furthermore, tortoises feature in endless stories, legends, and fables, the most celebrated being those of La Fontaine. But the tortoises were not just subject to mere admiration and observation; they were also heavily exploited, and this exploitation has continued to such an extent that their very future is in serious doubt.

Protection. For a long time, *T. graeca* has been transported and exploited by humankind. In ancient Greece these tortoises were sometimes used to wedge or support amphorae aboard ships. They were displayed in cabinets of curiosities throughout the Western world. During the colonial wars they were gathered by soldiers on service in European theaters of war, and until their protection under the Washington Convention (CITES)—about 1976 in France—they were sold in fish markets and pet shops, both as a source for soup with "special" health benefits, as well as for (allegedly) ridding gardens of snails and slugs. These practices caused some populations to be reduced to vestigial levels.

Today tortoises are still gathered and sold in the souks throughout the Maghreb. The city governments pay little attention to the requirements of CITES, even though they are voluntary signatories to this convention, and this commerce is denuding their own countries of an important component of its fauna. Moreover, the tortoises are offered to tourists under the bleakest conditions, in tiny cages, from sacks lying on the floor, or crowded into baskets exposed to full sun, engendering pity on behalf of the visiting tourist, who then proceeds to buy the animals to "rescue" them. When they are taken across international borders, they are often seized by customs officials, or even if they are "lucky," they end up in suburban gardens in areas that are too cool and too humid for their survival. The sale of Mediterranean tortoises has subsequently created and developed a demand for "garden tortoises" among hobbyists and pet keepers, and this turns out to be disastrous for chelonians in general. Controlling these chaotic developments requires that a permanent education program be launched, sales brought to a halt in each range state, and detailed studies and enlightened field conservation programs established throughout the range. The species is placed in CITES Appendix 2 and Annex C1 of the European Community, and this status also applies to all the "formerly *graeca*" taxa that are discussed below.

Testudo (graeca) anamurensis

Weissinger, 1987

Distribution. This species occurs along the southwestern coast of Turkey, from the Bey Mountains in the west to the plain of Mersin in the east, including the site of Anamurium, which gives the species its name.

Description. This form differs from *T. ibera* (the former *T. g. ibera*) by the following traits: carapace trapezoidal, more narrow, and strongly widened posteriorly; females reaching a length of 260 mm. The color is generally greenish yellow, with moderately distinct black markings. The plastron is often entirely covered with an irregular black figure, but in old animals this may fade to greenish or yellowish, with some random markings.

Natural history. This tortoise appears to be active year-round, even in January and February, but is usually observed in the early morning and in the later afternoon. One nest included 7 eggs, but clutches may contain as many as 15 to 19 eggs, each being about 35 × 26 mm. Incubation lasts for 58 to 83 days.

Protection. Illegal collection continues, and there is still no area where this species is protected. It is placed in Appendix 2 of CITES. The Turkish authorities sometimes seize illegally gathered tortoises and release them back into the wild, but nothing more concrete is being done to protect this species.

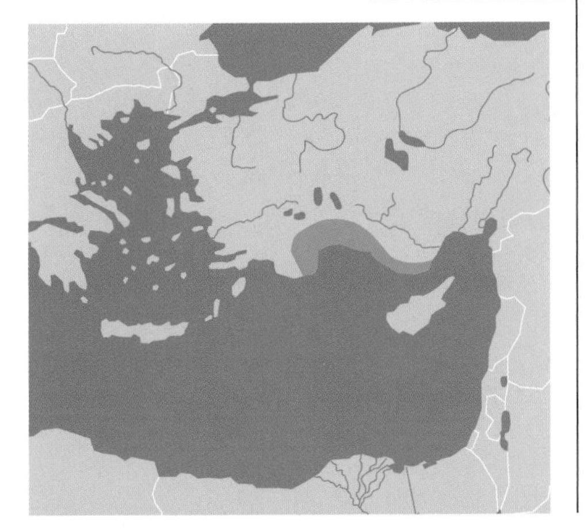

Common name

Anamur tortoise

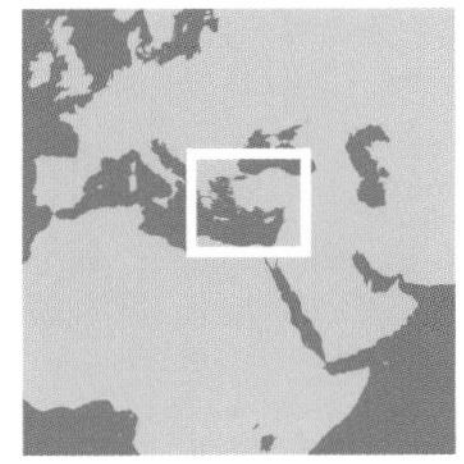

This individual has a more diffuse, highly irregular pattern, as well as spread-out marginal scutes.

© M. Rogner

Testudo (graeca) antakyensis

Perälä, 1996

Distribution. This tortoise is still poorly known, but it is has been found east of the Amanos Mountains in Turkey. It is possible that it also occurs in Syria, Lebanon, Jordan, and Israel, although the tortoises in Israel are larger than typical *T. antakyensis* and may be referable to *T. terrestris.*

Description. Certain authors consider this name to be a junior synonym of *T. (graeca) terrestris* or a subspecies of *T. graeca.* It is distinguished by its smaller size (to 162 mm), the yellow head coloration, and the posterior marginals not being broadened. A detailed morphometric study also notes the construction of the suprapygal scute, the length of the fourth costal scute, the neural bone configuration, and the rounded tubercle at the tip of the tail. The coloration is yellowish, with strong, regular black markings, although the

Common name

Antakya tortoise

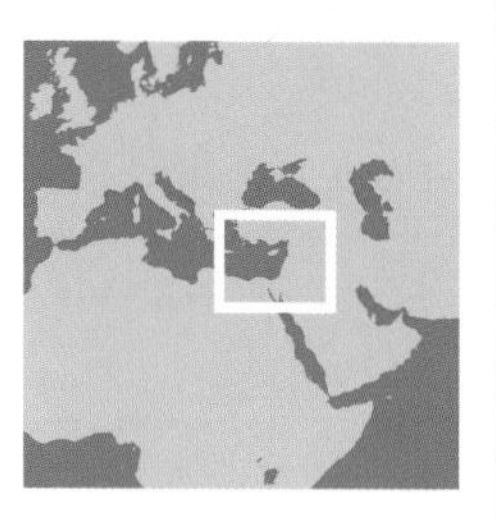

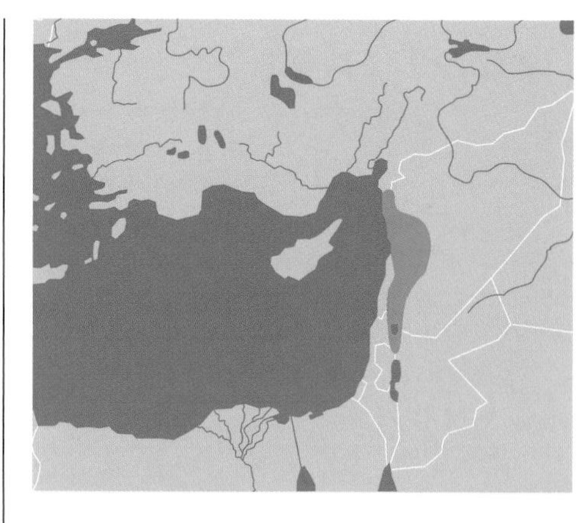

latter may be faded or lacking contrast in some individuals.

Natural history. This tortoise is certainly active all year round and is out and about even in January and February. Clutch size may be as small as four or five eggs, which are rounder and smaller than those of *T. anamurensis.* Incubation lasts for 64 to 90 days.

Protection. This tortoise does not appear to be threatened at present, but there are no special protected areas in which it is found. It is listed in Appendix 2 of CITES, under the name *T. graeca.*

Testudo (graeca) armeniaca

Chkhikvadze and Bakradze, 1991

Distribution. This species is found in southern Armenia, Turkey, Azerbaijan, and northern Iran.

Description. Close to *T. graeca,* this species also shows some characters of *Agrionemys (Testudo) horsfieldii,* such as a very rounded carapace, looking like a large pebble, and a completely rigid plastron. It may reach a length of 230 mm, with a maximum of 257 mm. The nuchal is narrow and sharp-pointed. The marginals are separated from the costals by a wide, very visible seam. The gulars project in a distinctive fashion, and in most individuals they bifurcate anteriorly. The hinge between the hypoplastron and the xiphiplastron is almost nonfunctional. The carapace is brown to chestnut, almost immaculate but sometimes with sketchy markings. The entire animal—plastron, carapace, and soft parts—is of the same earth tone.

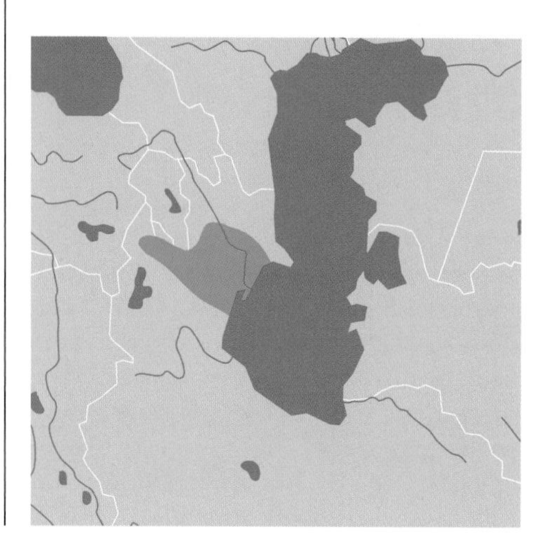

Common name

Armenian tortoise

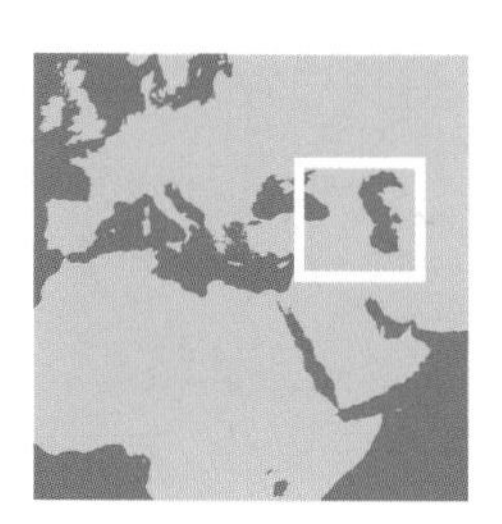

Natural history. This tortoise occupies semidesert habitats up to 3,000 m above sea level, where the vegetation is scant and the stony hillsides support a modest growth of spiny bushes. In summer, activity is confined to the early morning and late in the afternoon; at other times, the animal hides under bushes. It may nest three times per season, with three to five eggs per clutch.

Protection. This tortoise is occasionally consumed in the Megri area in Turkey, but there is no intensive exploitation. There is no specific protection in place for this animal, although it is included in the IUCN *Red Data Book* and also in Appendix 2 of CITES (as *T. graeca*).

Testudo boettgeri Mojsisovics, 1889

Distribution. The range is located east of that of *T. hermanni,* of which it was formerly considered to be an eastern subspecies. *T. boettgeri* occurs from Venice and Istria in Italy, through the Balkans, and as far as Dobruja at the mouth of the Danube. It also occurs on many of the Adriatic and Ionian islands, the Peloponnese, and European Turkey.

Description. Very similar to *T. hermanni,* this species also has a conical spur on the tip of the tail, a divided supracaudal (usually), black markings on the plastron, narrow vertebral scutes, small scales on the anterior face of the forelimbs, and strongly contrasting black and yellowish coloration. However, it differs in the following points: the yellowish color has a greenish rather than an orange tinge; the interfemoral seam is equal to, or shorter than, the interhumeral seam (reverse in *T. hermanni*); and in males the shell is significantly broadened at the level of the hind limbs. It is also larger, reaching up to 340 mm (record), although some of the Peloponnesian populations seemed to be stunted or dwarfed and do not exceed 140 mm (Bour, 1995). The plastron of the female may have a modest degree of kinesis at the level of the hind limbs, facilitating oviposition.

Natural history. This species lives in the Mediterranean garigue and maquis, in areas with some degree of tree cover but also with open areas, important both for thermoregulation (sun basking) and for nesting. The tortoises hibernate in natural niches and places of concealment, usually from October to March, but with some regional variation. Nesting occurs in May and June, up to three times per season, and with as many as nine eggs laid per clutch. The eggs measure about 38 × 28 mm. The diet is somewhat more carnivorous than that of *T. hermanni,* and small lizards, earthworms, other invertebrates, carrion, and feces may be consumed in addition to a wide variety of plant species.

Protection. This tortoise is now included in Appendix 2 of CITES. Before protective laws were passed, there were massive levels of collection and of exports to western Europe, which caused dangerous inroads into natural populations. Today there is much less commercial collection, but there are still problems with habitat degradation and urban development.

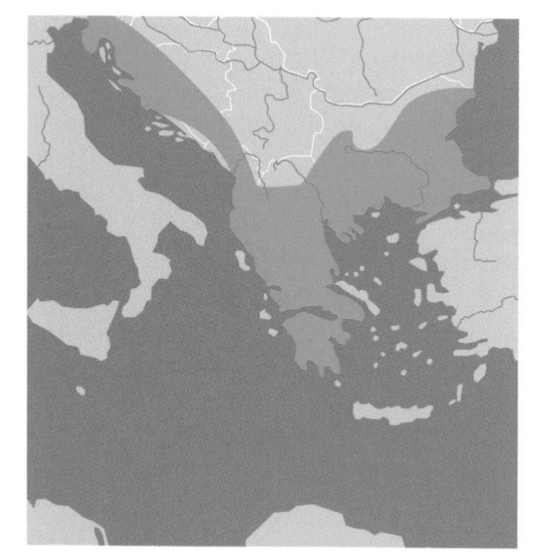

Common name

Eastern Hermann's tortoise

© B. Devaux

The coloration of this species appears very similar to that of *T. hermanni,* but the former is often lighter and sometimes has a greenish tinge. Males often have widely spread marginals in the area above the hind limbs.

Testudo (graeca) floweri Bodenheimer, 1935

Distribution. This species occurs in a narrow coastal band in the Gaza Strip, Israel, and Lebanon, as far as Beirut. It may also be found in Egypt, in the Sinai between El Arish and the Israeli frontier.

Description. This is a small tortoise, with a maximum length of 154 mm, but it still manages to have the appearance of a small tank (J. Perälä, 2004). Viewed from above, the carapace is somewhat elongated in the males and rounder in the females, which also have a flattened top. The males have a short plastron, exposing to sight the heavy, elongate, conical tail. The nuchal scute is short and normally is narrow in males and wider in females. The anterior edge of the carapace may be slightly serrated. The first vertebral has a rounded outline. In color, this species is distinctive: the background is pale yellow or beige or even greenish or brown, with very few, smaller black markings, usually present on the centers of the scales and sometimes taking on an elongate form. The plastron is greenish to yellowish, with diffuse but somewhat symmetrical markings, which tend to fade with age. Certain old males become totally melanistic. The large scales of the forelimbs are brown to yellow and sometimes have black tips. The head is yellow to dark brown and is lighter on top, around the nostrils. At the tip of the tail there is a large double scale.

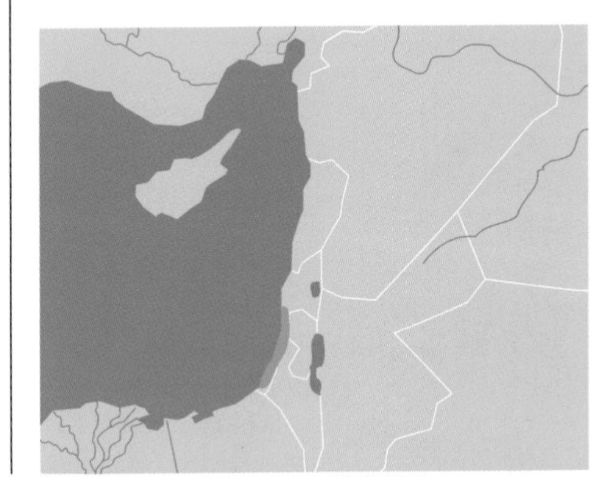

Common name

Flower's dwarf tortoise

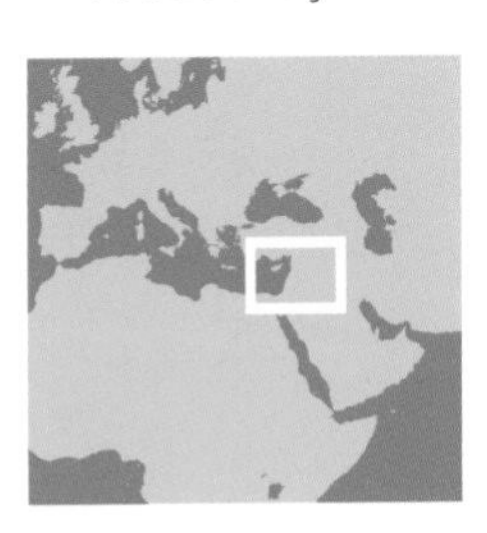

Natural history. The normal habitat is coastal plains where the climate is moderate to hot, and this tortoise is often active even in January and February. It buries itself in burrows to estivate and also retreats during the coldest times of the year. Nesting takes place in May to July, with three to five eggs per clutch. The eggs hatch in 90 days.

Protection. The range is largely occupied by people, and this species is progressively displaced by urbanization, warfare, and habitat destruction. All-terrain vehicles are also a menace, as is the introduction of predators, including grackles. However, commercial collection appears to be minimal. The species is classified in Appendix 2 of CITES. Populations do not appear to be abundant, and there is no reserve that offers them protection.

Testudo hercegovinensis Werner, 1899

Common name

Herzegovina tortoise

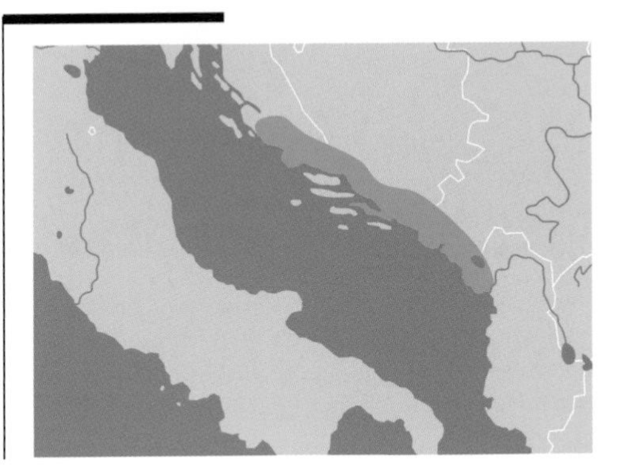

Distribution. This form lives in the Balkans—in the Zadar, Mostar, Bileća, and Trebinje regions of Croatia, Bosnia-Herzegovina, and Montenegro, along the shores of the Adriatic Sea.

Description. This species is often confused with *T. boettgeri* (or *T. hermanni boettgeri*); whether or not one considers the latter to be a distinct species, it is very close to the lineage of *T. hermanni* (Gmelin, 1789), which in many ways seems to be in a different clade from other tortoises in the

Testudo genus (see *Agrionemys horsfieldii* for discussion). *T. hercegovinensis* has no inguinal scutes, the space on each side being filled by an extension of the abdominals. It is small, not exceeding 147 mm. The coloration is yellowish, or beige with a greenish or reddish tinge, with symmetrical, bold black bands and well-marked growth rings. The plastron has two well-marked black bands, but they are frequently interrupted or discontinuous. The head is brown or blackish, without light spots, and there is only a limited yellowish area on the top and the back of the head.

Natural history. The habitat is one of forests or bushy country, like the cork oak forest near Trebinje (Werner, 1899) or certain dense mulberry and herbaceous woods surrounded by agricultural lands in Croatia. Three to five eggs are laid per nesting, and two clutches may be laid in a season. The eggs measure 32 × 16 mm, and incubation is said to be very fast—about 60 days.

Protection. Large numbers of this species were included in commercial shipments of *T. boettgeri* for many years, destined for export to western Europe and the USA, and populations must certainly have been severely affected by these episodes. War, urban development, and agricultural expansion have also weighed heavily upon this species, which today occupies only a fraction of its original range. Like all tortoises, it is listed by CITES, in Appendix 2, and should now be less subject to collection for any purpose, but unfortunately the habitats are severely degraded.

The shell markings are symmetrical, and the coloration is very contrasting, but darker than in *T. boettgeri*. The species can be identified by the absence of the inguinal scute.

Testudo hermanni Gmelin, 1789

Distribution. Hermann's tortoise occurs in western and southern Italy, southeastern France (central Var), and northeastern Spain. It also occurs on certain islands, including Corsica, Sicily, Sardinia, Majorca, and Minorca. The population in the French Pyrenees was eradicated by a wildfire in 1986.

Description. Formerly divided into two subspecies, one in the east and one in the west, Hermann's tortoise is now just plain *Testudo hermanni*. From the phylogenetic point of view, it is very close to *Agrionemys horsfieldii* and does not really fit into the *Testudo* complex, which may be considered a sister group. Nevertheless, *Agrionemys* is the name used for a group of tortoises specialized for digging extensive burrows, and *T. hermanni* is probably close to the common ancestor of the two groups and sufficiently distinguished to merit generic separation, according to F. Lagarde, X. Bonnet, and R. Bour (2004). But we shall keep it in the genus *Testudo* for the time being, as M. Cheylan recommends.

This tortoise has a spur on the end of the tail, most developed in the adult males; a divided supracaudal scute; two continuous black bands running from the front to the rear of the plastron; an

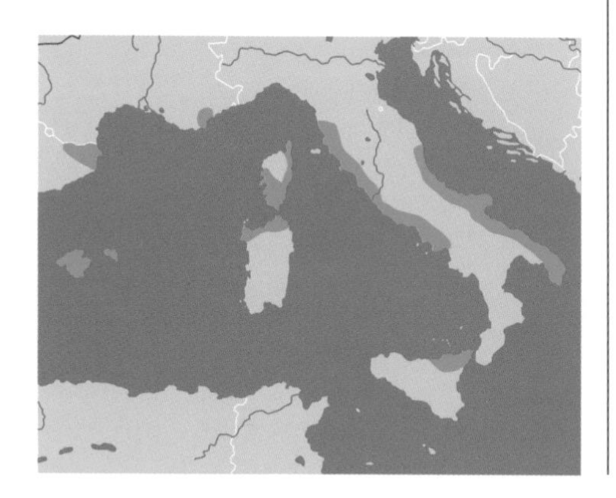

Common name

Hermann's tortoise

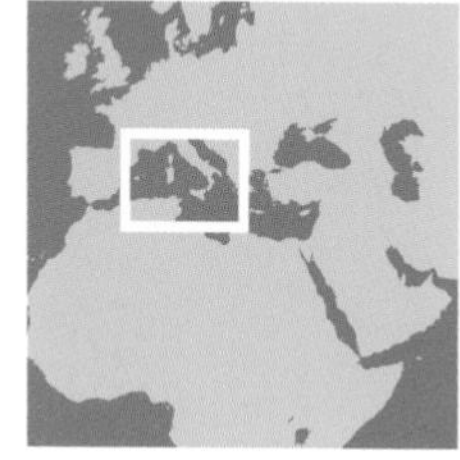

interfemoral seam longer than the interhumeral seam; and strongly contrasting colors, orange-yellow on the background and with intensely black markings. There is a standard design decorating the areola and the rest of each scute. The carapace is moderately domed, less so than in *T. graeca.* There are no spurs on the thighs. The length does not usually surpass 190 mm. The head often bears a yellow spot, low down and behind the tympanum. The limbs are grayish to brown, with bold yellow markings. There are differences between mainland populations and those of the islands. Those of the Balearics, Italy, and the Spanish Pyrenees are as described above. Those of Corsica are somewhat variable, suggestive of former mixing with *T. boettgeri;* thus the coloration is somewhat more greenish, with fewer black markings. The interfemoral seam is about equal to, or less than, that between the humerals. The males are especially different from those of other areas: they are expanded laterally in the rear part of the shell, having the overall shape of a massive triangle, with widened and somewhat recurved marginal scutes, and a coloration leaning toward gray-green. The size may also be greater than is usual for the species and may reach 250 mm in females.

Natural history. Hermann's tortoises live in the garigue, the maquis, and the scattered forests of the Mediterranean zone. They like semi-open country, where they can find shade and hiding places for estivation and hibernation, as well as open areas for insolation and oviposition. They emerge from hibernation in late February and embark upon courtship and mating immediately. These behaviors follow the same pattern as those of *T. graeca* but are less violent. In Provence, nesting starts around May 15 and ceases at the end of July. The photoperiod decides the length of the nesting season, more than the actual temperature does. The females dig a bean-shaped nest about 6 to 10 cm deep, and therein they deposit one to five eggs, about 32 × 24 mm in size. There may be a second nesting, about three weeks after the first. Incubation lasts about 90 days, and hatchlings emerge after the onset of heavy rains, from mid-August to September. If the hatchlings have failed to emerge because the rains never came or because the nesting was late in the season, the young will emerge from the egg but remain underground, not appearing until the following spring. They weigh about 10 g at hatching. Nest success in nature is about 80%, but the ensuing mortality of hatchlings may be very high. For every 1,000 tortoises hatched, only 4 or 5 are still alive three years later. The young face numerous predators: rats, raptors, badgers, magpies, hedgehogs, snakes, foxes, wild boars, and even ants and wasps. Only after 6 to 8 years, when ossification of the shell is complete, can the tortoises look forward to a somewhat less tormented life. The temperature of sexual differentiation is 28.5° C, and if the eggs are kept at 26° C for the first few weeks,

In the garigue of Provence, a Hermann's tortoise looks for an open spot to warm itself in the sun. The yellow-and-black pattern is highly contrasting. In this young tortoise, the growth rings are easily discerned.

only males will be produced. At 30° C, on the other hand, only females are produced. Sexual maturity is reached at about 8 to 12 years in males and 10 to 14 years in females. Hibernation in Provence starts around mid-November. The tortoise buries itself in a bed of dead leaves in a cavity dug under a bush or under old rotting planks. It will spend the three winter months there and not reappear until late February or early March, but there may be major differences between individuals. Some go into hibernation quite late and may emerge several times during the winter, while others remain securely buried and asleep from late October to the end of March. Thermoregulation studies by C. Huot-Daubremont have revealed that the lowest and highest temperatures tolerated by the internal organs of this species are 3.7° C and 35° C, respectively. Temperatures outside this range are lethal. The natural longevity in the wild is probably around 30 years, but some individuals may reach 60 years of age and a few may become centenarians. This species is 90% herbivorous. It favors clover, dandelions, strawberries, yellow and white flowers, and numerous plants and herbs of the maquis, but it will nibble on an earthworm or a snail from time to time, and even upon small animals found dead (such as lizards, baby rabbits, and amphibians).

Protection. This species, like *T. graeca,* has been subject to intensive commercial exploitation and export at least until 1976, when protective laws were passed at last. It is currently listed in Appendix 2 of CITES. Because of food rationing during the last war, this tortoise was habitually consumed by some Provencal villagers. Up to the middle of the last century, it was consumed in convents and monasteries on fasting days, being considered (like otters, frogs, and snails) to be "neither meat nor fish." Today the tortoise continues to suffer from urban development, habitat destruction, ongoing collection from the wild, and the fires that strike the Mediterranean region from time to time, as happened in Var in 1990 and 2003. But it is the sheer multiplication of humanity that has reduced tortoise populations and today threatens their very survival. At the beginning of the last century, the range of *T. hermanni* was a virtually continuous coastal band from Spain to somewhere north of Rome. Sixty years ago, these tortoises were still to be found in Bouches-du-Rhône and in the Maritime Alps, as well as at the foot of the Abruzzo Mountains. Today the French distribution is limited to a small area in central Var and parts of Corsica, and in Spain and Italy the distribution has become fragmented. An association, SOPTOM, was created in 1986 as a result of biological investigations by Marc Cheylan and ecological work by David Stubbs, and in 1987 the Program for the Protection of Hermann's Tortoise in France was started, followed by the opening of the "Tortoise Village" in May 1988. Activities pursued since 1986 have included education of the general public, numerous field studies, and efforts to reconstitute natural populations by liberating captive-hatched young tortoises raised at the Tortoise Village.

Major concerns remain, especially resulting from the growing urbanization of Var, which has greatly restricted the areas available for tortoises, and also the expansion of pet shops and wildlife dealers in France and elsewhere who supply the growing demand for pets. This concept of "garden tortoises" is catastrophic for tortoises in our part of the world and also for chelonians in general; it encourages the public to obtain and keep these animals, usually in terraria or small enclosures, and this leads to ever-increasing commercial pressure on the surviving wild populations. Mediterranean tortoises (not to mention other species) will not be safe until all of the following steps have been taken: habitat protection (by outright purchase, declarations of reserves, environmental protection); an ambitious program of studies on wild tortoises to get a better understanding of their ecological needs and biology; education of the general public, and particularly children, about the need to protect tortoises; reinforcement of wild populations by means of captive-raised animals from specialized rearing centers (as with the Tortoise Village in Gonfaron); cooperative plans with other centers and other countries (international congresses, international publications, joint study programs, and so on); and above all a changed mind-set so that tortoises will be thought of and respected as wild animals and not just as pets. This species is classified as endangered by the Turtle Conservation Fund.

© I. Arndt

Most tortoises of this species have a well-developed yellow spot under the tympanum. This one has a strong, cutting beak, which enables it to crop the materials that make up its vegetative diet.

Testudo (graeca) ibera Pallas, 1814

Common name

Iberian tortoise

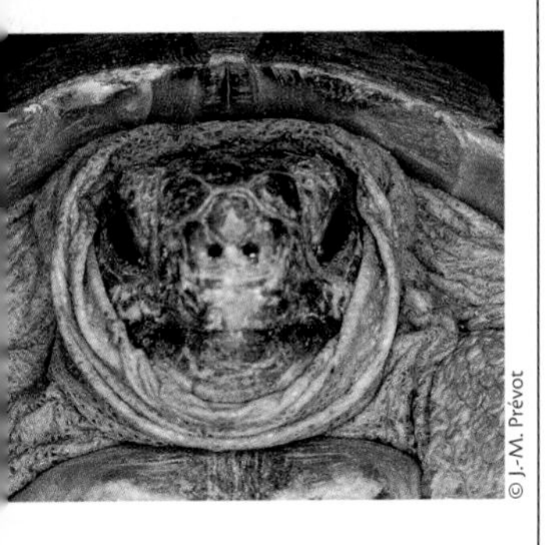

© J.-M. Prévot

This rather large tortoise has a solid shell, somewhat wider toward the back. The color ranges from blackish to chestnut and is quite uniform.

Distribution. This form occupies a limited range in the drainage basin of the Kura River and the low country of Lenkoran in Azerbaijan, eastern Georgia, Armenia, and northwestern Iran.

Description. Formerly considered a subspecies of *T. graeca,* a name that covered many other Middle Eastern tortoise taxa, it is now considered to be a species in its own right. The common name refers to Iberia, as eastern Georgia was once called. This species may attain a length of 260 mm and has a relatively wide, low carapace. The hind marginal scutes fan out somewhat and may have denticulate edges. The coloration, formerly reported to be uniformly dark brown, is in fact more contrasting, with an olive to beige background and very bold black markings radiating from the areolae. However, some old individuals do have an almost totally dark brown or black carapace. The head is dark olive or brown-black, occasionally with some yellow spots.

Natural history. In summer these tortoises reduce their activity levels, restricting daily activity to early morning, when they may concentrate around the few water holes. They hibernate from November to February. They are somewhat omnivorous, consuming worms and insects in addition to vegetable matter. Nesting has been observed from May to mid-June, and the clutch consists of four or five eggs, 35 × 25 mm in size. There may be two nestings in a season. Incubation takes about 70 days.

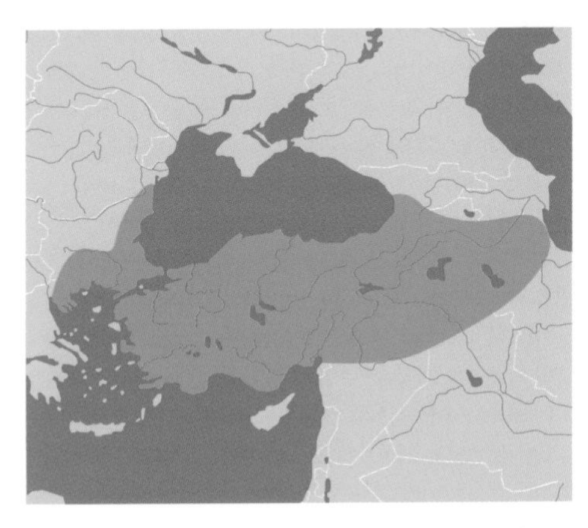

Protection. This tortoise is placed in Appendix 2 of CITES. Populations are declining in Azerbaijan, as a result of extensive new agricultural development. In former times, large numbers of these tortoises were collected and exported to Western markets.

© B. Devaux

Testudo kleinmanni Lortet, 1883

Distribution. This species is restricted to a limited coastal strip, about 50 km wide, from Libya (Tripoli, Cyrenaica) as far as the Nile in Egypt, and to rare isolated individuals and populations.

Description. This is the smallest tortoise of *Testudo,* with a maximum length of 144 mm. The carapace is strongly domed and almost cylindrical, with the marginals recurved at the sides, forming a rounded edge at the bridges. The general color is light yellow to beige, very similar to the color of the sandy substrates where the animal lives, with perhaps some traces of black around the edges of the scutes in younger specimens. The plastron is also diagnostic, with a reasonably kinetic posterior lobe (hence the Spanish vernacular name) and with a black triangular marking on each of the abdominal scutes (note that *T. marginata* has such markings on all of the plastral scutes). The limbs, head, and neck are also uniform yellow. The eyes are large and almond-shaped, with a black iris. There are no spurs on the thighs. On the forelimbs the scales are very large and flat, a trait that assists concealment in the sand. This species can dig in the sand to hide but also often uses existing niches or cavities, perhaps at the base of a bush. The limbs are relatively long, allowing the tortoise to do a high-stepping walk, no doubt to minimize contact with a substrate that may be extremely hot in the desert sun.

Natural history. The habitats are in sandy desert regions, with areas of pebbles and stones and near enough to the sea to gain some humidity advantage; these tortoises are usually found a maximum of 30 to 50 km from the Mediterranean itself. Some minor, scrubby vegetation, dominated by *Thymelaea hirsuta* bushes, may be present. The small body size of this species may be explained by the sheer aridity of the habitat. This tortoise digs into the substrate at the base of bushes, getting under leaves or actually into the sand to protect itself from heat and desiccation. It estivates for a major part of the year but may emerge very early in the morning and at the end of the day. In diet this is an omnivorous species with herbivorous tendencies, eating grass and fallen fruit, organic detritus, insects, and dead animals. It is never seen to drink, and its water requirements must be extremely limited. A single egg represents the typical clutch, laid in early spring, and it does not hatch until the end of summer. The egg is rather large, and this significant size presumably explains the necessity for mobility of the hind lobe of the adult female's plastron. Nevertheless some authors have reported clutches of up to four eggs (Hobbs and May, 1993). The young weigh between 10 and 14 g, which is rather large for such a small species. Predators, including monitor lizards, raptors, and jackals, wipe out numerous hatchlings of this species.

This tiny Egyptian tortoise appears to enjoy eating yellow or white flowers. The brown lines on the carapace disappear with age, and the shell becomes a uniform sandy yellow color.

Protection. Now listed in Appendix 1 of CITES, this is considered to be the rarest of all the *Testudo* species, along with the recently described *T. werneri.* Recent studies (Baha El Din, 2004) have shown that *T. kleinmanni* has been almost totally

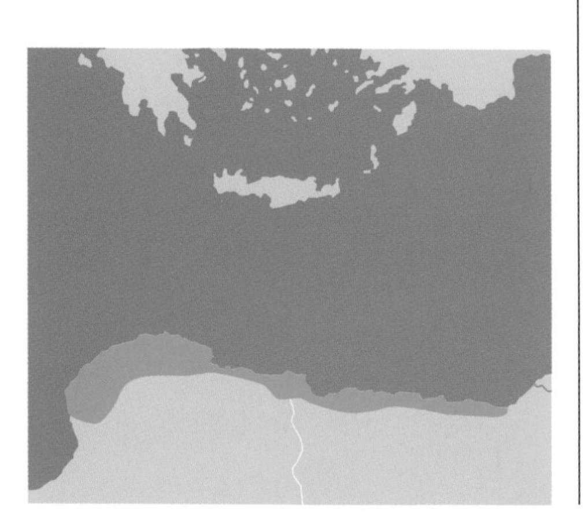

Common name

Kleinmann's tortoise

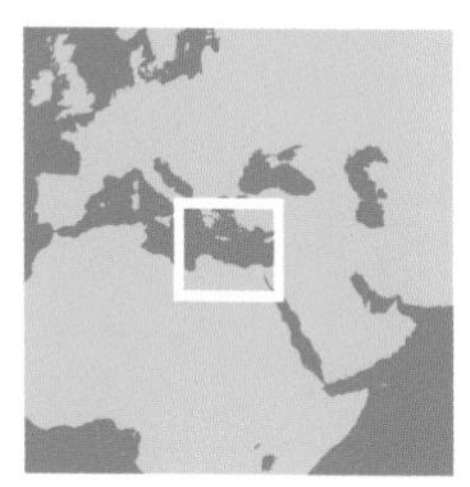

eradicated throughout its entire range with the exception of relict areas such as El Omayed Protected Area. It has suffered greatly from commercial collection, and its habitats have been ruined by overgrazing, spread of farms, and urban development, as well as by military maneuvers. It is not uncommon for batches of these tortoises to be seized by customs agents in Cairo. Researchers who specialize in this species are convinced that it is on the fast track to extinction and that urgent measures need to be taken to save it. There is a single rearing station at the University of Cairo, with the goal of providing young tortoises for reintroduction to natural habitats, but the actual productivity of the operation is very small. The Baha El Din team (2004) has proposed creating a series of protected areas, including Salum and Al Qasr and bringing together all parties interested in the welfare of this species to plan an integrated field research project, public education in both Egypt and Libya, and release of captive-raised tortoises into protected areas. This species is classified as endangered by the Turtle Conservation Fund.

© H. Fiello

Exit from the egg is always a difficult time. The caruncle allows the tortoise to cut the eggshell, and then the clawed feet assist in bursting open the restrictive confines of the natal prison.

Testudo marginata Schoepff, 1795

© B. Devaux

This photo presents a good view of the wide posterior "skirt," the midline constriction of the marginal scutes, and the contrasting black-and-yellow coloration.

Common name

Marginated tortoise

Distribution. Principally found in Greece, this species also occurs in the southern part of the Balkan Peninsula as well as in Sardinia (where it was certainly introduced by the Ligurians) and in Tuscany (also an introduced population).

Description. This tortoise is the largest species in the genus, with large females reaching 400 mm and 6 kg), and it is the only one that has a wide expansion of the posterior marginals, spread out like a fan or a skirt and present in both males and females. (To a lesser degree, the same "skirt" is present in *T. weissingeri.*) The shell is elongate and oval, with some degree of median constriction, somewhat as in males of the South American *Chelonoidis carbonaria.* The growth annuli are usually very clear, at least for the first 20 years of life. The overall coloration is brown to blackish, sometimes with beige, orange, or pinkish areas, and with beige seams. Old animals lose the colors and become gray-brown to greenish, perhaps with a yellow blotch on the areola of each carapace scute. The plastron is light yellow, with a bold black triangle on each scute (present, but only on the abdominal scutes, in *T. kleinmanni*). There are no horny spurs on the thighs, nor is there a cornified tip to the tail. The head is grayish to blackish, without light markings, and the forelimbs are armed with heavy scales on the anterior face. Females are larger than males, and the latter show a strong plastral concavity and a thickened area on the xiphiplastra, as in certain *Gopherus* species. This buttress allows the male to bash himself against the carapace of the female ("xiphiplastral ramming") at the moment of insemination. The young, being beige and black, look like little *T. graeca,* but the former are somewhat more elongated and already have the black triangles on the plastron.

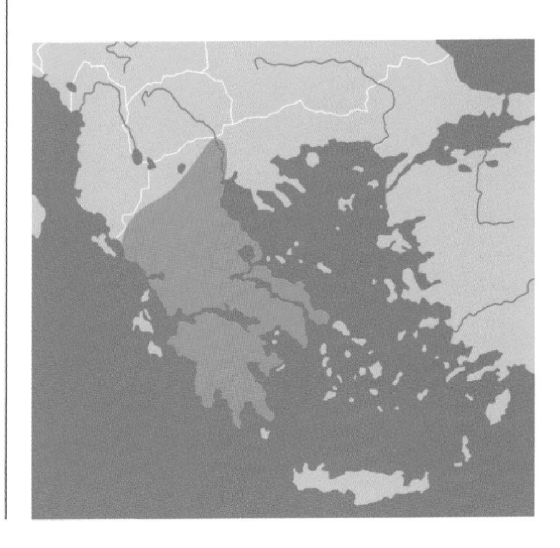

Natural history. This tortoise is most often found in hilly or mountainous areas, up to 1,300

m altitude, but it is also sometimes found at sea level. The extensive posterior "scute" does not seem to be of any benefit during copulation, and its presence forces the male to position himself vertically behind the female. He extends his hind limb until the female raises the posterior end of her shell to facilitate intromission. But all too often the male falls off laterally and has to start again from scratch. The eggs are quite large (38 × 32 mm), somewhat elongated, 6 to 9 per clutch (the record is 15). The juveniles begin to appear after the first thunderstorms of August or September; they weigh 15 to 20 g. These are herbivorous animals for the most part, but they feed also on dead insects, earthworms, small lizards, and snails.

© B. Devaux

Females may reach 400 mm, and it appears difficult for the male to copulate with a partner that has such a rigid skirt. But he manages in the end, and his courtship and mating zeal may be intense and prolonged.

Ethnozoology. This species plays a role, along with *T. graeca* and *T. hermanni,* in ancient Greek bestiary. It was represented on stater coins several centuries before Christ, and it may be seen also in numerous frescoes or sculptured on pediments at the foot of fountains or on architectural friezes. Some marginated tortoises, carried by raptors to the tops of vertical peaks in the mountains, have found themselves isolated on these remote, rocky platforms and have survived for many years, maintaining small populations in these locations, their only companions being holy men who use these lofty refugia as places for prayer and meditation.

Protection. This species is still abundant in the Peloponnese and is rather rarely collected, being completely protected by law. During the last century it was exported in large numbers, and it is still quite well represented in European collections. Nevertheless, intensive agriculture and urban sprawl have reduced its available habitat considerably. There is no study program or active conservation currently in practice for this species although it is one of the most spectacular members of the genus *Testudo.* It is, however, listed in Appendix 2 of CITES.

Testudo (graeca) nikolskii

Chkhikvadze and Tuniyev, 1986

Distribution. This species is found in the northern part of the Black Sea area of Caucasia, from the town of Anapa as far as the Pitsunda Peninsula.

Description. Formerly considered a subspecies of *T. graeca,* this tortoise is now elevated to the rank of a full species. This is a very large tortoise, reaching a length of 296 mm, with a wide, rounded shell, several small spurs on the thighs, and a single supracaudal scute. The growth rings are numerous, and the coloration is yellowish, with few dark markings (restricted to the front part of the scute and on the areola). In the young, yellow is dominant, although adults will darken with age and become heavily marked with black. Some specimens have black claws on the forelimbs (A. Leontyeva, 2004), although others may have yellow claws. The large scales on the forelimbs are often more rounded than triangular. The head is brown to blackish, with few light markings.

Common name

Nikolski's tortoise

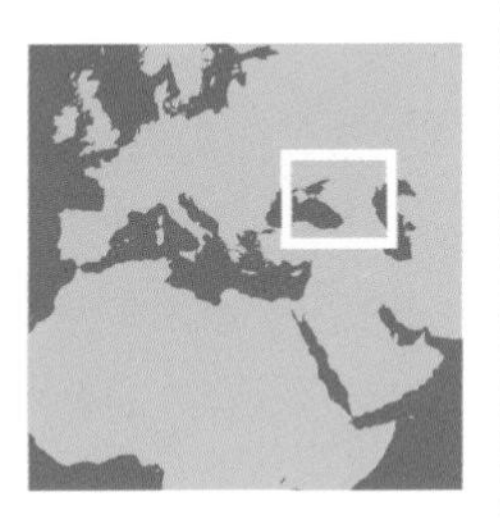

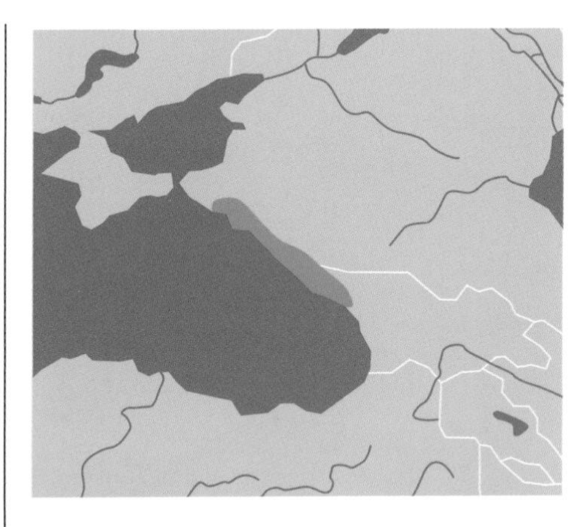

Natural history. This species undergoes a lengthy hibernation, from October to the end of April. Nesting takes place in May, and four to six are laid, measuring about 32 × 28 mm. In July and August the tortoises are less active and leave the dry hillside areas to descend to the valley floors, where the humidity is greater. Many of these tortoises are heavily infested with ticks.

Protection. This species' listing in Appendix 2 of CITES has not stopped heavy collection and habitat destruction. Today urbanization and agriculture are the worst problems, as well as feral dogs and cats, which abound in the area. One reserve has been established to protect the species at least in part of its range, but the reserve is much too small. The species is considered endangered by the Turtle Conservation Fund.

Testudo (graeca) pallasi

Chkhikvadze and Bakradze, 2002

Distribution. This tortoise is found north of the Caucasus Mountains in Russia, in the republic of Dagestan between the buttresses of Makhachkala and the low country of Primorskaya. It can also be found in Azerbaijan.

Description. Formerly considered a subspecies of *T. graeca,* this tortoise has been elevated to species rank. The maximum length is 247 mm. It has slight carapacial bossing and very dark coloration. Usually yellowish to greenish brown, with wide dark areas around the edges of the scutes, it may also be almost completely black. The plastron is also marked with dark areas, on a beige to greenish background. The posterior marginals are moderately broadened above the limbs and are lightly serrated. This species could be confused with *T. ibera* or *T. nikolskii* but differs in that the seam between the humeral and pectoral scutes traverses the posterior part of the entoplastron. The suture between the hypoplastron and the xiphiplastron is close to the anal notch.

Common name

Pallas' tortoise

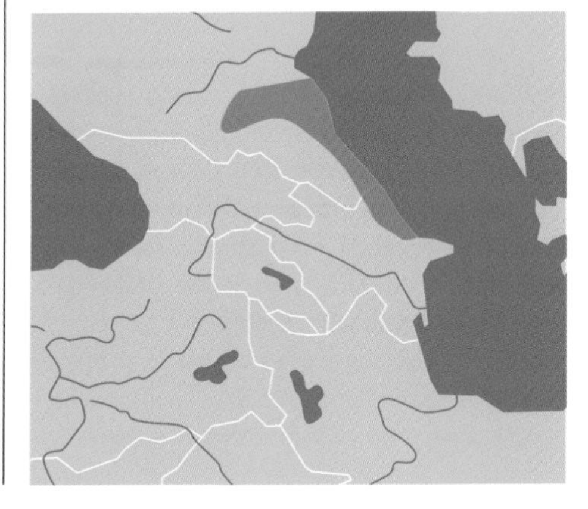

Natural history. Hibernation is prolonged, from October to mid-March, but for the remainder of the year this species is usually active all day long. In addition to vegetation, it seems to enjoy eating snails of the genus *Helicella.* Nesting may occur three times in a season, with three to six eggs each time. The eggs are almost round, measuring 36 × 35 mm. Hibernation lasts for 75 to 105 days.

Protection. Not only placed in CITES Appendix 2, this species also appears on several *Red Data Book* lists in Russia and Dagestan. It was formerly heavily collected and exported. Today human encroachment has severely reduced its range. Reserves in certain habitats—the sandy dunes on the coast of the Caspian Sea in particular—appear to be essential, and captive breeding programs are recommended.

Testudo (graeca) perses Perälä, 2002

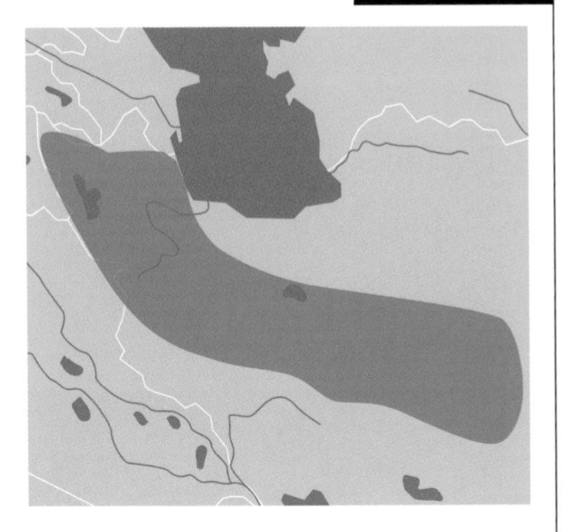

Common name

Zagros Mountain tortoise

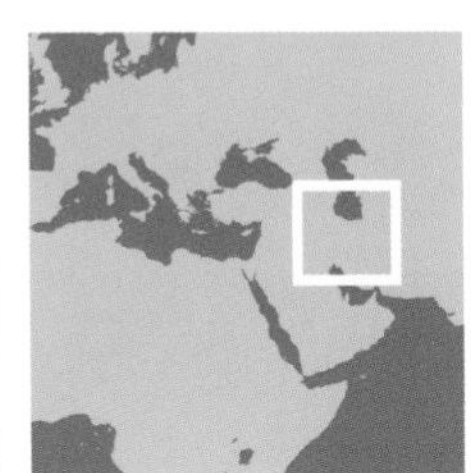

Distribution. This species occurs from southeastern Turkey (Zagros Mountains) into Iran (Lake Urmia) and northeastern Iraq.

Description. This is a tortoise of medium size, up to 193 mm for males and 161 mm for females, and with a robust, squat appearance. Individuals as long as 257 mm may possibly exist (Pritchard, 1966). The species is distinguished by the following traits: the first vertebral is wider than long, bordered in front by an acute angle; the fourth vertebral is very wide, and the fifth is short; the supracaudal is conspicuously widened; the anal scutes are very long and wide; and the interanal suture is longer than the interfemoral. The marginals are expanded, especially toward the front. The gulars are projecting, extending slightly beyond the anterior lobe. The forelimbs are massive and thick, with long triangular or rectangular scales. The tip of the tail is equipped with an enlarged double scale. The horny spurs on the thighs have wide bases. The color is distinctive: the background is sometimes greenish yellow, but more often light brown or reddish, with extensive black markings, and the limbs and head are blackish brown.

Natural history. The aridity of the environment, which is a pebble-strewn desert with little vegetation, explains the squat, encarapaced appearance of the animal, similar to that of some of the South African tortoises. These tortoises estivate during the hot months, or they may venture forth briefly very early in the morning. The eco-ethology of this species is still poorly known. Copulation has been observed in August, and one egg measured 45 × 35 mm.

Protection. Classified in Appendix 2 of CITES, this species is also listed in the IUCN *Red Data Book* as a vulnerable species. In the past it has been caught for commercial export. At present it still suffers from human activities. Nevertheless it is confined to certain dry, relatively intact ecosystems, and reserves should be established in these areas in order to protect this very desirable species.

Testudo (graeca) terrestris Forskål, 1775

Distribution. This species is found in the northern part of the ancient Mesopotamian area, bridging the Turkey-Syria frontier, and from northern Iraq to the Tigris in Turkey, as far as Eastern Anatolia.

Description. Formerly a subspecies of *Testudo graeca,* this form is now considered a full species, but its distribution is more limited than that of the former subspecies. *T. terrestris* reaches a length of 190 mm, and occasional individuals may reach 250 mm. The carapace is oval, short, and compact, with a very pronounced dome. The lateral marginals are recurved toward the bridge and give the shell a massive appearance. The posterior marginals are slightly incurved at their lower edges in the males. The growth rings are clearly marked on the carapace, with prominent areolae. Vertebrals 2 to 4 are very wide, and the first is elongate, with almost straight lateral borders. The gulars project

Middle Eastern spur-thighed tortoise

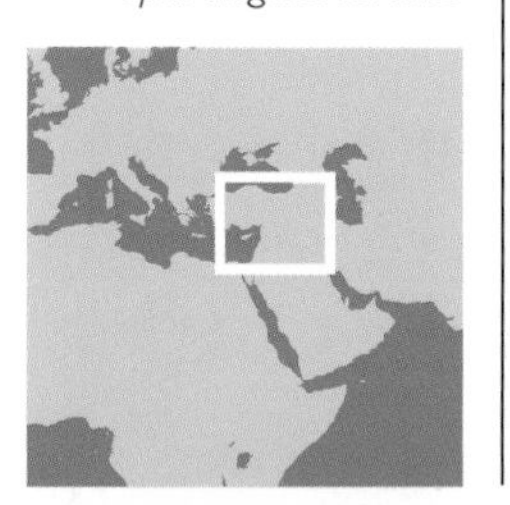

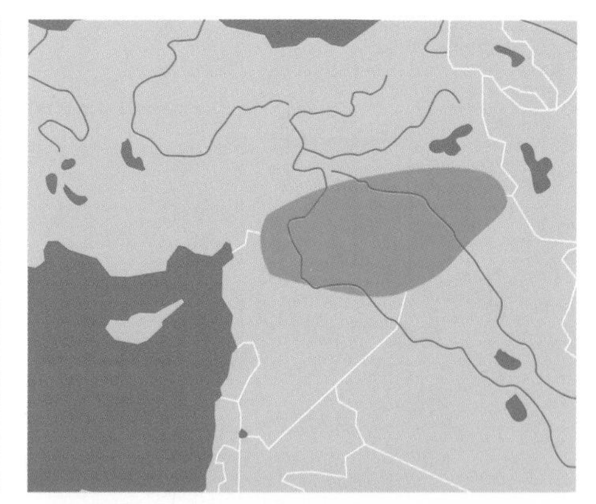

forward but are often quite narrow. The general coloration is usually yellowish to beige, without strong dark markings, just a small spot on each areola. Old animals may become darker or, on the other hand, may lighten up to a uniform orange-yellow or brown. The top of the head is covered with two large pentagonal scales. The head is beige to brown, with small yellow scales. There is black pigmentation on the outer rhamphothecal surfaces, giving the face a sort of mustache effect. There are five claws on each forelimb and just four on each of the hind limbs. The anterior face of the forelimbs bears 13 to 15 large scales, with granular skin in between. The plastron is orange-yellow, with a very large, symmetrical, brown-black central design.

Squat, wide, and heavy, often with monochromatic yellowish or brownish coloration, this species is easily recognized.

Natural history. No studies have been done on this species, but presumably the life history is very similar to that of *T. ibera.*

Protection. Placed in Appendix 2 of CITES, this species is also listed as endangered in the *Red Data Book* of the IUCN. The species has been heavily collected in recent decades and is commonly seen in European collections. Today it has much to fear from human activities and habitat destruction. No specific protection of this species is in force anywhere. It would seem that investigations into its biology and population status are necessary.

Testudo weissingeri Bour, 1995

Distribution. The range of this species is a small, narrow band of coastal terrain west of the Taygetos, in the southern Peloponnese, between Kalamata and Areopolis.

Peloponnese dwarf tortoise

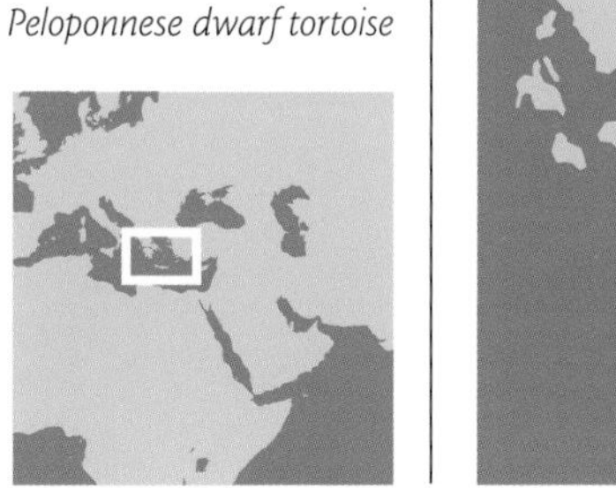

Description. This species was described quite recently and has the general appearance of a dwarfish form of *Testudo marginata.* The females have an average length of 210 mm (maximum 240 mm). The general form is very similar to that of *T. marginata,* but the "skirt" is much less developed and does not extend to cover the hind limbs. The growth rings are extremely fine and close-packed. The color is variable, according to the particular locality and the age of the indi-

vidual. The young are in general greenish or grayish, with dark contrasting spots and light areolae. The oldest animals may be entirely melanistic, or they may be a uniform orange-yellow, grayish, or marked with diffuse areas of yellowish or blackish. The plastron is yellowish with a black triangle on each scute, but the growth rings are finer, and the triangles perhaps more irregular and less developed, than in *T. marginata.*

© R. Bour

This tortoise is about to hide in a crevice to shelter itself from the intense summer heat.

Natural history. This tortoise occupies a particular habitat type—namely, hillsides with olive terraces and native maquis, from sea level up to 600 m altitude. In this demanding environment, with extreme summer heat and drought, it estivates in June and July, taking refuge among the boulders, sometimes digging burrows several meters long, or simply hiding in the undergrowth. Occasionally five or six tortoises may be found together in the same retreat. In winter this species remains active and does not reduce activity except on the coldest days. The main food is the leaves of the plant *Urginea maritima.* Sexual maturity may be reached at 11 to 13 years of age in males and 14 to 16 years in females. There seems to be more sexual activity in autumn than in the spring. The clutches contain about four eggs, elliptical in form and measuring 35 × 30 mm. The hatchlings are about 32 to 35 mm long.

Protection. The very limited range makes this species inherently vulnerable, and this region is also subject to heavy pressure from people, including urban development, tourism, and extension of cultivated areas. Many tortoises are found in a burnt condition after the peasants set fire to undergrowth and fallen wood, and others are injured by agricultural tools or land clearing. Furthermore, nests are pillaged by predators, such as martens, foxes, and wild boars. The worst threat is urban development. Numerous homes have been built in the olive fields and in the habitats of *T. weissingeri,* and the available habitat is being reduced rapidly. Commercial collecting is a problem too, and many hobbyists wish to add this tortoise to their collections or to do captive breeding so that they can supply other collectors. Moreover, the species is listed in Appendix 2 of CITES. Two other points may be made: the Greek authorities seem not to take much interest in the species, and no measures have been taken to protect the habitat. *T. weissingeri* is therefore a seriously threatened species that needs to be the focus of intensive conservation efforts.

Testudo werneri Perälä, 2001

Distribution. This species is restricted to a coastal band just 30 kilometers in length, north of the Negev in Israel, including Be'er Sheva, Be'er Mash'abbim, and Dimona, as well as the Sinai Peninsula as far as the Nile Delta. It may have been present in southern Sinai in the past, but the Egyptian distribution remains essentially anecdotal.

Description. These tortoises were formerly considered to belong to the species *T. kleinmanni.* Perälä's study, which included 140 specimens from Libya to Negev, have demonstrated that, east of the Nile, there is a distinct species, *T. werneri* (named after an Israeli herpetologist, Y. L. Werner, who collected the first specimen in 1963). It is basically very similar to *T. kleinmanni.* Key de-

Very similar to *T. kleinmanni*, this species is usually somewhat lighter in color—cream or yellow. The carapace is wider toward the rear, with expanded marginal scutes.

tails are as follows: The maximum length of females is 131 mm; males reach 106 mm. The carapace is quite rounded and very wide in the middle and also at the rear. The carapace is somewhat flattened in the back, in the region of the fourth and fifth vertebral scutes, and the supracaudal, which incurves toward the horizontal plane, is extremely long in males (shorter in females). The vertebral scutes are narrow, and the anterior lobe of the plastron is quite modest. Furthermore, the posterior marginals are more recurved and denticulated than in *T. kleinmanni*. The color is matte yellow, less bright than that of the western form, and sometimes tending to fade toward greenish. The dark areas are darker than in the other species, especially in young individuals, with brown seams along the edges of the vertebrals and costals, and triangular markings on the front sides of the marginals. The plastron is also darker than that of *T. kleinmanni*. The two triangles in the abdominal area are generally fairly clear, and there may be traces of the same pattern on the pectorals. The posterior lobe of the plastron is flexible in both sexes, but much less so than in *T. kleinmanni*. The head is also less dark yellow than in the western form and has more diverse grayish or greenish spots. The eyes are often bordered by dark markings, and there are also dark areas on the top of the head. In a way, these markings recall those of other Middle Eastern tortoises, including *T. terrestris*. The oldest animals of *T. werneri* are also marked with darks streaks and spots, whereas the majority of *T. kleinmanni* specimens become a uniform sable yellow over the entire body, including the carapace.

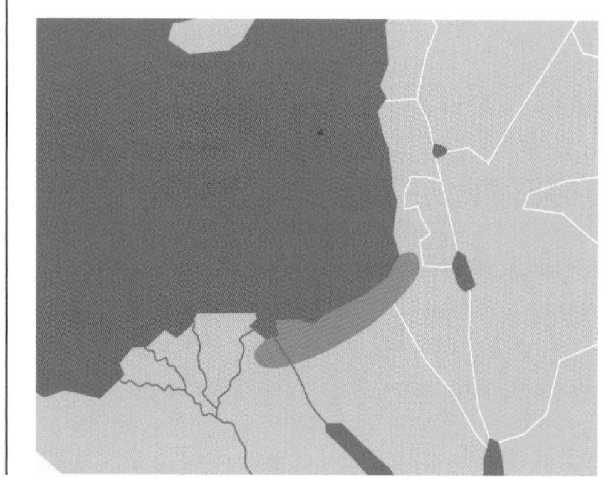

Common name

Werner's tortoise

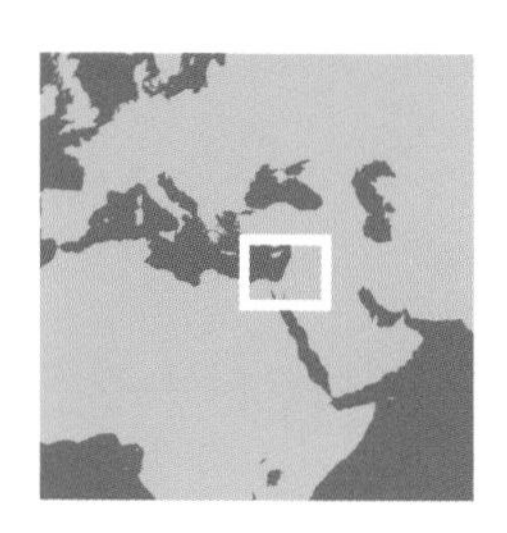

Natural history. Studies done in the Negev on what was then considered to be *T. kleinmanni* actually refer to this species. The humidity levels in the Negev are lower than those of western Egypt, where *T. kleinmanni* lives, and this factor may make the survival of *T. werneri* more challenging than that of its relative. This is a species that estivates for part of the year and emerges at the beginning of the day, especially after (rare) rains. Omnivorous by necessity, this animal eats small dead animals, invertebrates, fruits, and organic detritus. It lays a single egg, and the hatchling does not emerge until the summer rains have moistened the land.

Protection. This species is highly endangered. Recent studies (Baha El Din, 2004) show that it has almost disappeared from its Egyptian distribution, and only a few isolated relict populations survive in the Zaranik Protected Area. In Israel, populations are a little more substantial, and the habitats less degraded. These tortoises still suffer from overgrazing of the pastures by cattle, agricultural practices, urbanization, and habitat destruction (warfare, human activities, desertification), as well as from commercial collection. The act of naming or creating a new species also gets hobbyists excited, and they clamor to get specimens. Furthermore, it is possible that captive hybridization may occur with the western form, which will complicate the conservation of both of these *Testudo* species. The species must be placed in Appendix 1 of CITES, and it is essential to generate an emergency preservation plan, with extension of protected and monitored areas, initiation of field studies, restocking plans, and education of local peoples.

Testudo (graeca) zarudnyi (Nikolski, 1896)

Distribution. This species is found only in the plateau of central Iran, at altitudes between 1,000 and 3,000 m.

Description. Formerly considered a subspecies of *T. graeca* and sometimes thought of as a subspecies of *T. terrestris*, this form is now considered to be a species in its own right. It is a large form, up to 275 mm in length, with an elongated and dorsally elevated carapace. The anterior and posterior edges are strongly curved upward and are somewhat serrated. The front and rear shell edges are characterized by horny spines located between each pair of marginals. The general color is yellowish, with dark scattered and irregular markings, often concentrated around the areola of each scute. Old animals may be uniformly dark, with a reddish tinge. The anterior carapace opening is narrower than in other *Testudo* species.

Natural history. The ecology of this tortoise is poorly known. Hibernation occurs from October to March. Nesting takes place from mid-April to late May, and the typical nest contains four eggs.

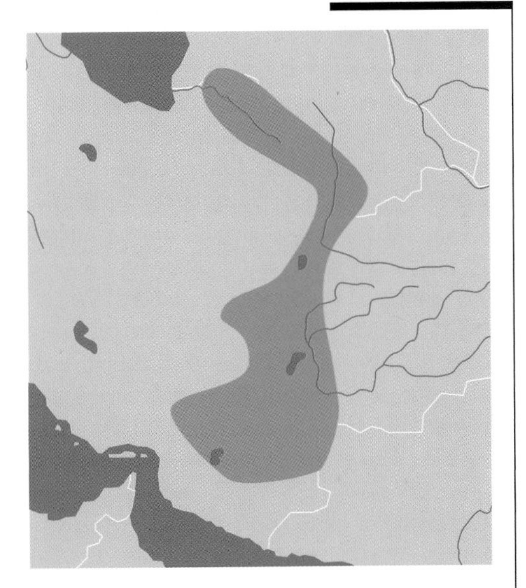

Common name

Central Iranian tortoise

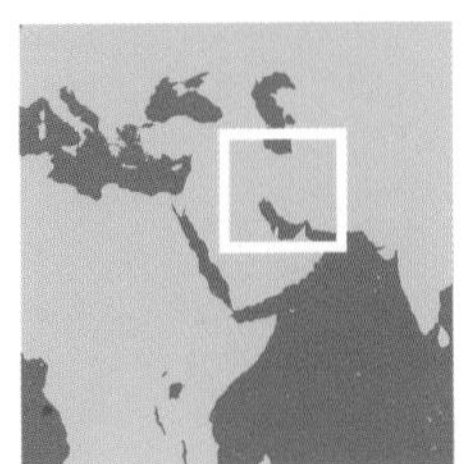

Protection. This is a rare species with a limited distribution. Commercial collection has occurred in past decades, and these tortoises have made up some minor fraction of shipments of *T. ibera* and/or *T. perses.* Studies on the status and biology of this species are necessary before the conservation needs can be evaluated

Xerobatinae Gray, 1873

The Xerobatinae are primitive land chelonians with a robust appearance, medium to large in size, and adapted to the arid areas of North America (where they dig burrows) and the humid environments of Southeast Asia (where they hide on the surface).

Gopherus polyphemus (Daudin, 1801)

Distribution. The range of this species is centered in the state of Florida (USA), with populations spreading as far west as New Orleans (Louisiana) and north into southeastern Georgia and narrowly into South Carolina.

Description of the genus *Gopherus*. The four species of *Gopherus* are endemic to the southern United States and northern Mexico. These are the only completely terrestrial chelonians of these regions and are the largest herbivorous land reptiles on the continent, although the maximum size of *G. flavomarginatus* remains debatable. The ancestral form was very close to *Stylemys,* which was widespread on the North American continent 50 million years ago. The special feature of the group is its burrowing ability, and the burrows created by one species *(G. polyphemus)* may be 10 m in length. The morphology is specialized along with these behaviors, and these tortoises share features with Old World burrowers like *Testudo horsfieldii* and *Centrochelys sulcata:* low, flattened form, with heavy limb and neck bones; powerful limbs with spurs or pointed scales along the edges;

strong, rigid claws; and an ability to survive in very arid habitats (although *G. polyphemus* lives in the humid environment of Florida). They have spade-like gulars, which are blunt and flattened and used for combat between males. The outer faces of the limbs are covered with conical overlapping scales, and when the tortoise digs its burrows, these scales assist in loosening the substrate material. On the other hand, when the limbs are retracted, the scales form a sort of sheet of armor across the anterior opening of the shell, totally protecting the head and rendering the opening impregnable. The coloration is quite uniform, tending toward sandy beige or earth gray. Males and females both possess odor glands under the chin and have a specialized pointed scale on each forelimb to distribute the odorous secretion during copulation.

Description of *G. polyphemus*. The maximum length of this species is about 38 cm; most adults are 25 to 30 cm. The carapace is the broadest and most flattened of those of the four species, and in some individuals there may be some constriction at the midline. The gular extension is well developed, flattened, and spadelike. The background color is bronze or coppery, with dark markings around the scutes. As the animal grows and ages, it tends to become a uniform dark gray-brown. There is a flattened area along the second to fourth vertebral scutes, but the fifth vertebral is strongly curved. In other words, this tortoise looks like a big flat rock and also reminds one of *Testudo horsfieldii*, from the other side of the world. The forelimbs each have five digits, with squared-off, thick, rigid claws that are highly adapted for digging. Sexual dimorphism is most evident in the old males, in which the plastral cavity is easily visible and the xiphiplastra are greatly thickened. Males are larger than females. The constant entering and exiting of burrows abrades the shell, and in adults it takes on a smooth, almost metallic look. The head is very wide (for a tortoise), much more so than in *G. agassizii* or *G. berlandieri*, and the neck vertebrae are also uniquely massive. The head is gray-yellow, with small, thick scales. The plastron is uniformly gray-yellow.

Common name

Gopher tortoise

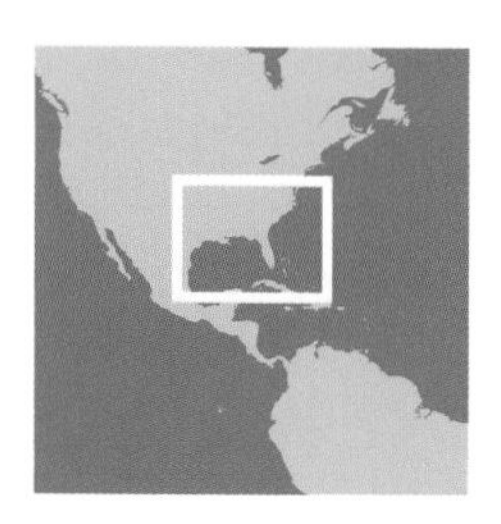

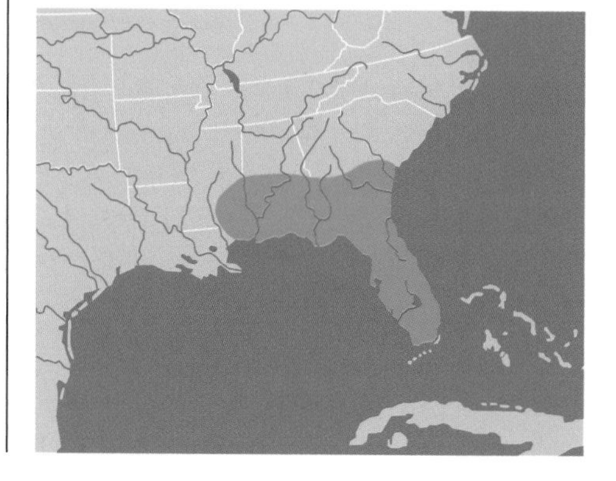

Natural history. This species reaches farther north than any of its congeners, and it lives in dry, well-drained, sandy areas, including dunes just a short distance from the sea. In Florida it occurs in pinewoods and on dry sandy soils with moderately open vegetation, but it also occurs in urban areas and in private gardens. It digs long burrows in the sand, which may be several meters long. The burrow is just the right size to allow the tortoise to turn around, and thus the dimensions (width and height) of the burrow give a good index to the size of the tortoise within. Some burrows have two entrances and may also have various underground bifurcations. Some tortoises dig several burrows in one season, which they occupy successively. On Egmont Key (Tampa Bay), an isolated population has become quite dense, and we have observed a burrow every 10 m or so. The burrows are always dug at the base of a bush or other vegetation. Sometimes the tortoise will sit in the mouth of the burrow, soaking up the morning sun on cool mornings and retreating quickly when approached. But during the months of great heat (July and August), these tortoises tend to estivate and may rarely be seen out in the open; in general they spend a great deal of their time, at all seasons, in the burrow. The burrows also serves as a retreat for a great variety of other animals, whether or not they are actually occupied by a tortoise, and one sometimes finds snakes, foxes, possums, and even wild dogs in them, as well as a frog species found nowhere else but in these burrows. When an animal dies in the burrow, the carcass may be a source of food for the tortoise, which is somewhat opportunistic in diet. This species seems to be more sociable than other members of the genus, and one often sees groups of *G. polyphemus* grazing on grass within sight of each other. Male combat occurs but is less violent than with the congeners. Clutch size is extremely variable, but the average is close to seven. Nesting occurs from March to June, and the nest may or may not be close to the parent's burrow. The eggs are slightly prolate and measure 45 × 40 mm on average. Incubation lasts for 80 to 110 days, with emergence of hatchlings occurring in early autumn. The young tortoises are brightly colored, usually yellow-orange with dark circles on the scutes, leaving a light area around the areola.

Protection. This tortoise shares it range with extensive human populations, and intense urbanization usually spells destruction for local tortoise colonies, although less intensive development, after a number of years, may create excellent habitat for this species. Such situations include old fields and unmaintained orange groves, the edges of airports and golf courses, and so on. Large numbers of tortoises also find refuge in controlled areas such as Eglin and Patrick air force bases in Florida, as well as the Kennedy Space Center itself. Gopher tortoises also profit from periodic burning of the habitat, which intercepts the formation of a closed canopy of trees (this species requires at least some open sky in its habitat), and they can easily escape actual damage or death during these burns by retreating deep into their burrows. The gopher tortoise is good to eat and has been consumed by American Indian groups for a long time, and much more recently by rural African Americans as well as such communities as the Minorcans in the St. Augustine area. Today it is protected by the individual states, and the western populations are listed by the federal government (U.S. Department of the Interior) as threatened.

A challenge to gopher tortoise survival is a respiratory disease, possibly spread to wild tortoises through the unauthorized release of sick captive animals. These mycoplasmas can now be found virtually throughout the range of the species and cause periodic die-offs of populations even in protected areas, such as the numerous state parks in Florida. This has led to a policy forbidding purchase, sale, capture, or release of gopher tortoises, at least in Florida, the heart of the range. This policy does not in itself solve the problem of habitat destruction by urbanization, and formerly it was common practice to transport displaced tortoises to "safe" habitat elsewhere. But this practice ran into many problems, and now it is usual in Florida to charge land developers a monetary fee for developing gopher tortoise habitat; this money is then earmarked for the purchase of much larger acreage of habitat in counties where land is cheaper. The species has been intensively studied, and much has been learned.

The photographer surprised this tortoise just coming out of its burrow in sandy ground.

Gopherus agassizii (Cooper, 1863)

Distribution. The range of this species includes the Mojave and Sonoran Deserts in the southwestern United States (California, Arizona, southern Nevada, and narrowly into Utah) and much of Sonora, Mexico, just entering Sinaloa.

Description. The carapace is oval and is wider, shorter, and less flat than in *G. polyphemus,* and the head is much narrower and the soles of the feet much wider. Also, a light eye-ring is present, whereas in *G. polyphemus* the entire eye appears

Usually, this species has gray-black coloration, like the specimen on the right, but a Mexican subspecies (shown on the left) is smaller and lighter in color.

© F. Bonin

This tortoise, found north of Alamos, Sonora, Mexico, has a well-sculptured shell and long legs that facilitate rapid walking through stony areas.

dark. There is a slight keel on the fourth vertebral scute, and some tortoises have an almost vertical fifth vertebral. The usual size of adults is about 250 to 300 mm, with a maximum around 400 mm. The gular area is quite projecting and is slightly forked. The color ranges from gray to sienna or even black, without the contrasting light areas of other species. The head is gray, as is the plastron, which is somewhat rectangular.

Natural history. A wide range of habitat types is used: dry canyons, rocky valleys, and arid plateaus dotted with rare cacti and spiny bushes. The short burrow in the desert makes life tolerable, with just a touch of coolness and humidity, as well as possible food in the form of deceased prey animals. These tortoises eat a great deal during the wet season, building up their reserves. The diet is largely vegetarian, but they also take insects, birds, lizards, and rodents. The tortoises estivate for part of the summer, but in spring they are active by both morning and evening. They seem to be somewhat nomadic and sometimes make long journeys. They dig multiple burrows, which they may use at different times of the year. Males are often highly aggressive, and when they encounter each other, there may be a fierce battle, the gular "crowbar" being brought into play, as each tortoise tries to overturn his opponent. The female digs a nest about 20 cm deep and therein deposits about 6 eggs (on average), the maximum being 14. There may be two or three nestings per season. The eggs are spherical, very white, and somewhat soft-shelled at the time of laying, but they quickly become hard. Nesting occurs from May to July, and hatching ensues three months later. The young have wide light-colored areolae, and it is only at the onset of maturity (8–10 years) that the color becomes a uniform gray.

Protection. These tortoises are threatened by many human activities: vandalism, vehicular accidents, bad tourist conduct, and the spread of industrial zones and military bases. The Bureau of Land Management established a major conservation program to study and protect this species in 1980, under the direction of wildlife biologist Kristin Berry, and this program is still in effect. Reserves, closed to the public, have been established, including the Desert Tortoise Natural Area (98 km^2) and the Chuckwalla Bench Area (368 km^2) in California. This species is the most intensively studied of its genus, and numerous congresses have been held on the biology of this single species (on physiology, captive rearing, pathology, etc.). During 1980–1990, this species was affected by a disease that has spread widely and reached the other species of the genus. This chronic bronchitis is often fatal, and has been studied at the University of Florida by E. Jacobson. It is caused by mycoplasmas and is amplified in its effect by herpesviruses. No good cure exists, and the specialists recommend total isolation of affected animals, no transportation of seropositive individuals, and even euthanasia. This pathological outbreak killed two-thirds of some populations, but then it stabilized and, by the end of the 1990s, had started to regress. Today it still occurs, but the surviving tortoises have perhaps developed resistance to the infection. Nevertheless an ongoing monitoring of the health of the populations is absolutely necessary to prevent further mortality and population loss.

Common name

Desert tortoise

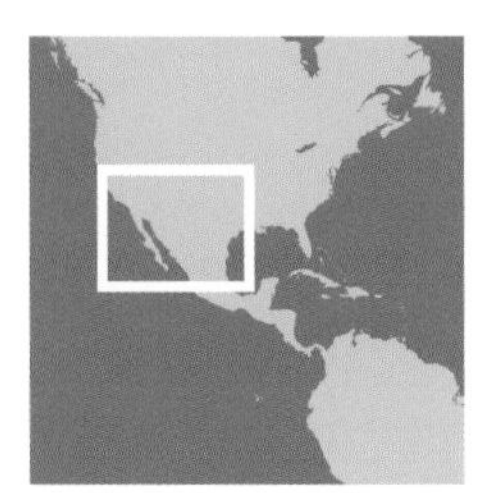

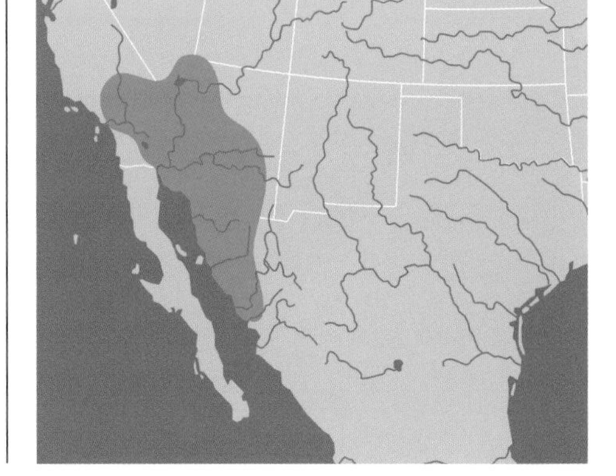

Gopherus berlandieri (Agassiz, 1857)

Distribution. The range of this species is rather small, extending from San Antonio, Texas, southward to Ciudad Victoria, Tamaulipas, Mexico. The populations are not continuous, but near Brownsville, Texas, densities as high as 23 tortoises per hectare may be reached.

© J. Maran

Description. This is the smallest of the *Gopherus* species (250 mm for males). The carapace shape is nearly square, in contrast to that in the other species, and the width is almost equal to the length. The carapace is quite domed, giving the correct impression that this tortoise is less of a burrower than its congeners. The gular region is bifid and well developed beyond the front of the plastron, and there is major sexual dimorphism. The general coloration tends toward dark gray or even red-brown. The background is cream or a dirty beige, and the annuli on the scutes are the color of anthracite. The anterior limbs are wide and flat and are not specialized for digging. The head is small, narrow, and grayish, sometimes with a slightly reddish tone. The plastron is cream to grayish. The limbs lack well-developed tubercles. The young are brightly colored, being greenish gray illuminated with bright yellow areolae. There are dark edges along the seams, and the marginals are bordered with yellow.

Natural history. This tortoise is less adapted for burrowing than the other *Gopherus* species. The burrows are rarely as much as 1 m in length, and often the retreat is simply a sandy niche or a spot under dense vegetation, without any attempt at burrowing having been made. The Texas tortoise lives in low hills in Texas, where climatic conditions are less severe than in the Mojave Desert. It is rather resistant to elevated temperatures, and it is said to be able to tolerate 42° C, although this may be just an anecdote. A favorite food is *Opuntia lindheimeri,* a local cactus. Nesting occurs from April to November in well-drained areas free of vegetation, often at the base of a bush. The tortoises may use these scarce sites in communal fashion. The eggs are more elongate than those of other *Gopherus* species and are quite large (50 × 35 mm). The number of eggs in a clutch ranges from 5 to 12, and incubation takes 88 to 118 days.

Protection. The Texas tortoise has been gathered for commercial purposes for a long time, although apparently not for export. The species is severely affected by agriculture along the Rio Grande Valley. Quite large numbers are killed on highways. Human activities, more and more intensive in this region, threaten the survival of the species. Formerly it was consumed by indigenous peoples and also by colonists, and sometimes it has suffered the indignity of being stuffed for tourist souvenirs, in questionable circumstances. It is theoretically protected by law in both states where it occurs (Texas in the USA and Tamaulipas in Mexico). No special reserves have been created for it, but it does occur in a couple of refuges—namely, Laguna Atascosa National Wildlife Refuge and Welder Wildlife Refuge.

© F. Bonin

In this *Gopherus,* the very dark seams between the carapace scutes can be seen.

Common name

Berlandier's tortoise

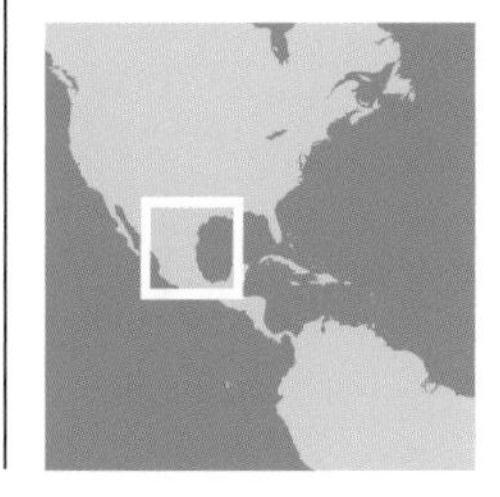

Gopherus flavomarginatus Legler, 1959

Distribution. This species survives only in a very small range in north-central Mexico, at the intersection of the states of Chihuahua, Coahuila, and Durango, and especially in the Desierto del Silencio at the foot of Cerro San Ignacio. It lives in zones of semidesert or dry savanna, up to 1,400 m in altitude.

Description. This large tortoise species was overlooked by scientists until it was described by John Legler in 1959. It is morphologically closer to *G. polyphemus* than to the geographically closer *G. berlandieri* and is the most southern representative of the genus. It is also the largest of the gopher tortoises and the largest tortoise in North America. Sadly, it is also the rarest. The vernacular names are appropriate—it is indeed a "giant," and the carapacial flanks of many specimens are yellow. How large individuals may get is uncertain, but 400 mm specimens are not unusual, whereas this would be an exceptional size for *G. agassizii* or *G. polyphemus.* The overall color is orange-beige to brown, with dark markings on the areolae and around the scutes. In very old animals the pattern dissolves, and the tortoise takes on a uniform tint. The "yellow borders" are mainly visible in the juveniles; most adults no longer show this feature. The overall shape is that of a flat, wide tortoise, with a powerful form. The plastron and the bridges are also robust and are colored a uniform dull yellow. In the Durango part of the range, the plastron may be adorned with black markings. In the males the posterior marginal scutes are very enlarged. The gulars form a wide, flat shovel, like that of *G. polyphemus.* The forelimbs are quite wide, with large rounded scales that are barely overlapping and that have contrasting colors: some are yellow-gray, and others black. The head is especially strong and is similar to that of *Centrochelys sulcata;* both species are powerful, sand-colored animals and, when old, have carapaces abraded smooth by a lifetime of burrowing. Sexual dimorphism is not marked, and the males do not have any major plastral concavity.

© F. Bonin

The burrow is often constructed under a bush, in order to enhance the humidity within.

Natural history. Considered very close to *G. polyphemus,* this taxon has a number of features (apart from geographic isolation) that justify its recognition as a full species. In the hot, dry areas of Mexico where it lives, it may be found at somewhat elevated altitudes, up to 1,400 m (Morafka, 1975). It digs its burrow at the foot of a hill or on flat ground at the base of a bush. Some burrows may reach 5 m in length. In the Desierto del Silencio, there may be as many as 20 burrows per hectare, some of them abandoned. The density of tortoises themselves is more like 7 per hectare. Morafka's studies demonstrate a temperature difference of no less than 14° C between the temperature inside the burrow and outside. The young tortoises hide under mats of dead leaves or in preexisting crevices and niches; we have seen no actual burrows of less than 15 cm width. Within the burrows, one may often encounter a bed of vegetation that may serve as both an actual bed and a source of food. These tortoises are opportunistic feeders, eating numerous vegetative species and particularly *Hilaria mutica,* an arid-zone herb, and *Opuntia rastrera,* a cactus. But they also appreciate dead animals, invertebrates, and the droppings of domestic animals. Maturity is reached rather late (about 15 years of age), but the species presumably has a very long life, judging by the great size of some of the animals found. Fossils similar to the living *G. flavomarginatus* may be a full meter in length.

The biology of the species has been studied mainly in captivity. About 20 eggs are laid on average, with several nestings yearly and an incubation time of 90 to 120 days. The young tortoises are more colorful than the adults, with the famous "yellow sides" that give the species its name. They are preyed upon by lynx, desert foxes, pumas, coyotes, and peccaries (Morafka; Rivera Garcia). In addition, burrows may be damaged or destroyed by deer, horses, and cattle. It seems that this *Gopherus* species is the most sociable of the

Common name

Mexican giant gopher tortoise

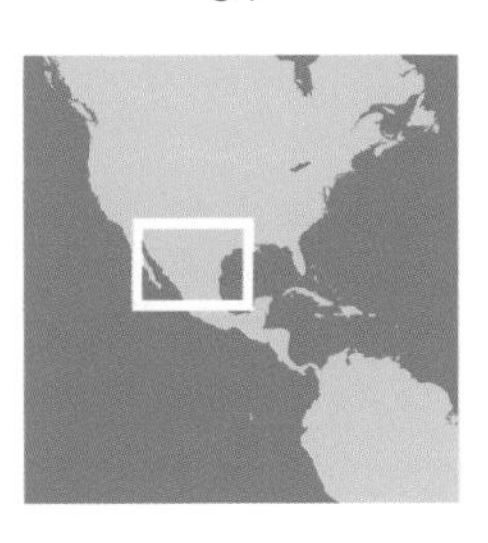

four, sharing territories and having a certain hierarchy among the males.

Protection. Several millennia ago, this species occupied all of northeastern Mexico, and it territory was contiguous with that of *G. polyphemus* of the southern USA. Today it is a relictual species in an isolated habitat and is the rarest member of the genus. Furthermore, its great size and rarity are inevitably appealing to hobbyists, and there have been ongoing raids on specimens in the wild. Since 1961 the "tortugueros" have done business with this animal, not hesitating to destroy an ancient burrow in order to extract its occupant. They also use a mirror on the end of a long pole and equipped with a long metal hook to extract the animals from their burrows. Even though placed in Appendix 1 of CITES (i.e., banned from international commerce), the species continues to be of interest to dealers. Happily, a good part of the habitat is essentially inaccessible, but the density is very low in the state of Durango, and it is possible that collectors do cover the entire range of the species. There are two protected areas: the Mapimi Biosphere Reserve and the Desierto del Silencio, but these areas are not large, and it is important that far greater expanses of habitat be protected within the Coahuila-Chihuahua-Durango triangle if this species is to have any chance of long-term survival.

© A. Dupré

This beautiful animal, observed in the Chihuahuan Desert of Mexico, has a perfectly rounded form and uniform coloration. Its burrow was several meters long (see preceding page).

Manouria emys (Schlegel and Müller, 1840)

Distribution. This species is found in Burma (hence the vernacular names), southwestern Thailand, Malaysia, Sumatra, and the northwestern part of Borneo. Iverson includes Vietnam, but these records may have been relocated animals; southern China also should not be considered a confirmed record.

Description. This is the largest tortoise of Asia, with a length of 600 mm and a weight of 30 kg. The form is massive, somewhat rectangular, and rather flat. In shape it is rather similar to *Centrochelys sulcata,* but the color is much darker, ranging from brown to virtually black. The scutes are rather thick and show strong growth annuli; they are also rather flat, with angles between them, giving the shell some of the qualities of a cut diamond. There is also a flattened area from the second to the third vertebral. The marginals are serrated and recurved at front and rear, remaining vertical in the bridge area. There is a rather wide, short nuchal scute. A small bunch of large spurs is present on each thigh, very similar to those of *C. sulcata.* The plastron is the same color as the carapace. The head is rather narrow and is gray to

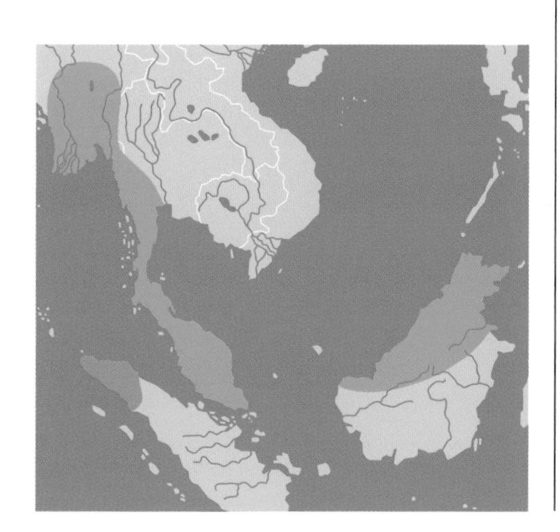

Common name

Asian brown tortoise

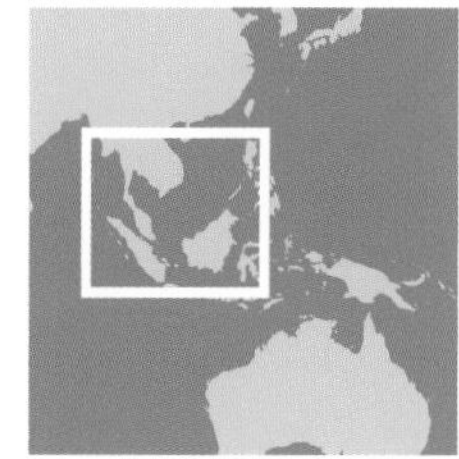

This tortoise, photographed in Malaysia, is somewhat similar to *Centrochelys sulcata,* but it needs forested, high-humidity environmental conditions and avoids open, sunny areas.

© F. Bonin

blackish; the lower jaw is very deep. The limbs are blackish, with large flat scales. The young are much more colorful, with a beige or sometimes cream background and brown to chocolate seams. Some subadults remain quite light, with a beige tone, while others darken up considerably. The adults become dark brown, the shade depending on the subspecies.

To differentiate *M. emys* from the other species in the genus (namely, *M. impressa*), the following points are important: The carapace is more domed; the scutes are flat but not concave; and the top of the head is covered with small scales, whereas *M. impressa* has large symmetrical scales toward the back of the head. The size is much larger (600 mm in *M. emys,* 350 mm in *M. impressa*); the nuchal is narrower; and finally, the coloration is quite varied and more contrasting, tending more toward brown or chocolate than in *M. impressa.*

Subspecies. The typical race, *M. e. emys* (described above), occupies the southern part of the range. It is the smaller of the two subspecies, reaching about 500 mm. It is also lighter in color, with beige to yellowish tones or even light brown. The easiest way to tell them apart lies in the plastron: in *M. e. emys,* the pectoral scutes are triangular and do not touch at the midline. *M. e. phayrei* (Blyth, 1853) is the more northern race, and it is also larger. The color tends toward dark brown, with some animals being totally black. The pectoral scutes are somewhat rectangular and have a major line of contact along the midline.

Natural history. This tortoise has certain behavioral traits unique among chelonians: it constructs an aboveground nest and also defends it from potential predators. As do many tortoises, it vocalizes during mating. These large tortoises actually occupy a variety of tropical ecosystems, at altitudes above about 500 m. They dig burrows in friable soil, but these burrows are very shallow and quite unlike those of *C. sulcata.* Some authors (e.g., Wirot Nutaphand) have observed these tortoises in dense forests and reported that they are partially aquatic. Others (e.g., N. de Rooij) consider them to be specialized for dry areas and stony savannas. Perhaps *M. emys* is actually somewhat ubiquitous, and its range is sufficiently large for it to be obliged to adapt to different milieus. Nevertheless, it certainly prefers a much higher level of humidity and moisture than *C. sulcata* does, and it knows how to hide itself in the forest litter as well as in burrows made by other animals and in rocky retreats. In Borneo it lives in rain forests with very high levels of humidity. It is basically herbivorous and enjoys both grasses and broadleaf plants as well as water plants (and even watermelons). It may also eat frogs, insects, and dead animals.

The reproductive behavior is characteristic. During courtship and mating, the males vocalize in a unique manner, and according to the late Sean McKeown, the partners seem to hold a sort of conversational exchange. The female, moreover, is less chatty than the male. Most of these observations were made in captivity, in the Honolulu Zoo, but in the field Burchfield has also noted the peculiar sounds emitted by the males, not unlike the roaring of a lion. The vocalizations have been carefully recorded and compared, and it is apparent that each individual has its own unique call, and this is especially true of the males. During the initial phase of courtship, wherein the male swings his head from one side to the other, he emits sounds of low frequency. Subsequently, the female responds with short, regular groans, while the male emits brief, high-frequency cries. During copulation itself, the male continues to produce short cries. Researchers plan to create a catalog of these sounds and to study their specific structure. Possibly these exchanges are of importance for communication between individual tortoises in the deep shade of the primary forests where they live.

At the time of nesting, in April to May and also in September to October, the female selects a location at which to prepare her nest. She will clear an area of about 100 m², pushing together all the twigs, dried leaves, and detritus around her. This operation takes several hours—and sometimes several days. The chosen site needs to be dry and to receive full sunshine. To construct the nest pile, she will use all four limbs as well as her mouth. When this work is finished, she digs a nest cavity in the middle of the pile and deposits about 40 large eggs (65 × 50 mm). Indraneil Das has reported a nest of 51 eggs, the largest clutch on record for any terrestrial chelonian. Incubation is quite short in this elevated, ventilated, and very warm nest, and the hatchling may emerge after just 70 days. During incubation the female indulges in a behavioral trait unique among turtles and tortoises and somewhat reminiscent of that of some of the crocodile family: she guards the nest and chases away predators. She remains close to the pile for several weeks, and when a monitor lizard or a human comes close, she advances upon the intruder and launches an actual attack, attempting both to bite the invader with her mouth and to bang him with her carapace. Her determination is great, and she often succeeds in chasing off the predator. After several days, however, she will abandon the nest. Even so, some have reported that as incubation proceeds, she will add leaves and twigs to the nest pile, possibly to increase the insulation of the nest or perhaps to make repairs and prevent gradual decay or collapse.

© B. Devaux

M. e. emys has a yellowish plastron with separated pectoral scutes (on left), while *M. e. phayrei* has a more grayish plastron, with the pectorals joining at the midline (on right).

Ethnozoology. This very widespread species is well known to the peoples of Southeast Asia. *Manouria emys* is worshipped by certain ethnic groups but is also widely consumed. The species was found in China thousands of years ago, but it is not found there today, apart from individuals that have been imported from South Vietnam. It is still captured for food in some countries and represents an abundant and much appreciated food resource. The flesh is supposed to have some curative properties, or at least almost all the cultures that consume this species give the same excuse. The shells of very large specimens have been used as cradles for young infants and may serve also as kennels or shelters for dogs and pigs.

Protection. Today this species is still collected and exported to China for food. Easy to raise and maintain in captivity, it is also exported in the direction of Western zoos. Listed in Appendix 2 of CITES, it is subject to national export quotas, but no smuggler has ever been intercepted. One assumes, therefore, that populations must be diminishing, and certainly this species is difficult to find in the wild. Habitats are being degraded, and human exploitation is constantly reducing the occupied range. Those who specialize in Asian chelonians are thus demanding good status surveys and evaluations of distribution, so that this tortoise can be better protected. The species is listed as endangered by the Turtle Conservation Fund.

Manouria impressa (Günther, 1882)

© B. Devaux

One can see the very flat scutes and the serrated marginals of this tortoise, photographed in Vietnam, at Cuc Phuong. The nuchal scute is very wide.

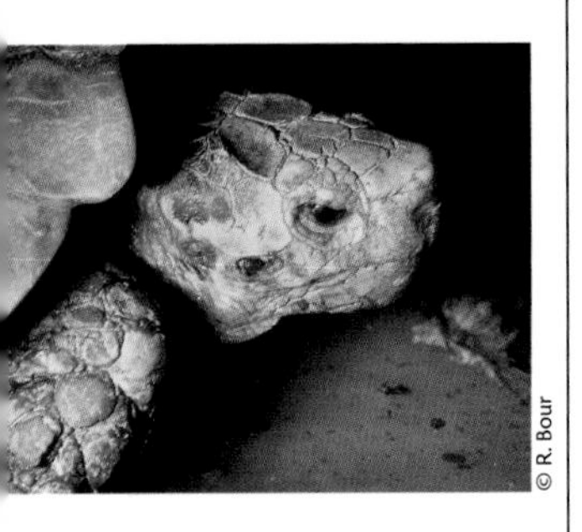
© R. Bour

The head is somewhat square and massive, with conspicuous scalation. This turtle can bury itself underground and has certain parallels with *C. sulcata*.

Distribution. The range of this species is distinctly interrupted, and although it overlaps with the range of *M. impressa,* the actual distribution is much more limited. It occurs in southern Myanmar, in Thailand and Malaysia, and in southern Vietnam. Observations in northern Vietnam and in China almost certainly correspond to individuals that have been transported by commercial interests.

Description. Although this species is similar in a general way to *Manouria emys,* there are many points of difference: the length does not exceed 330 mm; the nuchal is very wide; the coloration tends toward orange or reddish; and the principal scutes of the carapace have a concave face—hence the name *impressa.* There are two large, wide scales, oval or quadrangular in shape, at the back of the head. The back is quite flat, with a depression on the fourth vertebral scute, and the rear of the shell falls away sharply. The scutes are beige with brown markings and with distinct growth annuli and light lines along the seams, giving the impression of linked or conjoined shields. The marginals are strongly denticulated toward the rear, and less so in front. The head is yellowish or reddish, with brown scales on the sides and the rear. The plastron is beige to brown, with brown markings. There is a single large spur on each thigh. The limbs are protected with large round, flat scales, often yellow but most often brown to black. In young individuals, the coloration is bright and contrasting;, with a beige to light yellow background and with fine brown rings.

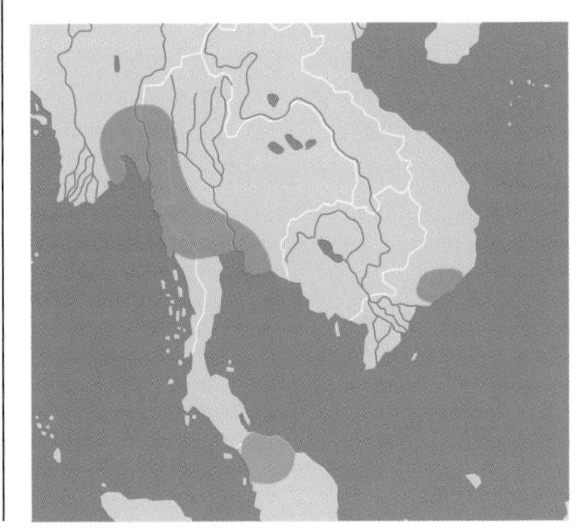

Common name

Impressed tortoise

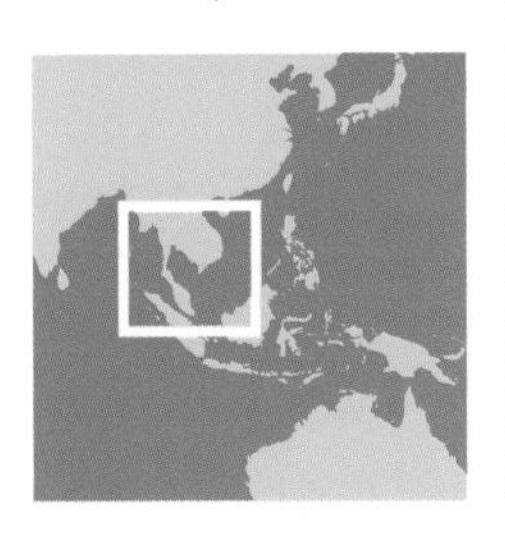

Natural history. This species is much less studied than *M. emys*. It is rarely seen in zoos, but sometimes specimens are seized by customs authorities and placed in rearing centers, such as Cuc Phuong in Vietnam. In the natural environment, this species may be found at altitudes of up to 1,200 m (Boulenger). Other authors have observed it in bamboo forests (Wirot). Certainly it does not burrow; instead, it hides by simply pushing itself under vegetative debris. It is most active at dawn and dusk and consumes bamboo and high-altitude grasses, as well as berries, fallen fruit, and dead animals. The behavior seems to be very different from that of *M. emys*. This species lives in humid zones and bushy thickets and has never been reported to construct a nest or to vocalize.

Protection. *M. impressa* occurs with growing frequency in wildlife confiscations at Asian borders and airports. China is the most frequent destination, where it is sold for food as if it were a domestic animal. Some of these tortoises reach as far as the Western nations. The species is also consumed by certain ethnic groups in Malaysia and is used for certain alleged medicinal virtues. On the other hand, the habitats are less degraded than those of *M. emys* and are less occupied by humans. Thus this species probably maintains reasonable abundance in discrete areas within its overall range, although it is seen only rarely in the wild. Some concerned parties consider it to be rare, and they insist that status surveys and protective efforts be undertaken, as well as placement in Appendix 1 of CITES.

GEOEMYDIDAE Theobald, 1868

This family includes a great number of freshwater species. It is of relatively recent origin (Tertiary) and has met with great adaptive success. It is found in the Orient, with extensions into the Palaearctic (China and the western part of the region), as well as in the American tropics (Rhinoclemmys). There are two subfamilies, Batagurinae Gray, 1969, and Geoemydinae Theobald, 1868.

Batagur baska (Gray, 1831)

Distribution. This species is scattered through certain river systems in Southeast Asia, from India into southern Bangladesh, the coastal plains of Myanmar (Ayeyarwady River), Thailand, the Malay Peninsula (Perak River and others), and south as far as Cambodia, Vietnam, and Sumatra.

Description. The carapace length may reach 560 mm. The shell is slightly domed and has smooth scales and an even, nondenticulated posterior margin. All the vertebral scutes are wider than long. The general coloration is gray to olive-brown, sometimes with a tendency toward black. The plastron is well developed and has a slight anal notch. The bridges are wide and very robust. The anterior and posterior lobes are both shorter than the bridge. The color of the plastron is uniformly yellow to cream. The head is somewhat small but quite elongate, with a pointed snout and a large mouth, the corners of which curve upward, giving the animal the appearance of a supercilious interrogator. The head is olive-gray on top and sometimes a lighter gray at the sides and underneath; the jaws are cream. When the males reach breeding condition, however, all of these colors change. At that time, the males become darker, taking on an intense black hue. The head also becomes jet black, with the only spot of light coming from the large white iris, sometimes encircled by a white ring. The anterior limbs are well webbed, and each has four claws. All four limbs have a large transverse scale on their anterior border. The skin of the limbs is olive-gray. The young have a strong median keel, with a small spine on each vertebral scute. The spines and the keel itself disappear with growth. The juveniles are olive-brown to gray black, with a serrated posterior margin. They grow very fast, a good adaptation for outgrowing predators, and at the age of two years they are already 120 mm in length.

Natural history. This species lives in the estuaries of great rivers and in major watercourses with strong current and sandy or clay bottom. It also occupies mangrove areas. Typically omnivorous, it has a particular fondness for the fruits, rhizomes, and leaves of a mangrove *(Sonneratia)*, as well as certain mollusks, crustaceans, and fish. The males are smaller than the females and, when copulating, assume the seasonal dark coloration mentioned above. Nesting occurs from late December to early March and takes place on sandbanks and also on banks somewhat distanced from the water. The females may swim as much as 80 km to nest. There are three nestings per season, with 15 to 35 eggs per nest, and these measure about 70 × 45 mm. The females often make a preliminary wide excavation that may serve to draw attention away from the real nest. Incubation takes 70 to 112 days. In Burma the females nest at high tide, as do many sea turtles, and thus take advanage of the current to reach their nesting grounds.

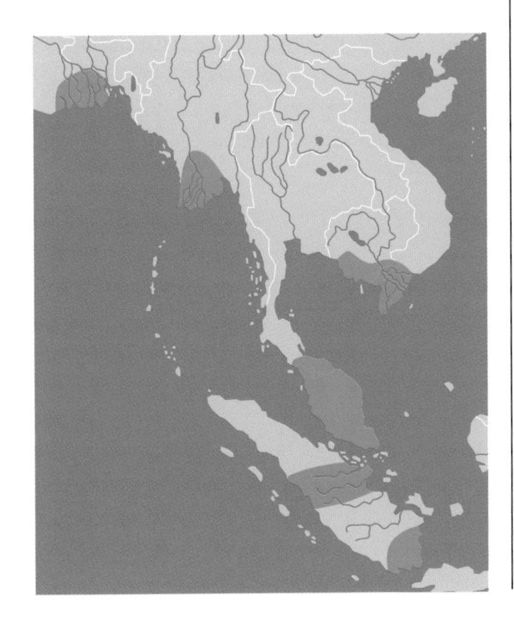

Common name

River terrapin

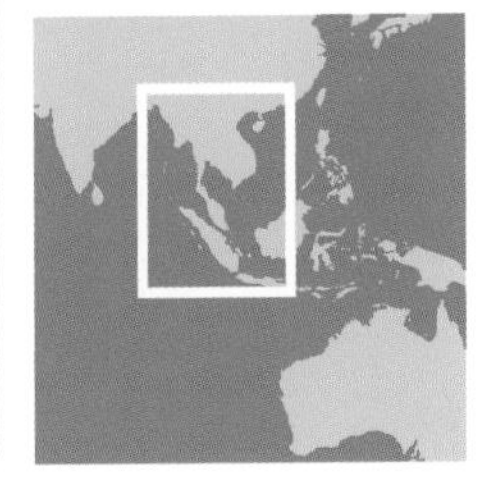

© B. Devaux

Protection. This species is listed on Appendix 1 of CITES and is diminishing fast throughout its range. Not only is it extensively eaten, but its habitats are heavily degraded as well. Furthermore, the construction of shrimp ponds has caused the rapid destruction of mangroves. Twenty years ago in Malaysia, as many as 20 nesting females might be seen per night in high season on the Perak River; nowadays one per month might be seen at Bota Kanan. Ambitious programs for captive rearing and reintroduction have been undertaken in the Malay Peninsula, at Bota Kanan, Bukit Pinang, and Kuala Behrand. At Bota Kanan, more than 4,000 two-year old juveniles are released each year into the Perak River. Unfortunately, this river offers protection to the turtles for only 15 linear kilometers, and if the species is to be saved, it will be important to set aside much larger sections of the original habitat. This species is classified as endangered by the Turtle Conservation Fund.

These tortoises take on brighter colors during the mating season, especially on the head and mouth, no doubt to make themselves conspicuous to their partners in the deep shade of the rain forests.

Callagur borneoensis (Schlegel and Müller, 1844)

Distribution. The overall range of this species, now mostly found in Borneo, includes southern Thailand, the Malay Peninsula, and northeastern Sumatra.

Description. This is one of the most colorful turtles in existence, and the color of the adult males during the breeding season is truly spectacular. The females may reach a length of 600 mm and weight of 25 kg. The males are much smaller than the females, reaching only 300 to 400 mm in length. The shell is oval, relatively flattened, and very smooth in the adults. There is no posterior serration, but the juveniles have a slight median keel that disappears with growth. The color of the shell varies with sex. In the males, the shell is greenish or olive, with three bold, black longitudinal stripes on the vertebrals and costal areas and with black markings on the marginals. The shell of the female is plain brown. More domed than the males, the females look very similar to *Batagur baska* females. The plastron is well developed and has a slight anal notch. The bridges are wide and thick. The inguinals are larger than the axillaries. The anterior and posterior lobes are shorter than the bridges. Overall, the color is uniformly yellow or cream. The head is rather small, with a pointed, projecting snout and prominent, round nostrils at the tip. The overall physiognomy is similar to that of *Batagur baska* and gives the animal a proud, smiling demeanor. The edges of the jaws are finely serrated. Large transverse scales are present on the anterior aspect of the limbs, which are well webbed, and the forelimbs have five claws (in

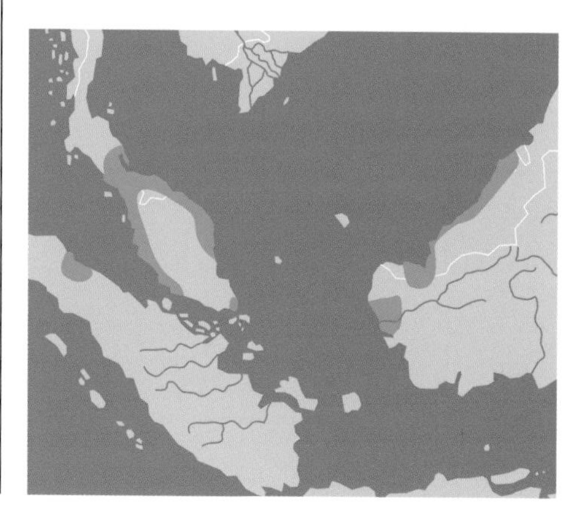

Painted terrapin

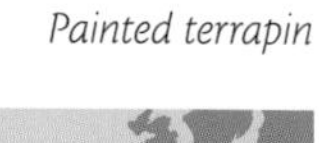

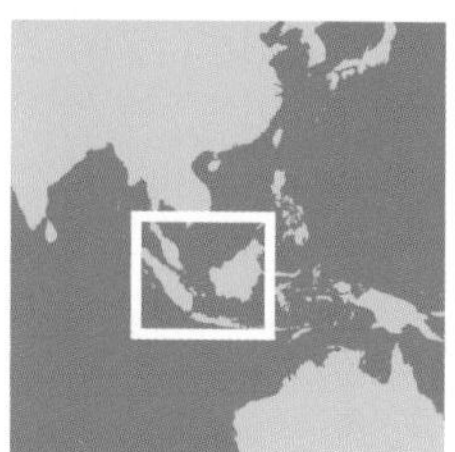

© B. Devaux

This extraordinary species shows a great variety of color patterns, according to sex and time of the year. Here, a male has a light carapace and a red and white head, showing that it must be the mating season.

contrast to four in *Batagur baska*). The young are almost circular and are light gray, marked with dark spots on the costals; a light line runs along the keel and serrations on the posterior marginals. During the season of love, the head of the female takes on a reddish tone, but the male undergoes a spectacular transformation: his shell becomes light in color, turning light gray or even white, which serves to emphasize the contrast of the three black stripes. The head is usually bluish gray with an orange band on top, becoming stark white and developing a sort of helmet with an almost vegetable-like appearance, resembling a large whitish lychee that encircles the cheeks, the eyes, and the sides of the head and extends halfway down the neck. On the top of the head, a red streak, the color of bright paint, completes the gaudy ensemble. These colors may allow the female to recognize her partner in the gloomy shade of the mangrove zone.

Natural history. These turtles, which live in estuaries and mangroves, may tolerate a salinity of about 50% of that of seawater. Dunson and Moll (1980) demonstrated that the juveniles may survive in pure seawater for two weeks. They nest on beaches and thus may tolerate the marine environment for many days before regaining the estuarine environment. They are mainly herbivorous, consuming mangrove fruits, fallen vegetation, and certain mollusks. In contrast to *Batagur baska,* they are rather irascible, and the males strenuously attempt to bite when picked up or molested. The females nest up to three times per season, each time producing 15 to 25 large, flexible-shelled eggs, 72 × 40 mm in size. Copulation occurs from January to February in the Malay Peninsula. The females migrate as far as 3 km from their place of residence. From June to August, one may see these turtles on the same beaches used by true sea turtles such as *Dermochelys coriacea, Chelonia mydas,* and *Lepidochelys olivacea*. Incubation of the eggs takes about 90 days.

Protection. *Callagur borneoensis* features in Appendix 2 of CITES, but it would seem to be a good idea to put it in Appendix 1. The truth is that this species is disappearing rapidly throughout its entire range. The eggs are widely consumed, as is the meat. Furthermore, its exceptional beauty is extremely appealing to hobbyists, who thus maintain a permanent drain on populations. A *Callagur borneoensis* spotted in the wild by a human being ends up in a casserole or in the hands of a middleman who steers it toward a Western market. Yet it does not survive well in captivity, and all those who try to raise it report difficulties. Furthermore, the mangrove habitat is heavily impacted by the creation of shrimp ponds, and the estuaries and rivers have become polluted or been channelized. These factors explain the decline of the species and justify its placement in Appendix 1 of CITES as soon as possible. A hatchery has just opened at Kedah, Malaysia, with the objective of assisting in the conservation of the species. This species is listed as endangered by the Turtle Conservation Fund.

The female (top) is black, with light-colored lips. The male (bottom) can have variable coloration associated with the reproductive season.

© F. Bonin

© F. Bonin

Chinemys nigricans (Gray, 1834)

© F. Bonin

This young tortoise is already very black, with a well-developed median keel. But the head is a more colorful yellow and brown. The plastron is very light, with several irregular spots.

© F. Bonin

Common name

Red-necked pond turtle

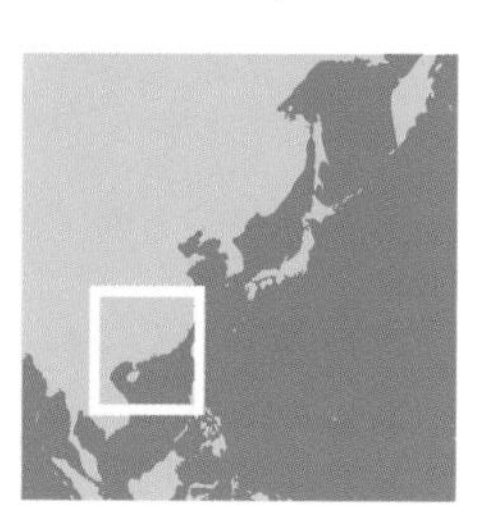

Distribution. This turtle occurs in southern China (Guangdong Province), as well as the extreme north of Vietnam.

Description. The vernacular name is appropriate for the species. This turtle is generally blackish, but the front part of the head is chestnut to reddish. It was formerly called *Chinemys kwangtungensis*. The carapace does not exceed a length of 269 mm. The shell is somewhat elongated, with a well-developed median keel even in old animals. The posterior marginals are very smooth and unpatterned, and the keel is marked by a lighter band. The seams between the marginals are sometimes yellowish. The plastron has an anal notch and is yellowish, with irregular dark brown markings on each scute. The axillary scutes are smaller than the inguinals, and the bridge is narrower than the posterior lobe. The head is strong and pointed, with a very smooth anterior aspect and wider, more granular scales toward the back. The general color of the head and neck is brown-black, but the snout may be more yellowish. The sides of the head are marked with small yellowish scales and light lines that extend from the eye across the tympanum and as far back as the base of the neck, where they fade out. The jaws are light (cream or yellowish), with fine dark markings. The limbs and tail are gray to dark brown, but there are scattered yellowish scales on the forelimbs. Males are often much more reddish than the females, which are basically dark gray.

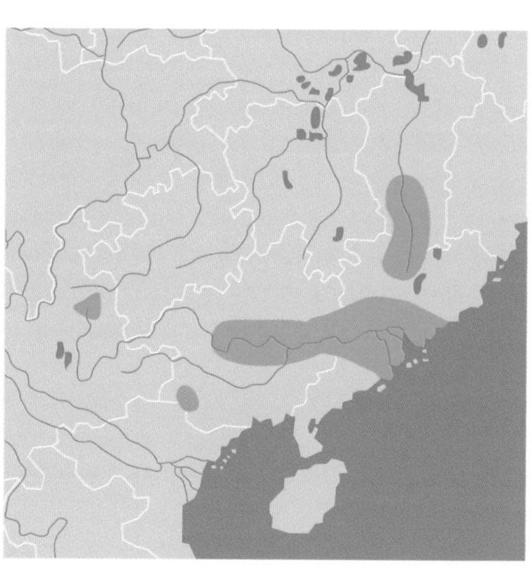

Natural history. This turtle is found in watercourses in hills and mountains, up to 1,200 m altitude, as well as in calm, mud-bottomed waters and sometimes fast-flowing, clear torrents. In diet it is omnivorous, eating aquatic plants, insects, amphibians, larvae, and small fish. The clutch size ranges from two to six. The eggs are large (48 × 26 mm) and take at least 90 days to hatch.

Protection. This species is caught for human consumption. Its current status is poorly known. It has been listed as endangered by the Turtle Conservation Fund.

Chinemys reevesii (Gray, 1831)

Distribution. This turtle occurs in China, south of the Yangtze and as far east as Shanghai. It also occurs in Japan (Honshu and Kyushu), as well as in Korea, Taiwan, and Hong Kong.

Description. The oval carapace does not exceed a length of 235 mm. It is not serrated posteriorly but has a median notch on the supracaudal scute. There are three well-developed keels. The two lateral keels are located about two-thirds of the way up the costal scutes. The color is light to dark brown, uniform or with the keels darker than the rest. The plastron has an anal notch and is yellow, as are the bridges; each scute bears a single large, dark brown blotch. The head is small and bullet-shaped, and the snout is protruding. The upper jaw is dark brown to black, with numerous yellow stripes, curved or straight, long or short, on the sides. A yellow streak, often incomplete, surrounds the tympanum. The neck is sometimes black or gray-brown, with narrow yellow bands along its entire length. The skin and the limbs are entirely olive or brown. The males are much smaller than the females and have a very large, thick tail. Often, they are less than 120 mm in length and are an unrelieved black in color.

© F. Bonin

This tortoise displays a very smooth shell, oval in shape, with keels almost nonexistent and light seam lines on the reddish background of the back. The head has light marking on a brown-black background.

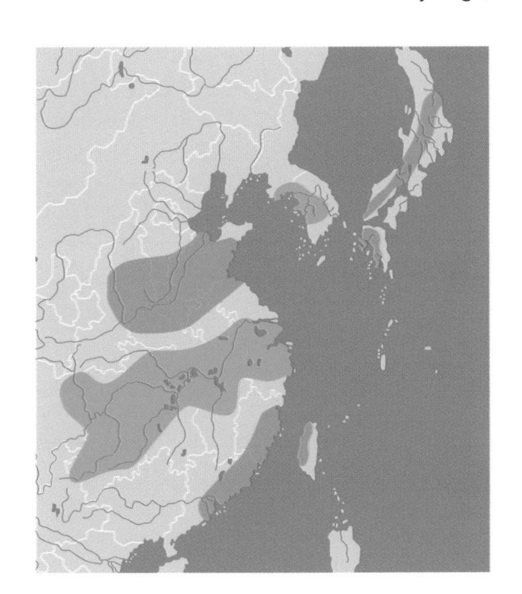

Natural history. This turtle occupies small, slow, shallow waterways, marshes, and irrigation canals, but it also occurs in large rivers as long as the current is slow and the bottom is clay. They bask for much of the day. This is an omnivorous species, eating larvae, other invertebrates, and vegetative material, including algae and aquatic plants. Courtship and copulation occur in spring, the males pursuing the females in the water, circling around them, and rubbing their snout against that of their partner. Nesting occurs in June and July, with four to nine eggs being laid. The eggs are elongate, white, and about 40 × 32 mm in size. Three nestings per season may occur, and incubation lasts for 90 days.

Common name

Reeves' turtle

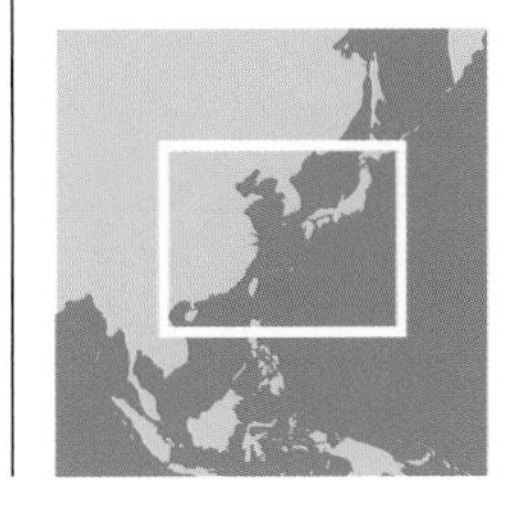

Geoclemys hamiltonii (Gray, 1831)

Common name

Spotted pond turtle

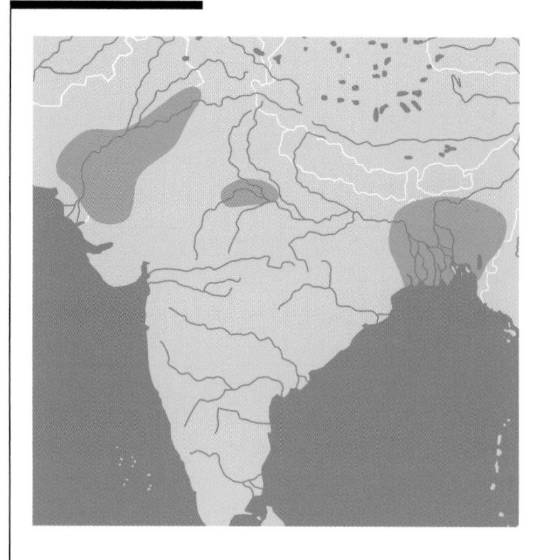

Distribution. The range of this species is very wide, although discontinuous, and includes southeastern Pakistan, northern India (Assam), and Bangladesh, in the Indus and Ganges Rivers.

Description. The common name is apt: the skin is decorated with white spots on a black background. This is a rather large turtle, reaching 350 mm in length and 5 kg in weight, and the shell is well domed and elongate-oval in outline. There are three dorsal keels, the vertebral keel being especially well developed. The keels are interrupted by a single protuberance on each vertebral and costal scute. The carapace is black, with multiple light streaks and spots, cream or yellow in color and well separated, on each scute. These ornamentations are especially clear in the juveniles and tend to disappear with age. Old individuals are often entirely black. The plastron is yellow and has no hinge; it is covered with numerous dark streaks. The bridge is always wide and strong. The anal notch is well developed. The head is large; the snout is short and covered on top with small scales. The ground color is black on the head and grayish on the neck. All of the skin is ornamented with large white or yellow spots. The tail is rather short, and the digits well webbed. The hatchlings are very brightly colored, with brilliant yellow or white spots.

Natural history. These turtles live in major rivers, but also in forest lakes where the waters are tranquil and well shaded and there is dense aquatic vegetation. In summer, they bask for long hours on emergent and floating wood. In winter, they hide in burrows or under vegetation. This is mostly a carnivorous species, feeding on snails, fish, amphibian larvae, and earthworms. The females mature relatively late, when they have attained a carapace length of 270 mm. Nesting has been observed from May to June in India and in December and January in Bangladesh (a second nesting may occur in February or March). The nests contain 19 eggs on average; they are oval and white and measure about 50 × 22 mm.

Protection. This species is included in Appendix 1 of CITES and is considered endangered in most of its range, because it is heavily collected for food. Furthermore, it is so attractive in appearance that it appeals to hobbyists and thus to dealers. The increasing penetration of humans into wetlands is rapidly reducing available habitat.

© F. Bonin

The head and limbs are generally spotted with white, giving this species a clownlike appearance, accentuated by the pointed snout. The carapace is a plain gray-brown.

Hardella indi Gray, 1870

Distribution. This turtle occurs in the Indus River in Pakistan and in associated and adjoining water bodies.

Description. This species is very close to *H. thurjii* but is smaller (up to 350 mm).

Protection. In this Muslim country, this species in not heavily exploited, but certain ethnic groups may collect it for consumption.

Common name

Indus crowned turtle

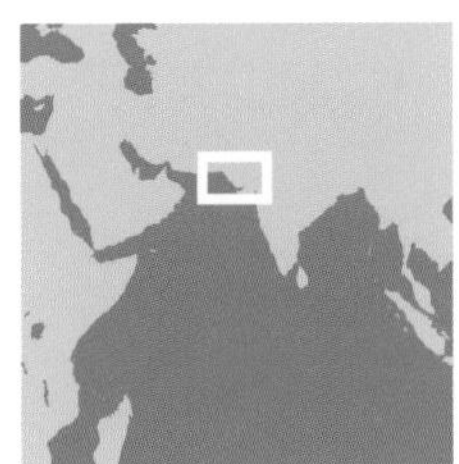

Hardella thurjii (Gray, 1831)

Distribution. This turtle is found in northern India and Bangladesh, in the effluents and tributaries of the Ganges and the Brahmaputra.

Description. Females of this species may attain a length of as much as 530 mm. The carapace is modestly domed, and the posterior marginals are not serrated, but there is a supracaudal notch. The median keel is weak. The color varies from dark gray or brown to black, with a yellow border between the costals and the marginals. The keel is black. The plastron is very solidly attached to the carapace and is unhinged; the posterior lobe is rather narrow, and an anal notch is present. The plastron is yellow, with a large dark blotch on each scute and others on the bridges. Some individuals are uniformly yellow, dark gray, or black. The head is brown or black and is ornamented with numerous yellow spots; the snout is rounded and rather short. One of the yellow bands starts below the nostrils and extends posteriorly, above the eyes, before curving downward behind the orbits to reach the neck. Another stripe forms a line on each side of the nostrils, on the upper jaw. Finally, a third band extends from the corner of the mouth along the lower jaw and along the neck. Some individuals show additional yellow markings on the dorsal area of the head. A single wide scale covers the nose and the top of the head, while the posterior part of the head bears a large number of scattered, small scales. The eyes have black irides, with golden dots. The jaw margins are serrated. All four feet bear extensive webs, and the limbs are brown or dark gray, with yellow or cream edges. The males do not exceed a length of even 180 mm and are thus one-third the length of the females. The young are more colorful than the adults, with wide marginals and often serrations around the posterior margin. Vertebrals 2, 3 and 4 are much wider than long, although they become narrower with age.

Natural history. This turtle lives in watercourses with a muddy bottom, abundant aquatic vegetation, and a feeble current. It also occurs in certain lakes and ponds with extensive vegetation. Primarily herbivorous, it feeds on aquatic plants and vegetative debris that falls into the water. However, it may also sometimes feed on

© T. Zhou

The carapace of this species is ordinary enough, being flat and modestly keeled. But the head is highly distinctive, with its bright yellow spots spaced irregularly over the snout and neck.

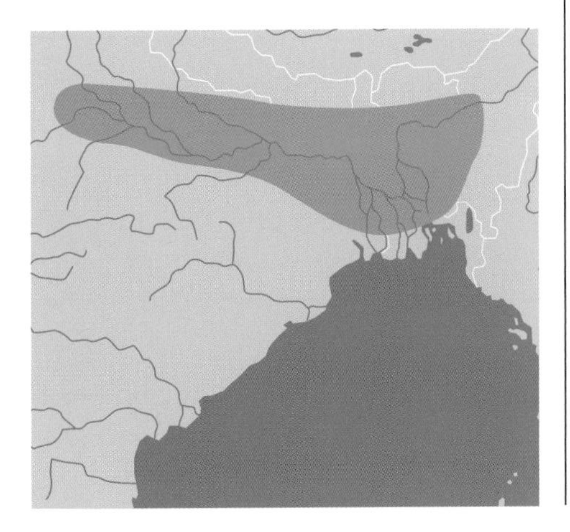

Common name

Ganges crowned river turtle

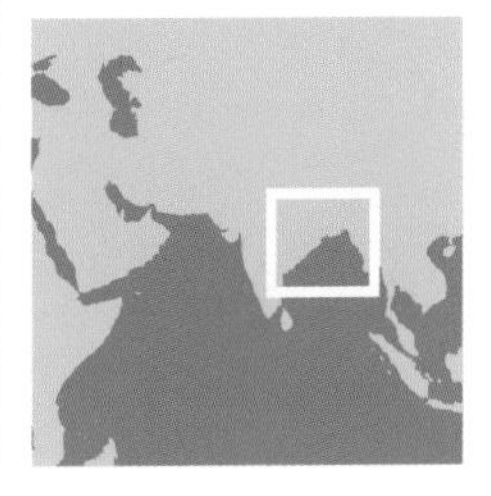

freshwater shrimp or small fish. It rarely comes forth from the water to bask, usually remaining at the surface, far from the banks, or at the bottom of the water, where it rests immobile. If it feels itself to be threatened, it buries itself in the mud. In Bangladesh the nesting season starts in November and lasts until January. A single clutch is produced each year, consisting of 12 to 16 eggs measuring about 60 × 34 mm.

Protection. This species may be found in all the markets in eastern India and in Bangladesh, being of large size and having a long history of human consumption. It is sometimes collected for export to China. Populations are becoming increasingly decimated.

The almost black carapace is encircled by a yellow line, and the keel is minor. On the other hand, the throat and the snout are brightly marked with yellow.

Hieremys annandalii (Boulenger, 1903)

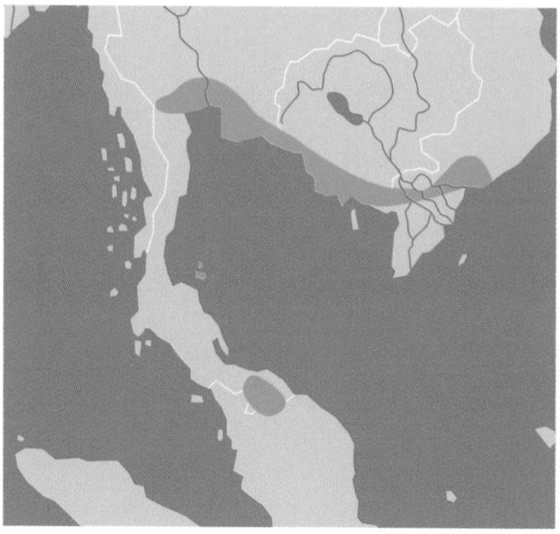

Common name

Yellow-headed temple turtle

Distribution. The range of this species is fragmented and includes central Thailand, northern Malaysia, southern Vietnam, and southern Cambodia.

Description. This is a very large turtle, reaching a length of as much as 600 mm. The shell is elongated and slightly domed, with a modest vertebral keel. The posterior marginal scutes are serrated, with a deep notch between the twelfth and thirteenth marginals. The general color is dark brown to black. The plastron is attached to the carapace by a strong band of connective tissue. The bridges are well developed, and the inguinals and axillaries are large. The plastron is shorter than the shell opening and lacks hinges. The anterior lobe has angled edges and is larger than the posterior lobe. The anal notch is always wide. The plastron and the bridges are yellow, with a large black marking on each scale. All these details of pattern tend to run together with age, and the plastron becomes entirely black. The lower jaw has a serrated edge. The snout is not raised nor protuberant, and the head as a whole is blackish to olive, with several gray marks. Major yellow bands are present on the snout and the sides of the head, as well as a wide stripe on the jaws and along the mouth, and the throat is also yellow, sometimes with dark dots. The yellow markings on the head are particularly striking and give the turtle its vernacular name. All the limbs bear extensive webbing, and the fifth digit has just two phalanges. The forelimbs are covered with large scales. The anterior face of each limb is dark gray, while the underside is light gray. The tail, also gray, is quite short. In the young, the three yellow bands on the head make for a particularly colorful appearance: the first extends from the snout to the neck, circling around the tympana; another starts at the jaw and

continues along the sides of the neck; and the third starts at the lower jaws and extends posteriorly. Sometimes, all the bands run together and form a wide yellow area behind the eyes.

Natural history. This is a species of wetlands, inundated fields, wet forests, and swamps and is not a riverine turtle per se. Nevertheless, some very large specimens have been observed at the mouth of the Chao Phraya River, near Bangkok (Smith, 1931), and the species tolerates saline waters. An herbivore, it consumes fruits and aquatic plants, as well as land plants and flowers. Nesting occurs from December to January. Four eggs, on average, constitute a clutch, and they are very large (60 mm long), with hard shells.

Protection. This species is collected and consumed, but fortunate individuals may escape this destiny by being deposited in Buddhist temple ponds, especially in Bangkok. The fragmented range suggests that the species' distribution is already relictual.

© F. Bonin

Note the long, yellow stripes on the neck and jaws and along the head.

Kachuga kachuga (Gray, 1855)

Distribution. This species occupies the basins of the Ganges and the Brahmaputra in Bangladesh, northern India, and southern Nepal, and it is also found in the Krishna and the Godavari Rivers.

Description. The name comes from the Hindi *khechuwa,* or *kachua,* which simply means "turtle." It may reach a length of 560 mm. The carapace is elliptical and somewhat domed, with a robust appearance, like a rounded stone. The bridges extend into massive buttresses within the carapace. The rear marginals are widened and sometime lightly serrated. There may be a vertebral keel, but it diminishes at a young age. The surface of the shell eventually becomes completely smooth, without seams, sutures, or rough areas, and is a uniform gray or black. The plastron is long and narrow, follows the curved configuration of the bridges, and is entirely orange-yellow. There is a slight anal notch. The humeropectoral seam is angled at the midline, whereas it is transverse in *K. dhongoka*. The bridges are yellow but adorned with dark markings. The head is rather large and wide, with a short, elevated snout typical of all species of *Kachuga*. The eye, with a curious, inquisitorial aspect, is located very close to the front of the head. The mouth is short, and the jaws are both bright yellow, with numerous serrations, al-

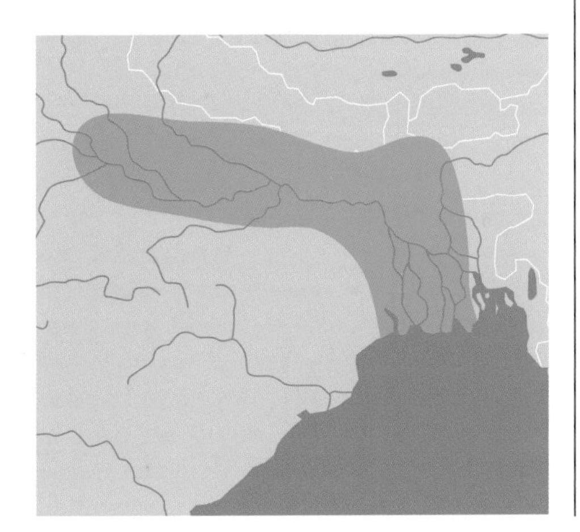

Common name

Red-crowned roofed turtle

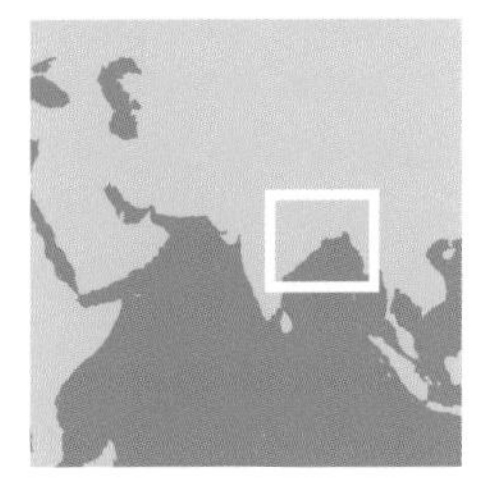

Adults have wide light-colored lips and a blackish body. The young are more colorful, with a strong median keel.

though the rest of the head is gray or brown, giving the animal a "smiling" appearance. The top of the head is covered with fine, soft skin and bears a reddish crown, although in some individuals the head is simply brown. The underside is light gray or cream. The sexual dichromatism is exaggerated during the breeding season: the head of the male becomes midnight blue, with a red area that extends from the nostrils to the top of the head, whereas that of the female remains dark brown to black, with yellow jaws. In young individuals, one may see 10 yellow lines along the sides of the head and 6 red lines on the neck, but all of these colors disappear with age. The young grow rapidly and have a strong keel with a blunt spur at the rear of the second and third vertebrals, as well as serrations along the hind margin. The head is light gray and marked with a reddish spot behind the eyes, and the jaws are yellow or cream. The jaw surfaces are finely denticulate.

Natural history. These turtles live in medium to large rivers and are powerful swimmers. One may spot them in clear water, not far from the surface, thanks to their almost luminous mandibles. They like to sun themselves on rocks, floating wood, or banks, but they are very skittish and dive at the least provocation. Mainly herbivorous, they eat fruit and aquatic grasses. They nest in March and April, depositing up to 25 ovoid eggs of large size (68 × 45 mm). The nests are excavated not far from the water, on sandbanks or on islands in the middle of the rivers. Incubation lasts for 80 to 86 days.

Protection. This large species is appreciated for its meat and is heavily fished, especially in Bangladesh. According to R. J. Rao, who heads up the breeding program at Gwalior and has studied the species, the populations are generally in a low state and remain vigorous only in certain reserves and protected rivers. Reintroduction programs will be necessary to restock some of the former colonies. Some of the habitat has been degraded by human activities, barrages have been built on the rivers, and sand mining has taken place at excessive levels. Water pollution has also occurred.

The species is classified as endangered by the Turtle Conservation Fund.

Kachuga dhongoka (Gray, 1834)

Distribution. This species is found in the northern part of India and in southern Nepal (Chitwan). Good populations exist in the Chambal River, in the state of Madhya Pradesh.

Description. This species attains a length of 480 mm and has a modest vertebral keel, which reaches its highest development on the second and third vertebral scutes. In the young, the vertebral keel is very well developed, and traces of lateral keels may also be seen. The nuchal scute is very small. The width of the carapace is greatest at the level of the fourth vertebral scute. The general color is grayish to brown or olive, with a black band along the keel, as well as two narrow lateral bands along the traces of the lateral keels. The plastron is long and narrow and is much smaller than the carapace openings. There is an anal notch. The width of the bridge is considerably greater than the length of the posterior plastral lobe. In the adults, the plastron is uniform, without any pattern, and varies in color from light yellow to creamy white. In old individuals, the plastron becomes darker, especially in the males, whereas in

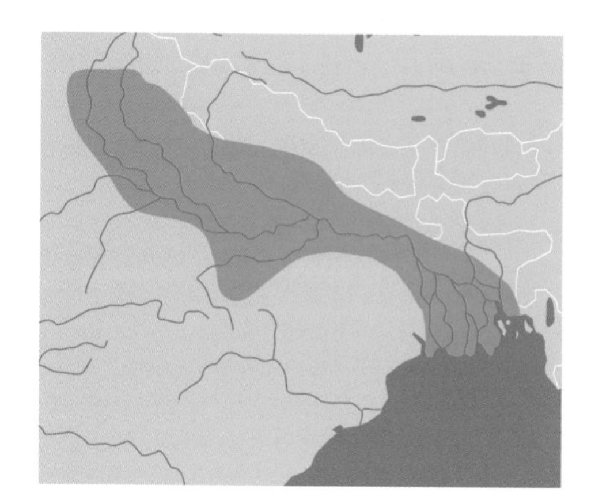

the young it is usually yellow, with brown or reddish areas on each scute. The head is small, with a somewhat elevated snout. The skin on the top of the head is smooth in front, but toward the rear it consists of a large number of small irregular scales. The upper jaw possesses a median notch, flanked by two well-developed cusps. The head is olive-gray, with a somewhat dark upper surface and, notably, a vivid yellow or cream-colored band on each side, which extends from the nostrils, passing above the eye and dropping down toward the tympanum. This band is very bright in the young and is a good character for identifying the species. The iris is grayish or sometimes brown. The digits are extensively webbed, and all the limbs have transverse scales, which increase the functional surface area and assist with swimming. This species is an especially adept swimmer. Sexual dimorphism is well developed, the females attaining 430 mm and a weight of 8 kg, whereas the males do not exceed 190 mm and 1 kg.

Natural history. This species occurs in the great permanent rivers and leaves the water only to bask or to nest. In the northern part of the range, these turtles hibernate, buried in the mud at the bottom of the rivers. Omnivorous in diet, they enjoy freshwater mussels and various aquatic plants. The fertilized females may retain oviductal eggs for months during the coldest season. They lay 20 to 35 eggs per clutch in March to April, along the edges of the rivers or in midriver sandbanks. The eggs are hard-shelled and elongate and measure about 57 × 36 mm. Incubation in the ground takes about 55 days. The young make haste for the water as soon as they emerge.

Protection. Although thoroughly protected by Indian law, these turtles are extensively collected for food, and they have disappeared from some areas. In Nepal they are almost gone. In India they benefit from the protection offered by certain reserves and parks, including a sanctuary on the Chambal River, where the populations are still strong. Programs of study and reintroduction have been conducted at the Jiwaji University in Gwalior, under the direction of R. J. Rao. This species is classified as endangered by the Turtle Conservation Fund.

Common name

Three-striped roofed turtle

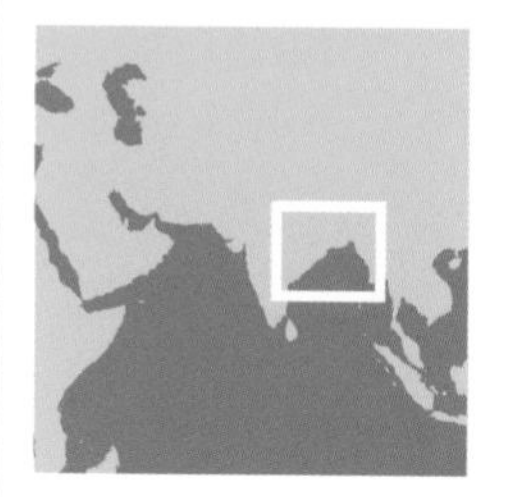

© I. Das

This old specimen has a flat, keel-less, beat-up carapace. The head bears a light line from the extreme front of the nostrils to the beginning of the neck.

Kachuga trivittata (Duméril and Bibron, 1835)

Distribution. This species is found only in Myanmar, in the Ayeyarwady and Salween River systems.

Description. This is the largest of the *Kachuga* species, reaching as much as 580 mm, but is the least colorful of the three. The carapace is elliptical, with a very strong keel that becomes less prominent with age. The young have a protuberance at the rear of the first, second, and third vertebral scutes. Old specimens have a very smooth, domed shell, resembling a large, polished rock. The posterior marginals are serrated in the young, smooth in the adults. The color is brown to grayish, uniformly so in the females, but the much smaller males (up to 460 mm in length) have three black lines in place of the keels. The plastron is long, narrow, and quite thick and is attached to the carapace by means of extensive buttresses. It is yellow or orange, and the bridges are edged with black. There is an anal notch. The head is of medium size and rather wide, with a raised snout, a "smiling" mouth, and light-colored mandibles with serrated jaw surfaces. The top of the head is covered with smooth skin, while the neck bears numerous irregular, rounded scales. The head and neck are brown to olive, tending toward black on top. The top of the head may be marked with a very dark line in certain specimens. The feet are gray to black on their distal parts, the proximal shaft being lighter. The powerful limbs are equipped with extensive webs and with wide scales on the anterior aspects.

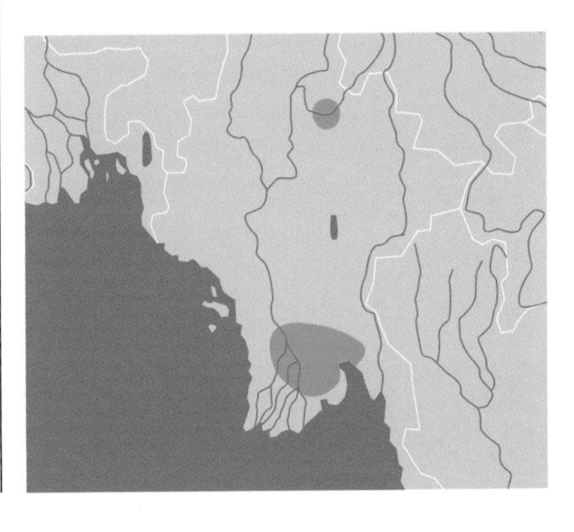

Common name

Burmese roofed turtle

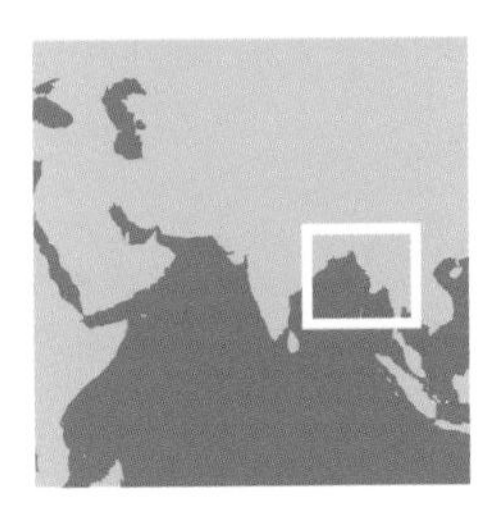

Natural history. This species occurs in large, deep, slow, muddy tropical rivers. It is mostly herbivorous. Nesting occurs in December and January on the riverbanks, several meters from the highest water. Clutch size may range from 18 to 25 eggs, each being about 67 × 42 mm, and incubation lasts for about 80 days.

Protection. Turtles of this species are occasionally caught by fishermen, but it is an elusive species and difficult to catch. The present status is poorly understood, but populations are threatened by humans and by habitat degradation and pollution. It is covered by general regulations protecting turtles of all kinds in the Ayeyarwady, but these are usually merely "on paper" prohibitions. The species is classified as endangered by the Turtle Conservation Fund.

Malayemys subtrijuga (Schlegel and Müller, 1844)

Distribution. The range is somewhat discontinuous and includes southern Vietnam, southern Cambodia, southern Thailand, southern Myanmar, and the northern part of the Malay Peninsula.

Description. The carapace is oval and modestly domed, and three discontinuous carapacial keels form a small knob on each of the larger scutes. The median keel extends the length of the five vertebral scutes, whereas the lateral keels rarely reach as far as the fourth costals. The carapace length is less than 200 mm. The posterior marginals are not serrated. The dark to light brown color of the carapace tends toward chestnut, and a fine yellow line borders the carapace. The small knobs on the keels are darker than the rest of the carapace. The plastron is unhinged. It is narrower than the carapace and has a strong anal notch.

The posterior lobe is shorter than the bridge. A large dark brown blotch is present on each of the plastral scutes, and two more dark blotches lie over the bridges. The head is rather large and is black, with several light or even whitish stripes. The first stripe extends from the nostrils and passes over the eyes on its way to the neck. Another extends from each corner of the snout and curves downward to pass below the eye and rejoin the neck. Two more light bands cross the tympana behind the eyes. A large scale covers the crown of the head, and the area posterior to this is covered with numerous small scales. The limbs and the tail are gray to black, with a yellow edge on the outer borders.

Natural history. This turtle may be found in wetlands, including canals, rice paddies, small lakes, and marshes, with a muddy bottom and abundant aquatic vegetation, but it is also often found on land. It is carnivorous and eats large numbers of small snails (hence the vernacular name), earthworms, aquatic insects, crustaceans, and small fish. The clutch size ranges from four to six eggs, which are white, elongate, and about 45 × 25 mm in size.

Protection. This turtle is often to be seen at Buddhist temples. Its small size limits local consumption, but large shipments are sent to China from time to time, and populations continue to drop as a result of habitat deterioration, and the range becomes ever more limited.

© F. Bonin

The shell has a central keel and modest lateral keels. The head is strongly striped with yellow on a black background.

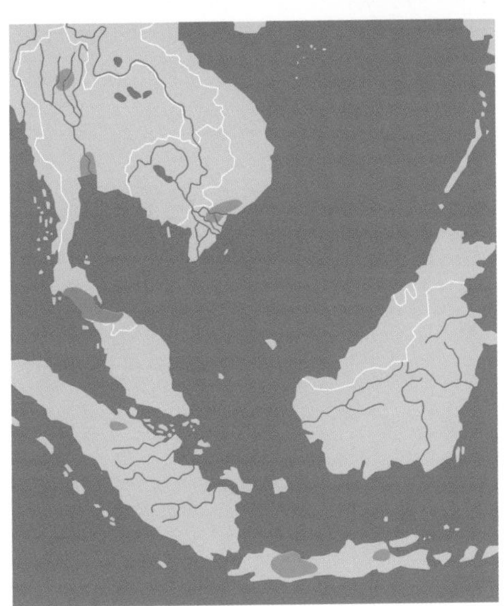

Common name

Malayan snail-eating turtle

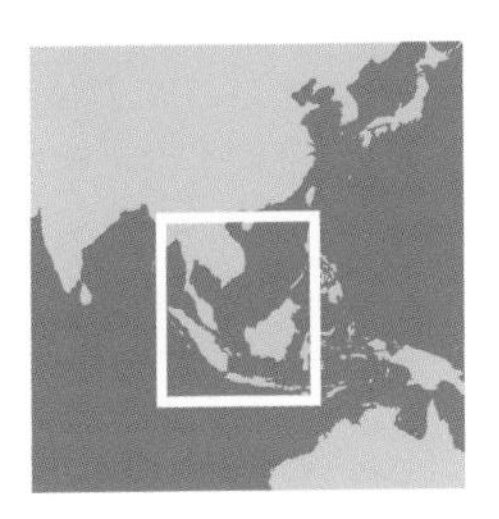

Morenia ocellata (Duméril and Bibron, 1835)

Distribution. This species is found in southern Burma, from Toungoo to Mergui, in Tenasserim.

Description. This species has a rather domed shell and may reach a length of 220 mm. The carapace has a median keel, made up of a series of elongate knobs. The posterior margin of the shell is smooth. The nuchal scute is wider than long, the width being equivalent to one-fourth of that of the first marginal scute. In color the carapace is olive, dark brown, or black, and it has a yellow black-edged ring with a black center on each of the vertebral and costal scutes. This adornment is less distinct in old animals. The plastron has a strong anal notch. The posterior lobe is narrow, although the bridge is wider than the length of the posterior lobe. The axillary and inguinal scutes are large. The overall color of the plastron is a uniform

Common name

Burmese eyed turtle

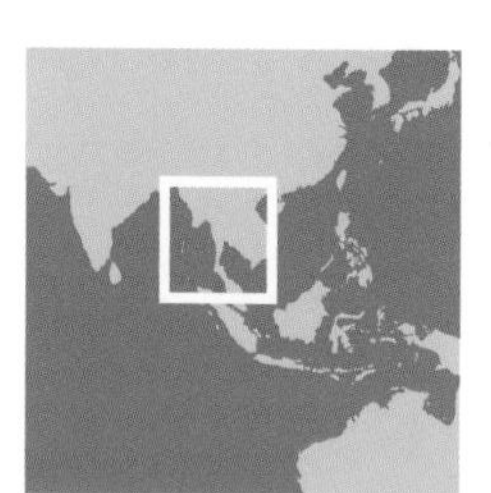

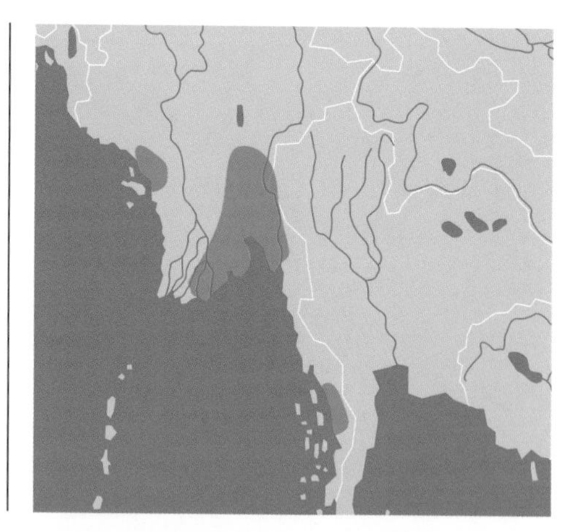

yellow. The head is rather small, with a short, nonprotruding snout. The upper surface and the sides of the head are covered with a single large scale, behind which is an array of numerous small scales. The color is olive to brown, with a yellow band on each side that extends from the rear of the snout and passes over the orbit on its way to the neck. A second yellow band occurs from behind the eye to the neck. The tail is very short and is of the same color as the limbs—brown or olive.

Natural history. This is a strictly aquatic species. It is found in permanent rivers, swamps, lakes, and marshes, usually in well-vegetated sites. It often basks on tree branches, rocks, or banks. It is highly herbivorous and consumes aquatic plants. The female lay five to eight eggs, about 48 × 22 mm in size.

Protection. This species is included in Appendix 1 of CITES. It is caught intensively on a local basis, especially during the dry season, and large numbers may be seen in Chinese markets. Conservation measures are necessary to prevent populations from continuing to decline.

These shy turtles, with little color, often form assemblages in lakes, ponds, and basins, and as they disappear into the water, only the yellow triangles on the tops of their heads remain visible.

Morenia petersi (Anderson, 1879)

Distribution. This turtle is found through Bangladesh and also in a small enclave of southeastern India, from Bihar to Bengal.

Description. The shell is moderately domed and in females may attain about 220 mm. A median keel, in the form of a series of elongate protuberances, extends along the vertebral scutes. The posterior marginals are smooth-edged. The overall coloration is olive to dark brown or black; the costals, like the vertebrals, have yellow to whitish cream borders. Each marginal scute has a vertical white streak. There is a strong anal notch. The bridge is wider than the length of the posterior lobe of the plastron. The axillary and inguinal scutes are large. The plastron is uniform yellow or orange, and dark areas may be seen on the bridges and the underside of some of the marginals. The head is small and has a protruding, pointed snout. The upper jaw is provided with a pronounced median notch and extends considerably beyond the lower jaw. A large scale covers the top and sides of the head, with small scales behind it. The head color is olive, with numerous yellow bands on each side. One of the bands extends from the tip of the nose backward, while another extends from the orbit to the neck. The tail is short, and the limbs have yellow edges. They are covered with narrow scales resembling lamellae. The digits are completely webbed. Males are smaller than females (195 mm) and have a more pronounced keel, whereas the females are more highly domed. In the young, each vertebral scute has a light L-shaped mark, opening toward the front. Each costal has a pale ocellus close to the marginal seam. These markings disappear with age.

Natural history. This species occurs in sluggish rivers, swamps, and lakes. It is essentially diurnal and spends long periods of time basking on floating objects or swimming at the surface. In diet it is herbivorous, eating floating vegetation such as *Eichhornia*. Nesting has been reported in December to January in Bangladesh and is carried out at night. The clutches consist of 6 to 10 elongate eggs, measuring about 45 × 20 mm.

Protection. Collected heavily and for a long time for export to China, this turtle is now so reduced in numbers that this trade seems to have come to an end on its own. Surveys are needed.

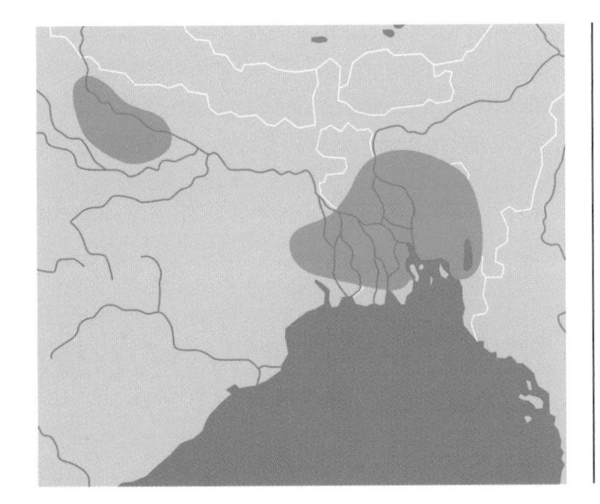

Common name

Indian eyed turtle

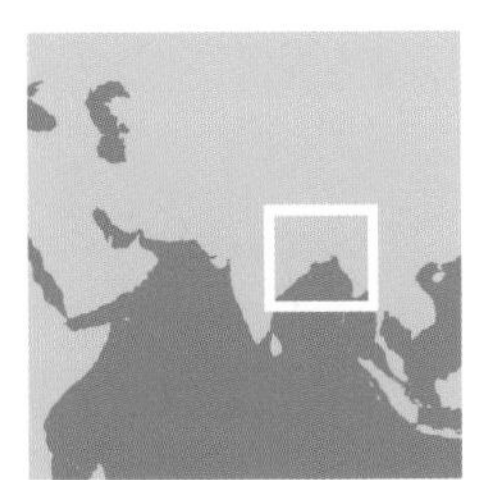

© I. Das

This small turtle, oval and domed in form, has a carapace with conspicuous ocelli and a characteristic yellow triangle on the top of the head.

Ocadia glyphistoma McCord and Iverson, 1994

Common name

No common name; this is possibly a hybrid.

Distribution. This turtle is found in the southeast of Guangxi, China.

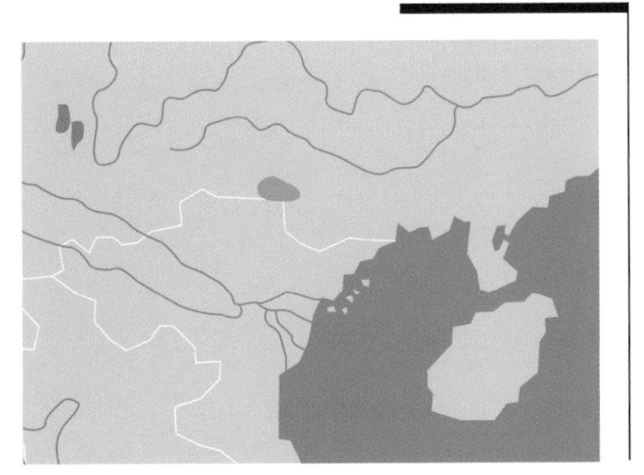

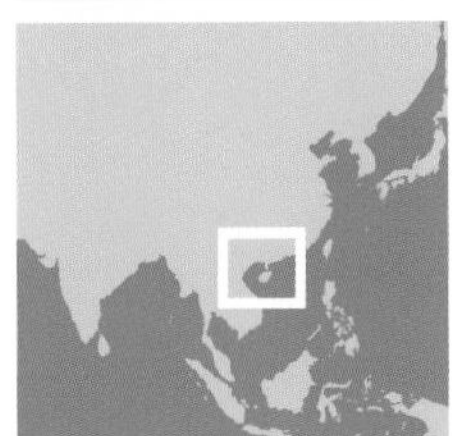

Description. This is a controversial form, which may be a hybrid between *O. sinensis* and *Mauremys annamensis*. It is close to *O. sinensis* but differs in the following points: fewer light stripes on the neck and head (usually four along the side of the head, whereas *O. sinensis* has eight); stronger black markings on the ventral surfaces of the hind limbs, a notch at the middle of the upper jaw (from which the name *glyphistoma* is derived—*glyphi,* meaning "notch," and *stoma,* meaning "mouth"). The plastron and carapace are wider, the posterior lobe of the plastron is longer, the interpectoral seam is shorter, and the interhumeral seam is longer. The plastron is cream to yellowish, with wide quadrangular dark markings on each scute. Large round blotches adorn the bridge, the axillaries, and the inguinals. Maximum size is 199 mm for males and 180 mm for females.

Ocadia philippeni McCord and Iverson, 1992

Common name

No common name; this is possibly a hybrid.

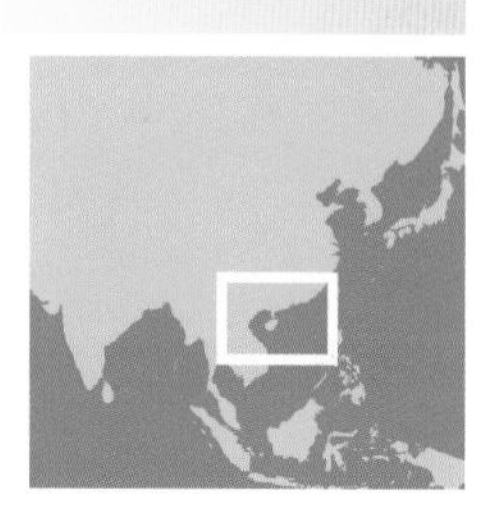

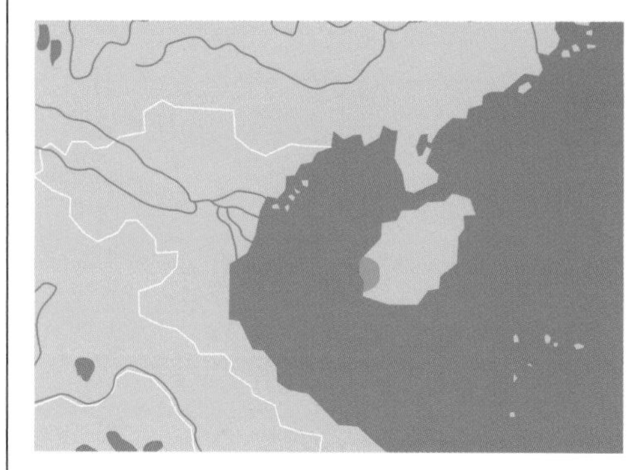

Distribution. This turtle is known only from "near Kancheng," on southern Hainan Island in China.

Description. This is a controversial form that may be a hybrid between *O. sinensis* and *Cuora trifasciata*. Close to *O. sinensis,* it differs in the following points: fewer light lines on the head and neck (the lines are well spaced, discontinuous, sometimes rounded off at the ends), a wider plastron, orange to pinkish overall tinge, a shorter interfemoral seam, and a longer intergular seam. This species may reach 360 mm in length and 2.7 kg in weight.

Ocadia sinensis (Gray, 1834)

Common name

Chinese stripe-necked turtle

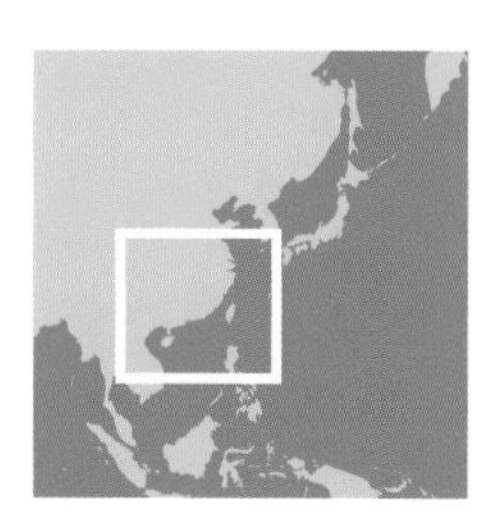

Distribution. This species has a fragmented range in southern China (Fujian, Hangzhou, Suzhou, Kwangju, Shanghai, and Hainan), into North Vietnam, and including Taiwan.

Description. This small species does not exceed a length of 240 mm. The shell is elliptical, slightly flattened toward the back, and highest between the second and third vertebrals. The carapace is lightly serrated posteriorly. Juveniles have three dorsal keels, which smooth out or disappear in adults. The carapace scutes often show clear annual growth rings. The overall color is reddish brown to black, according to the individual, with yellow seam lines in the young and sometimes yellow or orange bands following the original keel lines. There are no plastral hinges. The plastron is large, with well developed bridges and an anal notch. The plastron, bridges, and undersides of the marginals are yellow, with a brown to black blotch, large and quadrangular, on each scute. The head is narrow, olive on top, and yellow below. Fine, very straight yellow and black rays decorate the sides of the head from the snout to the base of the neck and are the source of the vernacular name. The snout is slightly pointed. The irides are pale green. The upper jaw, white or cream in color, is inwardly curved. The limbs are olive, with numerous very fine yellow stripes that are not as straight as those on the neck. Large scales cover

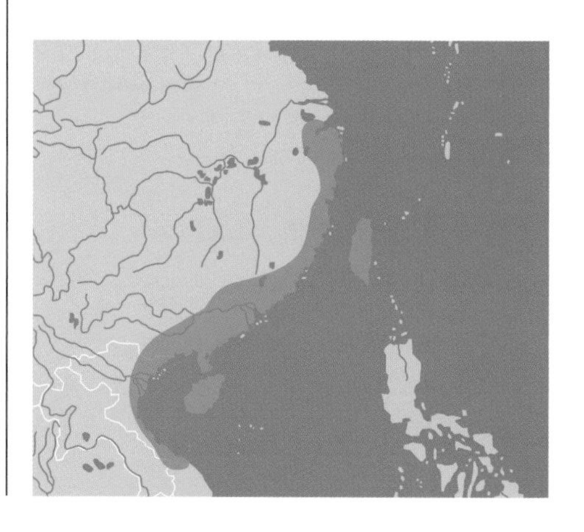

© B. Devaux

This very beautiful turtle, here seen in Vietnam, is marked with fine lines on a greenish to blackish background. The head has a narrow snout. The throat and the neck have an attractive pattern of white and black lines.

most of the skin of the forelimbs, and the remainder of each limb is covered with small, flat scales. All of the limbs are fully webbed. The hatchlings have three conspicuous keels, but these disappear quickly with growth.

Natural history. This turtle lives in watercourses and creeks with slight current, as well as in lakes, marshes, and canals—wherever it may encounter a mud or clay bottom. It is omnivorous in diet, preferring aquatic plants and fruits that fall into the water. Nesting has been observed from April to June, and usually just three eggs are laid, rather large in size (40 × 25 mm). When first captured, this turtle is liable to eject a powerful jet of cloacal fluid.

Protection. This species is classified as endangered by the Turtle Conservation Fund.

© F. Bonin

Orlitia borneensis Gray, 1873

Distribution. This species is found in the southern part of Sarawak and in a small area of Indonesian Borneo (Kalimantan), as well as the western part of peninsular Malaysia and northeastern Sumatra.

Description. This is a very large turtle, with males reaching a carapace length of 800 mm. The shell is narrow and elongate and is somewhat elevated in the vertebral region. The second vertebral scute has a mushroom-like shape that is very characteristic. The posterior marginals are quite small in relation to the lateral and anterior marginals. The fourth costal scute is very small. All the scutes are somewhat rugose, and the general

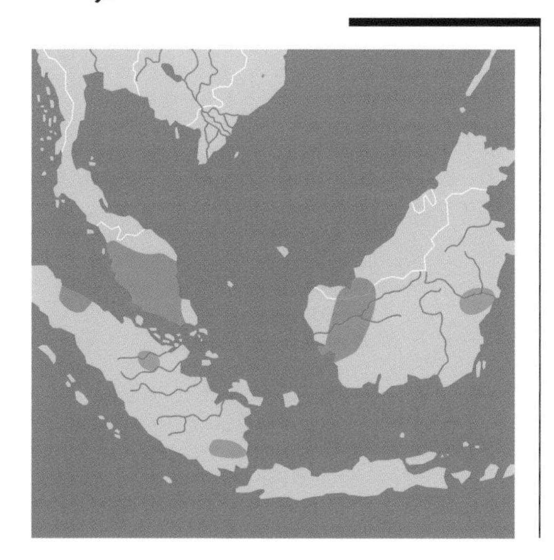

Malaysian giant turtle

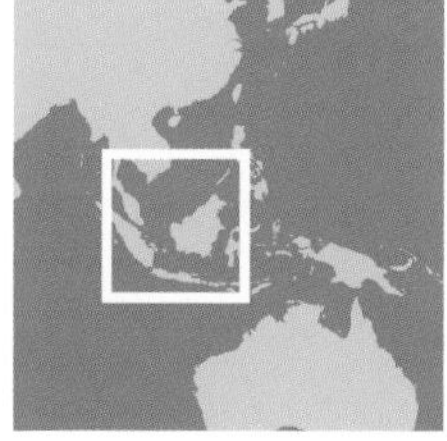

This large, dark turtle is flat and massive, without any major unusual features. The head is large and strong, and the snout is well developed.

color ranges from dark gray to brown but may also be black. The plastron is long and narrow and shows a moderate anal notch. Its edges may form a slight keel. The anterior and posterior lobes are narrower than the pectoral and abdominal scutes. The bridges are wide and thick, encircling the axillary and inguinal scutes. The overall color of the plastron is yellowish. The head is wide and large and has a raised snout. A series of granular scales is present between the eye and the tympanum. The skin of the top of the head is broken into small scales. The head is rather dark on top and has a dark yellow line on the side; the underside of the head and neck are cream. The limbs are gray, brown, or black. The forelimbs have wide horizontal scales on the upper edge, and all the digits are well webbed. In the young, the posterior marginals are serrated, and the carapace is very elevated between the second and third vertebral scutes. The head is decorated with dark shading and has a light streak from the corner of the mouth to the base of the neck.

Natural history. This turtle lives in major rivers and may be encountered in estuaries. Omnivorous in diet, it mainly eats fish and has been observed attacking a snake, but it also enjoys vegetation and fruit such as banana and papaya. It feeds equally well in water and on land. Very calm in disposition, it may rest for hours on a riverbank without moving. The females use sandbanks of the great rivers for depositing their clutches of 12 to 15 elongate, hard-shelled eggs, each about 80 × 40 mm in size. Upon hatching, the young measure about 60 mm in length.

Protection. This species has been heavily exploited for its meat, and populations are collapsing, especially in the Klang River. Details on the exact current status are lacking. The species is classified as endangered by the Turtle Conservation Fund.

Pangshura (Kachuga) smithii (Gray, 1863)

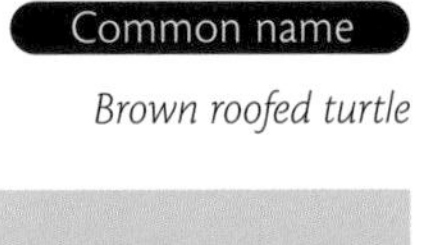

Brown roofed turtle

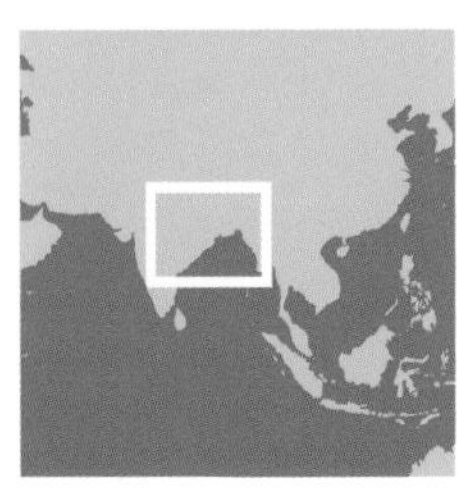

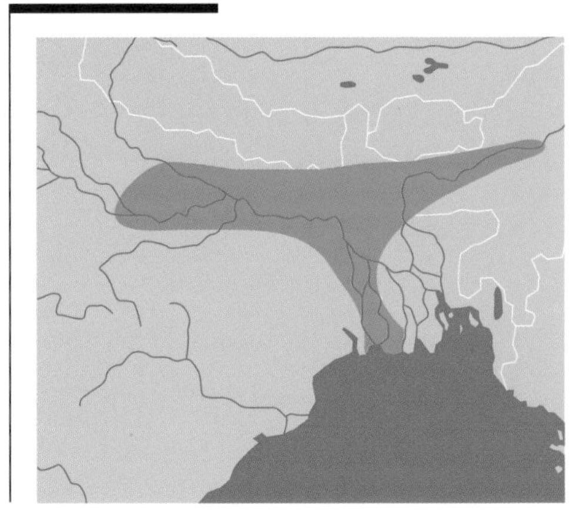

Distribution. This species occupies an elongate, narrow band of territory in the north of the Indian subcontinent, in the basins of the Ganges and Brahmaputra, from Pakistan to Bangladesh and southern Nepal.

Description of the genus *Pangshura.* The generic name is based on the Bengali word for "turtle" *(pangshura).* Formerly included within the genus *Kachuga, Pangshura* includes four species and several subspecies. All are aquatic and herbivorous, occupying slow-moving waters, swamps, and fast watercourses. These are

© H. Schleich

This is a small, shy species. The keel is well developed. The general coloration is brown-black, with a light line around the periphery. There may be a number of light, even pink, markings on the front of the head.

medium-sized turtles, with the females significantly larger than the males. The carapaces are roof-shaped, or tectiform, with a high keel and sometimes with blunt spines at the rear of the anterior vertebral scutes.

Description of *P. smithii.* This species carries the name of Sir Andrew Smith, naturalist and director general of the Army Medical Board of British India. This turtle is the color of toasted bread, as its vernacular name suggests. The females may reach a length of 227 mm; the males, just 108 mm. The elliptical carapace is widened toward the rear, and the vertebral scutes are narrow and keeled. The first three vertebrals each have the form of a narrow rectangle. The third is often elevated to form a peak, the fourth is lozenge-shaped, and the fifth, wide and flat, makes contact with six of the posterior marginals. The shell is brown with a reddish background and sometimes has a darker vertebral line. The plastron is long and narrow, the posterior lobe is smaller than the anterior lobe, and there is a small anal notch. The plastron is yellow, with black spots that are large or small, according to the subspecies. The head is of medium size, with a slightly elevated snout, and is covered with large, irregular scales. It is grayish to yellowish in color, sometimes with a slight pink tinge. The neck is gray, streaked with yellow. The forelimbs are covered with large scales. The young have a second vertebral scute whose shape differs from that of the adults; instead of being rectangular, it is pentagonal.

Subspecies. Two subspecies are recognized. *P. s. smithii* (described above): This turtle occupies a large slice of the southern part of the range, from Pakistan to Bangladesh. It is the larger of the two subspecies and has black pigment on the sides of the head and on the anterior face of the limbs. There is a dark brown to reddish blotch behind each eye. The irides are pale blue-gray, and the mandibles yellowish. Black lines and triangles are present on the areolae of costals 2 and 3. The plastron is strongly marked with black, relieved only by the yellow edges. *P. s. pallidipes* Moll, 1987: This subspecies occurs in southern Nepal (Bardia) and in the center of the species' range in northern India. Its name refers to the light-colored limbs. This turtle differs from the nominate race by the very reduced pigmentation of the plastron and the much lighter coloration of the head and limbs. The shell is often less keeled, and the spur on the third vertebral scute is completely lacking. The head is light olive, light gray, or yellowish, and the reddish spot behind the eye is very attenuated or even absent. The irides are pale blue-gray. Altogether, the patterning of this species is quite variable, some individuals having reddish lines on the vertebral scutes and a more or less dark plastron.

Natural history. This turtle likes deep watercourses, canals, lakes, and swamps with a muddy bottom and abundant vegetation, but it is not found in the great rivers with major outflow. Activity levels are reduced during the cold months, and it hibernates in Pakistan and in Nepal, but it is often seen thermoregulating on promontories or bridge stanchions, and it may estivate in summer in the southern part of the range, hiding itself in the mud. This turtle is an excellent swimmer and also is able to burrow deeply into the riverbanks or mud in order to hide. It is strongly vegetarian but may also consume insects, amphibians, and fish. It also may nibble on carrion, and Prashad noted in 1914 that it would feed on decomposing human bodies. It is a nervous species, plunging back into the water at the slightest pretext and struggling when caught. The females mature after five to six years. In Pakistan, nesting occurs from mid-July to September. From five to nine eggs are laid in each clutch, and normally there are two nests per season. The eggs are rather large (44 × 23 mm). Incubation lasts for 150 days.

Protection. This species is heavily collected and consumed. In Bangladesh, it is often seen in market stalls during the cold months. A better knowledge of its status and populations would seem to be highly desirable.

Pangshura sylhetensis Jerdon, 1870

Distribution. This turtle's range is limited to small territory in Bangladesh and northeastern India, in Assam.

Description. This species has a different habitat from that of others of its genus, being found more in forested areas, and it is unusual in having 13 pairs of marginal scutes (because of the division of the supracaudals). It is a rare species, very difficult to observe in nature. It does no exceed 190 mm in females, and males are smaller still; it is thus the smallest member of the genus *Pangshura*. The carapace is oval, with a strong keel on the third, fourth, and fifth vertebral scutes, and a strong hook or spur at the posterior of the third vertebral. The marginals are strongly serrated toward the rear. The shell is brown to olive, with a lighter line marking the vertebral line. The plastron is rather narrow and more or less oval and has a modest anal notch. The plastron is yellow, with dark brown or black markings on each scute. Some individuals may have a dark plastron or even a totally brown one, except for a yellow line around the edge. The head is brown and of medium size, with a protuberant snout and a hooked upper jaw. The head is covered with irregular scales. Two light lines, yellow or cream in color, mark the sides of the head; one extends from the rear of the eyes to the front of the cranium; the other runs along the marginals and curves inward in the tympanic region. The neck is brown, with fine cream or whitish, rather diffuse lines, which disappear in old individuals. The forelimbs are covered anteriorly with large scales.

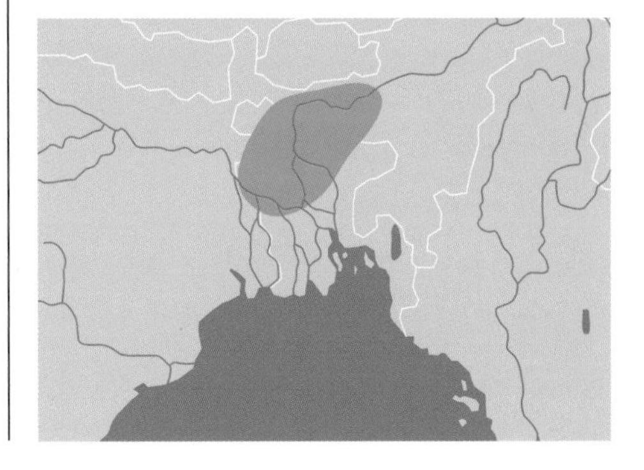

Common name

Assam roofed turtle

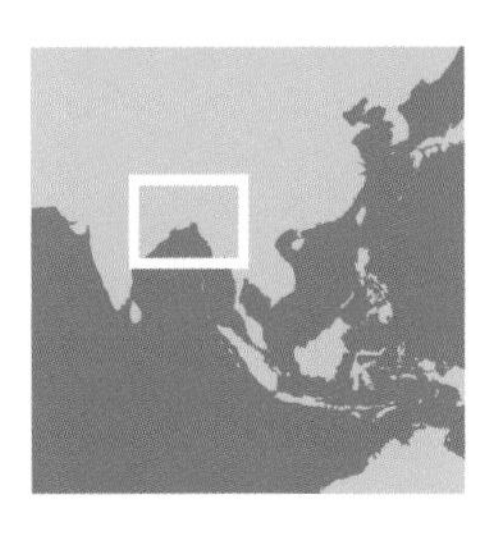

Natural history. This turtle buries itself in leaf litter and detritus in the warm, humid forest habitat and is active mostly at the beginning and end of the day. It occupies mountain streams and small rivers but also spends much time in the humid substrate. Primarily carnivorous, it eats insects, carrion, and small fish. Altogether, it is an opportunist and, when fully grown, may subsist on berries and vegetation.

Protection. The IUCN has recommended that this species be studied immediately to evaluate its true status. It is not hunted much nor consumed, perhaps because of its small size, but some of its habitats are degrading rapidly and it is possible that populations are very reduced. It is classified as endangered by the Turtle Conservation Fund.

Pangshura tecta (Gray, 1831)

Distribution. This species has a wide range in northern India, in the drainages of the Indus, Ganges, and Brahmaputra Rivers, from Pakistan to Bangladesh (and southern Nepal). Isolated populations exist in southern Pakistan and in west-central India.

© F. Bonin

The tectiform shell of this turtle identifies it as a member of the genus *Pangshura*.

Description. The Latin name refers to the rooflike, or tectiform, shape of the carapace. Females do not exceed about 240 mm in length, and the shell is strikingly elevated, the first two vertebral scutes being somewhat keeled and the third rising to a sharp point, as is characteristic of the genus. The fourth vertebral may also have a point at its posterior edge. The rear marginal scutes are not serrated. The color is brown, and along the upper edge of the marginals there may be a line of orange or yellow, visible especially at the sides. The plastron is very long—sometimes longer than the carapace—and is rather narrow and slightly elevated in front. There is a small anal notch. In color the plastron is yellowish or pink, with irregular black spots distributed over the scutes, the bridges, and the underside of the marginals. Each gular scute bears only a single black spot. The head is of medium size, and the snout is pointed and elevated. The skin of the head is covered with fine scales. The gray color of the head darkens on the top and is lighter below. Some very colorful individuals have reddish spots on the temples and jaws, with numerous cream-colored lines extending from the back of the cranium and running to the base of the neck. The eye is large and near the anterior of the head. The irises are somewhat reddish in the males and pinkish to greenish in the females. The postorbital spot is light, sometimes reddish, and extends like an arrow toward the back of the head. The males are often more greenish than the females, which are predominantly brown. In some individuals, the carapace seams are cream or yellowish. The limbs are gray-brown, with discontinuous series of yellow dots. The young are very colorful, with a light green carapace and orange lines along the vertebrals, with black rings. The neck is black, marked with yellow stripes; the spots behind the eyes are bright yellow. The plastron is dark orange, with numerous irregular black dots.

Natural history. This turtle inhabits all manner of calm and stagnant waters, although its congeners prefer deep rivers. It may be seen in ponds, lakes, swamps, and canals. It is a poorer swimmer than *P. smithii* and spends long periods of time basking on banks and floating wood. One often sees it extend its neck and limbs toward the sun. When disturbed, according to Indraneil Das, it reacts in the same way, holding out its limbs and lowering its head toward the ground. This is a nervous species and not at all aggressive. It is primarily vegetarian and only occasionally eats insects, carrion, or snails. The females mature in six to eight years. Courtship consists of the male turning circles around the female in order to get her attention. The nests are seen in February and March,

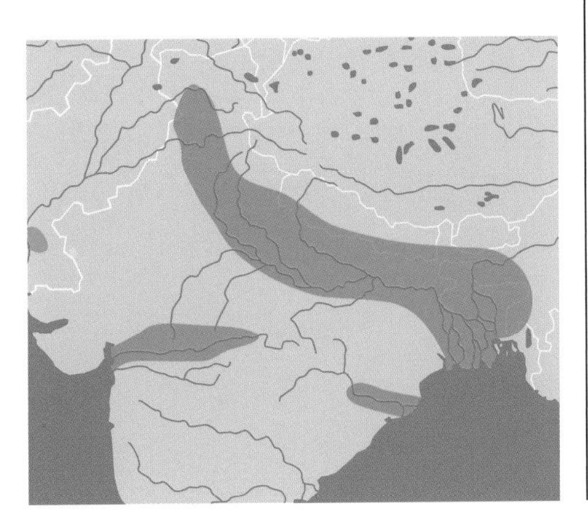

Common name

Indian roofed turtle

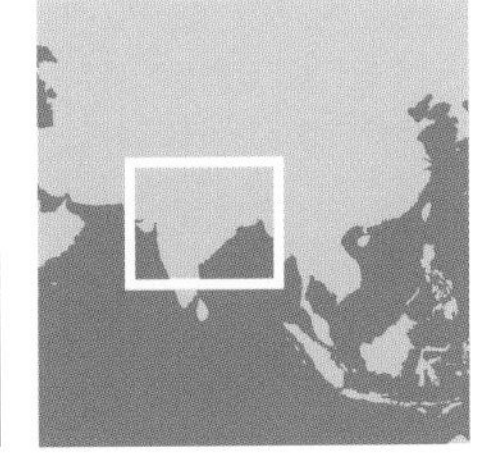

the females usually laying clutches of 3 to 12 eggs twice per season. The eggs measure about 42 × 27 mm. Incubation lasts for nearly 70 days.

Protection. This species is listed in Appendix 1 of CITES, even though populations are still reasonably abundant. And indeed, it is heavily collected for consumption and also captured for sale. Nevertheless, some other *Pangshura* species certainly merit protection, perhaps more so than *P. tecta*. The species is collected for its alleged curative properties, and it is said to cure numerous maladies. The eggs are also collected in large numbers. According to H. H. Schleich (2003), the species has become highly vulnerable in many parts of its range and suffers also from habitat destruction and the construction of dams and barrages. The nesting areas are also destroyed by agricultural practices that extend farther into the wetlands every year.

Pangshura tentoria (Gray, 1834)

Distribution. The three subspecies together extend over a wide expanse of territory in the central and eastern parts of India and in Bangladesh, as well as in southern Nepal.

Description. The Latin name refers to the tent-like form of the carapace, although it is somewhat less so than in *K. tecta.* Of medium size, the females reach 271 mm. The shell is elliptical and is wider behind than in front. The keel is more or less developed, according to the subspecies, with the first vertebral inconspicuous, the second having a slight posterior point or spur, and the third with a well-developed ridge and posterior projection. The marginal scutes are not serrated and are significantly widened posteriorly. The carapace is brown to dark beige, generally paler than in *P. tecta*. A reddish or brownish vertebral line runs along the central keel. The plastron is shorter and wider than is seen in other members of the genus and has a medium anal notch. It is yellow, with dark markings that vary according to the subspecies. The head is of medium size and has wide scales on the prefrontal and frontal areas, with smaller ones behind. The snout is somewhat elevated, and the upper jaw has a slight beak, unlike other *Pangshura*. The coloration of the head is rather pale, except in *P. t. flaviventer*. The head is gray, darker on the top, and lighter below and bears light spots behind the eyes and on the jaws, as well as light lines along the neck. The juveniles are more brightly colored, with brilliant yellow lines on the head, where the background in blackish gray to dark olive, and with a brighter plastron. The spots behind the eyes are more colorful, sometimes pink, and are quite bright. The libs are gray, with light cream-colored lines.

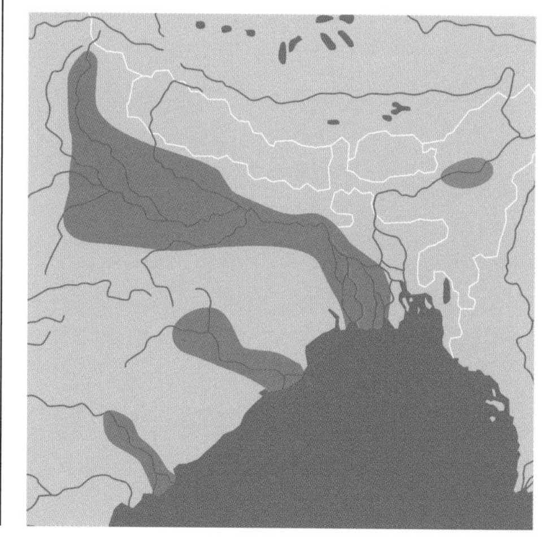

Common name

Indian tent turtle

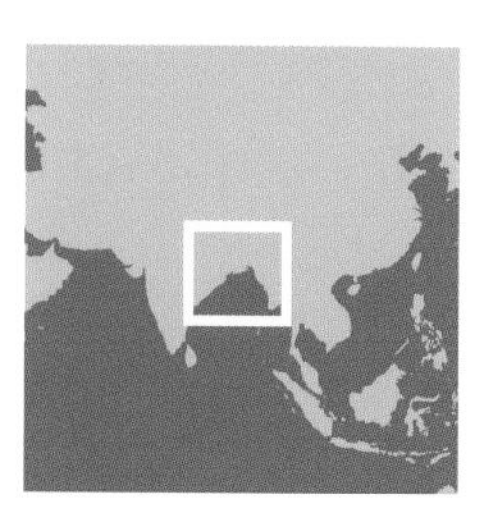

Subspecies. Three subspecies are recognized. *P. t. tentoria* (described above): This turtle occurs in the central part of India, in the Mahanadi and Krishna Rivers. The carapace measures up to 230 mm, and an amber line runs along the first three vertebral scutes. *P. t. circumdata* Mertens, 1969: This subspecies occurs farther north, in the northern and central parts of the Ganges drainage and in southern Nepal. It is recognized by a pink to reddish streak separating the costals and the marginals. It may reach 271 mm. The plastron has black spots occupying about half of the area of

© F. Bonin

Young *P. tentoria* display a highly tectiform carapace and a well-developed spine on the third vertebral scute. Fine white lines run from the rear of the orbit to the base of the neck. In the adult, the neck is brightly colored, but the head is blackish and loses its markings.

each scute. The bridges and the undersides of the marginals are black, with yellow rings. The head is olive-gray, with a pink or reddish spot behind the eye, and there are two reddish lines behind the head. *P. t. flaviventer* Günther, 1864: This subspecies is considered by H. H. Schleich (2003) to be a full species. It occurs in the northern tributaries of the Ganges, into western Bangladesh and southern Nepal. The Latin name defines it well: the plastron is yellow, and only the undersides of the marginals have black spots on each scute. This is the smallest of the three subspecies. It is very much roof-shaped, and the central keel, easily visible on the second vertebral scute, is well developed on the third one. In old individuals (as in all *Pangshura*), the carapace may flatten somewhat and completely lose the keels and vertebral spikes. The carapace is brown to olive, with a pale orange line in the vertebral keel, and sometimes a light yellow or orange area is present on the flanks, between the marginals and the costals. The head is of medium size, with a somewhat elevated snout, and there is no beak on the upper jaw. The coloration is rather variable, but in certain individuals the top of the snout is ornamented with a black spot that continues toward the rear of the skull, and with numerous cream-colored lines extending from the rear of the orbits and the cranium to the base of the neck, on a black background. Behind the eyes, this marking continues, cream-colored or lightly tinted with orange.

Natural history. This species occurs in large rivers, and it may also be seen even in very small tributaries. It spends long periods of time basking on banks and floating wood, especially in the Ganges itself. In may also be seen in temple ponds, where it is frequently kept. Its diet is primarily vegetarian, but it also eats animal material, including insects, and the juveniles are somewhat omnivorous. In areas frequented by crocodiles, it may also feed on the crumbs and fragments left by these giant reptiles. The nests, perhaps two per season, include 4 to 12 eggs, about 42 × 27 mm in size. Incubation lasts for about 130 days.

Protection. The natural predators of this species include hyenas and jackals. But these turtles are also often captured by fishermen for their own consumption, or the lucky ones may be placed in temple ponds. Nevertheless, Blanford reported that consumption of the flesh of these turtles may bring on a variety of maladies, and this may perhaps limit the numbers taken by people for consumption.

© F. Bonin

Siebenrockiella crassicollis (Gray, 1831)

Distribution. This species has a fragmented range in Southeast Asia, in southern Myanmar, Thailand, southern Vietnam, the Malay Peninsula, Singapore, western Borneo, eastern Sumatra, and Java.

Common name

Black marsh turtle

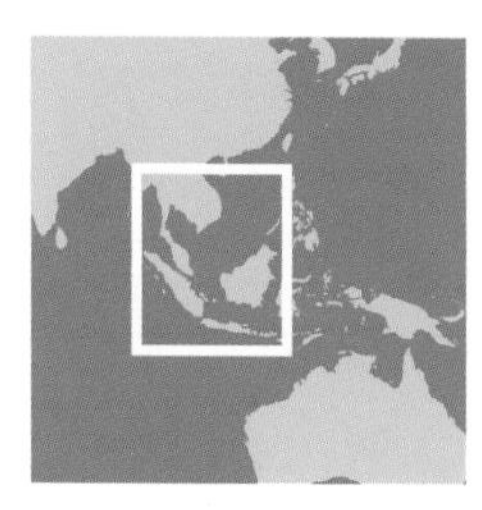

The carapace is black, sometimes relieved by light markings in young individuals. The head is large and wide and may be marked with light zones around the eyes. The neck is powerful.

Description. See page 325 for *S. leytensis.* This is the only species in the genus, which is named after Friedrich Siebenrock, a great turtle scientist at the Vienna Natural History Museum in the early twentieth century. The name *crassicollis* means "thick-necked," although this characteristic is true only of the poorly stuffed type specimen, not of the living animals. This is a small turtle, not exceeding a length of about 200 mm. The carapace is black, sometimes with slightly lighter areas, and the shape is elongate, almost quadrangular, and serrated at the back. There is a moderately developed median keel in adults. The plastron is well developed, lacks hinges, and is also black, with somewhat orangish zones along the seams and under the marginals—in some individuals the plastron is unrelieved black. The bridges are strong and very solid, dark brown to black in color or sometimes dark beige with darker markings. The head is wide, with a short, blunt snout. The upper jaw has a median notch. There are two wide scales on the front of the head, with finer scales behind it, as well as a band of small granular scales extending between the eye and the tympanum on each side. The head is blackish, with scattered clear spots in young individuals; these are white to pale gray and are located around the eyes, on the jaws, and on the sides. The limbs are black and well webbed. Males have a slight plastral concavity and lose the light spots on the head, although females keep them. The young have three keels, the two lateral ones disappearing early in life. The keels are marked by light lines, and the head has yellow spots.

© F. Bonin

Natural history. This turtle lives in slow watercourses with a mud bottom and in marshes, ponds, and artificial impoundments, as well as flooded prairies. Slow and nonaggressive, the turtles do not fear human presence. They may be seen crossing roads and wandering long distances across country, probably largely when seeking nesting sites, and they may move extensively by night. In the water, this species is not really a swimmer; rather, it should be considered a bottom walker. Sometimes the carapace is be infected with bacteria that leave permanent scars, as occurs with *Mauremys leprosa*. This is an omnivorous species, eating plants, vegetative debris, and fruits that have fallen into the water, as well as snails, amphibians, shrimp, and various other invertebrates, and it can feed equally well on land and in the water. When taken in hand, it may secrete a powerful musk. Nesting has been observed in April to June, and there may be three or four clutches in a season,

with two to five eggs per nest. These are elongate and rather large (45 × 19 mm), with hard, shiny shells. Incubation takes 60 to 80 days.

Protection. This species is sometimes consumed or sold in markets, but the flesh is little sought after, probably because of the musky odor. The progressive fragmentation of the habitat leads one to believe that populations may be dropping.

Cistoclemmys flavomarginata
(Gray, 1863)

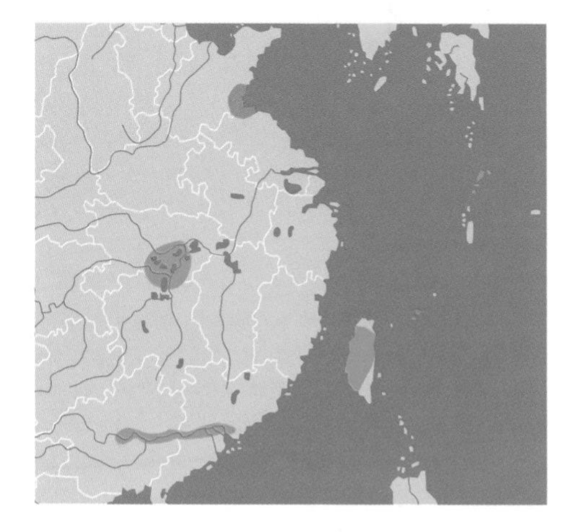

Distribution. The two subspecies occur in southern China (Fujian, Hunan, Szechuan) and in Taiwan, as well as in the Ryukyu Islands.

Description. This turtle was previously classified within *Cuora*. The carapace may reach 170 mm. The shell is rather domed and has a yellow vertebral band, which contrasts with the overall dark brown to black color. As its name indicates, there is a yellow line around the rim of the shell. Each vertebral scute is narrower than the corresponding costal scute. The carapace is wider in the posterior half than in the anterior. The vertebral keel is well developed. Annual growth rings are well marked, which gives the shell an engraved, sculptured appearance. The lower part of the marginal series is yellow, whereas the plastron is dark brown to black. The posterior lobe is very wide, has no anal notch, and can completely close off the rear of the shell. The seam between the anal scutes may disappear with growth in some individuals. The head has a gray to greenish dorsal area, with a wide yellow band that rejoins each orbit after looping around at the base of the neck. The upper jaw and the side of the head are brilliant yellow, and the chin varies from pink to yellow. The upper jaw overlaps the lower one to a considerable degree. The limbs are gray-brown on the exterior face, except for the yellow tips. The forelimbs are covered with large scales. The tail is short and gray, with a pale yellow dorsal band.

Subspecies. *C. f. flavomarginata:* The Taiwan yellow-margined box turtle is described above. *C. f. evelynae* (Ernst and Lovitch, 1990): The Ryukyu yellow-margined box turtle is found on the islands of Ishigaki, Shima, and Iriomote. *C. f. sinensis* (Hsü, 1930): The Chinese box turtle occurs in southern China in the provinces of Fujian, Hunan, and Szechuan.

Natural history. This turtle lives in rice paddies with very humid understory and at the edges of

Common name

Common yellow-margined box turtle

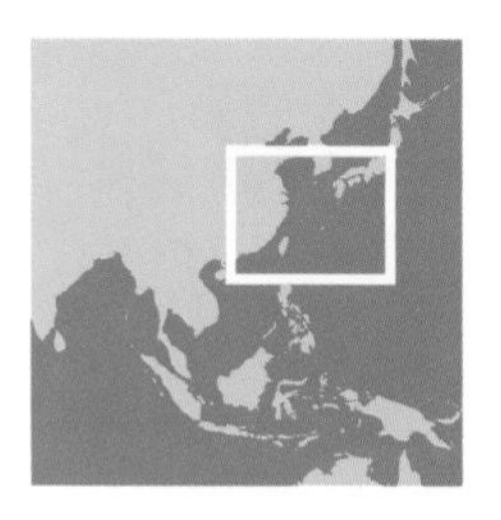

The carapace is rounded, with a central yellow band, especially visible on the posterior scutes. The head and neck are brightly colored, as are these parts of most of the species of the forest understory, perhaps for better recognition during the breeding season.

© J. Maran

slow creeks, and in Taiwan it may also be found in heavily vegetated hillsides. Ill at ease in deep water, it always swims on the surface and seems to be more terrestrial than aquatic. Long hours are spent basking in the sun, out of the water. This is an omnivorous species that mainly consumes fruits and vegetables. Nesting (a single egg, 42 × 23 mm) has been observed in July. There may be two nestings in a season.

Protection. This species is heavily collected in China, and it features more often on restaurant menus than in the wild. Its scattered habitats and the general isolation of the surviving colonies suggest that these turtles are fast disappearing. This species is classified as endangered by the Turtle Conservation Fund.

Cistoclemmys galbinifrons (Bourret, 1939)

© B. Devaux

The face of this turtle shows a truncated snout, expressive eyes, and gaudy head markings. Old specimens become uniformly beige or yellow.

Vietnam box turtle

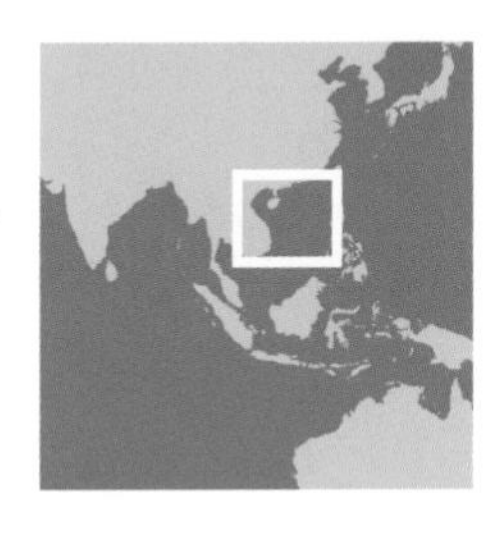

Distribution. The four subspecies occupy a fragmented range in southern China (Hainan Island) to northern and central Vietnam, and into Cambodia and Laos.

Description. Formerly included in the genus *Cuora,* this group of turtles has recently been reclassified into four subspecies and a new full species, *C. serrata*. *C. galbinifrons* has a very domed shell and may reach a length of 190 mm. It is wider behind than in front and the posterior marginals are smooth. The vertebral keel of the juvenile disappears in adulthood. A narrow cream or yellow stripe extends along the vertebral scutes, bordered by a broad, dark brown central band which extends beyond the vertebral scutes on to the tops of the costals. This band often has an elaborate internal pattern or design made up of small dark markings. The remainder of the costal scutes is generally creamy-white or yellow, but occasionally dark figures invade the light areas. The upper edge of the marginals is brown to yellowish, with lighter markings. The lower edge often includes a yellow spot on each marginal. The plastron is very large and can close off the entire carapace, front and back. There is no anal notch and the plastral lobes are smoothly rounded. The plastron is dark brown or black, but the seam between the carapace and the plastron is always edged with yellow. The head is pointed, and has a very short snout. Yellow to pale green or gray, it is modestly dark-spotted in some juveniles. The chin and throat are yellow to cream. The limbs and tail are greenish to gray. There are large scales scattered across the anterior face of the forelimbs.

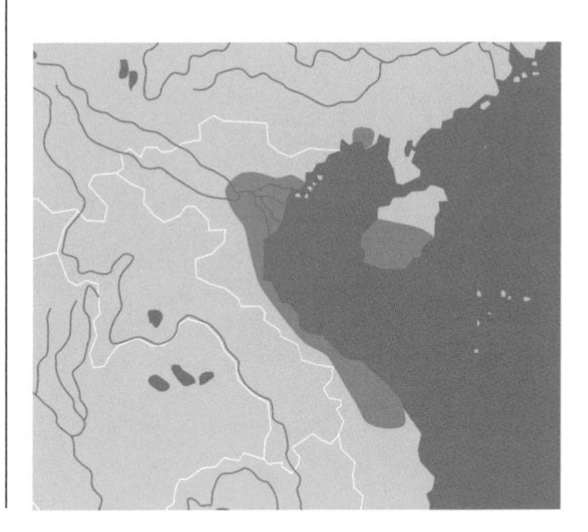

Subspecies. Four subspecies are recognized.

C. g. galbinifrons occurs in northern Vietnam and southern China. Its carapace is flatter and more elongate than that of the other subspecies. The plastron is almost entirely black, as is the lower edge of the marginal scutes. There are no wide brown bands on the sides of the carapace.

C. g. bourreti (Obst and Reimann, 1994) occurs in central Vietnam and in the regions adjacent to Cambodia and Laos. The dark brown central band on the carapace is very wide and reaches the midline of the costal scutes. The anterior marginals have dark patterns.

C. g. hainanensis (Li, 1958) only occurs on Hainan Island, in China. The very domed shell may reach 186 mm. It is partly light yellow, with a yellow vertebral keel. The vertebrals and the costals are brown to light chestnut, as is the entire edge of the carapace. The yellow region includes brown spots, and the light brown area includes several yellow bands. The plastron is chestnut to brown, with a number of irregular light yellow

spots. The tympanum is light yellow, and the lower jaw and the throat are gray-white. The neck is yellowish with dark brown bands on each side. The tail is yellow, with brown and black spots.

C. g. picturata (Lehr, Fritz and Obst, 1998) occurs in southern Vietnam and in the area adjacent to Cambodia. The dark brown band on the shell extends further down at the sides than it does in *bourreti,* to the top of the marginals. The first marginals are light and lack dark figures. The soft parts of the body carry an olive-gray, reticulated design which is very pronounced.

Natural history. This box turtle is more terrestrial than aquatic, and lives in humid places, in forests where it can bury itself in leaf litter and detritus. It is often active by night and after rains. Mainly carnivorous, it eats terrestrial invertebrates, including earthworms and varied insects, as well as carrion. Very shy, it usually retracts all extremities and closes up the shell when handled.

Protection. Unfortunately, this species is one of the most heavily exploited within its range, in China and in Vietnam, and is caught both for human consumption and for sale as a garden pet. It has also featured extensively in the international pet trade, although it requires specialized conditions in captivity, and often dies quite quickly.

© B. Devaux

Mostly collected in Vietnam and with China the principal destination, smuggled turtles are often seized by the authorities and placed in rehabilitation centers, as here at Cuc Phuong.

Cistoclemmys serrata

(Iverson and McCord, 1992)

Distribution. Confined to the central part of the Island and Province of Hainan, China.

Description. As its name indicates, the carapace midline is moderately serrated. The posterior marginals are also serrated, a feature unique in the genus. There are three keels, even in adults. The shell is darker than in other *Cistoclemmys,* the costals being almost black. The growth rings on both carapace and plastron are well marked, even in old animals. The plastron is often light in color, and never completely black, especially in the middle.

Some specialists consider this species to represent natural hybrids between *Cistoclemmys galbinifrons* and *Pyxidea mouhoti*.

Natural history. Probably similar to that of other *Cistoclemmys*. It lives in humid understory environments.

Protection. This species is less heavily collected than other members of the genus, but its scarcity and novelty make it very attractive to collectors. The status needs to be reevaluated and populations censused.

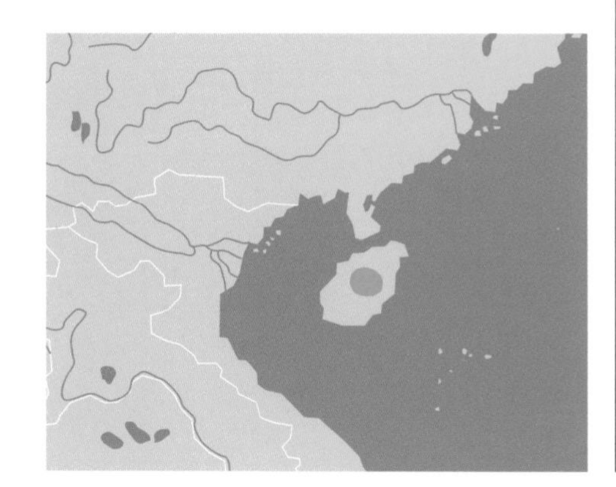

Common name

Hainan box turtle

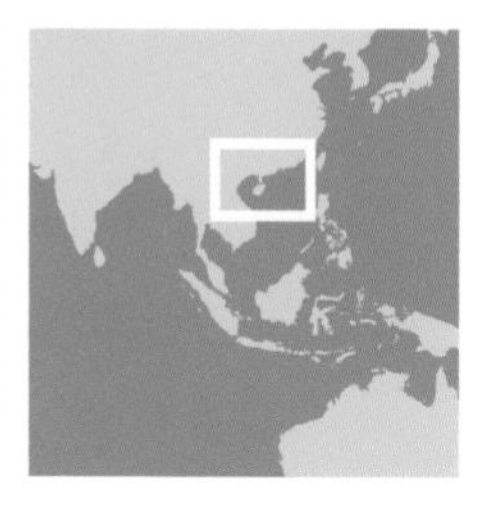

Cuora amboinensis (Daudin, 1801)

Distribution. Widely distributed in Southeast Asia, this species survives today in isolated residual areas, demonstrating a breakdown of the integrity of the overall range. It occurs in Bangladesh, the Nicobar Islands, and Assam, in the Kaziranga National Park. It also occurs in southern Myanmar, Malaysia, Sumatra, Java, and the Philippines and in Indonesia, as far east as the Moluccas.

The head is small and triangular, well marked with light lines and with a bright yellow patch from the snout to the base of the neck. This is one of the commonest and most heavily collected Asian turtle species.

© J. Maran

Description. The name of the genus is based on the Malay word *kura,* meaning "turtle." The specific name refers to the island of Ambon (or Amboina), one of the Moluccas Islands in Indonesia. The species is currently often called Asian box turtle, to distinguish it from the American box turtles; although the two groups are in different families, the large, hinged plastron is very similar in both. This species does not exceed about 250 mm in length, and it has a well-domed shell—although lower than that of the American species—with a vertebral keel in younger animals and a rounded back in more elderly individuals. The posterior marginals are somewhat widened but never serrated. The general coloration is light gray to uniform black, the keel often being light. The plastron is large and completely closes the shell openings. The hinge cuts across the middle of the plastron and allows the anterior and posterior lobes to be raised. There is no anal notch. The axillary and inguinal scutes are often lacking. The plastron is yellow to light brown, with a large black blotch on each scute, extending to the plastral border. The underside of the marginals is yellow, with a black blotch at the posterior extremity of each scute. The head is medium-sized and pointed, with a protruding snout and a slightly prognathous upper jaw. There are numerous fine serrations on the upper jaw. The head is lemon yellow, with strong black bands on each side. These designs give the turtle a characteristic "smiling" face. The limbs are grayish on the upper surface and have large horizontal scales on the front. The young are flatter than the adults and have three well-defined keels that are black on top. The head is yellow below, from the throat to the neck.

Common name

East Indian box turtle

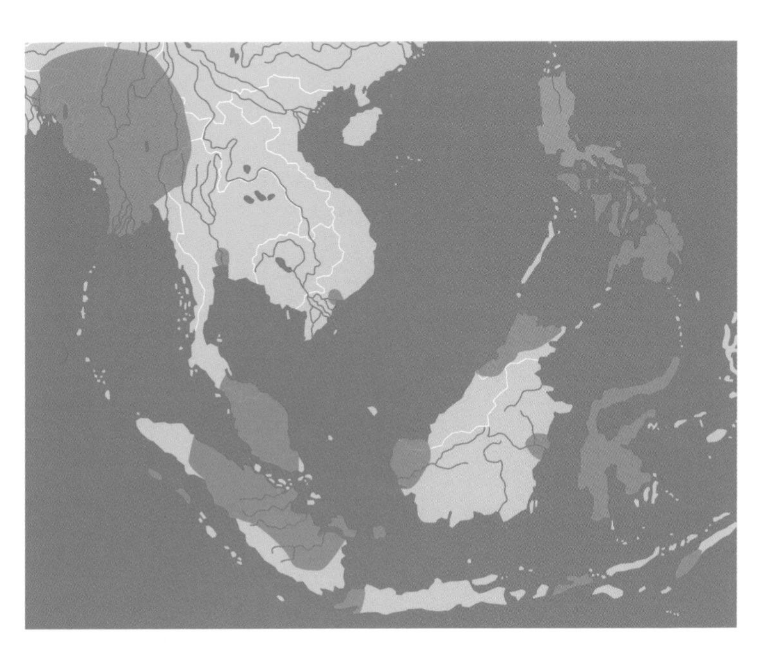

Subspecies. Four subspecies have been described.

C. a. amboinensis (described above) occupies the eastern part of the range, in the Philippines, Sulawesi, and the Moluccas. It is abundant on the island of Ambon, from which it gets its name. The carapace is rather flattened and wide and is expanded at the sides. The plastral pattern shows large dark blotches that are more spread out than those of the other subspecies.

C. a. couro (Schweigger, 1812): The Sunda box turtle occurs in Sumatra and Java. The plastron has a large dark mark in the center. The design extends along all of the plastral scutes, without being interrupted at the seams. The underside of the marginals is uniformly yellow, without dark blotches.

C. a. kamaroma Rummler and Fritz, 1991: The Asian box turtle occupies the north-central part of the range, in Thailand, Cambodia, Vietnam, and Malaysia (peninsula and eastern). The carapace is the most high-domed of the subspecies, is narrow, and has a modestly marked border.

C. a. lineata McCord and Philippen, 1998: The Burmese box turtle a recently named subspecies, has a very limited range in southern Myanmar, in the province of Kachin. The carapace is very domed and rectangular in shape. The vertebral keel is absent. The black spots on the plastron are symmetrical and sharply cut off at the seam lines. The gulars are almost entirely black. The underside of the marginals is marked by a large black blotch toward the rear of the scute.

Natural history. These aquatic turtles live in swamps and small watercourses, as well as in rice paddies, but they are capable of moving rapidly on land and can feed outside the water. Adults often spend the night on land; the juveniles are more aquatic. The turtles are not nervous around people and are often found in small water bodies or puddles near human habitation. Mostly herbivorous, they eat aquatic plants as their first choice but also enjoy mollusks and small crustaceans. On land, they eat mushrooms, earthworms, and certain plant species. The females usually stop eating about one month before nesting. The courtship is often violent, and the males bite the females very hard, sometimes inflicting serious wounds. Sometimes males attempt to mate with each other, and violent fights can result. Mating takes place from November to April, after some preliminaries in the course of which the male bobs his head in front of the female, perhaps 20 times or more. Finally, he bites her neck and holds on to her shell with his forelimbs. Three or four nestings may occur, between early April and the end of June, with two or three elongate, hard-shelled eggs per clutch, each measuring about 43 × 32 mm. Incubation lasts about one month—a remarkably fast development period for any turtle. Upon hatching, the juveniles measure about 45 mm.

Protection. This turtle is often eaten in China, where it is the most frequently consumed turtle species in restaurants. Furthermore, it is frequently captured and then released in ponds at Buddhist temples, especially in Malaysia. Finally, it has for a long time been gathered for export, and it is commonly seen in Western collections. These factors have led to progressive reduction of the overall distribution. The collection pressure is so severe that one is led to believe that a population collapse in the decades to come is more than likely.

© F. Bonin

The carapace is wide, with flattened marginal scutes and almost no median keel. The species may be recognized by the wide yellow line on the top of the head.

Cuora aurocapitata Luo and Zong, 1988

Distribution. An endemic Chinese species, this turtle is confined to a small area of central Anhui Province, in the region of Nanling, Yixian, Guangde, and Jingxian.

Description. The common name is self-explanatory: the head is bright yellow on top. The carapace does not exceed 120 mm in length and is rather flat, without posterior serration. The carapace is a uniform dark brown, and the plastron yellow and lacking an anal notch. The plastral hinge is well developed and fully mobile, and the raised lobes are able to close the shell openings completely. The head is of medium size, pointed,

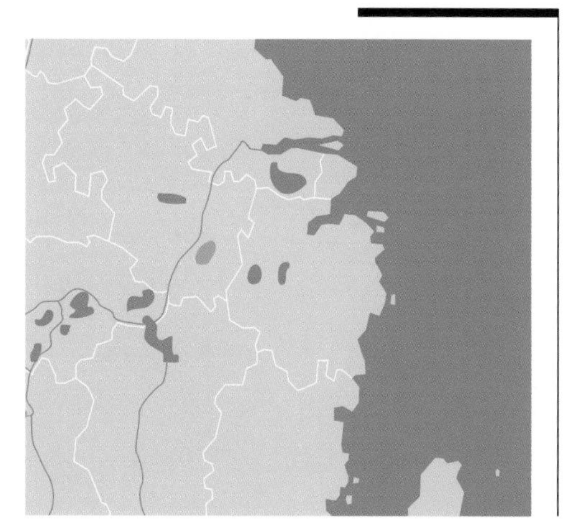

Common name

Yellow-headed box turtle

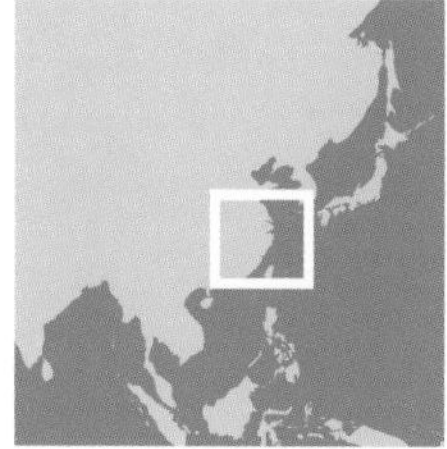

© F. Bonin

This turtle carries its name with pride: the head is golden yellow and very bright. In temple ponds and tanks, where it is often held in captivity, one may see hordes of these yellow triangular markings as they emerge from the depths of the greenish water.

black or grayish below and on the neck, with light areas on the mandibles and the sides and with a brilliant yellow shield or helmet on top. The forelimbs bear strikingly enlarged scales.

Natural history. This species lives in fast-running creeks and in marshes, in terrain with many valleys, but it may also be seen on land in moist areas. In diet it is omnivorous but seems to prefer aquatic plants.

Protection. This turtle is thought to be very rare and is considered to be "in critical danger" in China. A number of specimens have been seen recently for sale in markets, and the species is highly sought after by dealers who sell them to hobbyists. It is critically important that this turtle be added to Appendix 1 of CITES as quickly as possible and to establish conservation programs to save the last wild populations. The Turtle Conservation Fund lists this species as endangered.

Cuora mccordi Ernst, 1988

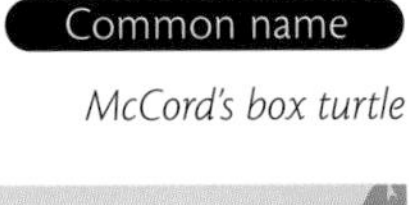

Common name

McCord's box turtle

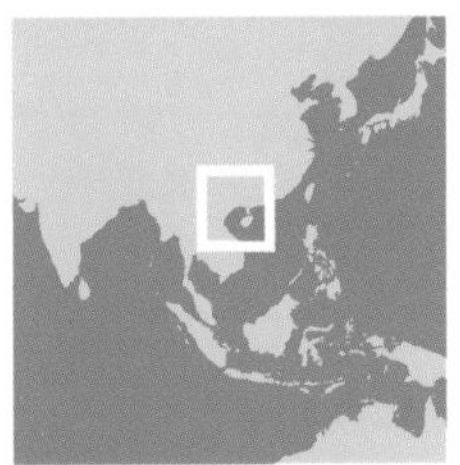

Distribution. This turtle occupies a limited range in the autonomous region of Guangxi, in southern China.

Description. The carapace length does not exceed 135 mm. The carapace is quite strongly domed, widest at the level of the eighth marginal scute, and highest between the second and third vertebral scutes. The marginals are expanded anteriorly and posteriorly and have a slight posterior serration. A modest median keel is present between the rear of the second and fourth vertebrals. The cervical scute is longer than wide. The color is reddish brown with dark seams, and there is a dark figure on each marginal. The carapace is bordered with yellow. The plastron is well developed and is yellow, with a large, black median area and also with two black markings on each bridge. The lower face of the marginal scutes is yellow. The anal notch is conspicuous. The posterior lobe completely protects the posterior part of the animal when it is raised. The head is narrow and pointed and extends anteriorly into a long snout. On the top of the head, the skin is smooth and yellow. The iris is yellow to yellow-green. A fine yellow band with a black border extends back from the snout, passing by the eyes and reaching the neck. The jaws, chin, and neck are light yellow. Brown to red-brown scales cover the upper surface of the forelimbs. The hind limbs are brown, but the rest of the soft parts are yellow. The tail is

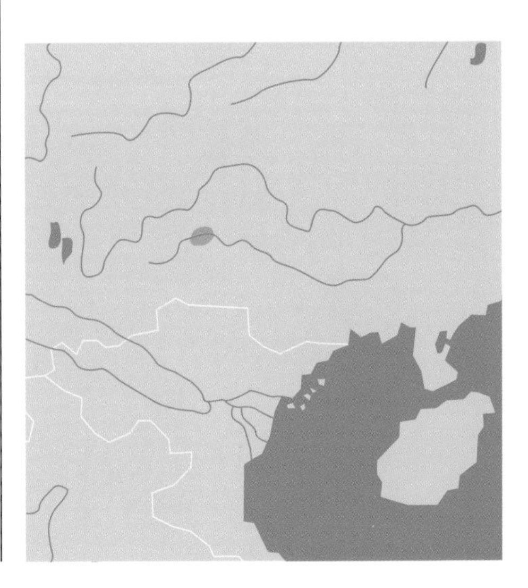

yellow and decorated with a brown and greenish band on its upper side.

Natural history. This is a semiaquatic species that walks through the shallow watercourses where it lives, under thick vegetation. It is most often seen in marshes and scattered patches of humid understory. In diet it is carnivorous and consumes earthworms, small insects, diverse plants, and flowers as well as fruit.

Protection. This species is classified as endangered by the Turtle Conservation Fund.

The carapace is flat and dark, without ornament, although the head and the underside of the limbs are very light, almost whitish. There are just a few black lines marking the snout and the top of the head.

© F. Bonin

Cuora pani Song, 1984

Distribution. This species is found only in the province of Shaanxi, in central China.

Description. This turtle can reach a length of 160 mm. The carapace is elongate and slightly domed. It is widest at the level of the eighth marginal scute. The marginals are widened, and a small anal notch is detectable. There is a median keel along the vertebral scutes, but this often disappears with age. The nuchal scute is long and narrow. The carapace is dark brown, with lighter coloration (sometimes reddish) along the keel and on the costals. The plastron is well developed, is yellow in color, and has black triangular blotches along the seams. A black bar also marks the bridge on each side, and the underside of the marginals is yellow. The posterior plastral lobe can completely conceal the retracted limbs, and the anal scutes are separated by a deep notch. The head is narrow and pointed and extends forward into a slightly raised snout. The top of the head is covered with very smooth skin. The iris is greenish yellow. The head is lemon yellow and carries a narrow brown band that extends from the posterior of the orbit to the neck. The jaws are yellow. The top and sides of the long neck are brown and covered with rugose scales. The chin and the underside of the neck are yellow. The color of the dorsal part of the limbs varies from greenish gray to brown, while the underside is orange-yellow. Wide horizontal scales cover the upper surface of each forelimb. The hind limbs are covered with small scales. The tail is yellow, with brown stripes on top.

Natural history. This turtle is mainly aquatic, living in marshy zones and slow, shallow water-

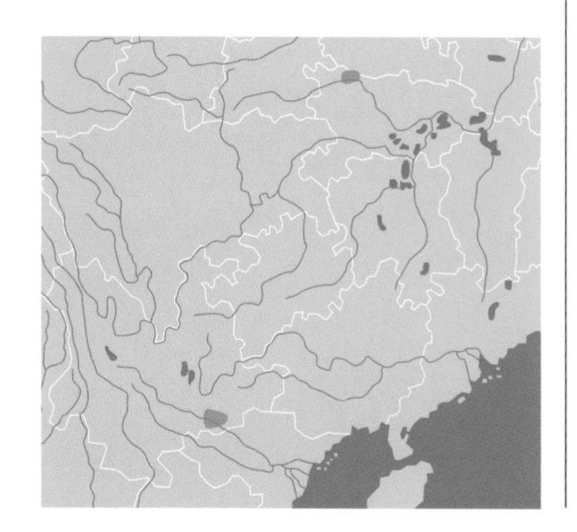

Common name

Pan's box turtle

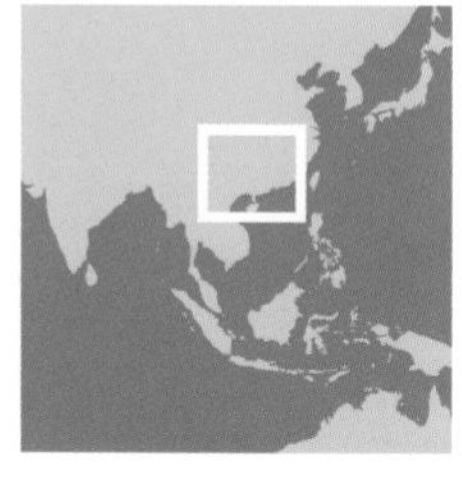

Widely spread marginal scutes and a flat, uniformly brown or greenish shell characterize this shy turtle, but the head is more vividly colored. The first vertebral scute has the form of a leaf.

© G. Cydirim

© F. Bonin

courses. It ventures onto land from time to time to walk in the wet forest. Omnivorous and opportunistic, it consumes earthworms, insects, and also fallen fruit.

Protection. The species is listed as endangered by the Turtle Conservation Fund.

Cuora trifasciata (Bell, 1825)

Distribution. This species is generally found in southern China (Guangxi, Guangdong), Hainan Island, and Hong Kong, as well as in northern Vietnam. It is sometimes seen in Laos.

Description. Usually less than 200 mm in length but sometimes up to 300 mm, this turtle has a moderately domed or flattened shell, wider behind than in front. A small notch separates the two supracaudal scutes. Adults have a long vertebral keel and two shorter lateral keels. The carapace is brown, sometimes chestnut, with black bands along each of the keels. The black band along the vertebral keel is longer than the others, which tend to fade out behind the third costals. The plastron is well developed, but the posterior lobe is not large enough to completely cover the retracted hind limbs. There is a strong anal notch. The hinge that allows the plastral lobes to be raised does not develop completely until a length of 90 to 100 mm has been reached. The plastron, a uniform dark brown or black in color, is bordered with yellow from the level of the pectoral scutes to the anals. Light streaks are often present along the seams. The lower face of the marginals is light orange to pinkish yellow or even bright yellow. It is marked with large dark spots, especially on the anterior marginals. The head is narrow and pointed, with a protruding snout and a rather flat upper jaw. The head is a very bright olive-green on the top and black at the sides. An olive, black-edged band circles around the nostrils, passes the orbits, and forms a large yellow area behind the eye. The

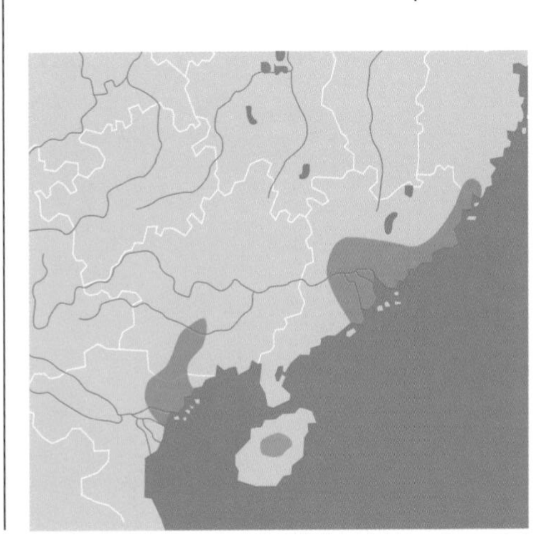

Common name

Chinese three-striped box turtle

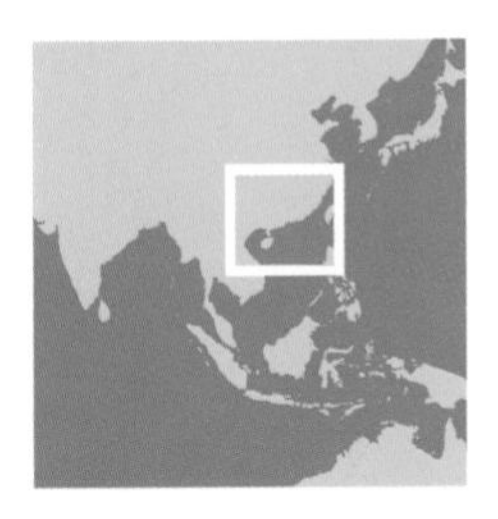

upper jaw is yellow, and a yellow band extends from the corner of the mouth, passing by the tympana and reaching the edge of the neck. The chin and lower jaw are yellow, and the upper surface of the neck, as well as the sides, are olive and covered with small scales. The throat is yellow to orange. The limbs are olive to brown on top and bright orange or pink-yellow below. Wide horizontal scales cover the front face of the forelimbs, while the front of each hind limb is covered with small scales. The end of the tail is olive, with two black bands, and the sides and the base of the neck are orange. The males, smaller than the females, have more understated colors. This is one of the most colorful members of the genus.

© F. Bonin

Natural history. This turtle leads a semiaquatic life, in small watercourses in hilly areas not exceeding about 400 m altitude in southern Guangdong. It enjoys long periods of basking. Mainly carnivorous, it eats earthworms, crabs, and small fish. The males are very active during courtship and breeding and can cause serious wounds to the females. Mating takes place on land or in the water, after a highly unusual courtship in which the male faces the female and vibrates his head and neck while holding his mouth open. These intense bouts of vibration can last for several seconds. Following penetration, the male hangs on to the shell of the female with all four limbs, his claws gripping the edges of her carapace. Nesting occurs in May, and usually just two or three eggs are laid, 50 × 26 mm in size.

© F. Bonin

Protection. The attractive appearance of this species has precipitated intense collecting for the export trade. Furthermore, in China this turtle is collected opportunistically whenever found, for medicinal purposes, and it has become quite rare in the wild. Currently, great numbers are raised in farms, especially on Hainan Island, and sold mainly for the establishment of secondary farming operations. The species is listed as endangered by the Turtle Conservation Fund.

The carapace bears three conspicuous dark keels. The head is flat and pointed, the throat is white, and stripes mark the jaws. This coloration also occurs in juveniles, which in addition have serrated marginal scutes.

Cuora zhoui Zhao, Zhou, and Ye, 1990

Distribution. This species, described by Zhao Er-Mi in 1990, has been observed in markets of Nanning and Pingxiang, in the Guangxi region of China. Later, additional specimens have been found in markets in Wuding and Yuanmou, in the province of Yunnan.

Description. The carapace measures up to 165 mm. The shell is rather flat and oval, with significantly expanded marginals, even at the sides. Some individuals have a trace of a median keel, but it is conspicuous only in juveniles. The first vertebral scute extends widely and covers the first marginal scutes. The carapace is brown-black, with irregular black areas. The plastron is also black, with a conspicuous white triangle at the center. The hinge is somewhat rudimentary, and the plastron cannot conceal the extremities of the turtle

© C. Zhou

This rare turtle is modestly colored, grayish to brownish, and has no keel. The head is dark, the only light marking being a cream-colored line on the side.

Common name

Zhou's box turtle

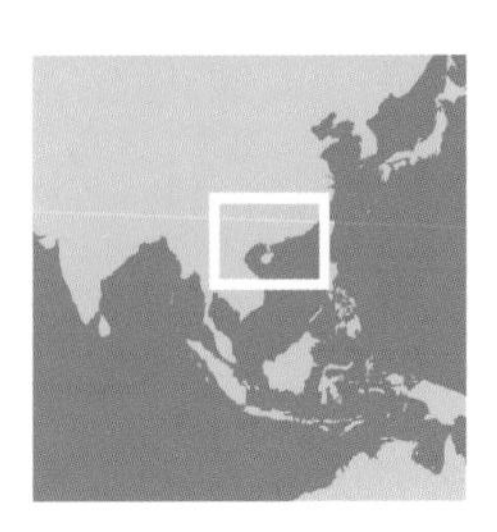

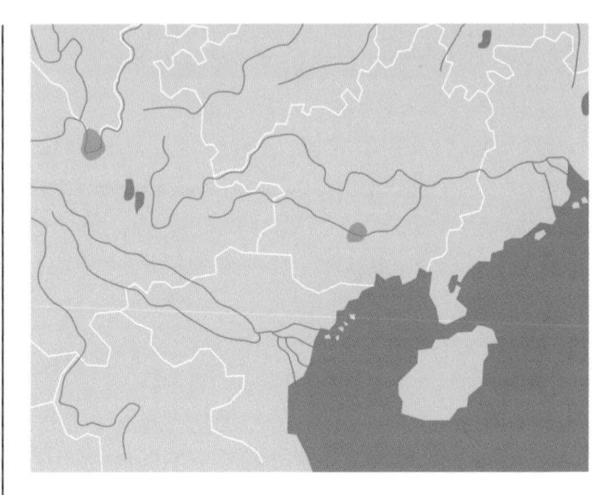

completely. The intergular seam is six times the length of the interhumeral seam. The head is rather small and has an elongated snout. The top and the sides behind the tympana are brown to olive, well separated by a yellow or cream section that starts at the nostrils, passes below the eyes, and drops down to the throat, extending along the underside of the neck. The neck is quite long and is covered with numerous fine, soft, rounded tubercles. The limbs are long and limber, with slight webbing. They are yellowish or beige, sometimes with darker edges.

Natural history. This species lives in forests near torrential streams and creeks, at medium altitudes. It does not seem to be very aquatic, preferring moist environments where it can bury itself in the substrate. It is mostly carnivorous.

Protection. The scarcity of this turtle makes its conservation challenging. Several specimens, obtained from markets, have been incorporated into live collections, while others form part of the great cohort of turtles earmarked for human consumption. Furthermore, the habitats of this species are subjected to deforestation and the construction of hydroelectric facilities. One of its discoverers, M. Zhou, has tried to breed this turtle in his facility in central Nanjing, and Elmar Meier has had some success with captive breeding in Münster (Germany). Field study and conservation programs continue to be urgently needed. Some authors have raised doubts about the validity of this species; others question whether it is possible to save a species that has become so rare. The Turtle Conservation Fund lists this species as endangered.

Cyclemys atripons Iverson and McCord, 1997

Distribution. The range of this species is limited to the hills of southeastern Thailand and the Cardamom Mountains region of southwestern Cambodia.

Description of the genus *Cyclemys*. The generic name refers to the circular form (Greek: *kyklos*) of the juveniles. These turtles have a wide range in northern India, in Nepal, and as far east as the western Philippines and southern China. The English name also describes them well: they burrow into wet leaf litter beside the creeks and marshes where they live, and their leaflike form and flat body shape complete the resemblance to dead leaves, making them very difficult to see. These turtles are quite adaptable, and they also vary greatly in color and form, according to their particular microhabitat, and this can make identification difficult. In certain species, the plastron has a somewhat mobile or kinetic anterior lobe, which can close so as to conceal the anterior extremities completely. The bridges are narrow and not totally ossified; instead, the plastron is attached to the carapace by a band of connective tissue and by plastral buttresses that are not sutured to the carapace.

Description of *C. atripons*. The identity of this species has been challenged (R. Dalton; see Fritz, Gaulke, and Lehr, 1997); it may be identical to *C. pulchristriata*. This turtle does not exceed 224 mm in length. The carapace is rather wide and flat, with three keels, of which the middle one is the most developed. The color ranges from olive-brown to black, at least in old individuals. Striated and radiating designs are visible on the juveniles, but they fade with age. The posterior edge of the carapace is strongly denticulate, especially in juveniles. The plastron has a line of flexion between the pectorals and the abdominals and is cream to yellow-brown, but the colors are never bright. Small dark radiations on the seams of the plastral scutes are frequent in the young but disappear with growth. The anal notch is very small. The bridges are darker than the plastron—hence the name *atripons* (*atri* = "black"; *pons* = "bridge"). The upper jaw of the adult, which has a pronounced horny beak, is cream to yellowish, with irregular black spots on the upper surface and wide bands on the sides, running as far as the neck. In the juveniles, these bands are separated by a fine light line that thickens with age, resulting in the presence of two prominent bands in some adults. The top of the head also darkens with age. In the young, the head is covered with little dots or black streaks. Juveniles have a cream to yellowish chin, and the skin of the soft parts may be a salmon color.

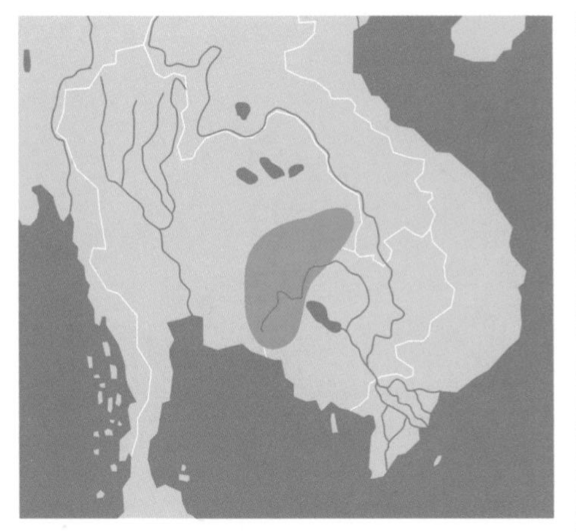

Common name

Cardamom leaf turtle

Natural history. There is no specific study of the natural history of this species, which was included within *Cyclemys dentata* until very recently.

Protection. See under *Cyclemys dentata*.

Cyclemys dentata (Gray, 1831)

Common name

Asian leaf turtle

Distribution. This turtle occurs in Thailand, Malaysia, Sumatra, and Java. It is also present in Borneo and the Philippines, on and near the islands of Palawan and Sulu.

Description. This species is distinguished from the others by the wide orange-red bands along the sides of the head, as well as the entirely yellow plastron; the interfemoral seam is short, and that between the anal scutes in long. In some individuals, there may be a few fine black rays on the plastron, but they disappear with age. The carapace does not exceed 126 mm. Oval in shape, it has a single vertebral keel. As mentioned above, the coloration is variable, but the tendency is toward dark brown or lighter olive. Certain individuals are mahogany in color; others, almost completely light beige. The scutes are very smooth. The posterior border of the carapace is saw-toothed. The skin of the back of the head is divided between two large scales, and the top of the head is a uniform brown. The surface of the forelimbs is covered with wide transverse scales, sep-

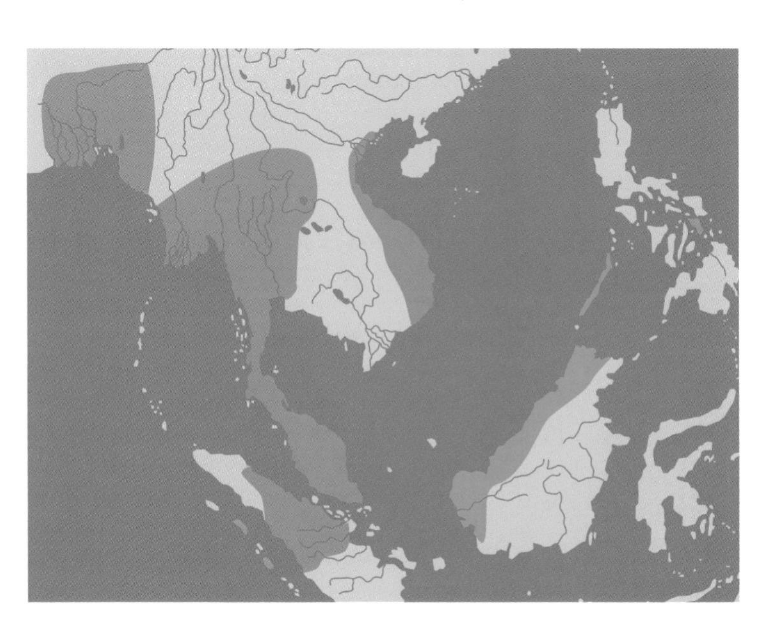

The general appearance of this turtle is one of massiveness, with a heavy carapace, well-developed marginal scutes, and dark, somber colors. The head is strong, as is the neck, and usually the skin is blackish, with sparse light dots.

arated by narrow bands of skin. The tail of the juvenile is, relatively speaking, much longer than that of the adult.

Natural history. This semiaquatic turtle occupies calm watercourses in well-vegetated hilly regions, up to 1,000 m altitude. The young seem to be more aquatic than their parents, which spend a great deal of their life out of water. This is an omnivorous, opportunistic species, which consumes aquatic plants, fruits, invertebrates, and sometimes carrion. When it is seized in the hand, it releases an abundant outflow of foul-smelling fluid. Sexual maturity is reached at about 8 years in males and 10 to 12 years in females. The courtship and coupling take place in the water. The males place their forelimbs on the bottom, then turn to face the female and carry out graceful movements of the forelimbs before the face of their partner. There may be four or five clutches per year, with three or four eggs laid each time. The eggs are quite large (57 × 33 mm). The young are also large—56 mm on average.

Protection. This species is widespread and, although exploited for food and pets, still appears reasonably abundant. It has no special protection, apart from inclusive national protection of all turtles.

Cyclemys oldhami Gray, 1863

Distribution. This species is found in northeastern India, Nepal, Myanmar, Thailand, Malaysia, Indonesia (Borneo, Sumatra, and Java), and southern China.

Description. Of moderate size (to 240 mm), this turtle has a rather flat, oblong carapace, with the anterior part raised in the middle, with a keel, and with serrations along the rear margin. The young have a well-developed middorsal keel and are almost circular in shape, again with a serrated posterior border. The coloration is quite variable, ranging from light beige, finely striated with black, to an almost uniform mahogany tint, sometimes touched with black. The plastron is quite elongate and articulates in the midsection, behind the pectoral scutes, permitting the anterior lobe to be raised to some extent, but some individuals show less plastral kinesis than others. The plastron ranges

Common name

Dark-throated leaf turtle

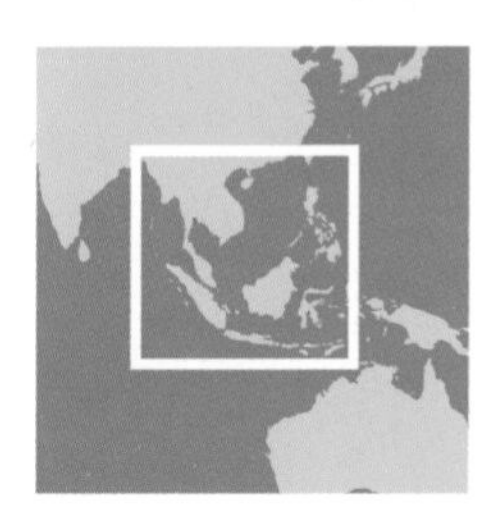

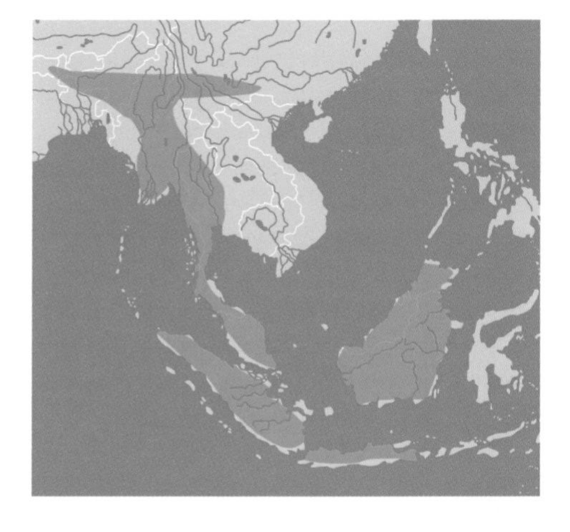

from chocolate to dark brown, with a light line along the hinge, or it may be very light yellowish, with dark concentric lines on each scute; many variations are possible. The seam between the femorals is much longer than in *C. dentata.* The anal notch is large and forms a right angle. The forelimbs each have five claws, and the hind limbs each have four; all digits are well webbed. Wide transverse scales are located on the front of the forelimbs. The head is brown to black and has no light band. The throat is dark—a defining characteristic of the species. In the young, the lower throat region has numerous small tubercles, lighter in color than the background. The young have no plastral hinge; this develops with age.

© F. Bonin

This species is sometimes confused with *C. dentata,* but it is generally more colorful, with light spots on a greenish background. The interfemoral seam is longer than in *C. dentata*.

Natural history. This turtle occurs in small, slow-flowing streams with no great depth, in hilly, well-vegetated regions. It often hides under dead leaves, not far from wet areas. In Thailand, it has been observed in altitudes up to 3,000 m. Mainly diurnal, it nonetheless may start to become active under cover of night. Omnivorous in diet and quite voracious, it eats just about anything it can find: vegetables, organic detritus, small animals. Very nervous in disposition, it pulls into its shell at the slightest disturbance and often emits a malodorous fluid. It may remain immobile in this fashion for several days. It nests four to six times per year, producing two to four elongate eggs (57 × 32 mm) each time. Incubation may take at least 80 days. The hatchlings have a dark throat and no light band on the head.

Protection. See under *Cyclemys dentata.*

Cyclemys pulchristriata

Fritz, Gaulke, and Lehr, 1997

Distribution. The range of this species is limited to a small area of Vietnam, not far from Hoi An.

Description. This turtle is the most colorful of the *Cyclemys* species, with creamy yellow stripes (never reddish) on the head and neck. The throat is a brilliant yellowish white. The light bands on the side of the head and neck are very wide, quite different from the appearance of the head and neck in *C. dentata.*

Natural history. This species lives in the upper reaches of dense highland forest areas, in small streams.

Protection. Extensively caught for food, this turtle is also sold to hobbyists. The smallest individuals are also bought by private individuals as "garden pets."

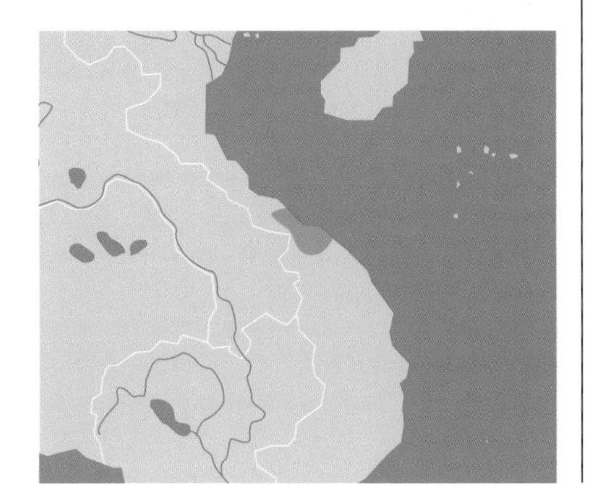

Common name

Yellow-striped leaf turtle

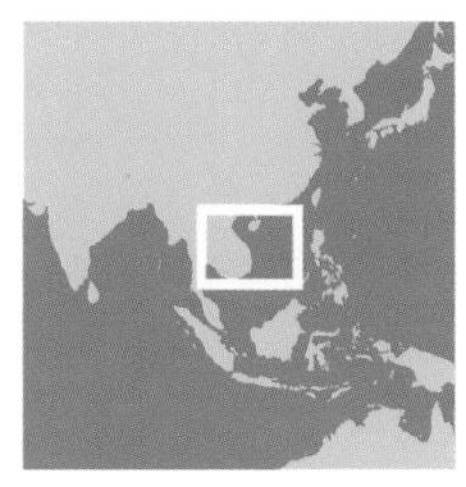

This *Cyclemys* is the most colorful of the group, with three beautiful white lines on the head and neck and with the carapace plain and unmarked.

© F. Bonin

© F. Bonin

Cyclemys tcheponensis (Bourret, 1939)

© J. Maran

The rough carapace is serrated and varies in color as the turtles age, becoming progressively more brown. Distinct dark and light lines are seen along the head and down the neck.

Natural history. This species lives in small watercourses in forest areas and hides in leaf litter. It is omnivorous and consumes worms, insects, larvae, dead animals, leaves, and fruits. It nests several times each year, with clutches of two to four elongate, hard-shelled eggs.

Protection. This is one of the four most heavily commercialized turtles in China, and the wild populations are becoming progressively more reduced.

Common name

Northern stripe-necked leaf turtle

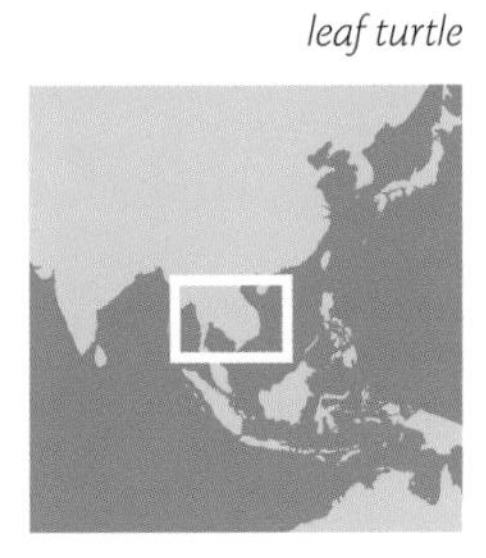

Distribution. The range of this turtle is limited to northern Thailand, near Chiang Mai; the old province of Tonkin, in northern Vietnam; and the frontier country between Laos and Vietnam.

Description. This turtle is very similar to *C. oldhami.* However, the top of the head is decorated with black dots and marbling, and the neck has light streaks.

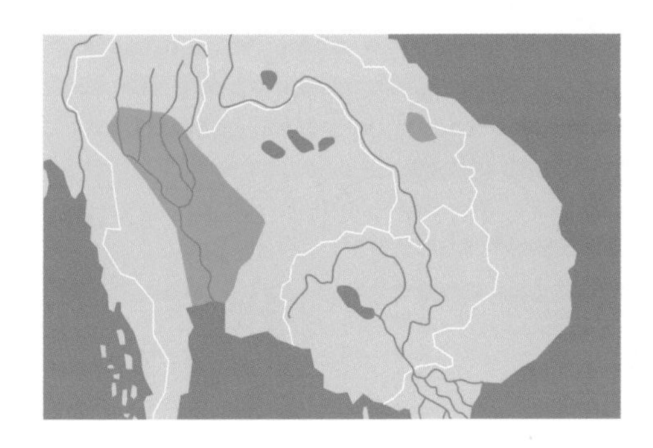

Geoemyda spengleri (Gmelin, 1789)

Distribution. This species is found in southern China (Guangdong, Guangxi, Hainan Island), in Vietnam (northern and central), and perhaps in Borneo.

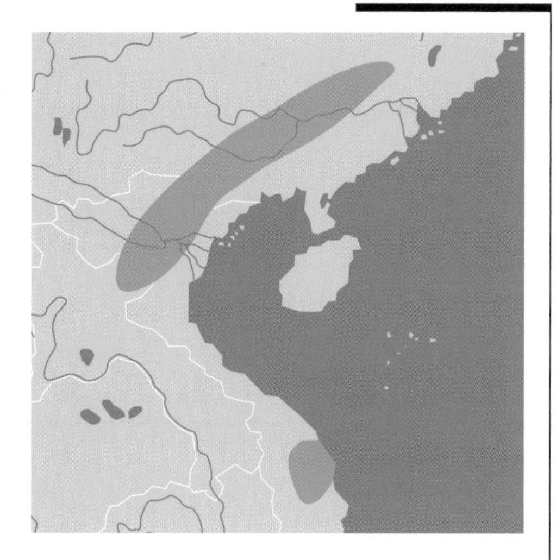

Description. This beautiful turtle is immediately recognizable. The carapace is somewhat elongated and rather flattened, with a maximum length of about 130 mm. It is widest in the posterior half of the body. It is especially remarkable for the three well-developed keels (especially the vertebral one) and the strongly saw-toothed margins of the posterior lobe. Each posterior marginal terminates in a sharp spine, pointing to the rear. The growth rings impart a rugose aspect to the carapace. The color varies from yellow-brown to gray-brown or just plain dark brown. The keels are darker than the rest of the shell, especially the vertebral one. The plastron is large and elongate and has a significant anal notch. There are no hinges or plastral rugosity. The bridges are as long as the posterior plastral lobe. The plastron is dark brown to black, with a yellow lateral border, and the bridges are entirely dark brown to black. There are no inguinal or axillary scutes. The head is flat, with the upper jaw recurved in the middle. The skin covering the top of the head is smooth and uniform. Its color is olive to brown, with a yellow band that extends from the orbits to the neck, passing over the tympana. Other yellow dots may appear on the sides of the head and on the jaws. The eye has a characteristic, striking white iris. The upper face of the forelimbs is covered with large pointed scales that overlap. Large scales may also be found on the soles of the hind feet. The base of the tail and the thighs are covered with small spurs. The hind limbs are not completely webbed. The color of the limbs and tail is a uniform brown or gray.

Natural history. This species is rarely seen in water and prefers wild, wooded mountain areas,

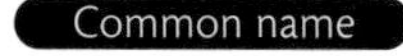

Black-breasted leaf turtle

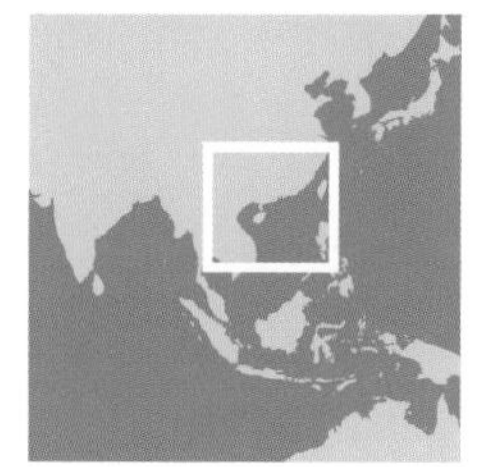

© F. Bonin

© F. Bonin

One could not confuse this species with any other. The carapace is quadrangular and well-keeled, and the marginals are strongly saw-toothed at the back. The head is small and marked with light lines on a brown to greenish background. The upper jaw shows a pronounced beak. The eyes are large.

where it is found close to small creeks and wetlands. Omnivorous in diet, it eats small invertebrates such as insects and earthworms, as well as various fruits. In captivity, newborn mice are eaten with great avidity. The clutch size is just one or two, and there may be up to three nestings in a season. The eggs measure about 43 × 18 mm, and incubation takes 66 days at 28°C.

Protection. This turtle is much sought after by hobbyists because of its attractive appearance and, perhaps, its small size. Populations are probably reduced because of this pressure. This species is classified as endangered by the Turtle Conservation Fund.

Geoemyda japonica Fan, 1931

Distribution. This species is confined to the Japanese Ryukyu Islands, on the islands of Okinawajima, Kumejima, and Tokashikijima.

© J. Maran

Description. This species has a carapace length of up to 160 mm, with a rather elongate, lightly domed shell with a flattened top. The widest point is reached behind the eighth marginal scutes. There are three well-developed longitudinal keels, the vertebral keel being the highest. The posterior border is strongly denticulated, and in young animals the anterior marginals are similarly shaped. This spiny appearance diminishes with growth. The surface of each scute is somewhat rugose, and the growth rings are very distinct. The areolae of the vertebral scutes and of costals 1 to 3 lie against the posterior border of the scute in question, along the keels. On the other scutes, the areola is located a little to the rear of the center of the scute and always on one of the keels. The coloration ranges from dark orange to dark yellow, brown, or even reddish. The keels are always darker than the rest of the shell. The plastron is large, unhinged, and somewhat elongate, with a significant anal notch. The bridge is short and is black or brown in color. The plastron has a light lateral line, yellow or yellow-gray, along each side. This species can be distinguished from *G. spengleri* by the presence of two large axillary scutes. The inguinals are small but usually present, whereas they are absent in *G. spengleri.* The head, of medium size, has a very smooth top and a well-developed upper beak, without a hook. The tongue is large, thick, and heart-shaped and is covered with numerous small excrescences. The color of the neck and head varies from orange-yellow to red-brown or sometimes even black. Irregular reddish or yellow lines, sometimes with spots of the same color, decorate the snout and the sides of the head and neck. The top of the head has a dark, irregular design on the front part. The throat is the same color as the light markings on the head, with darker spots. The exposed surface of the limbs is covered with large imbricate

© F. Bonin

The plastron of this species is completely black, with a yellow band on each side. The head is modestly colored, with a few light markings in the young that disappear with age (see photo on right). The snout is well marked, and the upper jaw is hooked.

scales. The limbs and the tail are always dark in color—red-brown or black, often ornamented with yellow or red lines and circles. In the young, the two stripes along the sides of the plastron are bright red or orange.

Natural history. This turtle lives in dense wetland forests, where the ground is covered with leaves and dead branches and twigs. Omnivorous, it eats small vegetable shoots as well as invertebrates such as earthworms and insects. Nesting occurs from June to August, and clutch size is four to six eggs, 45 × 30 mm in size.

Protection. This species is classified as endangered by the Turtle Conservation Fund. In Japan it is listed as a national treasure and is strictly protected.

Common name

Okinawa black-breasted leaf turtle

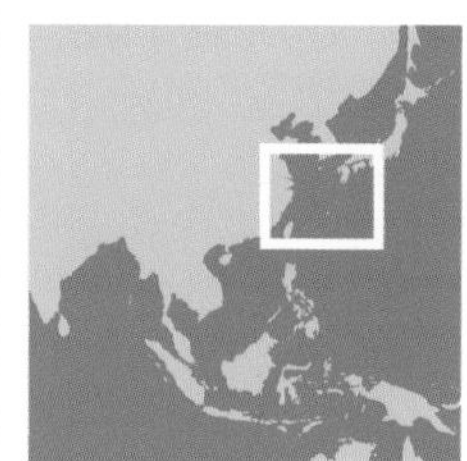

Geoemyda silvatica Henderson, 1912

Distribution. This turtle is found only in three small areas in southwestern India, in the Cochin region.

Description. This small turtle does not exceed a length of 130 mm and has an oval, moderately elevated shell with three low, indistinct keels. The median keel is the highest, extending along all five vertebral scutes, whereas the lateral keels are present only on the first three costals. All the vertebral scutes are wider than long, and the posterior border of the shell is lightly denticulate. The color is a uniform bronze, orange, or brown, with black areas. The plastron is rather large and has an anal notch. The bridge is quite long, and the axillary and inguinal scutes are usually present, although never large. The plastron is uniformly yellow or orange, with a black marking on each pectoral and abdominal scute. The head is small and is yellow in front, as are the jaws. An orange dot marks the tip of the snout. The iris of the eye is reddish, and red and pink pigment may also be seen on the upper eyelid. This coloration may extend to the top of the head, which is otherwise brown to black. The limbs and tail are light brown. Large scales cover the face of the forelimbs. The hind limbs are curved. The digits are unwebbed, but the claws are well developed. The males possess spurs on the rear of the hind limbs, and the head takes on a bright red color during the breeding season. The females are less colorful and slightly larger than the males. The plastron is somewhat mobile at the level of the junction between the plastron and the bridges.

Natural history. This species lives in dark dense forests below 300 m altitude. It uses burrows and hides itself in dead leaves and decaying tree trunks, as well as in rocky crevices. Crepuscular in nature, it is difficult to observe and seems to be rare within its habitat. It is omnivorous, eating fallen fruit, leaves, and grasses, as well as invertebrates such as millipedes and mollusks. Mating takes place from October to November.

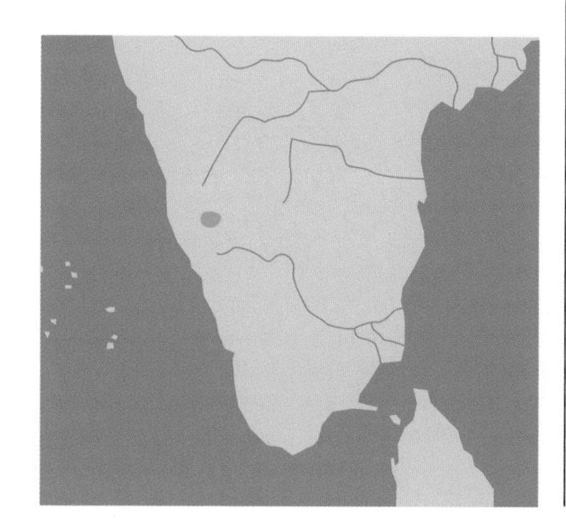

Common name

Cochin cane forest turtle

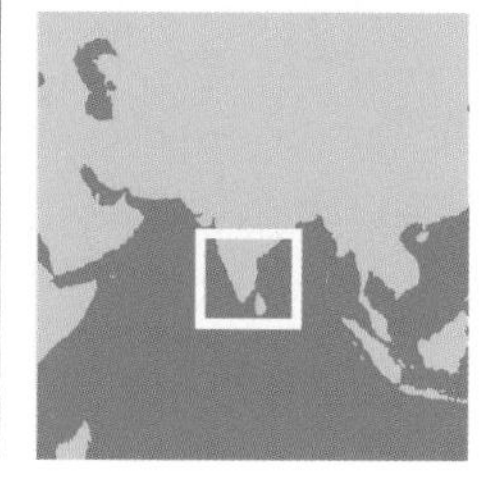

Protection. Only about 100 specimens of this turtle have ever been seen, and its status appears to be both reduced and rare. It needs to be studied and protected immediately. At press time, this turtle had been elevated to generic status (the monotypic *Vijayachelys*), named after J. Vijaya, an Indian turtle student who conducted field studies on the species in the 1980s, before her tragic death at the age of 28. The species is listed as endangered by the Turtle Conservation Fund.

Heosemys spinosa (Gray, 1831)

Distribution. This species is found in Southeast Asia, in southern Myanmar, Thailand, the Malay Peninsula, and Singapore and as far east as Sumatra and Borneo in Indonesia and Sulu and Mindanao in the Philippines.

Description. This unique turtle is well described by both its vernacular and scientific names. The marginal scutes are so strongly denticulate that they form a series of long spines. The carapace is wide and almost circular and is very flattened dorsally; it reaches a maximum length of about 220 mm. The vertebral keel is very wide and flattened, and the light color contrasts with that of the carapace as a whole. On the posterior edge of each vertebral, an outgrowth is present that contributes part of the keel. On each costal scute, a small spine marks the rear of the central areola. More strikingly, each marginal forms a sharp, angled spike that points in an outward, radial direction, giving the entire animal a spiny appearance, most evident in young specimens. As the turtle grows, the spines become less obvious, and fully mature individuals lose the distinctive appearance altogether. The carapace ranges in color from dark mahogany to a sort of "toasted" beige. The plastron is well developed, with wide bridges. The anal notch is well developed. Both axillary and inguinal scutes are present. The marginal undersides and the plastron are yellow, with a striking pattern of thin radiating black lines on each plastral scute. The head is rather small and has a median notch in the upper jaw, flanked by two cusps. The color of the head is reddish gray to brown, with a yellow area near the tympanum. The limbs are dark gray, with a number of yellow and red spots. The forelimbs have large, overlapping scales on the exposed face, whereas the hind limbs are covered with large, spiny scales. The thighs and the base of the tail also have spiny scales. The digits show little webbing.

The extraordinarily spiny shell of this species, with spines extending from the vertebrals to the marginals, probably serves to discourage predators, especially those that swallow their prey whole. The head is small, triangular, and modestly colored.

© F. Bonin

Natural history. These turtles live in very humid, warm environments, near shallow creeks with clear water, and usually in well-wooded, mountainous habitat. They frequently hide beneath the leaf litter on the forest floor. They are most active late in the day and in the early morning, so as to avoid the intense midday heat. They can best be regarded as semiaquatic; they swim poorly and have a rather jerky gait on land. Their diet is primarily herbivorous, and they can feed both on land and in the water. They consume plants, vegetative debris, fallen fruit, and also, from time to time, insects, carrion, and earthworms. The females have some degree of mobility of the hind plastral lobe, which facilitates the passage of the very large, white, hard-shelled eggs (about 60 ×

32 mm). Often only one egg is laid, and never more than three.

Protection. This species is listed as threatened on the IUCN Red List. It is consumed by certain ethnic groups in West Malaysia. But above all, it is collected for sale to Western hobbyists, who prize it for its unique appearance. It also suffers from deforestation and habitat destruction, and populations are dropping.

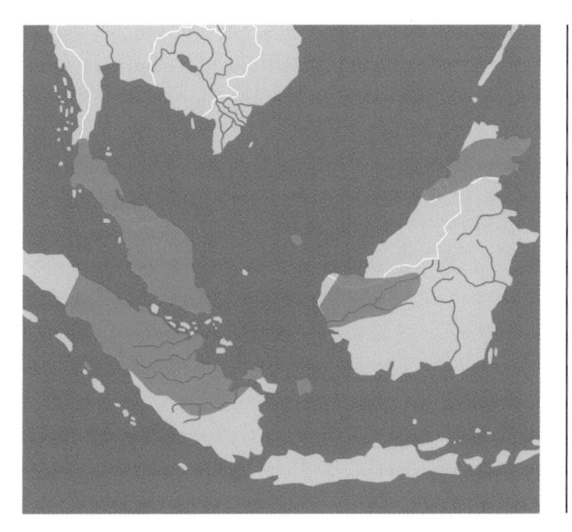

Common name

Spiny turtle

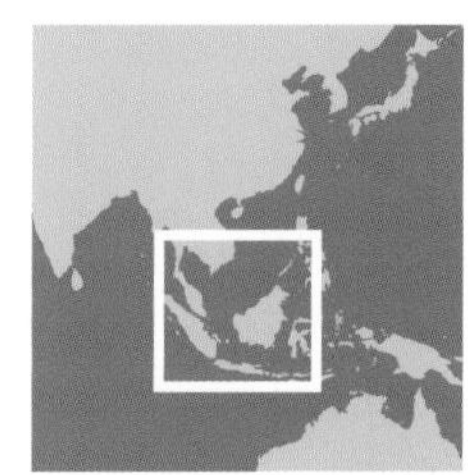

Heosemys depressa (Anderson, 1875)

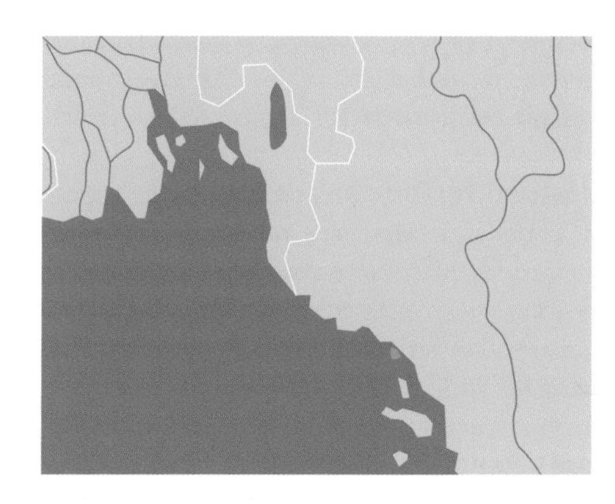

Common name

Arakan forest turtle

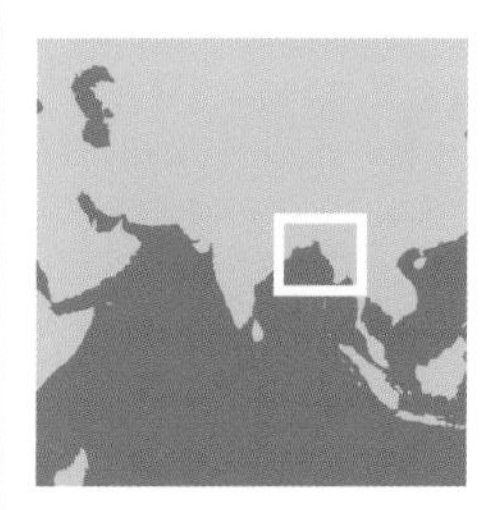

The carapace is very dark and rectangular, with a well-marked median keel bordered by two longitudinal depressions. The coloration is uniformly grayish to blackish.

Distribution. This species is endemic to western Myanmar, in the hill forests of the Arakan region, in Rakhine State, and possibly also in Kayah State.

Description. This turtle can reach a length of 250 m, and the carapace, as the name suggests, is depressed in the middle. There is a narrow dorsal keel on the first vertebrals, fading out in the middle but enlarged posteriorly. The posterior margin of the shell is strongly serrated. The overall color is light brown, but some individuals have black markings. There is an anal notch, and the posterior lobe is narrower than the length of the bridge. The plastron is yellow to brown, sometimes with dark brown or black streaks on the scutes. The bridge is uniformly dark or even black, and both inguinal and axillary scutes are present. The head is rather small and is brown or grayish, and the nostrils are slightly elevated. Large scales are present on the top of the head, toward the back. The iris of the eye is brown. The neck, limbs, and tail are beige to brown. The forelimbs have wide, pointed scales, sometimes square in shape, on the exposed face. They are partially webbed, and the claws are powerful. The posterior limbs are more flattened; are covered with small scales, except in the areas that contact the ground; and are similarly equipped with large claws and modest webbing. The males have very large hind limbs and a wide, thick tail; the tail and hind limbs are smaller in the females.

Natural history. This is a burrowing species that hides in wet piles of dead leaves, under

vegetation. It is omnivorous in diet but shows no preference for vegetable matter, fallen fruit, gastropods, or other invertebrates.

Protection. This species is consumed on a local basis, and its plastra are exported to Taiwan, where they play a role in traditional medicine.

Heosemys grandis (Gray, 1860)

© B. Devaux

This species has a wide head, marbled with black on a cream background.

Distribution. The range is somewhat fragmented but includes southern Myanmar, Thailand, southern Cambodia, southern Vietnam, and peninsular Malaysia. It may also be found in southern Laos.

Description. This is one of the largest semi-aquatic turtles of Asia, reaching a weight of 12 kg and a length of 480 mm. The shell is oval, rather wide, and massive. The carapace is moderately elevated but is flattened toward the rear and has a well-developed, wide median keel, pale yellow in color and thus lighter than the rest of the carapace. The hind marginals are moderately serrated. The carapace is brown to blackish. The plastron is large and has an anal notch. Both the anterior and posterior plastral lobes are narrower than the midsection. The bridges are well developed and are longer than the posterior lobe. Both inguinal and axillary scutes are present. The plastron is yellow, except in the young. The upper jaw has a shallow median notch and is bordered by two lateral cusps. On the top of the head, the skin is divided into irregular scales. The head is dark brown to orange, often with a vermiculate aspect that imparts a very decorative effect. This patterning is best seen in the young, for it fades with age. The jaws are wide and powerful and often are light in color. The forelimbs are very developed and have large scales on their anterior face. All the limbs are reasonably webbed. The plastron of the juveniles has a yellow background color with an attractive design formed by radiating dark lines around the areolae of each plastral scute. The growth rings are usually quite distinct.

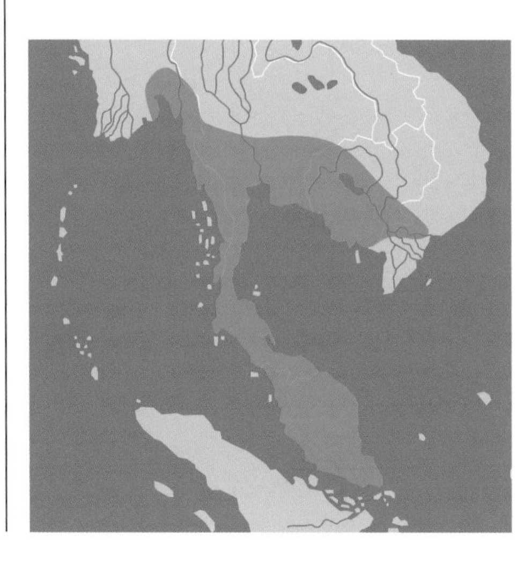

Natural history. These turtles live in or close to water, including rivers, swamps, lakes, creeks, and ponds, from sea level up into the mountains. But they spend much of their time on land, hidden under vegetation, and they are sometimes seen crossing roads not far from watercourses. This species is thus equally at home in water or on land, and it is a true omnivore, consuming aquatic plants as well as dead animals, insects, amphibian larvae, earthworms, and snails. The clutch size ranges from four to eight elliptical eggs, measuring about 65 × 35 mm but quite variable in size. Incubation lasts for 80 to 100 days.

Protection. The status of this species is poorly known, but this turtle is often caught and consumed, and its numbers seem to be dropping. In China it is currently imported extensively, its large size making it a desirable food item. In other countries it is captured and then placed in temple ponds. Between 1994 and 1999, great numbers were exported from Vietnam for sale to hobbyists.

Common name

Giant Asian pond turtle

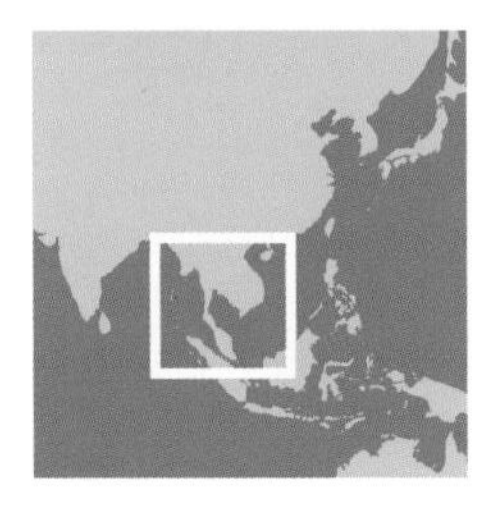

© B. Devaux

This animal seems quite bold and extends its neck frequently, as if with curiosity. The jaws are wide and strong, and the eyes expressive.

Siebenrockiella leytensis
Taylor, 1920

Distribution. This species was formerly considered to be confined to the island of Leyte in the Philippines, but that conclusion now appears to be an error. The actual distribution is on the northern part of the Philippine island of Palawan, notably in the municipality of Taytay.

Description. The scientific name refers to the island of Leyte, the reputed type locality for the species. No specimens of this species were found between 1920 and 1988, when a single specimen was found. Others were reported in 2004. The typical adult length is 210 mm, and the maximum 270 mm. The shell is oval and rather flat, with a slight median ridge sometimes present. The nuchal scute is triangular, tapering anteriorly. The first marginal scute on each side is large and extends forward beyond the nuchal scute. The anterior marginals present somewhat of a saw-toothed margin in younger specimens, and the posterior border is somewhat sinuous and feebly denticulate. The carapace is brown or red-brown, with no markings. The gulars may be large and somewhat divergent, and there is a deep anal notch. The bridge is wider than the length of the posterior plastral lobe, and both inguinal and axillary scutes are present. The color of the plastron and bridges is a uniform yellow-brown or reddish brown. The head is relatively large, the top being concave in profile and the nasal area swollen. The eyes are small, and the iris is white or pale greenish. The skin of the head is divided into scales on the dorsum and sides. The head color is brown, with several small black markings on the temples. A narrow yellow band extends around the head just behind the tympana. It passes below the tympana and rejoins on the neck. A small yellow blotch may be present on each side near the angle of the jaw.

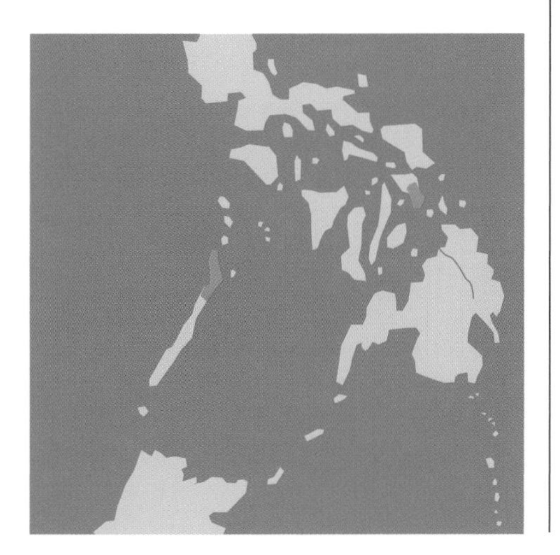

© P. Fidenci

This is a small species, with an elongate snout and several light lines on the snout and neck.

Common name

Philippine pond turtle

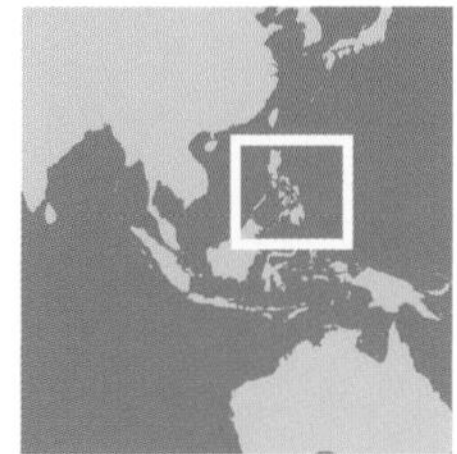

The neck is dark brown above and beige on the sides and below. The limbs are brown, very dark above and lighter below. The skin of the limbs, neck, and rest of the body has numerous small tubercles, which give the animal a rugose aspect. The forelimb has four large transverse irregular scales on the exposed face. There is only a single enlarged transverse scale on each hind limb. The feet are all well webbed and armed with powerful claws.

Natural history. The natural history data for this recently rediscovered turtle are not yet available.

Protection. This species is classified as endangered by the Turtle Conservation Fund.

© P. Fidenci

The carapace is oval, strong, and devoid of ornamentation. The upper jaw forms a strong beak.

Leucocephalon yuwonoi
(McCord, Iverson, and Boeadi, 1995)

Distribution. This species is endemic to Sulawesi (Celebes), where it is confined to the northern and central part of the island near Poso and Gorontalo.

Description. This turtle carries the name of Frank Yuwono of Jakarta, who found the type specimens. It is recently described and was considered close to *Geoemyda,* but it differs in being much larger in size (to 300 mm) and is considered by J. Iverson to be closer to *Notochelys.* The shell is flat and wide, with three keels, and is orange-brown to burnt sienna in color. The plastron lacks hinges and radiating patterns, and the bridges are well developed. The gulars are short, and the carapace is serrated behind. The head is rather large and, in males, has a strongly hooked beak. The coloration of the head varies according to sex and season; in females, the top and sides of the head, as well as the upper jaw, are blackish brown, with a creamy white zone extending from the eye to the rear of the tympanum. The males have a larger, more powerful head, cream to whitish in color, of which the top and sometimes the sides are marked with scattered blackish brown spots. The limbs have wide, thick scales on the upper sides and small scales on the lower edges and are brown, with various lighter (cream) markings. Small scales are also present on the tail.

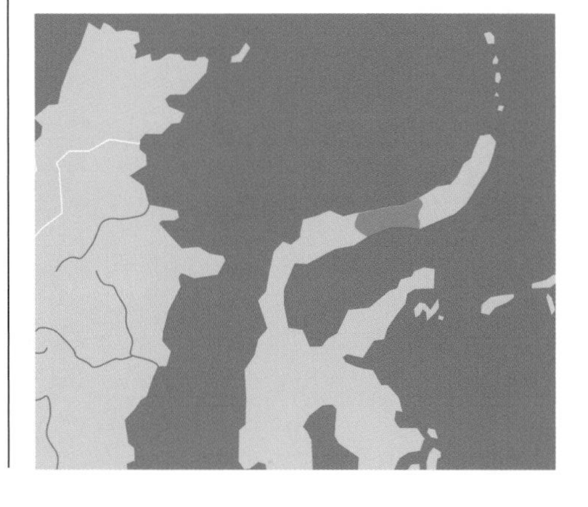

Natural history. This semiaquatic turtle lives in rocky, heavily vegetated habitats, close to watercourses and swamps, where the thick forest has a dense leafy substrate. It is primarily herbivorous and consumes leaves and fruits. Gentle in disposition, it flees rapidly when disturbed and can easily scale rocky precipices. A single egg is laid at a time, or very rarely two. The eggs are hard-shelled and well calcified; they average 65 × 45 mm.

Common name

Sulawesi forest turtle

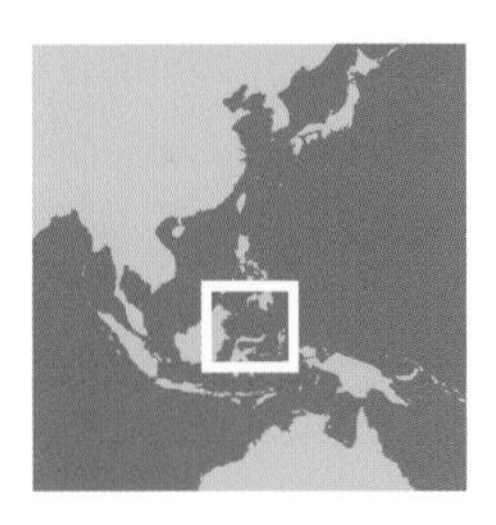

Protection. This turtle is one of the food resources for local fishermen, but it has also been collected for the wildlife trade in recent years. It is sometimes sold in markets, but the quantities collected from the wild are still unknown. Good densities might exist in a very restricted area of Sulawesi, but in view of the endemism of the species, the heavy collecting that has occurred, and the very low reproductive potential, it seems urgent to gain a more accurate picture the status of populations than is known today (S. Platt et el., 2001). Furthermore, the deforestation of Sulawesi has been particularly intense, and this turtle's habitats have already heavily degraded by humans. This species is classified as endangered by the Turtle Conservation Fund.

This turtle is rectangular and flat, and the head is adorned with wide white or yellowish spots on a blackish background.

© J. Maran

Mauremys annamensis (Siebenrock, 1903)

Distribution. This species is endemic to the central part of Vietnam (ancient Annam).

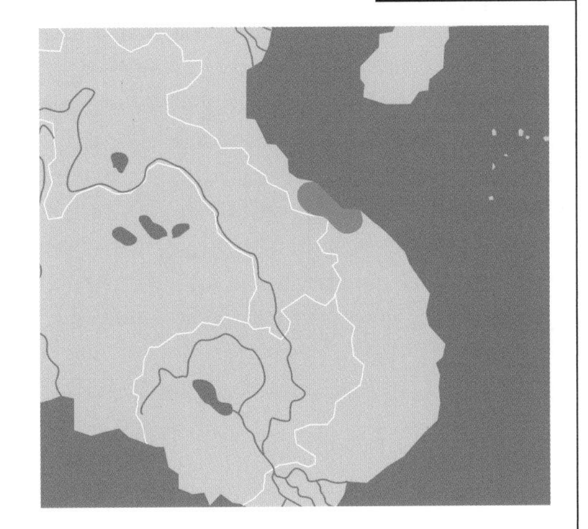

Common name

Annam leaf turtle

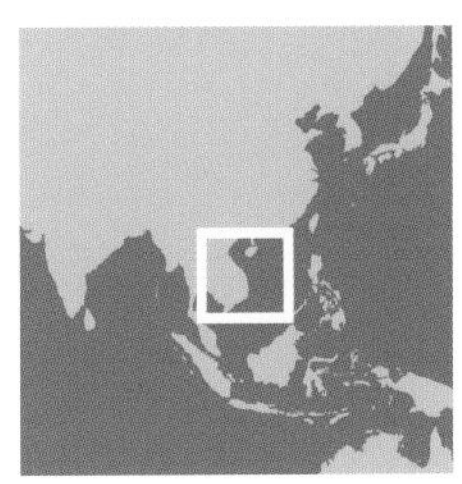

Description of the genus *Mauremys.* These are small aquatic turtles, 180 mm on average and 250 mm maximum, that occupy an enormous range in West Africa, southern Europe, and the middle East and as far east as Japan. Sometimes very dense populations may be encountered. In many cases these turtles are plain grayish in color, with little ornamentation, but along with *Emys* they often have yellow dots on the skin and carapace. Furthermore, in Morocco there are populations that have a remarkable, brilliant blue iris. In contrast to the characteristics of *Emys,* in *Mauremys* the plastron is unhinged, and the interfemoral seam is longer than the interanal seam. The nuchal is wider than long. The carapace is somewhat flattened, and the juveniles are almost circular, with three light dorsal keels, whereas the young of *Emys* lack the median keel and are more oval in shape.

Description of *M. annamensis.* Formerly known as *Annamemys annamensis,* this turtle is usually observed in vegetative mats at the bottom of lakes and swamps, which give it the name "leaf turtle" in English. Its length does not exceed 170 mm. The carapace is uniform brown in color, with three reasonably developed keels, outlined in black to some degree. The vertebral keel is more developed than the others. The body is oval and elongate, wider posteriorly than anteriorly. The head has a prominent snout, bluish black and decorated with wide yellow bands: the first band, V-shaped, underscores the front of the head and continues behind the temples to the base of the neck; the second band, wider and brighter, extends from the nostrils to the base of the neck, tra-

© F. Bonin

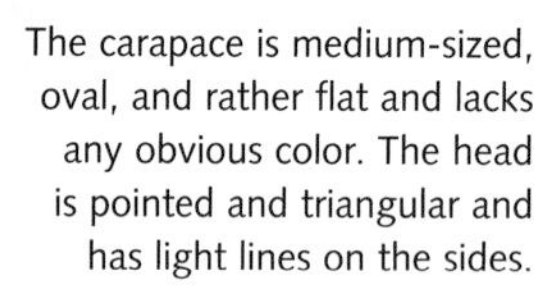

The carapace is medium-sized, oval, and rather flat and lacks any obvious color. The head is pointed and triangular and has light lines on the sides.

versing the eye and the top of the tympanum; the third starts at the middle of the upper jaw, passes below the eye, and drops down to the neck. The underside of the head and neck is yellow. The eyes are yellow, bisected with a horizontal black line. The plastron is well developed, with a strong anal notch and a truncated anterior lobe. The plastron is orange-yellow, with a large black spot on each scute, forming two lateral dark streaks running the length of the plastron. The black bands are also prominent on the bridges, which have both axillary and inguinal scutes, and on the underside of the marginal scutes. The digits are fully webbed and have five digits in front and four behind.

Natural history. This species lives in swamps, lakes, oxbow lakes, and marshy areas with thick vegetation. It hides itself below the substrate and is thus difficult to see. In captivity it is omnivorous, with herbivorous preferences. Reproduction in the wild has not been studied, but captive breeding is now undertaken at the Cuc Phuong Turtle Conservation Center (Vietnam) as well as in the USA.

Protection. The range of this species is poorly known, and most of lowland Vietnam is now committed to agriculture. Formerly, significant numbers of these turtles were caught for Chinese markets, and today there appear to be no wild stocks in existence. These animals used to be found in Vietnamese markets from time to time, but no more. The species does well in captivity, and good numbers of hatchlings have now been produced, so it is possible to envisage a restocking program if safe habitat can be identified. This species is listed by the Turtle Conservation Fund as endangered.

Mauremys caspica (Gmelin, 1774)

Distribution. The original distribution of *M. caspica* shrunk considerably when the western form, *M. rivulata,* was recognized as a separate species. *M. caspica* retains the eastern part of the range, in Turkey and Azerbaijan and passing through Iraq, Iran, Saudi Arabia, Georgia, and Turkmenistan.

Description. This is the larger of the two species formerly recognized as *M. caspica* and reaches up to 240 mm. The shell is oval, flat, and reddish brown to blackish. There is a slight median keel, conspicuous in the young and still present in some adults. The posterior marginals are very smooth and somewhat elevated in the caudal region. The plastron is well developed, with a strong anal notch, and is yellow with reddish or brown markings. The head is of medium size and olive to dark gray, with fine yellow or cream lines on the sides. Some of these lines end at the tip of the snout, after passing above the eye, and others extend from behind the eye and reach the base of the neck. Additional lines mark the mouth and the sides of the neck but fade with age. The neck, limbs, and tail are equally blackish and striated with fine yellow or cream stripes. The young are more colorful and have three keels, but the two lateral ones do not persist with growth.

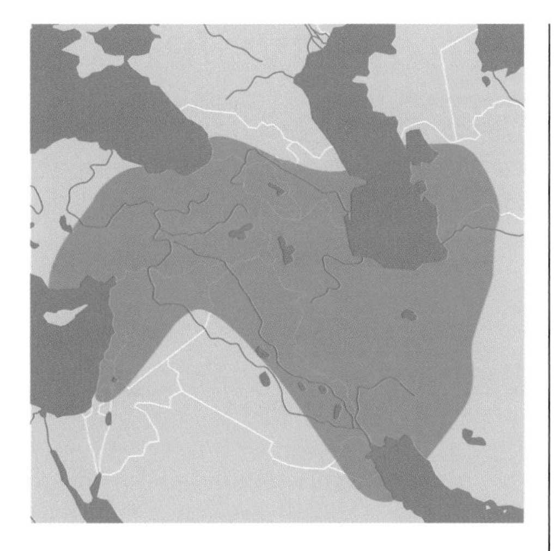

Caspian terrapin

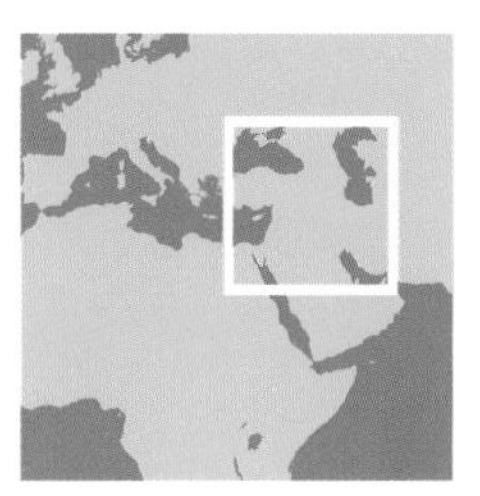

Subspecies. Currently, three subspecies are recognized. *M. c. caspica* (described above) occupies the entire northern and western part of the range. *M. c. siebenrocki* (Wischuf and Wischuf, 1997) occurs in the southern component of the range (i.e., Iraq). The plastron has large, symmetrical black markings. The overall coloration is the lightest of the three subspecies. *M. c. ventrimaculata* (Fritz and Wischuf, 1996) has a small range in southern Iran, in the Kor and Maharloo Rivers. The plastron is yellow, with irregular black markings.

Natural history. This is a rather abundant turtle in rivers and coastal swamps. During the rainy season it may be very active; during dry times, it estivates under a layer of hardened mud. It spends much time basking on land but is always wary and dives at the least disturbance. Mostly carnivorous, it feeds on aquatic insects, fish fry, mollusks, and amphibians. The males are aggressive and quite territorial, and this trait can lead to violent battles and quite serious injuries. The females often lay twice per season, not far from the water. Clutches consist of 3 to 15 eggs, which are white and hard-shelled and measure about 37 × 25 mm. The hatchlings, about 33 cm in length, emerge following heavy rains at the beginning of autumn.

Protection. This turtle suffers from habitat degradation and especially from the annual summer droughts and the destruction of wetlands. Sometimes eaten where it occurs and sometimes sold, it is also stressed by collecting pressure. Nevertheless, it remains abundant in certain areas and is less affected by overcollection than *M. rivulata*.

© A. Dupré

The general form is oval, with marginals somewhat expanded both anteriorly and posteriorly, and sometimes a light peripheral line is present. The head is barely marked with a few faint lines. The animal as a whole is a generally uniform brown to grayish.

Mauremys leprosa (Schweigger, 1812)

Distribution. This species occurs in the western part of the Mediterranean region, in Spain, Portugal, the south of France (Banyuls), and the northwestern part of Africa (Morocco, Algeria, Tunisia, and Libya). Records from Niger and Mauritania are not confirmed.

Description. The first specimens examined by Schweigger, which happened to be elderly ones, had infected carapacial seams and looked as if they had some sort of leprosy between the scutes; hence the name he chose to give it. In truth, even if these turtles are partially covered with algae and live in somewhat turbid or even polluted waters, they are not in the slightest bit "leprous," and one wishes that this negative name could be changed. This is one of the largest *Mauremys* species, reaching 250 mm in females. The carapace is oval and rather flat. The median keel usually disappears in adults, and the animal may even have a shallow depression along the midline. The marginals are very smooth, without any serration. The color ranges from red-brown to olive, with, at least in the young, fine orange and yellow markings on each scute. A great variation in the coloration of this species is found, correlating with locality and geography and especially manifested in isolated populations. The plastron is yellowish, with a large dark central figure, sometimes accompanied by a central yellow stripe. The unique colors are best observed in young specimens in captivity. In the wild, the animals become monochrome and rather plain. The head of *M. leprosa* is similar to that of some of the American turtles. The brown or blue-gray background is striped with vivid yellow bands that extend from the neck to the tympanum, some of them extending to the rear of the snout. But these stripes fade out almost completely in some individuals. In well-marked specimens, there is an incomplete yellow circle around the tympanum, and sometimes an orange or yellow spot is present between the tympanum and the eye. Most typically, the animal becomes a uniform brown or possibly greenish, except on the neck and the limbs, which remain arrayed with lighter bands of cream or yellow. The young are almost circular and measure about 32 mm in diameter. They have three keels, very bright coloration, and a yellow band around the rim of the carapace.

© A. Hell-Kevorkian

When young, these turtles are brightly colored and recall some of the American species. The yellow lines on the neck fade with age, and the carapace becomes uniformly brown to grayish.

Subspecies. Recent revisions describe seven new subspecies, but some of them are contested. The morphological differences are minimal, and it is possible that these groups are simply vicariant forms, identified only by where they live. All of them are found in Morocco, except for the nominate widespread form, *M. l. leprosa* (described above). They include the following.

M. l. atlantica (Schleich, 1996), from the wadis of Loukos and Makhazan, has a completely dark plastron and dark bars on the carapace, including a narrow dark border, and a tympanic spot.

M. l. erhardi (Schleich, 1996), from near Melilla, has a light plastron, dark bars on the carapace, no dark border, and a tympanic marking.

M. l. marokkensis (Schleich, 1996), from the plain of Hazouz to the region of Agadir, has a light plastron, no dark edge to the carapace, and a tympanic blotch.

M. l. saharica (Schleich, 1996), from the Noun wadi near Guelmin, has a light plastron, hieroglyphics on the carapace, no dark border, and no tympanic blotch.

M. l. vanmeerhaeghei (Bour and Maran, 1998), from the Draa wadi, is remarkable for the blue eyes shown by certain populations, as well as large black symmetrical figures on the plastron, well-developed stripes on the neck, a rather colorful carapace with light spots on a mahogany background, and a tympanic spot often present but sometimes not.

M. l. wernerkaestlei (Schleich, 1996), from the wadis of Serou and Oum er Rbia, has a light plastron, dark stripes on the carapace, no dark carapacial borders, and a tympanic blotch.

M. l. zizi (Schleich, 1996), from the Zizi wadi,

Common name

Spanish terrapin

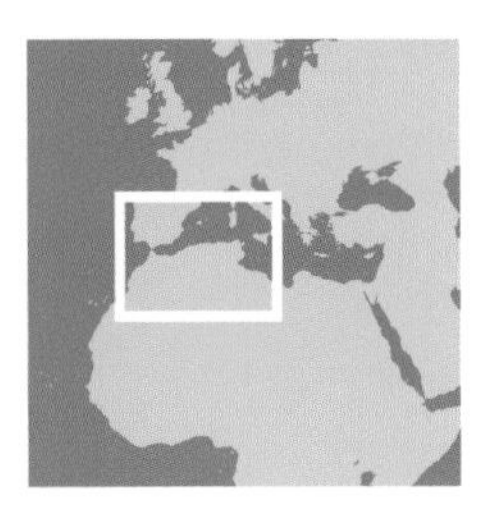

© B. Devaux

This very old turtle has a very flat, smooth, brown shell. Only the neck and head preserve some light spots and streaks on the greenish to brown background.

has a light plastron, hieroglyphics on the carapace, no dark border, and no tympanic spot.

Natural history. This is a cryptic turtle, often hiding in muddy lake bottoms or accumulations of vegetative debris. It lives in stagnant waters, lakes, ponds, and even brackish waters. In the southern part of the range, it occurs in oases and wadis that may carry water only on a seasonal basis, and sometimes how the turtles crossed the desert sands to get to these locations is far from clear. This is a basking turtle, often choosing an emergent branch or the sandy bank itself, but it plunges back into the water at the slightest disturbance. It is very tolerant of water pollution, and in situations where waters housing both *Emys* and *Mauremys* have become disturbed or polluted, *Emys* will succumb and disappear, while *Mauremys* will survive and maintain a population (R. Mascort). In the northern part of the range the turtles may hibernate for several months, and in their southern range (from central Morocco) they may estivate in mats of dry leaves or in dried mud.

The diet is primarily carnivorous: small fish, amphibians, tadpoles, gastropods, and insects. But this turtle also eats algae and aquatic plants, as well as vegetative debris. Mating occurs in early spring, especially in the south, and nesting starts in May or, in the northern areas, as late as June. Mating takes place underwater, rather rapidly, and the process is quite simple. The male will turn around several times in front of the female, then turn toward her head while stretching out his limbs to the fullest extent. Nesting takes place quite a long way from the water, on a suitably oriented part of the surrounds of the lake, with good drainage and exposed to full sun. A female may nest up to three times in a season, laying six to nine elongate eggs (35 × 21 mm). Incubation is short, and emergence starts 60 days after deposition. The young measure about 32 mm in length and weigh 8 g—slightly smaller than *Emys* hatchlings.

Protection. This species is not heavily collected and is very rarely eaten. There is no real commerce or traffic carried out with this species, although it may be seen in local souks from time to time. Resistant to pollution, it is not really threatened by destruction of wetland habitats, impoundments, lake dredging, or canal construction. In France the residual population is restricted and involves only about 100 individuals altogether (M. Franck), close to the Spanish border. The protection program includes modification of some of the local dams and barrages, cleaning of waterfront areas, surveillance by agents of the ONF (National Office of Forests), and a program of radio tracking. The population is the only one in France and is well worth preserving.

Mauremys japonica (Temminck and Schlegel, 1835)

Distribution. This species is found only in central and southern Japan, on such islands as Honshu, Kyushu, and Shikoku.

Description. This species, with its rather elongate, wide shell, is somewhat rectangular, and the carapace is slightly raised into a ridge along the vertebral scutes. The maximum length is 209 mm (for females). The posterior marginals are slightly serrated, and the scutes show persistent and clear annual growth annuli. The general color is brown, with some light beige areas and also blackish ones forming a somewhat repeating design on the scutes. The marginal scutes at the sides may be somewhat upcurved. The plastron is well developed, very flat, dark brown to blackish in color, and without any light areas. There is a large anal notch. The head is rather small and is light brown to gray-black in color. The upper surface forms a sort of uniform, fine-textured shield, often lighter than the rest of the head. The eyes are rather large and are placed high and well forward. On the neck are discontinuous light stripes, which are not always clear. The limbs are brown, with yellow to cream lines on the exposed surfaces. The young are more olive or light brown than the adults and also show light streaks along the seams between the costal and marginal scutes.

Natural history. This species lives is clear-water rivers with a modest current and a mud bottom, as well as in lakes and even in flooded prairies in mountain foothill areas. It is omnivorous in diet, eating amphibians, young fish, aquatic larvae and insects, snails, crabs, amphibian eggs, and algae, as well as fruits that have fallen to the ground. Although strongly aquatic, it likes to bask in the sun on emergent snags or on the riverbanks themselves, and it hibernates underwater in aquatic detritus or under the mud. Each animal is imprinted upon a given site, which it frequents year after year. Sexual maturity of this species seems to be achieved rapidly, and it has been estimated that reproduction may occur at just five years of age. Mating has been observed from September to April, with a peak in January and February. Males are especially active at this time, as they wander about seeking females. Such activity may take place even at quite low temperatures (13°C–14°C). Cloacal temperatures taken by Takashi Yabe have included surprisingly low values of 3°C for the males and 5°C for the females. Nesting has been observed from mid-May to early July. There may be two nestings in a season, with five to eight eggs each time. The eggs are white and hard-shelled and measure 36 × 22 mm. Incubation is quite short—70 to 80 days—and emergences occur from about the middle of August, if the soil has been sufficiently softened by rain. The hatchlings measure about 35 mm and weigh 8 g.

This turtle has a trace of a median keel on the first vertebrals. The shell is boldly sculptured with growth annuli. The head lacks color, and the species in general is very plain.

Protection. This species is protected at the national level, but certain populations suffer from degradation of their natural habitat. The extension of rice culture and the management and alteration of flowing waters have been harmful to this species. It also suffers from a high level of highway vehicular mortality, and it is gathered for the wildlife trade or just to show to children. These circumstances are progressively lowering populations, and in recent times this drop has gained momentum.

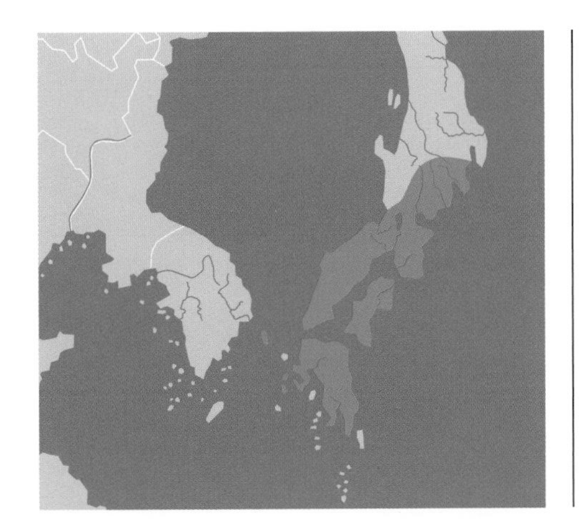

Common name

Japanese pond turtle

Mauremys mutica (Cantor, 1842)

Distribution. This Asian turtle occurs in a wide coastal strip in central and southern China—in the provinces of Yunnan, Hubei, Anhui, Jiangsu, Zhejiang, Jiangxi, Hunan, Fujian, Guangdong, Guangxi, and Hainan—as well as in northern Vietnam, in eastern Laos, and on Taiwan, where it has been introduced by humans. It is also found on the Japanese islands of Yaeyama.

Description. The curious name *mutica* refers to the short, flat snout of these turtles. They are oval, elongate, flat-shelled, and wider posteriorly than anteriorly, with females reaching a maximum of 170 mm. The median keel is weak, and the two lateral keels disappear with age. The margins may be somewhat recurved in old individuals, and the posterior ones may be slightly serrated. The overall color is brown to grayish. The plastron is orange-yellow, with a black spot along the border of each scute. The bridges, also orange, are each marked with two prominent black blotches. The head is of medium size and grayish to brown-black in color, and it sports a short, almost truncated snout, which gives the species its name. A wide yellow band extends from the snout, traverses the eyes (which are large and brown), and ends at the base of the neck. The lower edge of the upper jaw, as well as the underside of the head and neck, are a faded yellow color. The limbs are grayish to olive above and yellowish below. The tail of the male is significantly longer and thicker than that of the female. The young have three keels, as well as some light spots, on the carapace.

Subspecies. This turtle varies greatly throughout its wide range, according to the different habitats. Some animals are steel gray, with shells and plastra that are almost black. Others have a beige carapace with dark edges to the scutes, and a yellow plastron with dark markings on reach scute. There are two subspecies: the nominal race, *M. m. mutica* (described above); and *M. m. kami* Yasukawa and Iverson, 1996. *M. m. kami* has a lighter brown carapace, smaller plastral blotches, an indistinct (or absent) postorbital stripe, and unusual sexual dimorphism, with the male larger than the female.

Natural history. This turtle frequents river basins and tributaries of major waterways, as well as marshes, lakes, and flooded fields. It is omnivorous in diet, eating all the usual forms of aquatic

The short, obliquely truncated snout justifies the name *mutica*. Otherwise, this species resembles other *Mauremys* species. The light lines on the sides of the head and neck are indistinct and poorly contrasting.

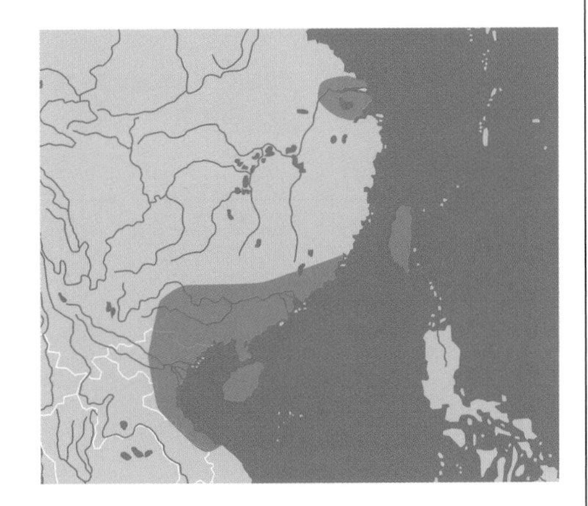

Common name

Yellow pond turtle

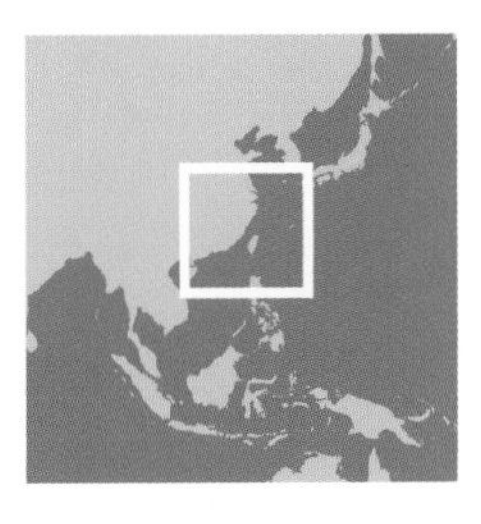

© F. Bonin

This aged turtle has a uniformly light-colored shell that is somewhat domed. The keel is still present on the posterior vertebrals. The head is also light, modestly relieved by yellow stripes on the sides.

prey, as well as algae, grasses, aquatic plants, and fallen fruit. Females appear to mature at an age of six years (length 130 mm). Mating occurs in March and April, and the nesting occurs later, in June and July. Eight eggs are the maximum; they are elongate and medium-sized (35 × 20 mm). Incubation time varies according to the locality, ranging from 64 to 96 days. The young are very circular in form and very brightly colored, weighing about 8 g. In captivity, males may be extremely aggressive.

Protection. Even within China, this species is recognized as being in danger. This is one of the most heavily harvested and traded turtle species in Chinese markets, and furthermore, much of its habitat has been taken over and converted for human use. Widespread use of pesticides has also been very harmful. The species is probably rapidly becoming rarer throughout its entire range.

Mauremys rivulata (Valenciennes, 1833)

Common name

Eastern Mediterranean pond turtle

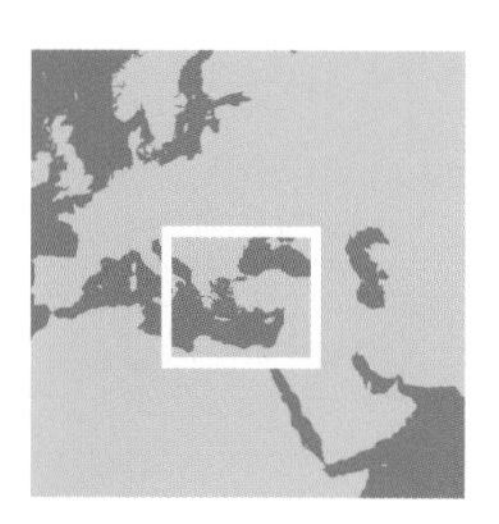

Distribution. This species occupies the western part of the range of the former *M. caspica* (in the old sense). This range extends from Bosnia and Herzegovina to Bulgaria to Syria and includes parts of Greece, Turkey, Lebanon, and Israel. The species also occurs on Cyprus and Crete.

Description. The length of this turtle does not exceed 200 mm. The shell is oval, flat, and rather rectangular, with a general anthracite gray or blackish color or even with a blue tinge. There is a central keel and two lateral keels, which disappear in adults. The plastron is well developed and black or bluish in color, with lighter areas on the periphery and on the underside of the marginal scutes. The head is blackish to bluish, marked with fine yellow lines that form a triangle on the head, and with reticulations along the neck. The lower jaw is light, and the underside of the neck is lined with small yellow streaks. The limbs are also blackish to bluish, with fine yellow lines.

Subspecies. *M. r. rivulata* is described above. *M. r. cretica* Mertens, 1946, which is found only on Crete, has lighter coloration on the carapace as well as very brilliant, light-colored irides.

Natural history. This is a widespread turtle that occupies all available aquatic habitats within its range, including irrigation canals and brackish waters, as in the Salonica region. Males are aggressive and territorial. This turtle leads a life identical to that of *Emys orbicularis:* much solar basking, carnivorous diet, long period of winter hibernation. Two nestings occur each season, with about a dozen eggs per clutch; the latter are medium-sized (about 35 × 22 mm), elongate, white, and hard-shelled. Nesting occurs at the end of spring,

and the young appear after the rains at the end of summer.

Protection. This species is very threatened. On the one hand, it is commercially collected, with illegal sales continuing in a number of European countries (both for hobbyists and for human consumption), and sometimes it is even passed off as an "American" or "Asian" turtle. On the other hand, habitats are progressively lost to urbanization, wetlands are drained, and agriculture proliferates. Use of phytosanitary chemicals is also a problem.

This young turtle is brightly marked and colored and looks quite exotic. The colors disappear quickly with growth. The carapace will become black, and the keels only slightly evident.

© J. Maran

Melanochelys trijuga (Schweigger, 1812)

Distribution. The numerous subspecies of this turtle are distributed over an enormous range in Asia, and there is much disjunction and isolation of populations one from another.

Description. The maximum length of this turtle is around 385 mm (for females). The shell is somewhat flattened and has three keels; the median keel is very strong, but the lateral ones are present only on the first three costal scutes. The posterior margin is smooth, and the marginals may be recurved in old individuals. The color of the shell varies from red-brown to black. The keels are lighter, often yellowish. The plastron is characteristic: it is black, with a wide orange-yellow border. The plastron is quite large, and the anal notch is well developed. The head is of moderate size, and the snout is very short. In color, the head is brown to black, with yellow or orange spots and sometimes a fine reticulation. Some individuals have a large yellow to cream-colored blotch in the temporal region. The limbs are well webbed and are gray, dark brown, or even black. The forelimbs have enlarged scales on their upper face.

Subspecies. This species shows wide morphological variation, not unexpected in a species with such a wide range and so much isolation of key populations. Six subspecies are recognized.

M. t. trijuga (described above): The peninsula black turtle occurs in India, from Bombay in the north to Coorg (or Kodagu) in the west, Cuddalore in the east, and Vijayawada in the south. The length reaches 220 mm, and the carapace is lightly depressed on top. The irides

Common name

Indian black turtle

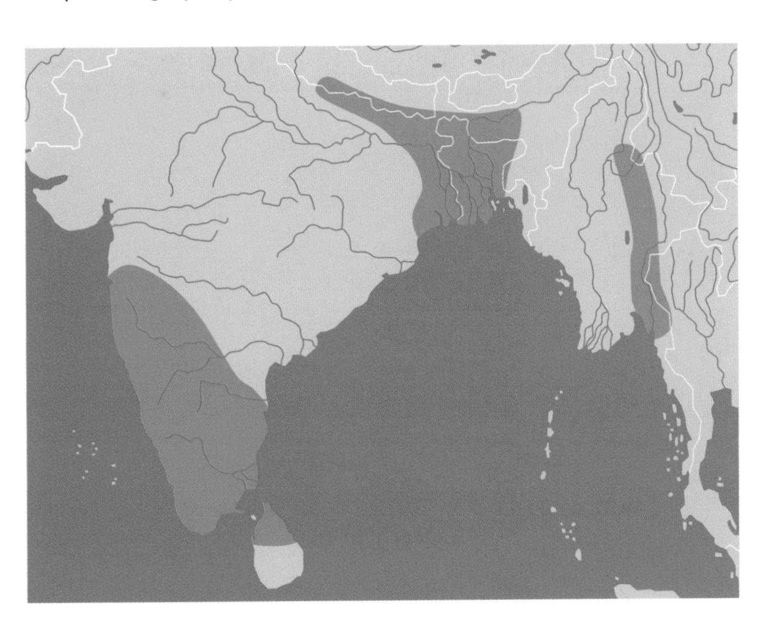

are white to light brown. This turtle is highly aquatic.

M. t. coronata (Anderson, 1878): The Cochin black turtle is found on the southwestern coast of India, from Calicut to Travancore (Kerala). The length reaches 230 mm, and the coloration is brown, with dark sutures, or the animal may be completely black. The plastron is black with yellow bands. The head is often black, with two large yellow patches on the temples. The irides are yellow.

M. t. edeniana Theobald, 1876: The Burmese black turtle occurs in Myanmar (Burma)—in Arakan and the hills of Karenni and Moulmein. The shell may reach 280 mm and is somewhat convex, with yellowish keels. The plastron is black, bordered with yellow. The head is brown, with yellow to olive reticulations.

M. t. indopeninsularis (Annandale, 1913): The Bengal black turtle is found in northern Bangladesh and southern Nepal. Its shell reaches 350 mm, is domed and black, and has whitish keels. The plastron is brown with a yellow border or is entirely cream. The head is brown to olive, without light spots, and may have a median black streak on top. The irides are nut brown.

M. t. parkeri Deraniyagala, 1939: The Parker's black turtle is found only in Sri Lanka and can reach a length of 385 mm (the largest size of any of the subspecies). It has a red-brown to black carapace and a dark brown plastron bordered with yellow. The head is olive to brown, with fine orange markings. The irides are yellow.

M. t. thermalis (Lesson, 1830): The Sri Lanka black turtle is found not only in Sri Lanka but also in the Maldives and perhaps southeastern India. The carapace reaches 280 mm. It is somewhat flattened and is dark brown with yellow keels, but sometimes the keels are lacking. The plastron is gray to dark brown or black. The head is black with yellow, orange, or red spots. The irides are brown.

M. t. wiroti (Nutaphand, 1979): The Thailand black turtle is found only in the eastern extreme of the range, in Thailand and Myanmar. It is lighter-colored than the other subspecies, mid-brown rather than truly black.

The three keels, outlined with a light stripe, are still visible in this specimen. The head is strong and marked with discontinuous white or yellowish stripes and blotches on a black background.

© F. Bonin

Natural history. These turtles are fundamentally aquatic, occupying calm rivers with clear water, as well as marshes and lakes. They spend a great deal of time basking (especially *M. t. thermalis,* as its name indicates). Some subspecies are more terrestrial than others, and *M. t. indopeninsularis* prefers inundated prairies and rice paddies. These turtles are primarily herbivorous, but *M. t. thermalis* is somewhat omnivorous and will also take carrion. When they are disturbed, they may emit a truly foul-smelling musk. During mating, the male is very active and aggressive and frequently bites the female at the base of her neck. Copulation is observed virtually year-round, except in the northern extreme of the range, and nesting also takes place throughout the year. The eggs are white, elongate, and rather large (48 × 35 mm). Incubation lasts for 60 to 65 days but varies according to latitude. A typical clutch consists of a dozen eggs. The juveniles are more colorful than the adults and measure about 43 mm.

Protection. These turtles are sought by collectors and, in some regions, are collected for the wild animal trade also. They are rarely consumed, and their populations are still quite strong, but they suffer from the effects of deforestation and the rapid degradation of certain habitats in Bangladesh, Thailand, Sri Lanka, and Nepal.

Melanochelys tricarinata (Blyth, 1856)

Distribution. The range of this species is restricted to Bangladesh, eastern India (Assam), and Nepal.

Description. Females of this species do not exceed 163 mm in length. The reddish brown shell of this turtle is lighter-colored than that of *M. trijuga* and has conspicuous yellow keels; the middle keel extends the length of the shell, while the lateral keels are on the first four costal scutes only. The plastron is large and is yellow to orange-yellow, with a strong anal notch. The head is red-brown and has a lighter narrow stripe along the sides; the stripe may be yellowish, orange, or reddish and extends from the nostrils to the neck, passing by the orbits and the tympana. A second band of similar color starts at the angle of the jaw and continues toward the neck. The limbs are red-brown to black, sometimes with yellow spots. The forelimbs are covered with enlarged scales on the exposed side, which may be pointed or square. The hind limbs may have some enlarged scales on the sole of the foot. The forelimbs are only about half-webbed, and as the English name implies, this trait is indicative of a largely terrestrial lifestyle.

Natural history. This is a species of tropical forests, but it needs access to wetlands where it may conceal itself in leafy substrates. It is essentially omnivorous and consumes insects, dead animals, berries, and fallen fruit. Its habits in nature are poorly known. Clutch size is at least three, and the eggs are rather elongate (44 × 25 mm). In old females, the posterior part of the plastron is flexible, which facilitates oviposition (Moll, 1985). Smith (1931) noted that old males often lose the light stripes on the sides of the head.

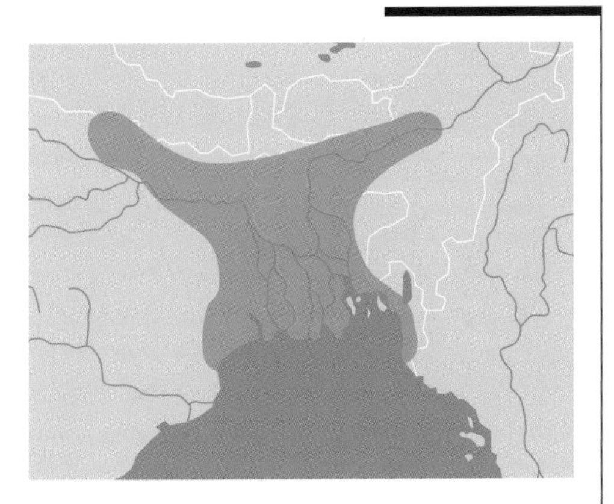

Protection. This species is included in Appendix 1 of CITES. Its populations appear to be low in density, and it is sought after by hobbyists. Furthermore, the habitats have been reshaped by humans and are degrading rapidly. This turtle is often seen in the tanks of Buddhist temples.

Common name

Tricarinate hill turtle

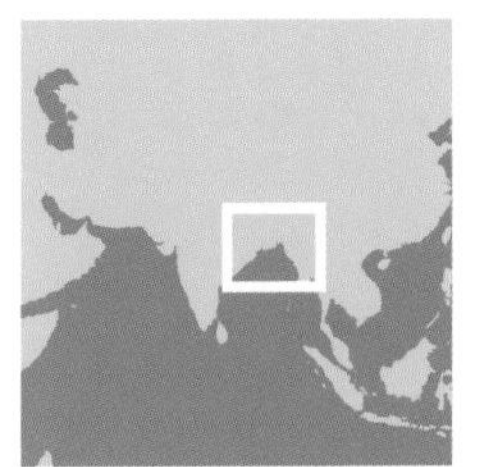

© I. Das

This small turtle is oval and well domed. The keels are still represented by light lines. The carapace is also adorned with a light line around the entire shall. The head is small, with a single light band on the sides.

Notochelys platynota (Gray, 1834)

Distribution. This species is found in southern Vietnam, Thailand, and Malaysia (East and West) to Java and Sumatra.

Description. This turtle may attain a length of 360 mm. The carapace is very unusual and indeed is unique among freshwater turtles, having an additional one or two vertebral scutes. The shell forms a keel on the first vertebral scute, then has a flat area covering vertebrals 2, 3 and 4, and has a posterior keel on vertebrals 5 and 6 (or 6 and 7 when two additional vertebrals are present). The overall color varies from reddish gray to mahogany brown. Each scute has strong growth annuli and radial streaks and has a single dark marking. The plastron is yellow to orange, with a black spot on each scute; is very flat; and is attached to the carapace by bands of flexible tissue. There are no plastral buttresses, despite the presence of well-defined bridges. The head is rather elongate and has a somewhat protruding snout, with two wide nostrils. The upper surface of the head is covered with small scales. The upper jaw has a median notch and fine serrations along the rami. The head, neck, and limbs are brown. There are wide transverse scales on the exposed face of the forelimbs. The digits are webbed, and the tail is of medium length. The young have a light-colored carapace, very circular and often green or yellow in color, with small symmetrical black spots on the middle vertebrals and on the costals. The carapace is strongly serrated behind and has a median keel. The head often shows a yellow band passing over each eye, as well as a second band from the angle of the mouth to the base of the neck. The plastron is orange in hatchlings.

Natural history. These turtles occupy slow waterways with a muddy bottom and dense aquatic vegetation, as well as marshes, lakes, and inundated tropical forest. Furthermore, they readily make long overland journeys, and their morphology is similar to that of many terrestrial species. They are sometimes kept in captivity in enclosures with little access to water. This is a primarily herbivorous species that eats aquatic plants and fallen fruit but also appreciates larvae, snails, and other invertebrates. The female lays two to six eggs, with hard shells and measuring about 56 × 27 mm. The juveniles measure between 50 and 60 mm in length. When picked up, this turtle will defecate frequently and remain fully retracted into the shell for a long time. Some writers have described it as aggressive, but this characterization does not seem to represent the animals that we have seen.

© I. Das

This turtle is remarkable for its flattened shell and for the presence of six or seven vertebral scutes rather than the usual five. The marginals are spread out, and the head is rather wide, with a pointed snout and with very light lines and spots on the black or brown background.

© F. Bonin

Protection. This is a scarce, poorly known turtle, still sometimes eaten by local people. It does not receive any legal protection by IUCN, mainly because of "lack of information." In Malaysia, it is often kept in Buddhist temples. Furthermore, it is sought by animal traders for commercial export. It would seem to be important to evaluate the impact of the commerce in this species before appropriate conservation measures can be drawn up.

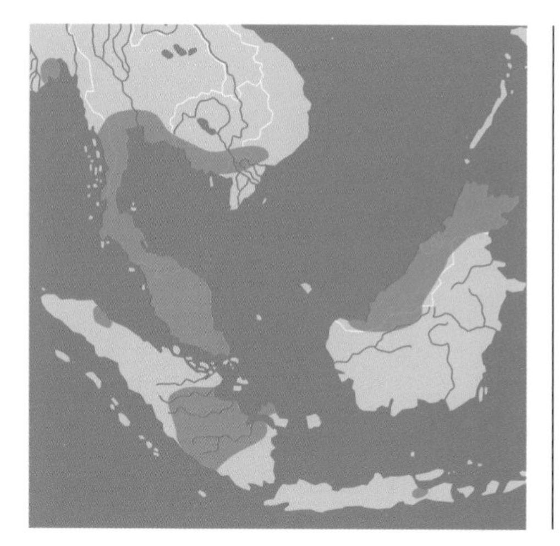

Common name

Malayan flat-shelled turtle

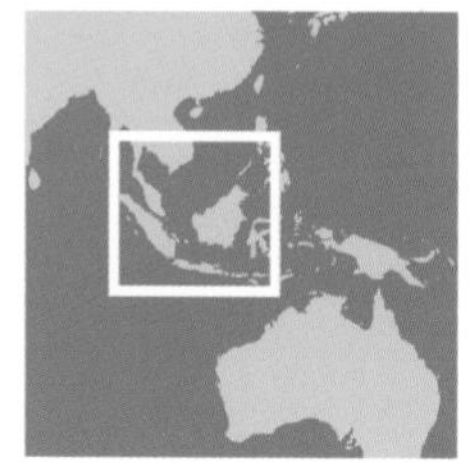

Pyxidea mouhotii (Gray, 1862)

Distribution. This species is found in southeastern Asia, ranging from India (Arunachal Pradesh, Assam, Meghalaya), southeastern Bangladesh, Myanmar, northern Thailand, Laos, Cambodia, and the Malay Peninsula to western Vietnam, southeastern China, and Hainan Island.

Description. The carapace is elongated and does not exceed 180 mm. Oval in shape, it is flattened on top, with three strong keels and major spiny denticulation along the hind margin. Less developed denticulations occur along the anterior margin. The median keel is slightly raised above the flattened vertebrals, and the two lateral keels coincide with a sharp falloff of the sides of the carapace, leaving the top of the shell forming a sort of plateau. The color is a uniform light or reddish brown to mahogany, with blackish areas running along the keels. The plastron is reduced and does not close the shell hermetically. A single hinge is located between the pectorals and the abdominals, and there is a significant anal notch. The plastron has narrow, poorly developed bridges, with only a ligamentous attachment to the carapace. The axillary scutes are often lacking. The bridge is yellow to light brown, with black areas on the scutes. The head is of medium size, and the upper jaw is provided with a strong horny beak, without a notch. The snout is short. The skin of the head is divided into large scales and is brown to mahogany in color, with dark lines and light dots giving the ensemble a vermiculated appearance. These dots sometimes form stripes bordered with black. The powerful limbs are dark brown, gray, or black. The upper face of the forelimbs is covered with large scales. The hind limbs have an inclined shape, like a golf club. All of the limbs are well webbed. The tail of the female has pointed tubercles at the base, and these may also be present on the thighs. The head of the males is dark brown, with red longitudinal lines on the sides, while that of the female is more gray-green, with well-defined yellow reticulations on the sides.

Subspecies. Two subspecies are recognized. *P. m. mouhotii* (described above): This race occupies almost the entire range of the species. The plastral

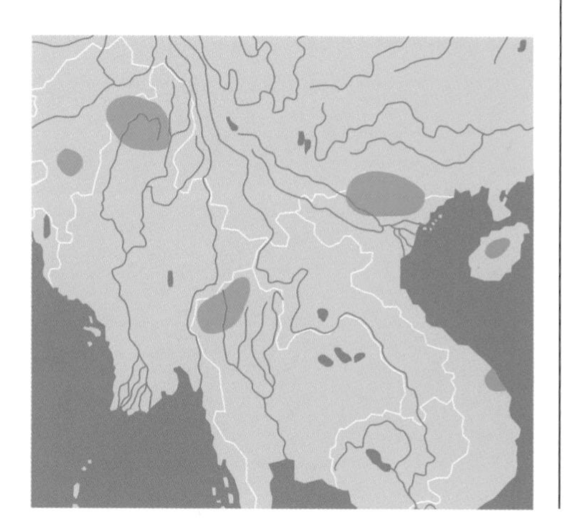

Common name

Keeled box turtle

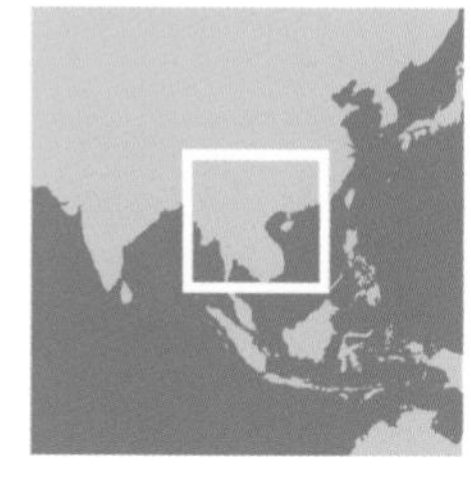

The carapace of this turtle is uniformly beige or chestnut and has scutes dotted with protuberances. The marginals are spread out and serrated posteriorly and, to a lesser degree, anteriorly. The head is strong, with a pronounced beak.

pattern is limited to black bars along the edges of the plastron. *P. m. obsti* (Fritz, Andreas, and Lehr, 1998): The Annam three-keeled turtle is found only in the ancient province of Annam, in the Phu Loc region of central Vietnam. In this subspecies, the design of the plastron includes radiating streaks, and the carapace is broader and rounder than that of *P. m. mouhotii.*

Natural history. This is a largely terrestrial species, but it occurs only in moist habitats in the forest or at the foot of hills, as on the island of Hainan. The habitat could be compared to that of *Terrapene carolina.* This species is mainly herbivorous but could also be described as opportunistic. The clutch consists of one and three elongate eggs, each measuring 40 × 25 mm.

Protection. These turtles, with their curious carapace shape, are collected for resale in Western nations. They do not do very well in captivity, however, and this poor outcome has somewhat reduced the volume of the trade. Today the main threat is collection for food, especially for the China trade, as well as habitat destruction. The species is classified as endangered by the Turtle Conservation Fund.

Rhinoclemmys punctularia (Daudin, 1801)

Common name

Labaria turtle

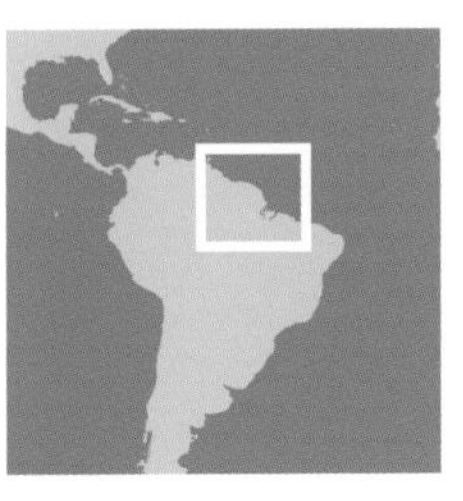

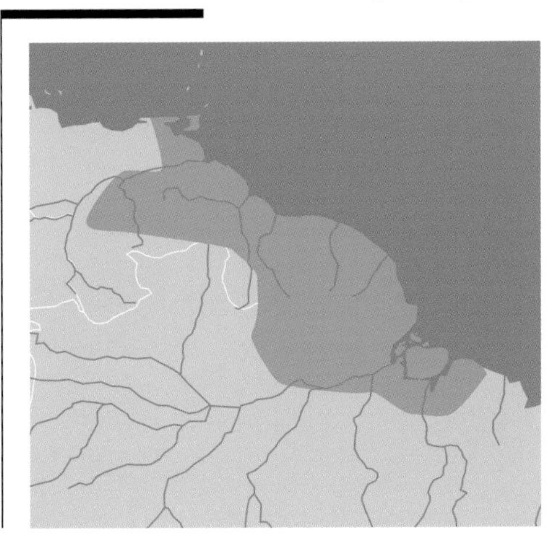

Distribution. This turtle is found in northern South America, in eastern Venezuela, Trinidad, Guyana, Suriname, French Guiana, and northeastern Brazil.

Description. The rather high carapace may reach a length of 254 mm and has a vertebral keel and a serrated posterior border, with a notch on the supracaudal. The maximum height and width are achieved just behind the center of the shell. The surface is smooth or perhaps has a slight rugosity. The color is uniform dark brown to black in adults, while the young are ornamented with yellow or bronze rays on each costal scute. The

plastron is well developed and raised in front, and the posterior lobe includes an anal notch. In color the plastron is red-brown to black, but the margins and the seams remain yellow. The bridge is yellow and has large dark spots. The head is small, with a somewhat pointed snout. It is black but has a motif on top made up of two orange or red stripes. The longitudinal bands reach as far back as the front of the nuchal. The eyelids are marked with a light stripe, and other light stripes are distributed between the eye and the tympanum. The irides are green or sometimes bronze. The forelimbs are covered with large red or yellow scales, spotted with black. The hind limbs are gray on the sides and mostly yellow with black spots toward the inside. The digits are completely webbed.

© A. Dupré

Subspecies. *R. p. punctularia* (described above) occurs in the Guianas, Trinidad, and the northern Amazon Basin in northern Brazil. *R. p. flammigera* Paolillo, 1985, occurs in southwestern Venezuela, especially in the area of the confluence of the Orinoco and the Ventuari. The head pattern includes numerous red spots that extend to the sides of the head and make up a semicircular flamelike design.

Natural history. This turtle occupies all types of freshwater habitats within its range, including marshes, ponds, savannas, and forest rivers, as well as lakes, and one often sees it basking on floating objects. It is carnivorous and feeds equally well on land and in the water. In courtship, the male pursues the female while sniffing at her cloaca, then turns toward the head of his partner for further love play. Mating takes place either in the water or on land. Nesting, which may occur at any time of the year, involves the production of just two large eggs, 65 × 34 mm in size. Incubation lasts for about 90 days.

Protection. This species is apparently common and relatively rarely exploited, in contrast to the sympatric *Geochelone denticulata,* which is eaten whenever it is found.

© J. Maran

The shell has the appearance of a smooth, blackish stone. The head is small, pointed, and decorated with light stripes and dots on the sides.

Rhinoclemmys annulata (Gray, 1860)

Distribution. This turtles occurs in Central America and northwestern South America, from southeastern Honduras, eastern Nicaragua, and Costa Rica to Panama, western Colombia, and northwestern Ecuador.

Description. This is a small species whose carapace length does not exceed 228 mm. The shell is yellow, brown, or black and is flattened in the vertebral region. Growth rings are evident on the dorsal scutes, and the posterior margin of the shell is somewhat serrated. The variable patterns on the shell range from uniform black to dark brown and may include orange markings on the vertebral and costal scutes. Sometimes the figures are light brown, with yellow touches. The markings on the costals are often in the form of radiating streaks that extend from the posterodorsal corner of each scute. Usually the vertebral keel is yellow. The plastron is well developed, with an anterior lobe that is somewhat elevated in front and with a distinct anal notch. The color ranges from black to dark brown, with a yellow border. In certain individuals, a groove is present between the scutes along the midline of the plastron. The bridge is black or dark brown. The head is small and has a slightly projecting snout. A wide yellow or red band extends from the orbit to the nuchal area.

Common name

Tropical land terrapin

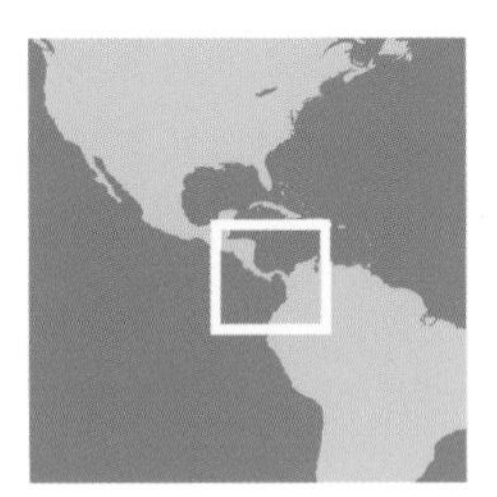

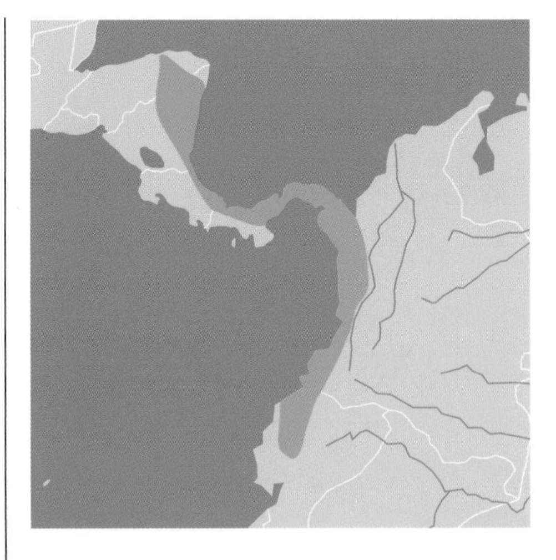

Another streak starts at the lower rear corner of the orbit and, at the level of the tympanum, reconnects with another stripe that starts at the upper jaw. The forelimbs have large yellowish scales and are decorated with dark bands or large black blotches. The digits are not webbed. There is much variation in color and decoration, according to the individual turtle.

Natural history. This is a terrestrial, diurnal species that lives in tropical forests, where it occupies the understory. It may be found at altitudes as high as 1,500 m, as it follows the penetration of the forest into mountainous areas. It is herbivorous in diet and eats various fruits and plants, including pawpaw, mango, and banana. It is mainly active in the morning and after rains, and it may be seen walking along roads and paths. During mating, the males salivate over the females. Nesting may take place at any time of the year, and the clutch consists of just one or two large eggs (70 × 40 mm on average).

Protection. This species is so cryptic in its forest environment that it is probably safe wherever the forest remains intact.

Rhinoclemmys areolata

(Duméril and Bibron, 1851)

Distribution. Mainly found in Mexico, in southern Veracruz, Tabasco, Chiapas, and Yucatán. It also occurs in northern Guatemala and Belize.

Description. The carapace is oval and somewhat domed, reaching a maximum length of 239 mm. It is always wider in the posterior half and has a light vertebral keel. The marginals are

© F. Bonin

This small turtle has a low keel that disappears with age. The carapace has distinct designs and ocelli during the first years of life, but these fade and it becomes dark. The head is relatively unmarked, but in young animals there are reddish lines on the neck.

widened or upcurved and are serrated posteriorly. In the young, the shell has a rugose surface, although it becomes smooth in old animals. The overall color is olive, with dark sutures. One sometimes also sees specimens with a lichenlike yellow reticulation. Some individuals have a light brown carapace; in others the carapace may be black. The well-developed plastron is always lightly upturned in front and has an anal notch. The plastron is yellow in color and has a large dark central figure. The seams are also dark. The bridges are yellow. The head is small and has an elevated nasal area. Behind the eye, there is a red or yellow band that reaches the side of the neck. The nuchal has two red or yellow elongate markings, and another yellow band passes between the eyes and the tympana. Additionally, there are black dots or circles on the chin and throat. The digits are slightly webbed, and large yellow scales, scattered with black dots, are present on the forelimbs. In the juveniles, each costal scute is decorated with a central red or yellow marking, often circled with black. These elegant adornments disappear with age.

Natural history. This species is mainly terrestrial, but it also frequents shallow swamps. It is also found in the savannas and dense forests that border the wetlands. It is omnivorous in diet, consuming insects and worms, as well as fallen fruit. At the time of oviposition, the plastron of the female becomes flexible, which allows for the passage of the rather huge eggs, measuring 60 × 31 mm on average.

Protection. Somewhat cryptic and not heavily exploited, this species is probably safe for the time being.

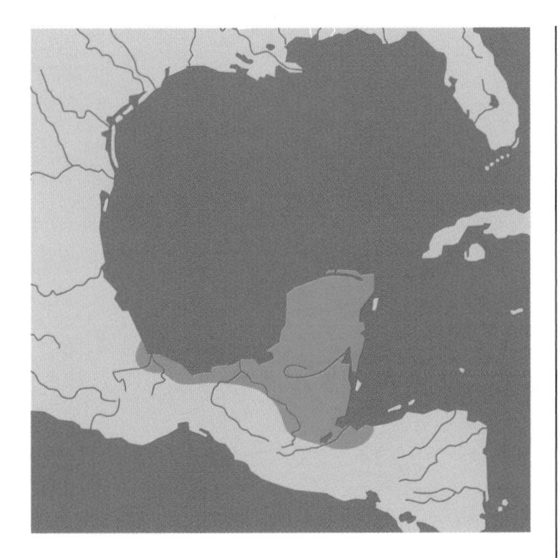

Common name

Furrowed wood turtle

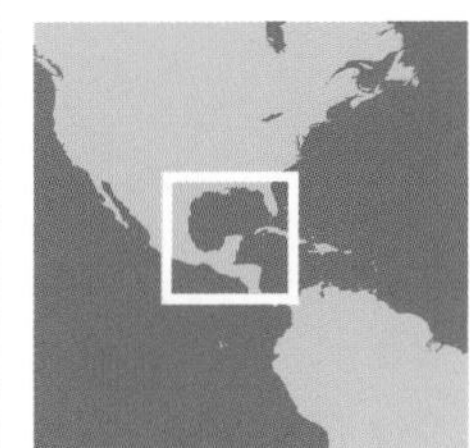

Rhinoclemmys diademata (Mertens, 1954)

Distribution. The range of this species is restricted to the western shores of Lake Maracaibo, in the Venezuelan states of Zulia and Táchira, and in neighboring Norte de Santander, Colombia.

Description. This species may attain a carapace length of 257 mm. The shell is quite domed and has a modest median keel and a serrated posterior border. The central keel is outlined with black. The plastron is well developed and has an anal notch. The bridge is strongly developed. The plastron is black, but the sutures and the edges are tinged with yellow. The head is of medium size, with a pointed snout. The top of the head has a creamy-white figure in the shape of a diadem—hence the species name. The rest of the head is uniform black. The forelimbs always have five claws, while the hind limbs have only four. The young have three keels, and the shell is rather flat and rugose (becoming smooth in the adults).

Natural history. This semiaquatic turtle lives in marshes, ponds, and wetlands. However, it also walks quite long distances from water and is sometimes tragically killed on nearby highways. It is omnivorous in diet and is equally happy feeding

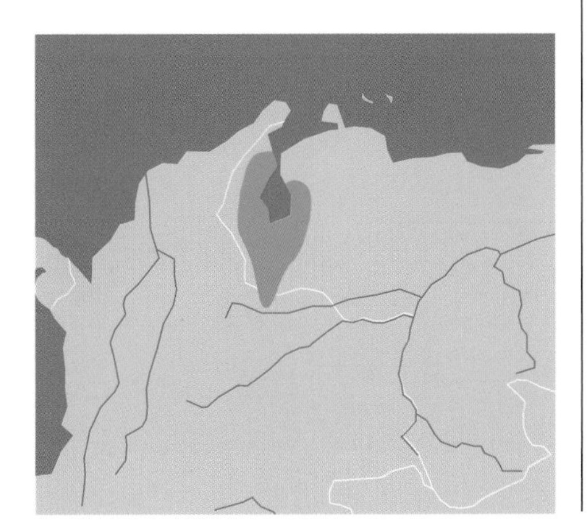

Common name

Maracaibo wood turtle

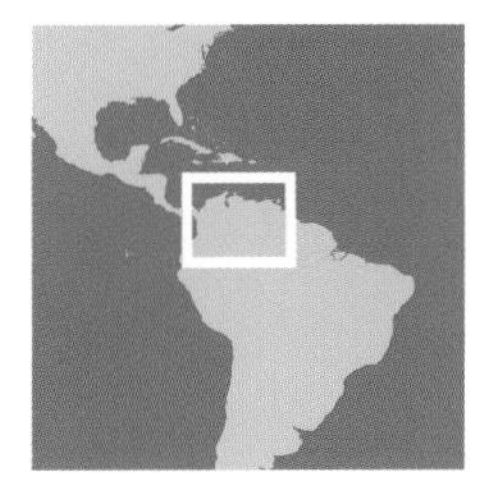

on small invertebrates (insects, snails), minor carrion, and aquatic and certain terrestrial plants. The female lays her eggs in natural cavities or small crevices, without digging a nest. The eggs have a white, hard shell and measure 46 × 31 mm on average.

Protection. Rarely exploited, this species is locally quite common in Zulia, but it is subject to road mortality on the main north-south highway.

This species is easily recognizable by the diadem-shaped marking on the top of the head. The carapace is grayish to blackish, with a black stripe along the keel.

© F. Bonin

Rhinoclemmys funerea (Cope, 1875)

© F. Bonin

Distribution. This species is found only in Central America, from the Rio Coco (frontier between Honduras and Nicaragua) south through Costa Rica to the Panama Canal.

Description. The name evokes the overall black, funereal coloration. This is the largest turtle of the genus and may reach 330 mm in length. The carapace is moderately domed and has a feeble median keel, and the shell is serrated behind. The maximum height and width are reached just behind the midpoint of the carapace. The color ranges from dark brown to black, according to individuals. The large plastron is upcurved in front and has an anal notch. It is brown to black, with yellow seams and a wide yellow band along the midline. The head is very black, like the rest of the animal, but the lower side is brightly colored with vivid yellow markings. A wide yellow band passes over the tympanum. Two other, narrower bands extend from the eye to the corner of the mouth, reaching as far as the tympanum. The chin and lower jaw

The name of this turtle derives from the black, or "funereal," color of the adults. The young have lighter coloration on the head and neck. In this young animal there is still an evident median keel, as well as a scattering of dark dots on the carapace.

© F. Bonin

are yellow, with large black spots. The limbs have numerous small yellow spots on the dark gray to black background. The hind limbs are well webbed.

Natural history. This turtle is conspicuous in the waterways along the Caribbean coast of Costa Rica, and unlike many of its congeners, *R. funerea* basks in large numbers on emergent branches and trunks and are easy to see. The diet is herbivorous, and broad-leaved plants are consumed, along with various herbaceous plants and sometimes fallen fruit. The female reaches sexual maturity at a length of about 200 mm. The male pursues the female underwater, undulating his head and neck up and down in a rapid rhythm. Each season the females may nest as many as four times, laying about three eggs each time. Each egg measures about 70 × 35 mm.

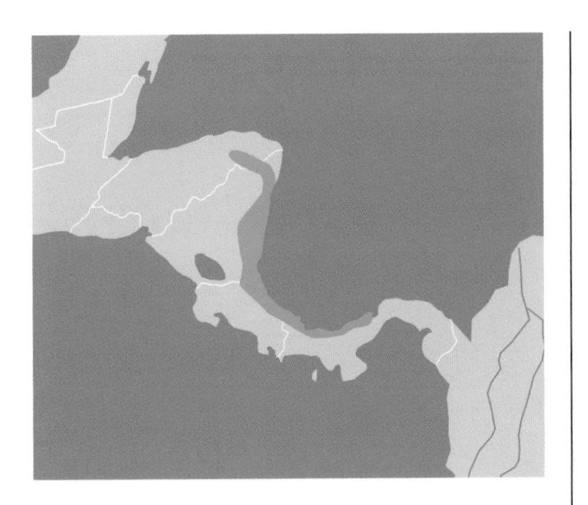

Protection. Not heavily exploited, this species is less popular as food than the sympatric *Trachemys scripta venusta.*

Common name

Black wood turtle

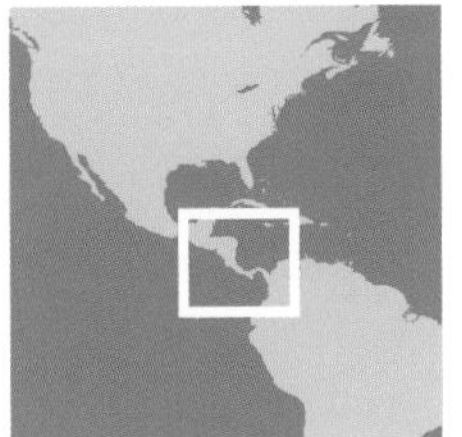

Rhinoclemmys melanosterna (Gray, 1861)

Distribution. The range of this species includes the Caribbean drainage basins of southeastern Panama and the Pacific drainages of western Colombia and northwestern Ecuador.

Description. Once again a species name refers to blackness, in this case the dark hue of the plastron. This form reaches a carapace length of 290 mm. The carapace is oval and well domed, widest at the level of marginals 6 and 7 and highest at the front of the third vertebral scute. The supracaudal is notched at the lower edge of the midline. The median keel is poorly developed, and some minor rugosity exists in the entire surface of the shell. The coloration is dark brown to black, without any pattern. The well-developed plastron is elevated anteriorly and includes an anal notch; in color it is red-brown, or more often black, with

© J.-M. Touzet

This turtle is still very black, especially on the plastron. Only the head and neck carry light lines, which disappear with age.

Common name

Colombian wood turtle

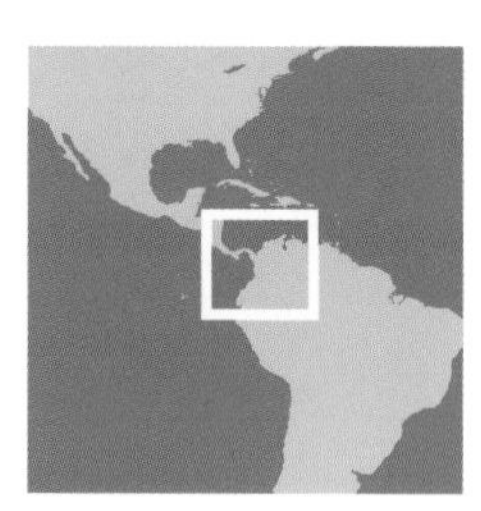

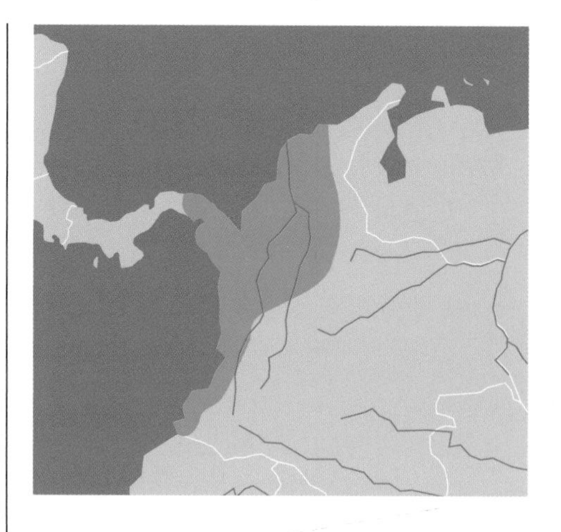

the seams and periphery yellow. The head is small, with a slightly pointed snout, and is dark or even black. On the top of the head is a pale gray—or alternatively orange or red—band that starts at the front of the orbit, passes above the tympanum, and drops down behind the eye and along the neck, where it becomes wider. The iris is yellow or white and is very bright. The forelimbs are covered with black dots, and the digits are well webbed. The males are smaller than the females (up to 250 mm). No subspecies have been described, but according to Medem, animals living in freshwater have red stripes on the head, whereas those that live in salty environments have green or yellow-green stripes.

Natural history. This species frequents marshes, lakes, ponds, and large freshwater rivers in the savannas and dense tropical forests, but it may also be found in brackish waters and in lagoons. It spends long hours basking in the sun. In diet it is herbivorous. The females nest year-round, with a peak between June and August and another in November, according to locality, and they lay just one or two very large eggs (about 60 × 33 mm). No nest hole is dug, and the eggs are simply covered with dead or decomposing leaves. Incubation lasts for 85 days or, it is said, for as many as 141 days.

Protection. This species is consumed by local people, but no information is available on its status.

Rhinoclemmys nasuta (Boulenger, 1902)

Distribution. This species is found in northwestern South America, from the Pacific drainages of western Colombia—including the Quito, Truando, San Juan, Docampado, and Bando Rivers—into northwestern Ecuador, near Esmeraldas.

Description. A length of 220 mm may be attained in the females; the males do not exceed 196 cm. The flattened shell has an inconspicuous dorsal keel, and the posterior margin is modestly serrated. The maximum height of the shell is reached around the middle of the carapace. The color is black to red-brown, with very dark seams. The surface of the shell is smooth in adults but more rugose in juveniles. The plastron is well developed and slightly upcurved anteriorly and has an anal notch. The background color of the plastron is yellow, with a large brown or black area on each scute. The bridge is also yellow, with two black markings. The head is of moderate size, and the snout is very protruding. A yellow or cream-colored stripe starts at the nostrils and connects to each eye. Another band starts behind the eye and reaches to the nuchal area, while a brighter line connects the tympanum to the lower edge of the orbit. The final stripe goes from the corner of the mouth to the tympanum. There are dark vertical bars on the lower jaw. The skin of the head and the limbs ranges from red-brown to yellow. The digits of all four limbs are well webbed.

Common name

Large-nosed wood turtle

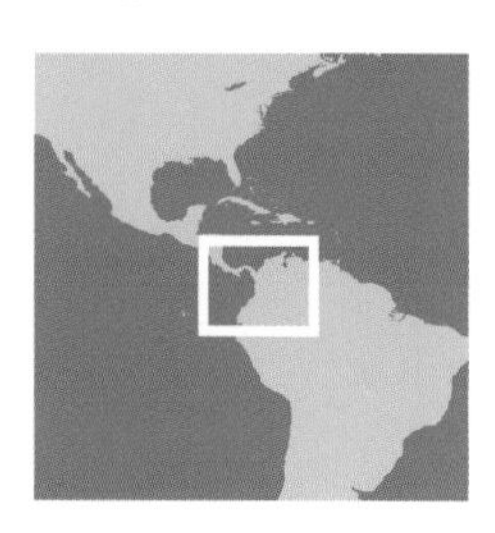

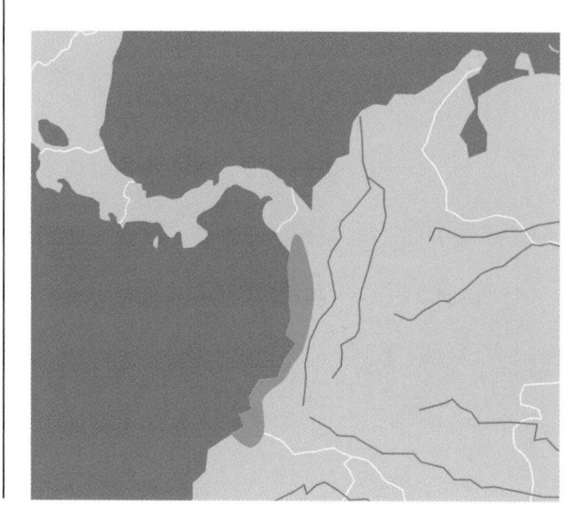

Natural history. This turtle lives in large rivers with rapid current, and it rarely leaves the water. Two nestings take place each season, from Janu-

ary to March. But in the southern part of the range, some females may nest all year round. They hardly cover their eggs, simply hiding them with a few leaves. The eggs number as many as four and measure about 55 × 22 mm.

Protection. This species is sometimes exploited, but it is hard to catch in it its fast-river habitat and probably is not seriously depleted.

The neck is extended, and light lines adorn very dark skin. The carapace is rather flat, without signs of a keel. The posterior marginals are sometimes lightly serrated.

Rhinoclemmys pulcherrima (Gray, 1855)

Distribution. The four subspecies occupy an extensive latitudinal range along Pacific lowlands of Mexico and Central America as far south as Costa Rica. Details are given under "Subspecies."

Description. This is the most familiar species of *Rhinoclemmys* and the most colorful. The carapace length does not exceed 214 mm in females (180 mm in males). The shell is quite rugose, with well-defined annual growth rings. There is a minor vertebral keel. The shell attains its greatest height and width just behind the midpoint of the carapace. Specimens from the northern part of the range are larger and flatter than those from farther south. The colors vary greatly between subspecies and even between individuals. In general, the carapace is dark to bright brown, with yellow, red, and black designs on each scute. Sometimes there are simple dots or yellow to red bands, but in others there are elaborate ocelli and rings of bright colors. The plastron is well developed and has an anal notch; it is yellow in color, with a very dark central figure whose width varies among individual turtles. The bridge is almost completely brown. The head is small, with a projecting snout and a notched upper jaw. The head is grayish and decorated with numerous orange and red bands and stripes. There are four main stripes: a midline one forms a U on the top of the snout, passes over the eyes, and rejoins other lines behind the temples; a second extends from the underside of the nostrils and reaches the tympana along the upper jaw; a third extends from the nostrils toward the eyes and continues behind them, broadening as it passes along the neck; and the last passes under the eyes and extends to the base of the neck. The jaws and the chin are yellow, often with ocelli, bars, or large black dots. The rest of the skin is olive to yellow. The forelimbs are covered with large red or yellow scales, with series of black dots. The limbs are slightly webbed.

Subspecies. Four subspecies are recognized.

R. p. pulcherrima (described above) is found only in the state of Guerrero, Mexico. The carapace is wide, not very domed, and brown, with dark markings. Each costal scute has either one or two small red or yellow spots, with a black center and encircled by a black ring. There are two or three light bars along the lower edge of each

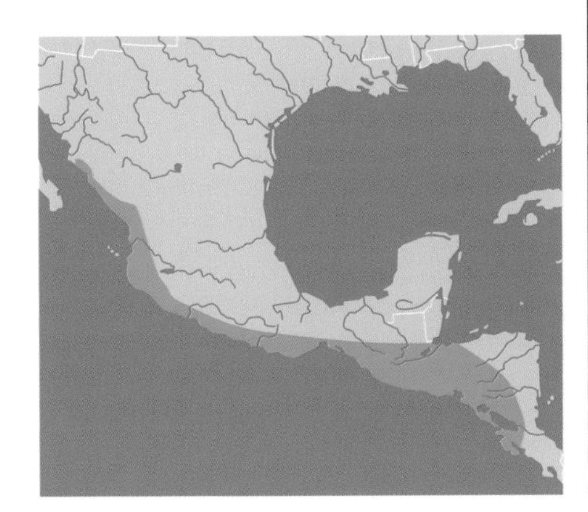

Common name

Tropical wood turtle

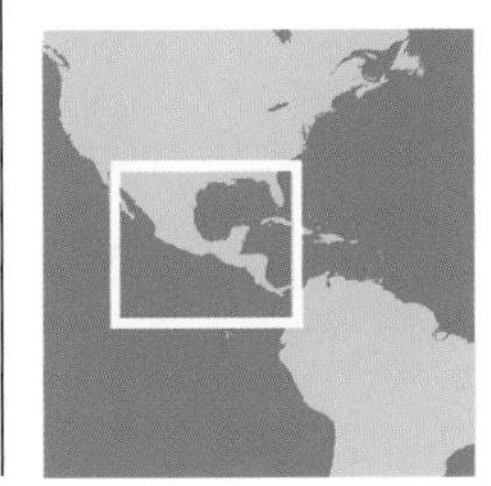

This small turtle is one of the most colorful of the *Rhinoclemmys,* with bright concentric designs on the carapace. The head and limbs are also marked with black and white lines.

© A. Hell-Kevorkian

The plastron is boldly marked with a broad black band along the middle, bordered by two yellow bands. The seams are also outlined with yellow.

© F. Bonin

marginal scute. The figure in the center of the plastron is narrow and is often forked at the gular and anal scutes. The bridge has two parallel stripes: one yellow and one black.

R. p. incisa (Bocourt, 1868) occurs in southern Mexico (Oaxaca), to Honduras, eastern Guatemala, and northern Nicaragua. The carapace is brown and only moderately domed in northern representatives, although toward the south, specimens tend to be highly domed. The shell has dark markings, and each costal scute is marked with a large yellow or red band or ocellus surrounded by a dark border. The marginals are decorated with a light bar on their lower face. The central figure of the plastron is narrow and unforked, and the bridge is brown.

R. p. manni (Dunn, 1930) occurs in southern Nicaragua and in northern and west-central Costa Rica. It is very brightly colored, with a well-domed shell in adults. The carapace has a brown background, and each costal scute is decorated with large red or yellow ocelli. On the lower face of each marginal there are two light bands. The central figure of the plastron is dark and narrow and sometimes forms a fork at the level of the gular and anal scutes. The bridge has a yellow bar and a black bar running from front to back.

R. p. rogerbarbouri (Ernst, 1978) occurs in Mexico, from southern Sonora to the province of Colima. There is no design on the pleural scutes, apart from a simple pale reddish band in certain rare individuals. The lower part of each marginal has a single light bar. The plastron has a large dark central marking that may be faded or lost. The bridge is always brown.

Natural history. This largely terrestrial turtle lives in humid forests and may be found moving on land in the vicinity of small streams. During the dry season, it may be seen entering water and swimming at the surface of ponds and puddles. It is especially active after rains. Omnivorous in diet, it shows a preference for vegetables and fruit but also eats earthworms, various insects, snails, slugs, and so forth. When a male encounters a female, he will evaluate her by olfactory means. He courts her by bobbing his head up and down, then touches her nose and sometimes bites her on the back of the neck or the limbs. Nesting occurs between September and December in the north and somewhat later in the south. The female lays between three and five eggs, each measuring about 45 × 28 mm on average. There may be as many as four nestings per season.

Protection. The main threat to these turtles is habitat deterioration and death by automobiles on highways.

Rhinoclemmys rubida (Cope, 1870)

Distribution. This species is partially sympatric with *Rhinoclemmys pulcherrima* in western Mexico but has a small, discontinuous range centered on the state of Colima in the north and Oaxaca in the south (see "Subspecies").

Description. This is a small species, usually less than 150 mm in length, although it has been reported to reach 230 mm. The flat shell has a median keel that widens and becomes higher in the rear half of the shell. The shell is always serrated or denticulate behind. The surface is rugose and has conspicuous growth annuli on each scute. The background color is yellow-brown, with darker seams and dark markings on each scute. Sometimes the animal is entirely brown. There is generally a yellow dot or spot at the center of each vertebral scute and also each costal. The plastron is well developed and slightly raised in front, and it has an anal notch. In color, the plastron is yellow, with a central brown figure and a brown bridge. The head is of moderate size, and there is a red or yellow figure on the top of the head that is unique and specific to each individual. There are light stripes around the nose, as well as a light band that passes between the eye and the tympanum. Another band originates at the tympanum and ends at the corner of the mouth. The chin and jaws are yellow, with small dark dots. The forelimbs are covered with large red or yellow scales that have black markings. The digits are unwebbed. At birth, the hatchlings are about 60 mm long.

Subspecies. *R. r. rubida* (described above) occurs in central coastal Oaxaca and as far east as Chiapas. The carapace is light brown with dark patches. The gulars are twice the length of the humerals. The marginals are somewhat widened at the sides, and an elongate light marking is present on the temples. *R. r. perixantha* (Mosimann and Rabb, 1953) occurs farther north, in the lowlands of Jalisco, passing through Colima and into Michoacán. The marginal scutes are light brown, without markings. The costals are brown and darker than the vertebrals and the marginals. The gulars are slightly longer than the humerals, and the marginals are strongly expanded at the sides. The temporal spot is oval.

Natural history. This turtle appreciates humid forests at the base of bushy hills. The eggs measure about 62 × 25 mm.

Protection. This species is protected by Mexican law and is so cryptic as to be rarely seen. It is very rare in European and American live collections.

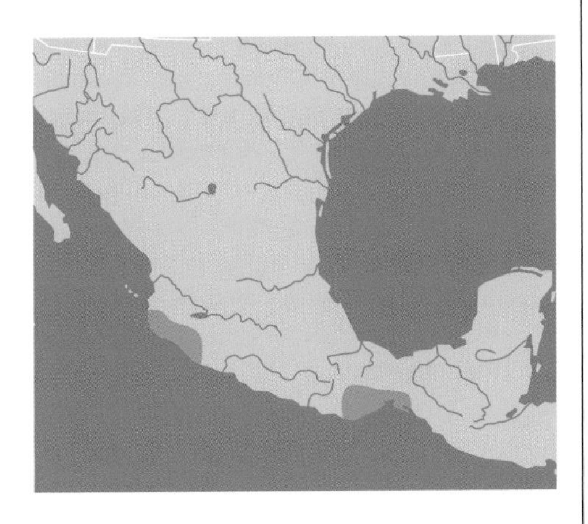

Common name

Oaxaca wood turtle

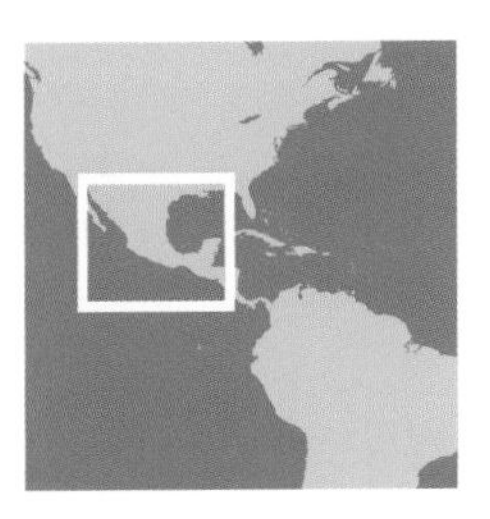

Sacalia bealei (Gray, 1831)

Distribution. This species is found in southern China (Hainan and Guangxi), as well as northern Vietnam.

Description. This small turtle reaches a maximum length of about 182 mm. The shell is elongate and smooth and has a very reduced median keel. The posterior marginals are smooth. The color is light to dark brown, with very small black dots. The plastron is large and unhinged and

© F. Bonin

The snout is short, and the sides of the head are rather dark, without much decoration. It is the top of the head that is astonishing in this species, as can be seen in this photo.

© F. Bonin

The *Sacalia* species have very bright ocelli on top of the head, which may startle a predator that momentarily mistakes them for the eyes of a dangerous opponent. Some butterflies can flash a similar pattern. Thin light lines extend back to the base of the neck. The carapace has a keel with distinct light coloration.

Common name

Beale's eyed turtle

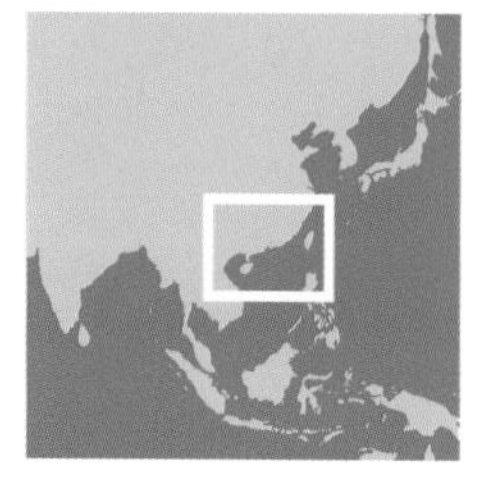

has a medium anal notch. The bridge is of the same size as the posterior lobe and has small axillary and inguinal scutes, although the latter are sometimes absent. The plastron is yellowish, with vermiculations and small dark spots. The head is small, the snout short, and the upper jaw unnotched. The skin of the posterior of the head is very smooth, but there are small granular scales on the sides of the head, between the orbit and the tympanum. The head color is yellow-brown, dark brown, or olive. On the top of the head, there are two to four yellow ocelli with black dots in the center, resembling a pair (or two pairs) of eyes. The two posterior ocelli are always present, but the front ones may disappear under secondary mottling or marbling. The jaws are brown to black and sometimes have fine vertical black lines. The neck is dark brown on top and may take on a reddish tinge below. On the top of the neck there are three thick yellow bands, bordered with black. The forelimbs are covered with enlarged scales on their upper face, and they may be dark brown, or sometimes yellow, reddish, or actually red. The males sometimes have bright green ocelli, with blood-red eyes, and the plastron is completely covered with a complex mahogany design, while the females have light yellow ocelli and a yellow plastron with fine radiating lines.

Natural history. This turtle lives in small watercourses, as well as in wetland areas in forested places that are not above 400 m altitude. It is very agile and climbs over rocks easily. Omnivorous in diet, it seems to prefer fruit. Nesting includes two to six elongate, white eggs.

Protection. Like the two other *Sacalia* species, this turtle is much sought after by hobbyists and is accordingly subject to heavy collecting pressure. It is also collected for food despite its small size. It is considered to be endangered by the Turtle Conservation Fund.

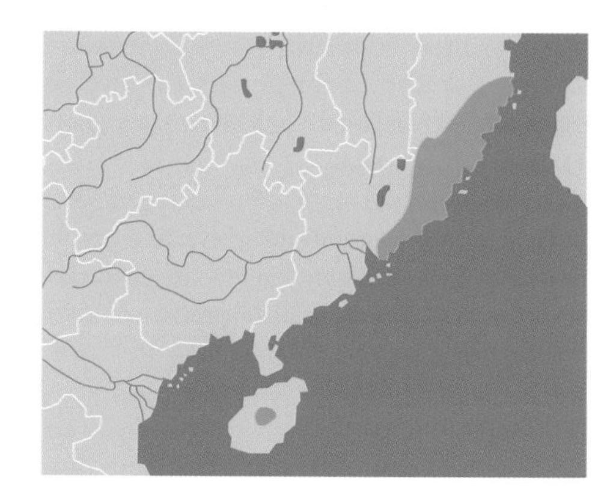

Sacalia pseudocellata Iverson and McCord, 1992

Common name

Possibly a hybrid.

Hainan eyed turtle

This is a questionable species that may be a hybrid between *S. quadriocellata* and *Cuora trifasciata.*

Distribution. Endemic to the island and province of Hainan, China.

Description. This turtle, quite recently discovered, may reach a length of 180 mm (in females). With a wider shell than its congeners, it has three small carapacial keels. The plastron is heavily pigmented, and is wider and longer than in other *Sacalia.* The gular scutes are narrow and rather long, and the interpectoral seam is very long, whereas the interfemoral is short. The anterior plastral lobe is shorter than the posterior one, and there is a small anal notch. The plastron varies from yellow to orange, with black pigment scattered over the entire surface. Only the edge of each scute remains yellow. The bridge has a longitudinal black band. The head is very narrow and there is no notch in the upper jaw. There are some very small tubercles behind the corner of the mouth and in front of the tympanum. The top of the head is uniformly yellow to olive-green. On top there are some designs, yellow to olive green in color, bordered with black, which are quite elongate but are interrupted in the posterior region of the head, from whence comes the name *pseudocellata.* A yellowish band, bordered with black, starts behind the ocelli and extends along the entire length of the neck and on each side. The chin and throat are uniform yellow. The lower surface of the neck varies from yellow in front to orange at the rear. The upper side of the forelimbs is covered with wide, imbricated scales, and the upper side of the limbs and tail has great numbers of minuscule scales. The ventral region of the limbs is orange to intense salmon, and the tail is dark olive brown on the underside.

Natural history. The lifestyle of this turtle is very similar to that of *S. bealei.*

Protection. No data are available.

Sacalia quadriocellata (Siebenrock, 1903)

Distribution. The specimens described by Siebenrock were found at Phuc Son, north of Annam in Vietnam. More recently the species has been reported to have a disjunct, interrupted range in suitable montane habitats from central and southern China—including the provinces of Guangdong, Hainan, Guangxi, Fujian, and Jiangxi—as well as in Laos.

Description. This species is very similar to *S. pseudocellata* and *S. bealei,* but with four well-developed, brilliant ocelli on the top and back of the head, showing variation between the sexes and to some extent between individuals. The carapace is oval, broad, and somewhat roof-shaped, or tectiform. The central keel is best developed on the fourth and fifth vertebrals. The carapace is brown to blackish or dark green, with black areas along the seams or, by way of contrast, widely scattered light markings on the remainder of the scutes. The head is dark beige or sometimes black or bluish, occasionally marked with mottling or light spots. The four ocelli are located behind the eyes and on the back of the cranium, forming devices of almost luminous intensity—often primarily yellow with a central black spot, or light blue encircled by white and with a central black point, somewhat like an aerial view of an atoll. Furthermore, the neck is marked with yellow lines, narrow but regularly spaced, that originate at the skin fold behind the scales of the head and finally disappear at the base of the neck. The ocelli in males are often brighter than those of the females and may even be a brilliant golden yellow. The digits are well webbed.

Natural history. This turtle lives in streams in mountainous regions and feeds on earthworms, mollusks, crustaceans, and small fish.

Protection. The beauty of the head markings makes this animal a favorite among collectors, and the small Hainan population is exploited for export. The species is considered endangered in China. The Turtle Conservation Fund (2002) also considers it to be endangered and in need of immediate protection.

Common name

Four-eyed turtle

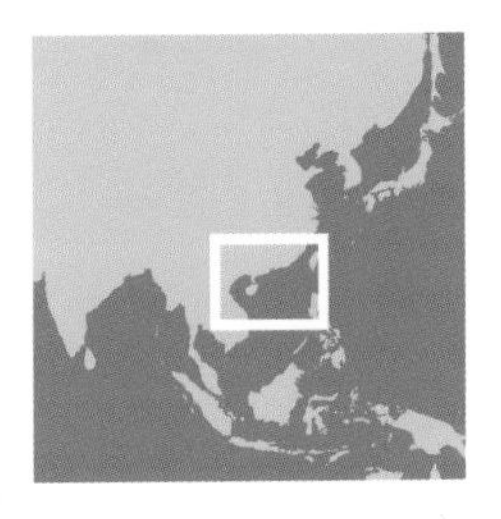

As its name implies, this species has four splendid, almost luminous ocelli on the back of the head, unique among turtles. They may serve to scare off predators.

© A. Dupré

EMYDIDAE Rafinesque, 1815

This is considered to be the most modern of the turtle families, with a known antixuity of just 80 million years; it also includes the greatest number of species. The genera show some one-on-one ecological parallels with batagurid genera in the Old World and include moderately large fully aquatic species to quite small, semiterrestrial or, in some cases, fully terrestrial forms. Apart from Emys orbicularis *in the Mediterranean region and the Near East, they are all confined to the Americas and are most diverse in North America. Shells range from domed to flattened, and some species have extremely decorative markings, which makes them sought after as pets—one of them, the famous red-eared slider, has been marketed (and introduced into natural ecosystems) on an almost global basis. Others are caught for food. The larger species are often quite strictly herbivorous as adults; the smaller ones may be omnivorous or even carnivorous. Some species have rather elaborate courtship rituals. Currently, two subfamilies are recognized, the Emydinae Rafinesque, 1815, and the Deirochelyinae Agassiz, 1857.*

Emydinae Rafinesque, 1815

Actinemys marmorata (Baird and Girard, 1852)

Distribution. Except for *Chrysemys picta bellii*, this is the only freshwater turtle in the Pacific Coast states of the USA. Its range extends from Seattle in the north to San Diego in the south (with a discontinuity south of Seattle as far as Portland, Oregon), along with scattered colonies in the northern half of Baja California, Mexico. Some populations, including those of the Truckee and Carson Rivers in Nevada, appear to be extinct.

Description. This species was formerly included within *Clemmys* and sometimes in *Emys*. The name *marmorata* refers to the marbled background of the skin and carapace. In truth, this marbled effect is mostly visible in turtles from the northern part of the range; elsewhere, the animal shows a remarkable superficial resemblance to the European pond turtle, *Emys orbicularis*, although the plastron in *A. marmorata* is more rigid. The carapace is flat and smooth, lacking keels and serration, and is wider behind than in front. The maximum size is about 210 mm. Most often the animal is greenish in color, sometimes olive or brown, with radiating series of yellow dots on the scutes. The soft parts have the same appearance, but the series of dots form actual lines and reticulations on the top and sides of the head. As these turtles age, some individuals become uniform brown or olive. The plastron is rigid, almost entirely black, but it may be cream or yellowish, liberally marked with blotches around the edge and along the seams. All in all, there is considerable chromatic variation, reflecting both different habitats and different populations. The head is rather small, with a somewhat protruding jaw. There is little sexual dimorphism. The tail, which is long in both sexes, is slightly wider at the base in males. The young are much more colorful and also display great chromatic variation, especially in the plastron, which may show ocelli or curved black lines forming graphic figures on the yellow background on the plastron. The young also have a strong median keel.

Western pond turtle

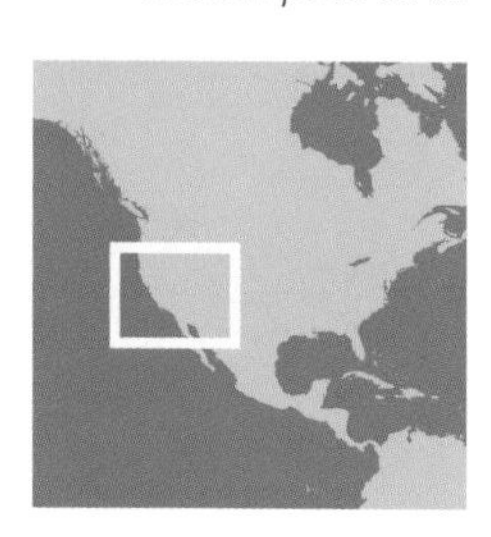

© F. Bonin

In many respects, this species resembles the European pond turtle, but it reaches a larger size, and the coloration is more marbled than dotted.

Subspecies. *A. m. marmorata* (described above) occurs in the north, from British Columbia to San Francisco Bay and western Nevada. The inguinals are often well developed and triangular. It is darker than the other subspecies. *A. m. pallida* (Seeliger, 1945), the southern representative, is found from San Francisco Bay to Baja California in Mexico. It is lighter in color and smaller than the nominate race, and about 60% of individuals lack the inguinal scutes. The head shows fewer yellow spots and tends toward a uniform grayish.

Natural history. This turtle lives in swamps, ponds, oxbow lakes, slow-flowing rivers, and even estuaries—indeed in virtually all freshwater habitats within its mostly arid range. It has also been observed at up to 1,800 m altitude. Although strongly tied to aquatic environments, it seems equally at home undergoing long treks across dry land, and turtles in some populations are supposed to be very active at night, as may occur in Oregon (J. Buskirk, 2002). In such places, the population density is high, and dozens of turtles may be seen taking the sun on snags and emergent branches, banks, and rocks. In the north, the turtles hibernate, sometimes for as much as six months, but those of the south do not do this. The diet is essentially opportunistic, and the adults are partially herbivorous. In general, the diet includes crustaceans, insects and their larvae, adult fish and fish fry, and amphibians, including tadpoles. Algae are also a major food source, especially for post-nesting females. Nesting occurs in June and July in the north, earlier in the south. Nesting may occur twice in a season, and each clutch may contain up to 13 eggs (35 × 20 mm, on average). Incubation takes 73 to 132 days (average 80). The young measure about 26 mm in length and weigh about 8 g. They have been observed in late summer in the south, but in the north they are seen only in spring, having passed the winter underground.

Protection. In recent years, populations of *Actinemys* in the Willamette Valley in Oregon and elsewhere have shown a serious decline. This is a result of direct and indirect impacts of human activities—urban development, agriculture, overgrazing, and highway construction. Many females have been crushed on highways while looking for nesting sites. More and more, wetlands are being reshaped or drained completely, and like the European turtles that they resemble so much, these turtles suffer greatly from having to share their habitats with humankind. Phytosanitary products (insecticides, etc.) are also a threat. Although this species was heavily collected and consumed by people at the beginning of the last century, these practices at least have virtually stopped. A number of researchers have centered their studies on this turtle, and conservation programs do exist, as in the San Joaquin Valley. This show of interest provides some hope that the status and populations may improve in the decades to come.

© F. Bonin

The young are decorated with regular designs on the scutes and black-bordered white lines on the sides of the head and neck.

Clemmys guttata (Schneider, 1792)

© J. Béhier

This small turtle is immediately identifiable. Its blackish to brown-greenish body is decorated with bright yellow spots, which presumably serve to break up the outline of the animal and make it harder to spot.

Distribution. This is the only species left in the genus *Clemmys,* and because it is the type species of the genus, the name has to be conserved. It has a wide range from southern Ontario and Maine southward into northern Florida. It may be found far inland in the southern Great Lakes region, western Ontario, New York State, Pennsylvania, central Ohio, northern Indiana and Michigan, and southern Illinois. It is also found farther south, in Georgia and a few scattered localities in Florida.

Description. This species is immediately recognizable by its black shell, boldly decorated with large, irregularly arranged yellow spots over the entire carapace, as well as on the head, neck, and limbs. In old specimens, the spots may fade never disappear altogether. This turtle is very small—not more than 125 mm in length. The shell is perfectly smooth, keelless, and unserrated, even in very young animals. The bridges each have an elongate black streak on a yellow background, which is quite distinctive and characteristic. The plastron is yellow or tends toward orange, with wide black markings on the edges. In old animals, these markings may spread, eventually covering almost the entire plastron. The head is of medium size, with a slightly extended upper jaw, like a parrot beak. Along the neck, fine yellow streaks sometimes form lines extending from below the eye to the base of the neck. In females the chin is often yellowish, whereas it is brown in males. The eyes are orange-yellow in the females, brown in the males. The juveniles are intense black, with an isolated yellow dot on each carapacial scute, except for the nuchal.

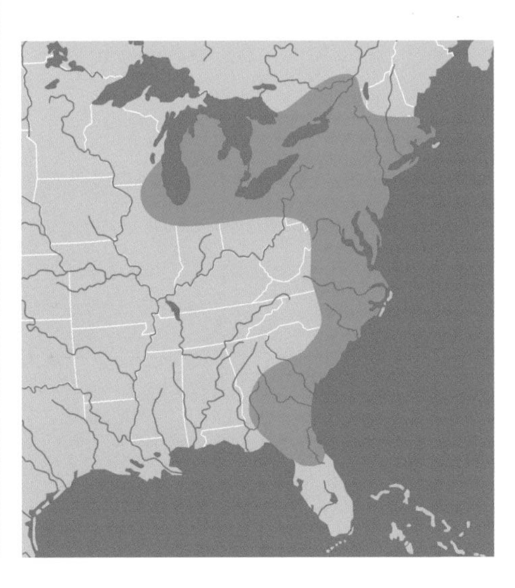

Common name

Spotted turtle

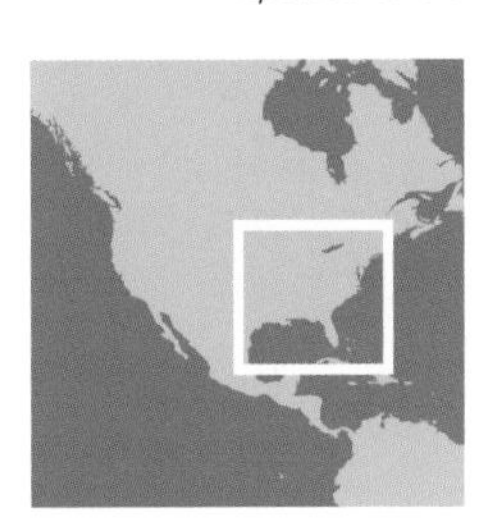

Natural history. This is a fairly aquatic species. It is found in marshes and small creeks, but it also is to be found in flooded meadows and friable soil, not far from rivers but deep in the woods. It is fully tolerant of the harsh winter temperatures that its northern habitat imposes, and Ernst has observed copulation at temperatures of 8°C

to 10°C. It hibernates for several months in the northern parts of the range and estivates in the south, utilizing burrows made by other animals or burying itself in the mud when the waters that it inhabits dry up. In diet it is omnivorous. Old specimens seem to enjoy algae and aquatic vegetation, but in general they feast on insects, mollusks, small amphibians, and freshwater crustaceans. Copulation is an energetic event and takes place at the edge of wetlands and marshes, in May to June in the north and in March to April in the south. The male pursues the female along the bank, then joins her in the water for actual pairing. Sometimes—but rarely—the male will fall perpendicularly onto his side during copulation. Two nesting per season are typical, occurring in May to July. The maximum clutch size is eight, and the eggs are small (32 × 17 mm). Incubation time is about 80 days. The young measure 30 mm and weigh 5 to 8 g. They hibernate in the nest in the northern part of the range and do not appear until spring.

© J.H. Harding

A young turtle has fewer spots, and they tend to be orange rather than yellow. Later the carapace will become darker and will lose some of the light spots.

Protection. Populations are dropping slowly, and the species is in danger of becoming rare. J. Behler considered that some populations were very threatened by intensive agricultural practices and modification of natural waters.

Emydoidea blandingii (Holbrook, 1838)

Distribution. This turtle is one of the most northern in the New World. It is found in southern Canada, from southern Ontario to the Great Lakes region, and west as far as Nebraska, Iowa, and northern Missouri in the USA. It also occurs in the New York State, Massachusetts, and southern New Hampshire and Maine.

Description. Superficially, this turtle resembles *Emys orbicularis,* at least in its yellow-spotted carapace and brown or olive background. However, the snout is more flattened, and the size is greater (females to 260 mm). The shell is narrow and rounded, unkeeled, and very smooth, with neither serrations nor rugosity. The vertebral scutes are wider than long. The first vertebral makes contact not only with the nuchal scute but also with the first four marginal scutes. The shell is black, with numerous yellow spots that are often arranged in linear series. With age, this decoration fades somewhat. The marginals carry an especially high density of yellow spots, which sometimes coalesce into irregular yellow blotches. The bridge is composed of flexible connective tissue and provides for a reasonable degree of plastral kinesis. The very large plastron is orange-yellow, with black markings on the outer parts; these markings form a symmetrical arrangement that, in the oldest animals, may extend over the whole plastron. The plastral hinge is located between the pectoral and abdominal scutes. The anal notch is slight. The head is small and flat, with a triangular snout. Brown to gray-blue in color, the head is some-

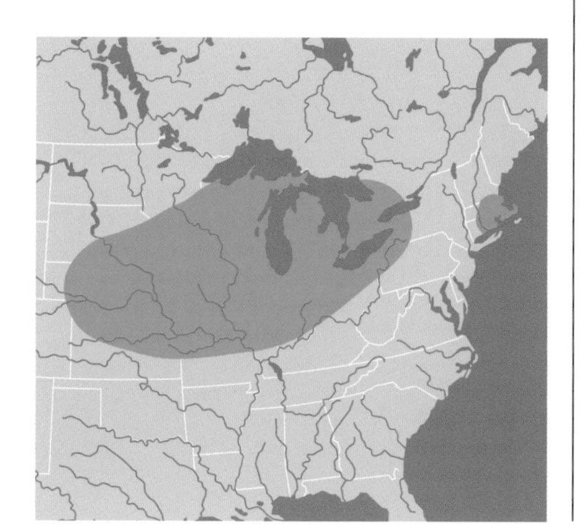

Common name

Blanding's turtle

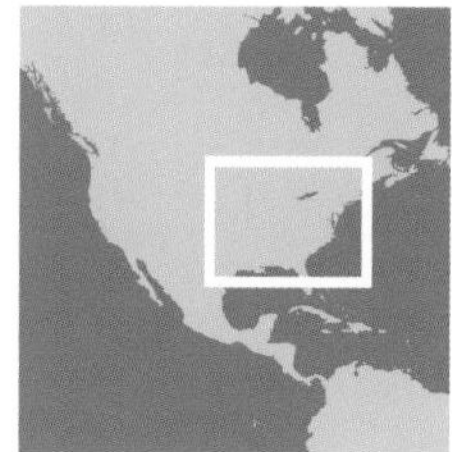

times finely adorned with yellow. The chin and throat are light yellow. The neck is very long. Sexual dimorphism is not greatly developed, but the plastron of males is slightly concave.

Natural history. This is an aquatic species, found in lakes, ponds, and mud-bottomed marshes, in which it loves to burrow. It may sometimes be seen basking on a bank or an emergent snag. Primarily carnivorous (eating amphibians, snails, slugs, and insects), this turtle does not disdain an occasional aquatic plant. Copulation takes place in the water, in late spring and early summer, and courtship is rather brief. Up to three nests are laid per season, from mid-June to late July. The maximum clutch is about 17 eggs, which have parchment shells and are about 35 × 25 mm in size. Incubation is short, about 70 days on average, but the young almost always overwinter in the nest and are ready to emerge the following spring.

© F. Bonin

This species looks like a large *Emys*, but the spotted pattern is more marbled, and old animals often lose their markings. The plastron is mainly orange-yellow, with black markings around the rim.

© F. Bonin

Emys orbicularis (Linnaeus, 1758)

Common name

Common European pond turtle

Distribution. This is the only representative of the Emydidae in the Old World, and it has a very wide range, from northern Germany and Poland in the north to Morocco in the south and the Caspian coast of Iran in the east. It also occurs in Lithuania, France, Portugal, Spain, Tunisia, Italy, Sardinia, Cor-

sica, all the nations of the former Yugoslavia, Albania, Greece, Turkey, the northern Black Sea nations.

Description. The name *Emys* apparently derives from an ancient linguistic error, from the time of Aristotle to that of Rondelet in the 1550s, who confused the Greek word *mus* (a mouse) with *Emus* or *Emys,* meaning a freshwater turtle. The French name, *cistude,* derives from the Latin *cistula,* a small box or jewel case somewhat like the shell of a land turtle. This species does not exceed 210 mm, and there is some geographic variation in size, with a tendency toward smallness in the southwest and toward large size in the northeast. The carapace is flat, well rounded, and covered with small yellow spots, which are scattered over the soft parts as well as the shell and sometimes form circles or radial designs. With age, specimens become uniformly brown, grayish, or greenish, but their skin and soft parts retain the attractive decoration throughout life. The eye is bright yellow, almost luminous. The plastron shows great chromatic variation, from uniform yellow to intense black, sometimes with yellow blotches and streaks, according to environment and location. There is a moderate hinge between the pectoral and abdominal scutes, with greater kinesis in the females. At the level of the bridges, there is a length of ligamentous connection that makes this possible. In some individuals, the plastron can move to a significant degree in response to pressure on the bridges. The head is medium-sized and somewhat elongate, with a triangular snout, a wide mouth, and rather large eyes, placed well forward. The skin of the neck and the limbs is very fine, with minute scales. The forelimbs are webbed, and the tail is often long and very slender—slightly shorter in females than in males. The males have a slight plastral concavity. Sexual dimorphism is manifest above all by the color of the eyes, which are brick red or scarlet in the males and yellow in the females. In the juveniles, the colors are brighter, and a slight median keel and modest denticulations along the marginals can be detected. The marginals are olive in color, with a yellow stripe along the marginals; the yellow spots appear during the second or third year of growth. The tail of the hatchling is very long and slender and is frequently truncated by the bites of its brethren (and elders). It is important to recognize the great plasticity of this species and the diversity of its color patterns, which has given rise to an array of subspecies.

This turtle has a shell with light radiating lines and very small dots on the skin. But there is great variation among the subspecies, the ecosystems occupied, the country, and the climatic zone.

Subspecies. Recent work by Uwe Fritz (1990–1997) has documented the genetic and morphological basis for recognition of a large number of subspecies, some of which are not always accepted. The variation is so complex and subtle at times that it becomes impossible to identify a specimen without knowing its origin. In general, northern and eastern turtles are larger than those to the south and west (210 mm and 140 mm, respectively, for females). In the south, plastra are often light yellow, while in the north they frequently show black spots. In the east, color tends toward red-brown, while in the southwest a greenish or beige tint predominates. The yellow pointillism is clearer in the west and southwest, while toward the east it is very understated. The following subspecies are recognized at present: *E. o. orbicularis,* the nominal subspecies and one of the original taxa described by Linnaeus (1758); doubtless the most widespread of the subspecies; found in northern Europe. *E. o. capolongoi* (Fritz, 1995). *E. o. colchica* (Fritz, 1994). *E. o. eiselti* (Fritz, Baran, Budak, and Amthauer, 1998). *E. o. fritzjuergenobsti* (Fritz, 1993). *E. o. galloitalica* (Fritz, 1995), present in Provence, Italy, Corsica, and Sardinia; small size. *E. o. hellenica* (Valenciennes, 1832), found in Greece, Turkey, and the southern part of Central Europe. *E. o. hispanica* (Fritz, Keller, and Budde, 1996), confined to Spain. *E. o. iberica* (Eichwald, 1831); *E. o. lanzai* (Fritz, 1995). *E. o. luteofusca* (Fritz, 1989), found only in central Turkey; coloration rather diffuse, consisting of a yellowish background with washed-out brown markings, distributed randomly. *E. o. occidentalis* (Fritz, 1993), found in Spain and Maghreb. *E. o. persica* (Eichwald, 1831), found in the extreme eastern part of the range.

Natural history. This species is often called mud turtle. In truth, it loves the muddy bottoms of slow creeks and rivers, marshes with flowing water, and ponds and canals choked with debris. Very aquatic, it nevertheless spends a large part of the day basking on emergent trunks, rocks, and banks during the warm season. It may stay immobile for several hours, still as a statue. But at the least disturbance, the slightest vibration in the air or the land, it dives into the water and does not show its head again for several minutes, and even then with great caution. Much of the winter is spent in hibernation, when the temperature of the ambient water falls below 14°C. It then hides itself in moist retreats at the edge of the water or sometimes under piles of leaves or in the mud and general detritus at the bottom of the river. In such a case, it will come back up to the surface at long intervals (several days or weeks) to take a new charge of air before returning to hide in the mud below. The flat shell and mud-covered carapace make the creature look like a smooth pebble, and this helps it pass unnoticed in the places where it spends its time. It is somewhat gregarious, and one often finds groups of a dozen or even a score of individuals. It will dash out from its hiding place at the bottom of the water to seize small prey—fish fry, tadpoles, aquatic insects. Carnivorous when young, it becomes more omnivorous with the passage of time and consumes aquatic plants and algae. In certain lakes, it lives in the reed beds, although in southern climes it seems to like water-courses with pebble bottoms. It is much more sensitive to pollution than its southern cousin *Mauremys*. When a body of water becomes polluted, *Emys* disappears rather rapidly, while *Mauremys* lives on. When attempting to hide itself, this turtle will fling mud or sand onto its carapace with its forelimbs, until it is covered and concealed from view (F. Bonin, 1995). When the mating season arrives (May–June, somewhat earlier in the south), the eyes of the male take on a brick red color. The male bites the female on the legs and neck and turns around her in a sort of ballet that has more in common with the behavior of some of the terrestrial turtles than with that of the familiar emydids of the New World. Mating always takes place underwater, and two or three extra males may join the couple, weighing her down so that, in some cases, she may be drowned. Nesting starts in early May in the south, and in early July in the

This turtle from Provence warms itself in the sunshine. The spotted design could make one confuse the markings with spots of sunlight reflected from the water, and it may aid in camouflage.

© B. Devaux

north. It may take place as much as several hundred meters from the water, and sadly, fatal incidents on highways frequently intervene. Nesting often starts late in the day, even at nightfall. The eggs are elliptical, white, and soft-shelled, measuring about 35 × 22 mm, and number 3 to 16 per clutch—most often 5 or 6. Incubation time varies with latitude but is rather short (60–70 days. Hatching in Provence occurs from mid-August to October, but some hatchlings (in the north) remain underground and emerge in the spring. The hatchlings, among the smallest of chelonians, weigh about 5 g, although some may be as small as 2 g. It was with this species that Claude Pieau (1974) demonstrated a relationship between the sex of a turtle and the temperature at which its egg was incubated. In the first weeks of incubation, if the eggs are kept below 26°C, only males result, and above 32°C the hatchlings are all females. In nature, of course, diurnal cycles of temperature, with cool nights and warm days, tend to produce about 50:50 males to females. As with most turtles, the mortality of the hatchlings is very high. After exiting the nest, the youngsters have to find their way to water—perhaps at a considerable distance—and while on land many are devoured by magpies, crows, rats, hedgehogs, and snakes. Up to the age of 5 to 6 years, they make themselves almost invisible, hiding in floating algae or under islands of vegetation.

Protection. This species is completely protected by law throughout its range, although up to the middle of the last century it was still sold and even consumed. Nevertheless, its habitats have been degraded rapidly as European lands become ever more urbanized. The wetlands are subject to draining and drying up, and throughout nearly all the nations where it occurs, it is undergoing fast retreat. It also suffers from the introduction of American turtles, escaped or released, which outdo the local species in both size and appetite. For the last 10 years, a series of congresses have been dedicated to this single species of turtle, so as to understand its status and natural history as fully as possible and to lays plans for its conservation. And more and more protective measures have been taken: reserves, parks, and protected areas have been declared, and the governments of most of the range states have done the best they can to prevent extirpation. Several raise-and-release programs have been undertaken in recent years (pioneered by the Swiss) and are starting to give encouraging results, as in France at the Lake of Bourget. Thanks to the mobilization of specialists, this turtle has become one of the most studied and best-protected species.

Glyptemys insculpta (Le Conte, 1830)

Distribution. This species occupies a wide range in the northeastern United States, extending from as far north as Nova Scotia southward to northern Virginia. To the west, these turtles are found in southern Ontario, New York State, northeastern Ohio, Michigan, Wisconsin, eastern Minnesota, and northeastern Iowa.

Description. The carapace reaches a length of 130 mm and presents a thickened and sculptured appearance, with strongly engraved grooves and striations that give it the name *insculpta*. The shell is rather flat but has a light keel and is wider in front than in the rear. The vertebral scutes and costals are large and bulge upward into a kind of dome under the pressure of the ribs and powerful shoulders. The general color is brown to grayish, with black and yellow on the costal scutes, radiating outward from the areolae. But these contrasting colors are best seen in young animals; the old ones tend toward uniform brown. The plastron is yellow, with a series of regular dark markings, one

© B. Devaux

Here the deep grooves in the shell that give rise to the name of this species can be seen. The head and neck are strong, and though the head is dark, the underside of the neck and the limbs are a brilliant light cream color.

on each scute. The soft parts are orange to beige. On the limbs, the enlarged flat scales are brown. There are four powerful claws on each of the limbs. The head is of medium size, with a beak with a strong hook, and the eyes are yellow and black. The top and sides of the head are brown-black. The males have protruding scales on the forelimbs, a concave plastron, and a very deep anal notch. The juveniles resemble the adults and are brown to grayish, without any orange on the neck and limbs. They have no keel, and their shell is about as wide as long (30 mm).

Natural history. These are forest turtles (hence the name "wood turtle"), living near shaded and marshy areas, but in the morning they like to warm themselves in the sun in the middle of open areas. In the course of hot, dry summers, they dig deep into the mud, and in the winter they hibernate. Mainly herbivorous, they eat grass, fallen fruit (bilberries, strawberries), willow leaves, sorrel, and filamentous algae. They also eat dead insects, earthworms, and gastropods, as well as dead birds and small mammals. Courtship occurs from late May to the end of September and takes the form of a dance. The male and female approach each other very slowly, heads held high and fully extended, in a posture of ecstasy. They bow their heads to each other, from right to left, for long minutes. Sometimes the males are more aggressive and bite the females. Sometimes the female has been observed to take the initiative in these preliminaries, which is exceptional among chelonians. Mating usually takes place underwater or in the mud of a marsh. The male immobilizes the female by gripping her firmly, with all four limbs. Nesting occurs in late May to June. The eggs, 4 to 18 in number, are rather large (40 × 26 mm). Incubation takes about 80 days. The young often remain underground for the winter, making their escape in the spring.

Protection. These turtles suffer from the destruction of their habitat and also illegal collection for sale to hobbyists. They have been placed in Appendix 2 of CITES and are considered to be vulnerable in certain states. In Quebec they are totally protected and may not be kept by private individuals. Capture is also forbidden.

Common name

Wood turtle

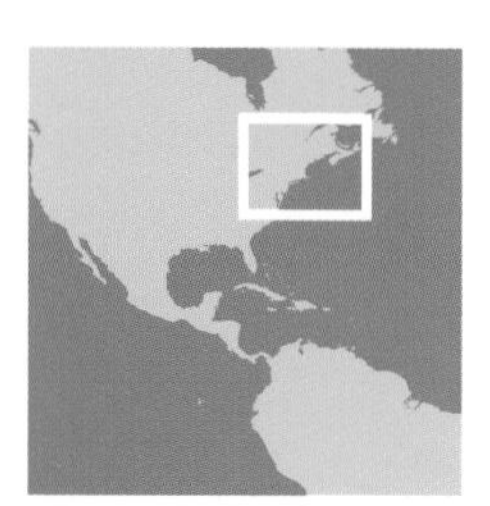

Glyptemys (Clemmys) muhlenbergii (Schoepff, 1801)

Distribution. This species occupies two separated areas in the northwestern United States. The first encompasses western Massachusetts, western Connecticut, and eastern New York. The second, to the southwest, includes Pennsylvania, New Jersey, Delaware, and Maryland. This turtle also occurs in southern Virginia and extreme northern Georgia.

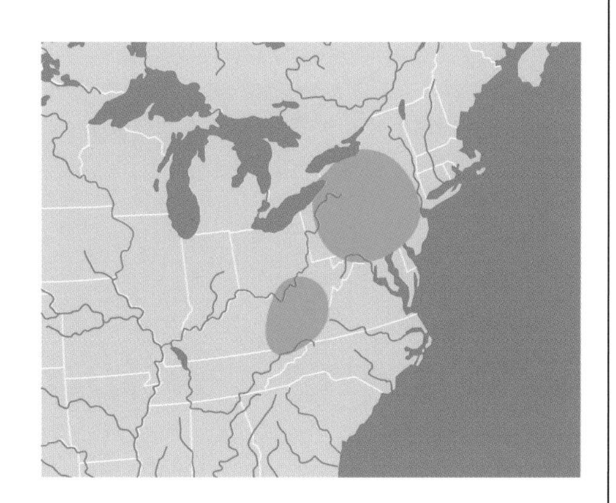

Common name

Bog turtle

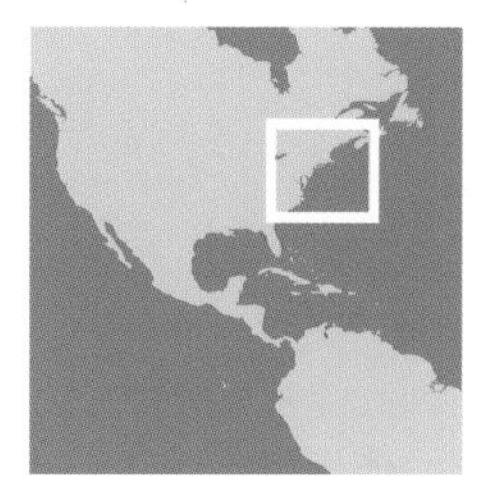

Description. At just 115 mm, this is the smallest of the "former *Clemmys*" species. It is rounded and rugose, with good growth annuli visible, as well as a light vertebral keel. Sometimes the marginals are slightly serrated. The general color runs from light brown to reddish black, with the areolae slightly lighter (beige to reddish). The keel is marked with a narrow straight line, beige to orange in color. The internal face of the marginals and bridges is the same color as the carapace. The plastron is dark brown, sometimes almost black, with irregular light areas that darken with age. The vertebral scutes are wider than long, the fifth being the longest of all. The costals are very large and heavily affect the narrow marginal scutes. The head is small to medium-sized, and dark brown to reddish, with a triangular, sloping snout and a large orange spot behind each eye. The back of the head, the sides, and part of the neck are lightened by a large beige to orange blotch. The upper jaw has a median notch. The soft parts are ornamented with lighter markings, yellowish or orangish in color. The limbs are brown to reddish, with light areas. The males have longer, stronger, and thicker claws on the forelimbs than the females do. The females have a slightly more domed and wider shell than the males. The juveniles are quite different: they have a brown carapace and a yellow plastron ornamented with an extensive dark figure in the middle. The head has light markings that are very well defined.

Natural history. This is not really an aquatic turtle, but it lives in flooded zones, stagnant and muddy waters, and small ponds with dense vegetation. It may be found at an altitude of up to 1,200 m in the Appalachians in North Carolina. It is most active during the warmest hours of the day and seems to have more need for heat than other members of the genus. Thus it will lie in the sun for long periods of time, on tree limbs or islands of vegetation, before it seeks to feed. It hibernates for a long time, October to March, in damp crevices along the riverbanks. It also uses holes made by other animals, including muskrats. It estivates in July and August and is mainly omnivorous, feeding equally happily on land or in water—perhaps a sign of its overall ambivalence. It eats insects, amphibians, and fish fry, as well as the bulbs of aquatic plants, berries that have fallen into the water, and algae. Mating has not been observed very often, because the females hide

© F. Bonin

The carapace is very rounded, with traces of a keel and a number of light spots on the blackish background. But one can always recognize this species, thanks to the large orange markings behind the eyes.

among dead leaves and it takes some time for the males to find them. Mating occurs only in the water, in May and June. There may be only one clutch per season, laid in mid-June to late July. The three to six eggs, are soft-shelled and rather small (30 × 16 mm). Sometimes the female lays them in sphagnum moss without digging a nest. Hatching starts in late August or early September. The young often remain on land, especially in the northern part of the range, not to appear until spring.

Protection. Placed in Appendix 1 of CITES because of its rarity, the species is still poorly known. Its status and the state of the various populations merit better evaluation. The species is listed as endangered by the Turtle Conservation Fund (2002).

Terrapene carolina (Linnaeus, 1758)

Common name

Carolina box turtle

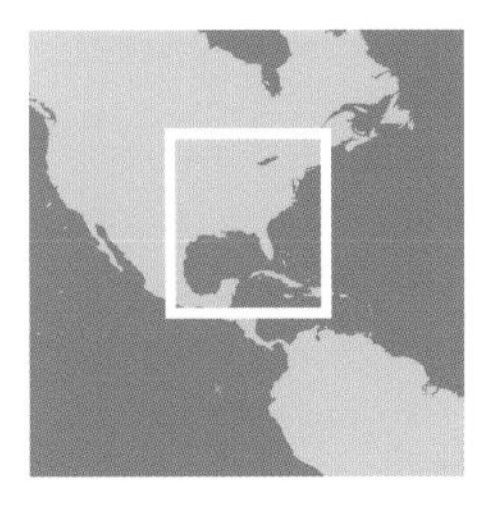

This species shows a wide variety of color patterns. The head is often brightly colored, with reddish, orange, or yellow markings, but in old animals the coloration may become uniformly beige, cream, or brownish.

Distribution. The wide range of this species encompasses virtually the entire eastern United States. Two isolated, quite restricted subspecies occur in Mexico. See details under "Subspecies."

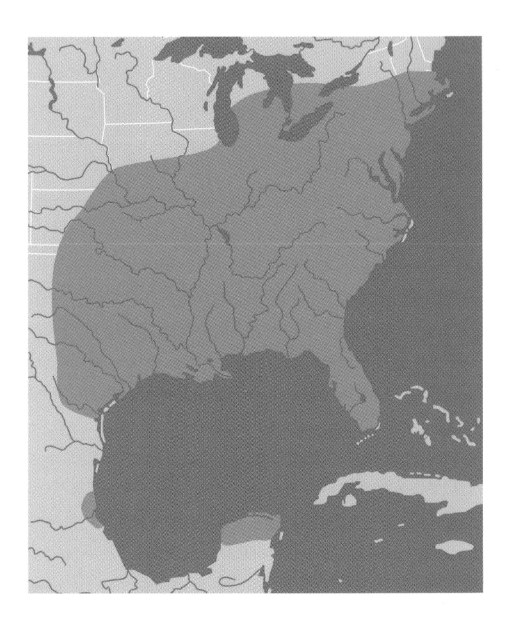

Description of the genus *Terrapene.* The name "terrapin" come from the Algonquians, the original indigenous people of the northeastern United States. Merrem recast the word as *Terrapene* in 1920. This genus is confined to the New World and includes four species, all known as "box turtles" because of the lidlike plastral lobes that can be raised to close the shell openings completely. None of them exceed 200 mm in length. Their morphology is closely parallel to that of the true tortoises, but they are more dependent on wet or humid habitats. Some, such as *T. coahuila,* are quite aquatic, but others, like *T. ornata luteola,* survive in very arid places. The hinge that allows them to close up so completely lies between the pectoral and abdominal scutes. The usual body form is rounded, with a strong dome, especially in *T. carolina* and *T. ornata*. They are able to extend their long neck vertically with the head angled horizontally, in a characteristic posture. They are brightly colored, at least when young. The head, covered with bright markings, the expressive eye, and their unique and often very elegant shell patterns have great appeal to collectors, and unfortunately they have been heavily commercialized in the past. They are omnivorous and enjoy insects, snails, small amphibians, and dead animals, as well as fruit and various vegetables. They are most active after rains and need to have frequent contact with

© J. Maran

This beautiful animal typifies the genus as a whole: a strongly domed shell, extensively spotted with yellow, and a head boldly patterned with very light blotches and marbling.

water or at least humidity. These small turtles can live as long as a century, and several cases of survival to 120 years of age have been reported, although such data are inevitably anecdotal rather than scientifically controlled. In captivity, they often develop respiratory maladies and ocular infections.

Description of *T. carolina.* The name *carolina* derives from the states of North and South Carolina, which in turn were named after King Charles ("Carolus") II. Since the seventeenth century, the species has been extensively imported into Europe, and countless pet chelonians in European hands have been named Carolina. The archetype of the genus, *T. carolina,* is short, round, and well domed, with almost vertical sides. The head has a well-rounded snout and an upper jaw with a pronounced beak. The coloration is quite variable, depending on the region, the individual, and the sex. In general the background is brown, with various hieroglyphs extending outward from the areolae or sometimes dispersed randomly. The forelimbs have wide, flat scales and are colored beige, brown, orange, yellow, or even red. The plastron is a uniform yellow, sometimes marked with dark spots in a somewhat radiating pattern. Some specimens, when old, become uniform brown, and this coloration is evident even in young specimens of *T. c. triunguis*. The head is often very colorful, with yellow or red markings on a brown or black background. The eyes are beige to red. The lower jaw and the underside of the neck are often very light yellowish or reddish, with dark spotting. Sexual dimorphism is not very obvious; the plastral concavity of the male is distinct in some subspecies but not all. The males do have a thicker tail, and their eyes are often reddish in the mating season, whereas the females' eyes are gray-brown or blackish. In some areas the males are larger than the females; in other areas the reverse in the case. The hatchlings, rather plain in appearance, do not have a plastral hinge. The carapace is rather flat and grayish, with a single yellow spot on each scute. There is also a slight vertebral keel. The plastron is cream to yellowish. The plastral hinge does not become functional for five or six years.

Subspecies. Six subspecies are recognized.

T. c. carolina (described above): The carapace is short, wide, and boldly marked. The marginals are almost vertical. There are four digits on each hind limb. Distribution is in the east: from southern Maine to Georgia and west to Michigan, Illinois, and Tennessee.

Often this species will extend its head and display a curious, inquisitorial face, which may explain why it was commonly sold, for a long time, as a "pet."

© B. Devaux

T. c. bauri Taylor, 1895: The coloration is less contrasting and less variable than in *T. c. carolina,* with radiating yellow lines on each carapace scute. The head is dark, with two light bands on each side. Three toes are present on each hind foot. This form is found mostly in Florida, as far south as the Keys.

T. c. major (Agassiz, 1857): This subspecies is the largest of the box turtles. The carapace is rather elongate, widely flared posteriorly, rather dark, and sometimes completely plain. On the other hand, some males are very brightly colored, with vivid markings on a black background. Four digits are present on each hind limb. Found in the south, this form ranges from eastern Texas to Florida.

T. c. mexicana (Gray, 1849): The carapace is long, domed, and beige to brown, sometimes pale yellow, with dark seams. Three digits are present on the hind limbs. This form is confined to a small area of Mexico in southeastern Tamaulipas, northeastern San Luis Potosí, and northern Veracruz.

T. c. triunguis (Agassiz, 1857): The carapace is narrow and rather elevated, with a tendency toward a midline keel. The background rather light, with dark designs. The plastron is uniformly yellow. The head and the limbs often have orange or yellow spots, and the head of some males is boldly marked with red. Three digits are present on the hind limbs. This form is found in the central part of the range—from Missouri to southern Texas and as far east as Alabama.

T. c. yucatana (Boulenger, 1895): The carapace is long and very elevated. The third vertebral forms a small knob, and the posterior marginals are slightly recurved. The background is beige to brown, with dark rays or with black borders on the scutes. The head may be very light, even whitish; the large eyes have orange irides. Four digits are present on each hind limb. This form has a small range in the Yucatán Peninsula of Mexico, in the states of Campeche, Quintana Roo, and Yucatán.

This individual is less domed than the turtle on the previous page, but the coloration is similar: yellow and cream streaks on the scutes, and bright markings on the head, neck, and limbs.

© F. Bonin

Natural history. These turtles live in wet areas, in forests or woods close to watercourses, but they can adapt themselves to dry, southern habitats, restricting their activity to the rainy months. They swim perfectly well and enter water frequently, but they are still terrestrial animals and can survive for long months without direct contact with the water. They also tolerate saline environments, as at Egmont Key, in west-central Florida. In the northern areas they hibernate for four to five months, but in the south they hide in the mud or under wet fallen leaves or they take over burrows excavated by other animals, but without becoming torpid. They are most active after rains. The young are mostly carnivorous, while old ones are frugivorous and herbivorous. At all ages, they show a fondness for mushrooms. The favored foods, especially of the young, include insects, small dead animals, earthworms, slugs, and small snails. K. Dodd has remarked that, on Egmont Key, *T. c. bauri* consumes lots of cockroaches that, without the turtles, would proliferate excessively. During the mating season, males will go long distances to find females. The male has red eyes and incurved claws on the hind limbs, which allow him to get a secure grip on the rear carapace edge of the female. She, by contrast, has yellow-brown irides. The mating ritual, which appears very clumsy, occurs in spring, in autumn, and anytime after rains. The male turns toward the female, bobbing his head vigorously and then biting her. Then he climbs on to her and comes down on her several times, gripping her firmly with his recurved claws. During copulation, he may continue to nibble at the female. The shell of his partner being nearly spherical, the male has to position himself vertically, standing on his hind feet. But if he falls backward while still penetrating the female, he may be dragged around unceremoniously. Nesting occurs in May and June. The eggs are elongate and small (30 × 21 mm) and number two to eight. Incubation varies from 70 to 110 days, with an average of 80 days. Hatching occurs from August to the end of October.

Protection. This species has been casually collected for three centuries, but today its export in controlled. In some areas the species is in decline, and one no longer sees it often in Western pet shops. Mortality in private hands is high, and raising these animals is very demanding. The species suffers mainly from human activities, the drying of wetlands, urban development, agricultural proliferation, and overgrazing.

Terrapene coahuila Schmidt and Owens, 1944

Distribution. The range is restricted to the basin of Cuatro Ciénegas, in Coahuila, northeastern Mexico. It has been found nowhere else.

Description. This is the rarest and least-known species of the genus *Terrapene*. It is thought to be essentially aquatic, but in truth it spends about equal times in the water holes themselves and in the surrounding grassy vegetation. In the marshes that constitute the preferred environment, these turtles walk along the bottom like a terrestrial species and rarely swim. This is the least colorful *Terrapene,* with an overall brownish, beige, or blackish color. Some may be mahogany or reddish in color, and certain individuals may have a few wispy dark markings on the head. The maximum carapace length is 165 mm. The shell is rather elongated and is rather flattened at the level of the second, third, and fourth vertebral scutes, and there are traces of a median keel. The marginals and vertebrals are very narrow, while the costals are quite high. The head is rather strong, with a short, round snout, with no beak on the upper jaw. The plastron is wide and well developed and lacks an anal notch. There is a large axillary scute on each side. The plastron may flex to an angle of about 45 degrees. In most males, there is a significant depression at the junction between the abdominal and femoral scutes. There are four digits on each hind limb and five on each forelimb. The juveniles are somewhat more colorful, with yellow streaks or light dots scattered over the dark brown background and a yellow band on each side of the head. This species possesses cloacal bursae, which it uses to moisten the soil at the nesting site. Sexual dimorphism is rather marked. In the males, the plastron is somewhat concave, the carapace more domed, and the iris brownish, spotted with yellow. The females have a slightly convex plastron and a yellow iris, spotted with brown.

Natural history. The only home of this turtle is a truly unique natural environment. In the valley of Cuatro Ciénegas, large round ponds called *pozas* receive water year-round from fresh or brackish springs. Sodium chloride is thus always present throughout the basin. The turtles occupy contrasting environments: water that is very pure or water that is charged with salt; wet savanna or dry savanna; and ponds that are cold (more or less) or warm. Each subpopulation of turtles adapts to the specific medium. Yet some individuals—mainly males—have the habit of wandering from *poza* to *poza,* and they may travel several kilometers in the course of these movements. In the principal lake in the basin, the turtles are covered with a white cloak of sodium chloride, which they seem to tolerate. The population density of the species may reach 20 individuals per hectare. At the site of a single *poza,* more than 400 turtles have been observed—all out of the water. According to the time of year, the atmospheric pressure, and the ambient temperature, one may find turtles either in the water or on a terrestrial substrate. They

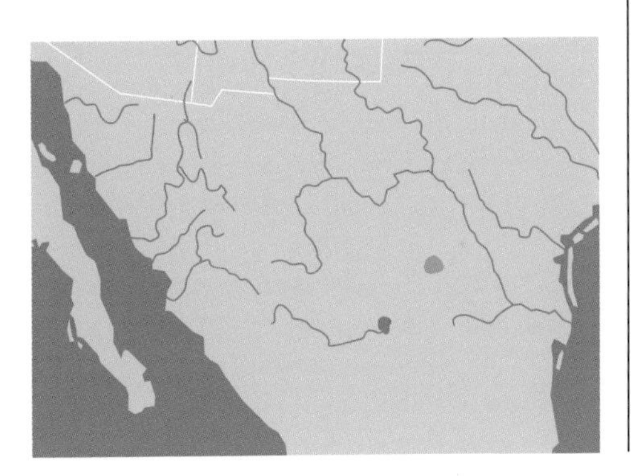

Common name

Coahuila box turtle

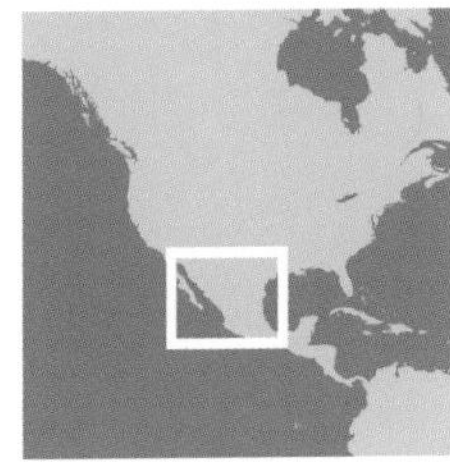

© F. Bonin

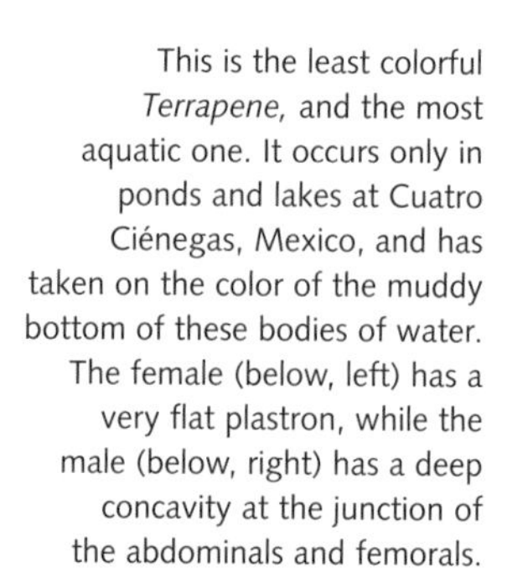
This is the least colorful *Terrapene,* and the most aquatic one. It occurs only in ponds and lakes at Cuatro Ciénegas, Mexico, and has taken on the color of the muddy bottom of these bodies of water. The female (below, left) has a very flat plastron, while the male (below, right) has a deep concavity at the junction of the abdominals and femorals.

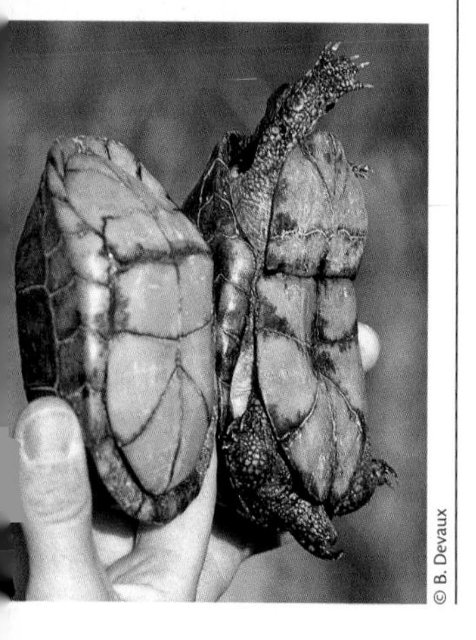
© B. Devaux

seem to appreciate both animal and vegetable foods, including insects and larvae, fish fry, algae, bulbs, and aquatic plants. They usually eat on land. When they are disturbed, they tend to flee into the water and then bury themselves rapidly in the bottom of the pond, digging with their forelimbs. They can remain underwater for half an hour without coming to the surface (Bonin, 2002). Mating takes place mainly from March to May and from September to November, usually out of the water. Nesting occurs from May to September. The eggs are hard-shelled, finely granulated, and elliptical (33 × 17 mm). There may be multiple nests in a season, but usually only two to four eggs are laid in a single clutch. The nest cavity is shallow and is excavated in the low dunes that encircle the *pozas*. Incubation time is about 60 days.

Protection. This is the only *Terrapene* placed in Appendix 1 of CITES. The guards at Cuatro Ciénegas have encountered several episodes of illegal collecting. Total protection of the whole basin seems to be necessary, since it is so rich in endemic life-forms, not just the turtles. Few chelonian species are adapted so completely to such a limited range, and anything that changes the environment, any degradation of this minuscule basin, may affect this species seriously. More and more, studies have been undertaken on this animal, with the purpose of understanding it better and planning effective protection. The species is classified as endangered by the Turtle Conservation Fund (2002).

Terrapene nelsoni Stejneger, 1925

Distribution. This species has a narrow distribution along the western side of Mexico, between the Pacific Ocean and the Sierra Madre, in the states of Sonora, Sinaloa, and Nayarit.

Description. This strange turtle with its unique shell pattern has rarely been seen by naturalists since its discovery two centuries ago. In the wild, it is a sort of Holy Grail of turtles: it is extremely rare for anyone to see one. It comes out just after rains but then hides for months, and we have little insight into its way of life or habits. The few known animals were found by luck or were given to zoos or to naturalists by local village people, but true observations of animals in the wild are essentially nonexistent. The local Mexicans call it "the turtle with little spots" because its carapace is spangled with bright yellow dots on a brown—or mahogany or almost black—background. The maximum size seems to be 159 mm. The shell is very domed,

elongate, and slightly flattened at the level of the third vertebral. The coloration differs greatly according to individuals and to geography, ranging from greenish to red-brown, and most of the time the beautiful spots fade away in old animals, especially males. Females are smaller, with a wider carapace and expanded marginals, especially at the rear. The head is rather strong, with a round snout and no beak on the upper jaw. Beige to brown or greenish in color, the head, like the shell, is covered with yellow dots. These dots coalesce to form blotches on the sides of the head, below the eyes, and at the base of the neck. The plastron is large and lacks an anal notch. It is yellow or beige, with dark markings on the sides, or is almost completely black, with a yellow border (see the "Subspecies" section). The hind limbs have four digits, as do the forelimbs, without traces of webbing. In contrast to the females, the males have a more massive head, red irides, a more pronounced upper jaw, a more elongate and higher carapace, less expanded marginal scutes, and fewer yellow spots on the carapace.

Subspecies. *T. n. nelsoni* (described above): This form occupies the southern part of the range, in Nayarit. The yellow spots on the shell are quite voluminous but less numerous. The first vertebral scute has a concave, scooplike profile. The plastron is black with a lighter border, sometimes spotted with faded yellow dots. *T. n. klauberi* Bogert, 1943: This subspecies has the larger range, which includes southwestern Sonora and northwestern Sinaloa. The yellow spots are smaller and more numerous and are dispersed on an olive background. The first vertebral scute is flat.

Natural history. This is the only species in the genus that is truly terrestrial. It occurs in various habitats, ranging from broiling deserts to chestnut forests at intermediate altitudes. But it is not seen during the rainy season (end of June, early July, and September–October). The main documented site is Guirocoba (410 m altitude), three hours by highway from Alamos, Sonora, where the animal occupies various environments ranging from semidesert to moist. Some turtles have been found by local people, but there is no information on their ecology. Captive specimens have survived for a short time, not counting those in the Ecological Center of Sonora in Hermosillo. In the latter case, three specimens kept in a cement bowl, along with various other animals, does not generate any useful information. This is an omnivorous species, feeding on insects, earthworms, dead animals, organic detritus, fruit, and trash of all kinds. Males seem to be particularly aggressive, and they court and mate with gusto. There is no information on the eggs, nesting, or incubation.

Protection. This species is an example of a rare turtle whose distribution and status are poorly known and that needs to be studied intensively in order to plan any useful conservation measures. Its environment is not destroyed or damaged, and its sheer "invisibility" prevents its collection. Nevertheless, its rarity and its narrow adaptations make it a remarkable animal and certainly one to save.

The shell and skin of this *Terrapene* are completely covered with yellow spots, on a brown or blackish background. The beak is quite pronounced and allows the animal to engage in a carnivorous diet.

© F. Bonin

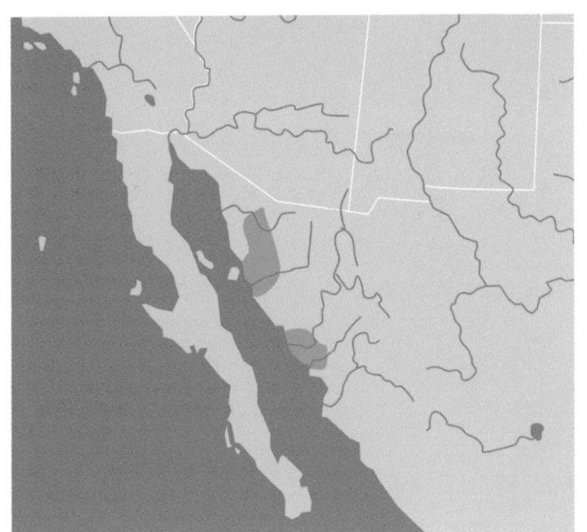

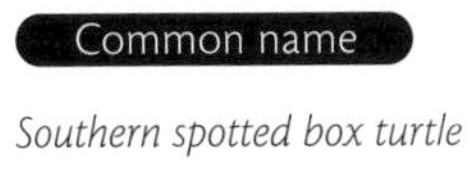

Southern spotted box turtle

Terrapene ornata (Agassiz, 1857)

If not for the radiating lines and brushstrokes on the scutes of the carapace, this *Terrapene* could be confused with *T. carolina*. This turtle is also flatter than the latter, and the plastron is colored like the carapace.

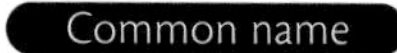
Common name

Ornate box turtle

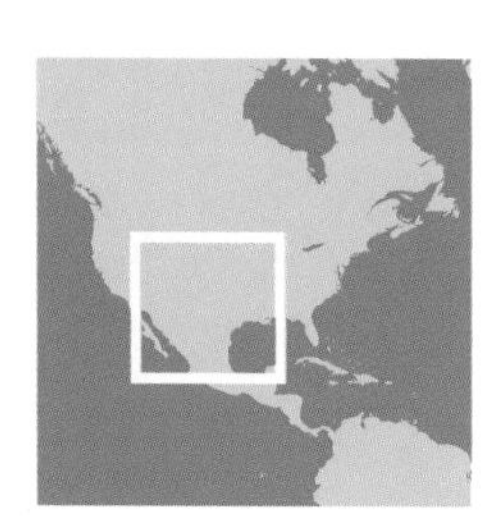

Distribution. This species is found in the central United States and northeastern Mexico, west of the range of *T. carolina*. Its range reaches up to the Great Lakes. It has been found in Illinois, Iowa, South Dakota, eastern Wyoming, southwestern Louisiana, Texas, New Mexico, southeastern Arizona, and Mexico, in northeastern Sonora and northern Chihuahua.

Description. Small is size (up to 140 mm), this turtle has a rather flat carapace, without a keel and with radiating and linear yellow designs on an almost black background. Often a vertebral stripe is present. The plastron is longer than the carapace and has the same patterning: broad yellow brushstrokes on a brown to black background. No bridge or axillary scutes are present. The marginals are vertical at the sides, expanded outward posteriorly. The head is rather small, with a nonprojecting snout and an overbite of the upper jaw. The soft parts are beige, light gray, or yellowish, and the head is brown, with light spots more or less organized into groups. The tail often has a yellow dorsal stripe. The hind limbs each have four digits, and the forelimbs have five.

Subspecies. *T. o. ornata* (described above): The carapace is somewhat flattened. The background is dark, and the yellow lines around each second costal scute number five to eight. This form occurs in the northern parts of the range, from Indiana and eastern Wyoming to New Mexico and Louisiana. *T. o. luteola* Smith and Ramsay, 1952: The carapace is lighter than in the type species, sometimes even with all colors faded out and a uniform beige predominating. One may see as many as 14 radiating lines on the second costal scute. This form occurs in the southern areas, in Texas to southeastern Arizona. In Mexico it attains the southern extreme of its range in Alamos, Sonora.

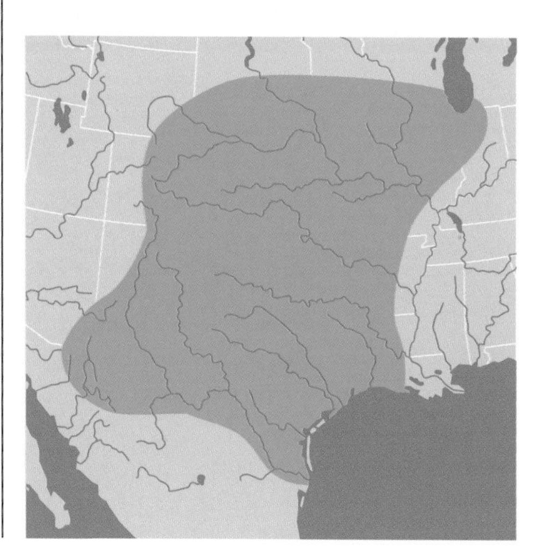

© B. Devaux

Natural history. This turtle lives close to wetlands but does not enter water voluntarily. It may be found up to 2,000 m altitude. It seems to prefer sandy, semiarid areas and can tolerate great heat. In the north of the range, it digs small burrows in winter, about 60 cm in depth, and buries itself from October to May. In the south, it slows its metabolism in summer and hides itself among leaves. It is most active after storms and rains. At such times it comes forth to catch and devour millipedes, beetles, dung beetles, and grasshoppers. It may also be seen catching rats and toads, and it loves berries and mushrooms. Some populations may reach high densities, the highest known for any *Terrapene*. Mating takes place primarily after spring rains. The male is recognized by his red irides and the moderate concavity of the plastron. The first digit of the forelimb is enlarged, thickened, and twisted inward, helping him maintain his hold on the female. Copulation seems to take a long time, sometimes lasting half an hour. Nesting has been reported from May to July, with one or two nestings per season. The eggs are fairly large (31 × 23 mm) and number two to six per clutch. Incubation takes about 70 days. In the young, the plastral hinge is not yet functional and does not develop until the third or fourth year.

The head and neck of this handsome specimen of *T. o. luteola* are brilliantly marked with three different colors. The carapace has typical brushstrokes of color.

Protection. This species was formerly exported in great numbers but is less so today. Some of the collecting pressure derived from the existence of legal protection for *Terrapene carolina* and all its subspecies (e.g., in Florida), without such protection being extended to *T. ornata,* which is still the case.

This photo of *T. o. ornata* shows the diagonal stripes and the keel outlined with yellow.

Deirochelyinae Agassiz, 1857

Chrysemys picta (Gray, 1844)

Distribution. This species has a very wide range in most of the United States, with the exception of the southeastern quarter, and also in southern Canada. Four subspecies share this huge territory.

Description. These are small turtles, not exceeding 250 mm, with very bright colors, as their name indicates, although they may fade a little with age. The shell is rounded, rather flat, and very smooth and lacks a vertebral keel. The color varies from olive green to black, with red or yellow stripes along the seam lines. The marginals have curved or straight red bars. The vertebral scutes are wider than long. The plastron is unhinged and is yellow, as are the bridges. The head, of medium size, is decorated with yellow stripes. In particular, one yellow line starts below the eye and extends

The carapace is wide, oval, and very flat, with just a light line along the keel and a light border around the edge. The head and neck are moderately striped with yellow.

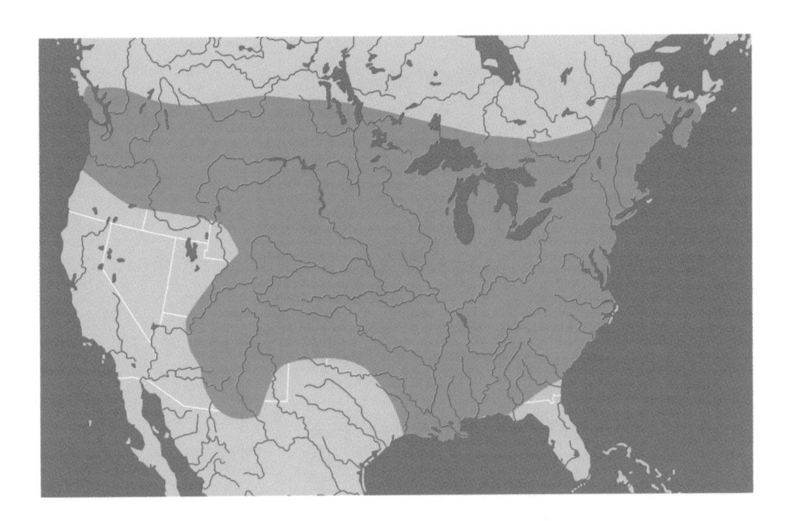

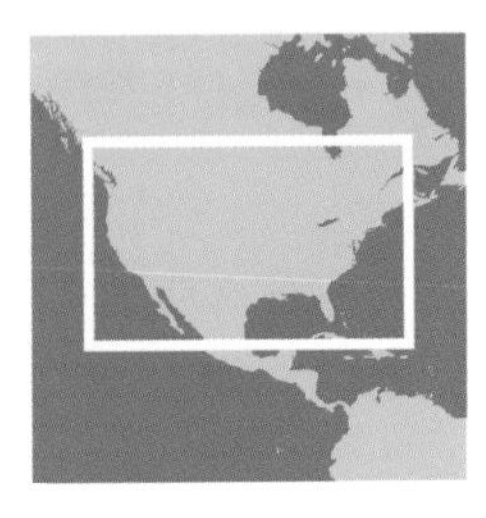

Common name

Painted turtle

posteriorly. On each side of the head, a wide yellow spot is located behind the eye. The chin is marked with two wide yellow lines, which reconnect posteriorly.

Subspecies. Four subspecies are recognized.

C. p. picta (Schneider, 1783): The eastern painted turtle is found from southeastern Canada to the state of Georgia, as well as in eastern Alabama and much of the eastern part of the United States. This race is easily recognized by its completely yellow plastron and the light seam lines on the carapace. The seams between the costals and vertebrals are aligned rather than alternating.

C. p. bellii (Gray, 1831): The western painted turtle is found in southern Canada (western Ontario to British Columbia) and in the United States in Missouri, northern Oklahoma, eastern Colorado, Wyoming, Idaho, and northern Oregon. This is the largest painted turtle. The carapace is almost always green, with numerous yellow markings. The plastron is marked with a dark design over almost the entire surface. The bridges and the undersides of the marginals, as well as the edges of the plastron, are often a bright orange-red.

C. p. dorsalis (Agassiz, 1857): The southern painted turtle ranges from the extreme south of Illinois and Missouri to northwestern Alabama and along the Mississippi to the Gulf of Mexico, in Louisiana. It has a very dark to black carapace, with a bright orange stripe along the midline. The plastron is uniform yellow. The largest individuals are only about 160 mm long.

C. p. marginata (Agassiz, 1857): The central painted turtle is found in Canada (southern Quebec) and in the central United States (Tennessee, northern Alabama, Pennsylvania, and Virginia). The seams between the vertebrals and costals are alternating. Along the seams are well-defined dark bands. A dark figure, of variable size, extends along the plastron. It is centered along the plastral midline and reaches the extremities of the plastron.

This photo brings to life the bright head coloration, as well as the light markings of the marginal scutes.

Natural history. These turtles frequent still waters or slow streams with a sandy or muddy bottom and with plenty of aquatic vegetation. Along the Atlantic coast they are also often seen in somewhat brackish waters. They are diurnal in habit and spend the night at the bottom of the water. Their activity period extends from March to October, but it varies with latitude. When they are basking, which occupies most of their time during the warm months, they may share their basking log with several other species. Hibernation occurs underwater. The turtles bury themselves in the mud and sometimes the water temperature may drop as low as 4°C. They are also able to stay out of water for long periods of time—for example, when they need to find a nesting site—and they may move long distances in search of alternative or better ponds. In some areas they spend a great part of the year underground. Courtship occurs from March to mid-June. The male places himself in front of the female, underwater, and caresses her with his long foreclaws. Nesting occurs from late May to mid-July, with 13 eggs per nest on average, each measuring about 35 × 28 mm. Incubation takes 70 to 80 days, and emergence starts after August 15. In the northern part of the range, the hatchlings remain underground after they hatch and do not appear until the following spring. These turtles are omnivorous and consume many species of plants as well as dead and live animals. Mobile prey, such as insects or earthworms, seem to be particularly attractive to them. Mostly carnivorous during their early years, they become herbivorous as they grow older.

Protection. One of the worst predators on these turtles is the raccoon, which is able to detect the scent of the fresh nest and to excavate the eggs without further ado. A great many painted turtles are killed by vehicles as they attempt to cross highways, and in some cases the habitat is being degraded rapidly.

Deirochelys reticularia (Latreille, 1801)

Distribution. This species is found in about 10 states of the southern and southeastern United States, including eastern Texas, all of Louisiana, and the southern halves of Mississippi, Alabama, Georgia, and South Carolina, as well as the whole of Florida, a large part of southeastern North Carolina, and Missouri. It reaches southeastern Oklahoma and the southeastern corner of Virginia and is also found in many of the barrier islands along the Atlantic coast.

Description. The strange name "chicken turtle" comes from past times, when this species was widely consumed and was considered everyday fare. It does not exceed 260 mm and has a long, narrow shell, rounded at the ends. There is no median keel. The sides are very smooth, with no notches, and there is a yellow band around the entire shell. The scales are rugose and colored a dark olive-brown. The vertebral scutes are wide, and the first is in contact with four marginal scutes. In most individuals, the scutes are adorned with a system of fine yellow lines that form an irregular network. The plastron is large and a uniform yellow in the eastern subspecies. The bridge is extensive and yellow and often has one or two dark markings, especially along the seams. The head is flat and elongate, and the neck is extremely long. The snout is pointed, and the upper jaw is smooth and sharp. The skin is more or less dark brownish, with brilliant light yellow bands on the neck and the head. The upper face of the forelimbs has a wide yellow vertical streak, on a very dark brown to black background. The webbed feet have fine, very sharp claws.

Subspecies. Three subspecies are recognized. *D. r. reticularia* (described above): The eastern chicken turtle is found from southeastern Virginia to the Mississippi River. It is identified by the presence of fine yellow bands forming a network over the scutes of the carapace, and there is often a dark marking at the seam between the femoral and anal scutes. *D. r. chrysea* Schwartz, 1956: The Florida chicken turtle is found only in Florida. The band around the shell is wider, and the color tends toward orange. The lines on the carapace itself are also wider, and their orange-yellow color is brighter. This is the largest subspecies. *D. r. miaria* Schwartz, 1956: The western chicken turtle is found throughout the remainder of the range—that is, west of the Mississippi River. The carapace is flatter, and an array of dark bands follow the seams on the plastron. The chin and throat are immaculate.

Natural history. This is a shy turtle that inhabits still waters—marshes, ponds, water holes, lakes. It prefers places where the vegetation is dense both on the banks and in the water. It

This large aquatic turtle is capable of extending its neck a considerable distance.

© B. Devaux

© F. Bonin

In this individual, the growth annuli on the scutes of the carapace are well marked. The head is decorated with a major line on the side and a finer stripe on the temples.

Common name

Chicken turtle

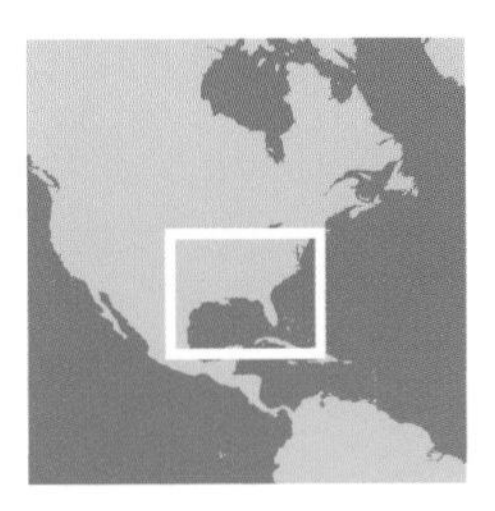

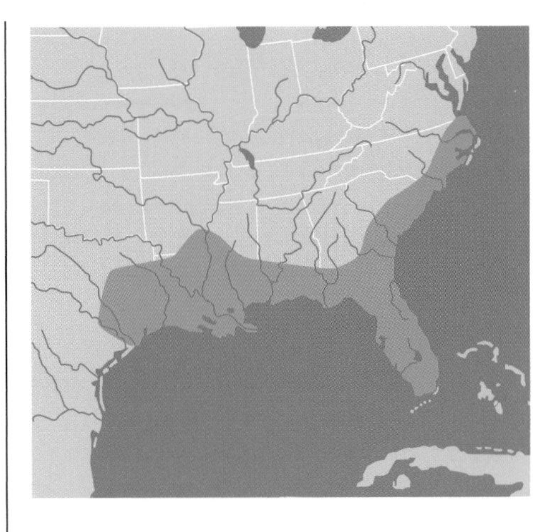

spends numerous hours out of the water, basking. It often walks overland, especially when drought dries up the marshes and it is forced to search for a new place to live. In the northern part of the range, it hibernates from November to early March. The male reaches sexual maturity very quickly (in three to four years). During courtship, he vibrates his elongate foreclaws rapidly against the head of the female. Nesting season depends on latitude. In Florida, it occurs from mid-September to early March. The eggs are white, soft-shelled, and about 36 × 20 mm in size and number from five to twelve. Incubation lasts for 84 to 150 days, according to latitude. The hatchlings weigh about 8 g. The young are mostly carnivorous, and the adults are also specialized for live food, especially crayfish, which they catch by means of "harpoon strikes" with their extremely long neck.

Protection. Raccoons, as well as alligators, are great predators on juveniles. In former times these turtles were also heavily predated by humankind, but this is rare today. They appear to be relatively short-lived even when protected, and Gibbons (1987) found that less than one individual in 100 survived to an age of 15 years.

Graptemys geographica (Lesueur, 1817)

Distribution. This species is found in the central United States, ranging from Canada (southern Quebec) to northern Alabama, and west to eastern Minnesota, and southern Wisconsin. It also occurs in northeastern Oklahoma, Arkansas, Tennessee, and northeastern Georgia. This is the only *Graptemys* species found in any river that drains into the Atlantic, rather than into the Gulf of Mexico.

Common name

Common map turtle

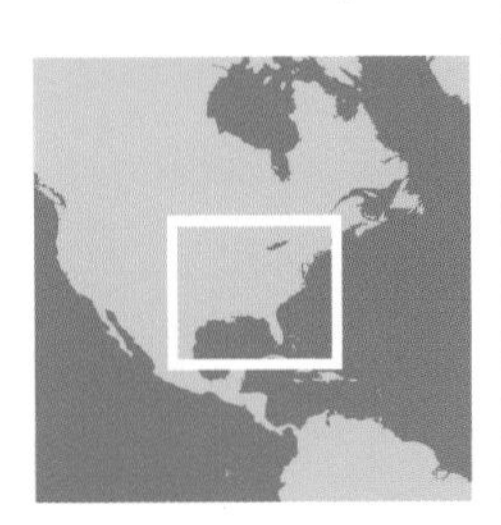

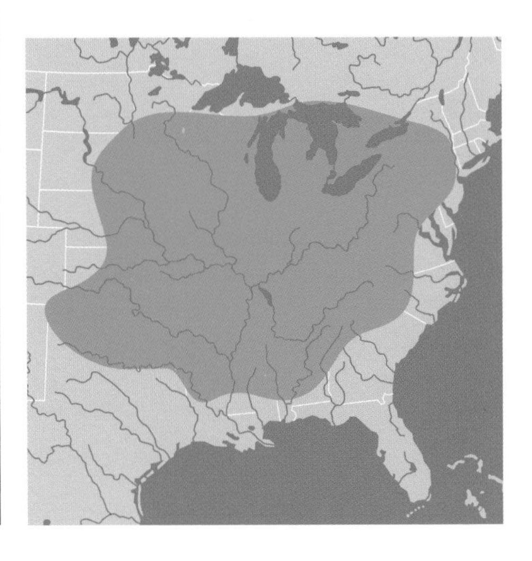

Description of the genus *Graptemys*. These turtles are often called map turtles because of the pattern, especially of the plastron, that indeed looks like the contour lines of a map. Small in size (males to about 160 mm, females to 300 mm or more), they lack a red stripe on the side of the head, but they have extensive dots and stripes on a dark background. They have a dorsal keel that is very pronounced in many juveniles but may nearly disappear in adults. Some species have strong spines on the second to fourth vertebral scutes. These are the most aquatic species of emydids, and apart from basking and nesting, they rarely leave the water. Females of some species (the "broad-headed" map turtles) are specialized for feeding on mollusks.

Description of *G. geographica.* This is a relatively large species (carapace to 270 mm in females) that lacks strong dorsal spines and has a very plain shell. The posterior marginal scutes are strongly serrated. The carapace is olive brown, with obscure designs made up of fine brown bands surrounded by black. The plastron is white or yellow in adults. The lower face of the marginals is yellow, with circular dark designs extending onto the bridge. The head is large (larger than that of G. pseudogeographica), with a smooth (unnotched) upper jaw margin. The yellow and black lines running along the head are very contrasting. The males are smaller than the females and do not exceed 160 mm. The females have a larger head and a much more rounded shell. Dorsal spines are present in the young, but they flatten out with age in the males and disappear completely in the females. The plastron has a gray-black figure that follows the seam lines.

Natural history. These turtles lives in major watercourses—rivers and large lakes. They may be seen basking along the edge of the water in major concentrations. In the northern part of the range, they hibernate from November to March. The largest populations are found where there are concentrations of water snails, and, curiously, natural populations seem to have a highly skewed sex ratio, favoring females by about 3 to 1. Basking animals are very skittish and dive at the least provocation. When grasped in the hand, they may bite quite hard. Courtship occurs in spring and autumn. The male swims around the female and places his snout against the muzzle of his partner, while bobbing his head up and down with neck extended. Nesting takes place from June to mid-July, according to latitude, and includes 9 to 20 white soft-shelled eggs measuring about 33 × 22 mm. There are often two nestings in the year. Incubation lasts for about 75 days. The hatchlings emerge in late August to the end of September, but they often spend the first winter still in the nest, appearing in March or April. The young measure 30 ×24 mm wide and are thus much wider, proportionally, than the adults. The diet includes aquatic mollusks, as well as insect larvae, crayfish, fish fry, earthworms, and a small amount of plant matter or algae.

Protection. Numerous predators—such as raccoons, opossums, and sometimes coyotes—destroy the eggs and the young. Other threats include water pollution, which kills off the mollusk prey species, and water management projects.

© F. Bonin

The state of the carapace of this animal indicates that it is old, and it has been through mycoses, fires, and bacterial attacks. The black head has very fine, irregular stripes.

Graptemys barbouri Carr and Marchand, 1942

Distribution. The distribution of this species is confined to three states of the southeastern USA—Alabama, Florida, and Georgia—in the elaborate interconnected river system comprising the Chipola, Apalachicola, Chattahoochee, and Flint Rivers.

Description. This species shows extreme sexual dimorphism, with females reaching 330 mm and males just 130 mm. The adult female has a huge head, and the shell is "swollen" in the anterior half and has very reduced vertebral spines. The posterior marginals are strongly serrated. The

© R. Bour

This is certainly one of the most decorative *Graptemys* species. The carapace is adorned with brilliant ocelli. A spectacular wide, light marking is visible behind each eye.

carapace ranges from dark brown to olive, with orange designs forming C shapes on the costal scutes. A very bold orange-yellow stripe, often with dark edges, marks the upper edge of each marginal scute. The plastron is yellow or white and often has dark stripes along the horizontal seam lines. The lower surface of the marginals is ornamented with round markings, gray to black in color. The skin is dark brown to black, with greenish bands on the limbs and yellow and black designs on the head. The greenish spot behind the eye is large and very striking in the males, and it encircles the rear of the orbit. This marking joins a wide yellow band located between the eyes, in a symmetrical fashion. The chin has a strong yellow transverse bar, and the limbs are well webbed.

Common name

Barbour's map turtle

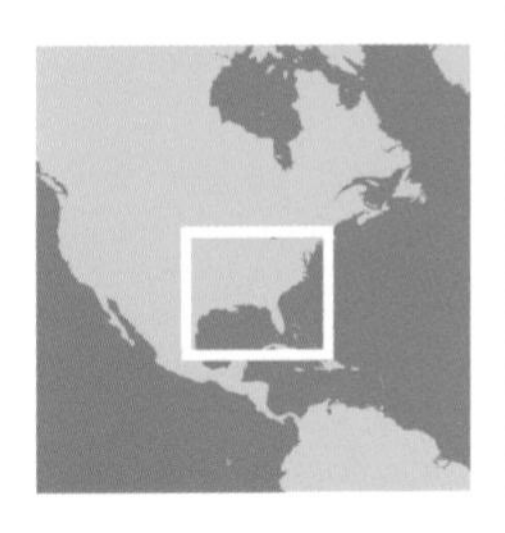

Natural history. This species is most easily seen in the Chipola system and its tributaries, the creeks having a limestone bottom and slow current, but it also occurs in great rivers like the Apalachicola. These turtles seek out protected banks or emergent snags on which to bask and will remain there for hours to sun themselves. They will move somewhat during the day to place themselves advantageously in direct sun, but basically they are sedentary turtles, not leaving their small home range. The males mature at about 70 mm, while the females must attain at least 180 mm. The courtship follows classic lines: the male moves toward the female, neck outstretched, snout against snout. The nesting season runs from June to late July. There may be two nestings in the season, each with 7 to 10 soft-shelled eggs, 35 × 25 mm in size. The hatchlings measure 35 mm and weigh about 11 g. The diet for the females includes aquatic mollusks more than anything else, their huge heads and powerful, flattened jaws and jaw surfaces being a major adaptation toward this diet, whereas the males have a more varied diet of insects, drowning butterflies, and some aquatic plants.

Protection. The eggs and young are heavily preyed on by raccoons, but even snakes will depredate the nests on the banks of the Chipola River. Furthermore, the species is still sometimes eaten by local people. While remaining numerous in the Chipola River (where the population is 60% male, 40% female), it is more and more threatened in its natural environment by pollution and collection for sale to hobbyists. This species needs comprehensive protection by federal law to control human take for any purpose.

Graptemys caglei Haynes and McKown, 1974

Distribution. This species is endemic to the Guadalupe and San Antonio Rivers and their effluents in central Texas, USA.

© B. Devaux

This small species is very shy. Note the serrated vertebrals and the S-shaped lines on the top of the head.

Description. This turtle does not exceed 210 mm. It is rather flat, is wider behind than in front, and has strongly serrated posterior marginals. The vertebral keel is interrupted by very high tubercles. The general color ranges from olive to light brown. Each scute is decorated with very contrasting yellow markings. The posterior part of each vertebral scute is dark. The plastron is well developed and uniform cream in color, with a dark pattern along the seams. The bridges and the lower surface of the marginals are cream and marked with four longitudinal light bars. The head is rather narrow and has a somewhat pointed snout. In color, the head is black, with seven cream lines on top. The midline stripe is the widest one. On the side of the head, the bands form a cross above the eyes. There is a yellow V-shaped mark on the dorsum and on the lower surface of the head. The chin is decorated with a cream-colored transverse bar, and the lower surface of the neck bears wide cream bands separated by fine dark lines. All the stripes form "hairpins" with the apex at the level of the chin. The neck, limbs, and tail are ornamented with well-defined black and yellow stripes. The male does not exceed 110 mm and has longer vertebral spines than the female. The carapace of the females is more rounded than that of the males.

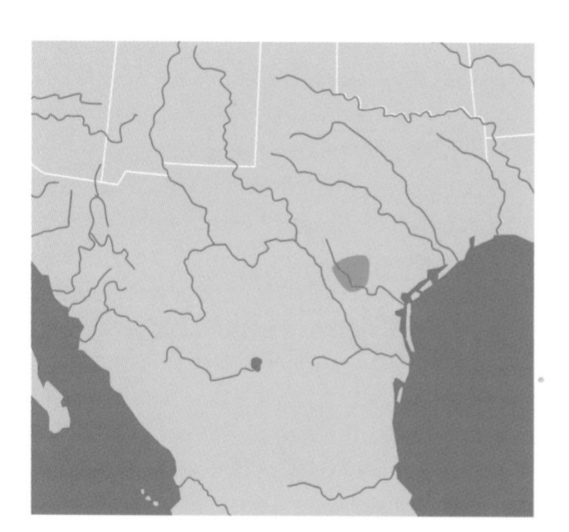

Natural history. This turtle lives in slow-flowing rivers and in lakes not exceeding 3 m in depth, with a muddy or limestone bottom. They bask a great deal, usually on rocks or trees, and rarely step on the ground itself except when nesting. Very nervous, they dive into deep water at the least disturbance. The female lays one to six eggs per clutch, between April and July. Two nestings per season are frequent. The critical temperature for sexual differentiation is around 29°C (100% males are produced at 28°C, and 100% females above 30.5°C). The diet is eclectic and includes aquatic plants as well as insects, larvae, and water snails (especially enjoyed by the females).

Common name

Cagle's map turtle

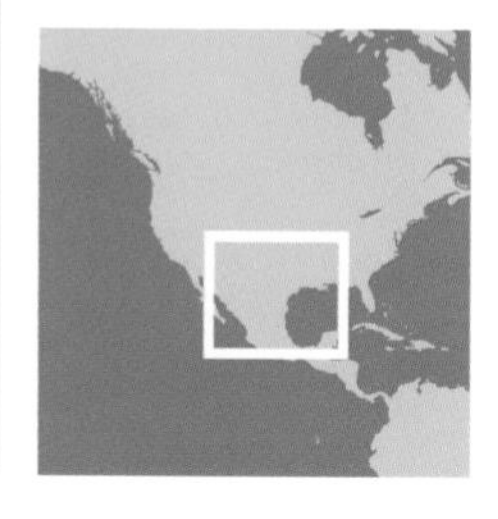

Graptemys ernsti Lovich and McCoy, 1992

Distribution. This species is found in Alabama and Florida, in rivers emptying into Pensacola Bay on the Gulf of Mexico, including the Yellow, Escambia, Conecuh, and Shoal Rivers.

Description. This was formerly considered a subspecies of *G. pulchra*. The name *ernsti* refers to turtle researcher and author Carl Ernst. The females may reach a length of 280 mm. The carapace is very high, with a strong median keel, but only the first vertebral scute has a slight spine on its posterior border. There is a deep notch above the head, around the nuchal and first vertebral scutes. The rear margin of the shell is saw-toothed. The costals are large and olive, with one or two long, light-colored bars bordered by a dark line that extends vertically between the vertebrals and the marginals. The plastron is cream and has dark elongate markings on the humerals and femorals, but these markings disappear with age. The head shows a large olive yellow area between the orbits, always separated by two marks of the same color located along the posterior rim of the orbits. The stripes on the neck are also wide, particularly the top ones. Females have a very wide head, with powerful jaws and strong crushing surfaces. The plastron is flat in both sexes, and the males attain their maximum size at around nine years of age.

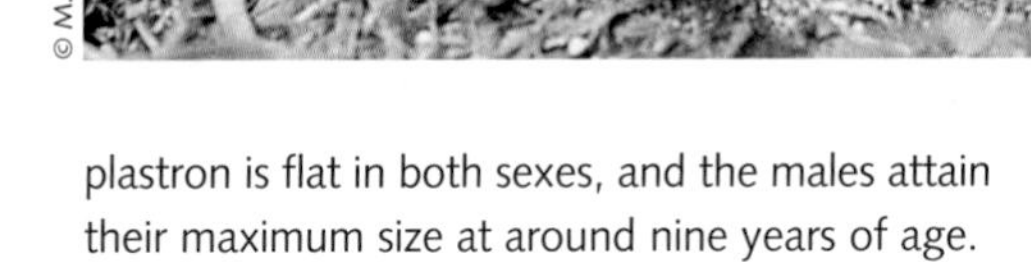

© M. Rogner

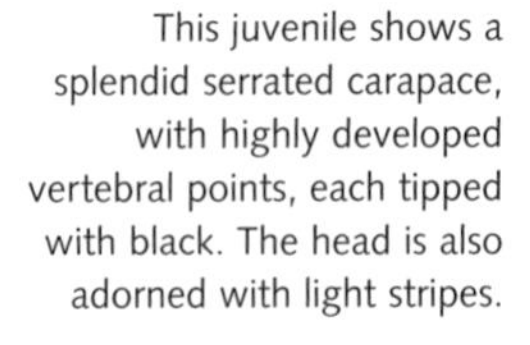

This juvenile shows a splendid serrated carapace, with highly developed vertebral points, each tipped with black. The head is also adorned with light stripes.

Natural history. These turtles live in major rivers with a sandy or gravel bottom. They prefer areas with plenty of rocks, tree trunks, or accumulations of vegetation that provide basking opportunities. They may be seen basking all year round, in places where the water is reasonably deep. The females may nest up to six times per year, especially when they are well advanced in years. Nesting takes place from late April to the end of July. It has been reported that the nests may remain underwater for several days if a flood invades the nesting beach, without the eggs being destroyed. The young never remain in the nest during the winter, and they move toward the water immediately upon emerging. They measure about 40 mm at birth. This species is mainly insectivorous, but the females also feed heavily on mollusks.

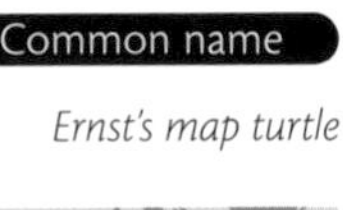

Common name

Ernst's map turtle

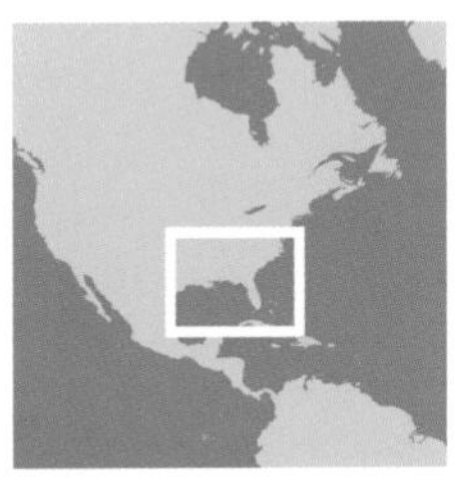

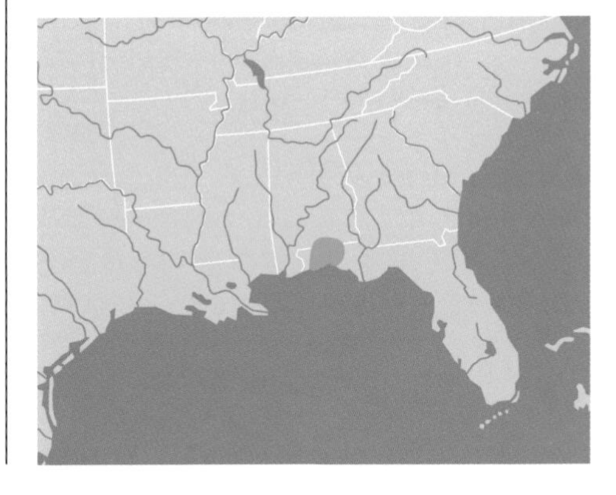

Graptemys flavimaculata Cagle, 1954

Distribution. This species is found only in southeastern Mississippi, in the Pascagoula River and nearby watercourses.

Description. The females do not exceed 180 mm. The posterior edge of the vertebral scutes is very elevated and forms strong spines on the first

three vertebrals in the males. The posterior border of the shell is strongly serrated. The vertebrals are wider than long, and the tips of the spines are black. The carapace is olive to brown, with a large yellow mark at the center of each costal scute (from which the species gets its name). On all of the marginals, there is a yellow marking in the form of a detached semicircle. The plastron is creamy white, with black bands that are prominent along the seam lines but fade with age. The head is narrow, the nose is flat, and the upper jaw appears poorly developed. The head is olive, with yellow stripes forming a rectangular figure behind the eyes, connected to a longitudinal band that continues onto the neck and is twice the width of the other lines. There are about 20 yellow stripes along the head and neck, separated by wide black stripes. The limbs have an identical design. The males do not exceed 110 mm, and they have elongated claws on the forelimbs.

This young animal shows bright yellow blotches on each of the carapace scutes.

Natural history. This turtle lives in major waterways with a moderate or sometimes rapid current. The bottom needs to be clay or sand, and the turtles bask on piles of vegetative debris. They also bask on roots and floating plant material. They will come out even when it is raining or when the temperature is low, and sometimes they climb as high as 3.5 m above the level of the water. They primarily eat insect larvae and aquatic mollusks. Courtship occurs in springtime: the male extends his head and neck to the maximum in the direction of the female, then agitates his forelimbs on each side of the head of his partner, caressing her with his claws. The female extends her head and repeats the gestures of the male. Sexual maturity is very precocious in the males and may be reached at three years of age.

Protection. Populations are dropping for three reasons: agricultural pollution, especially from the timber and chemical industries; commercial collection for illegal sale to hobbyists; and the mindless firing of guns at basking turtles by underemployed local youths for target practice. It features on the list of threatened species in Mississippi.

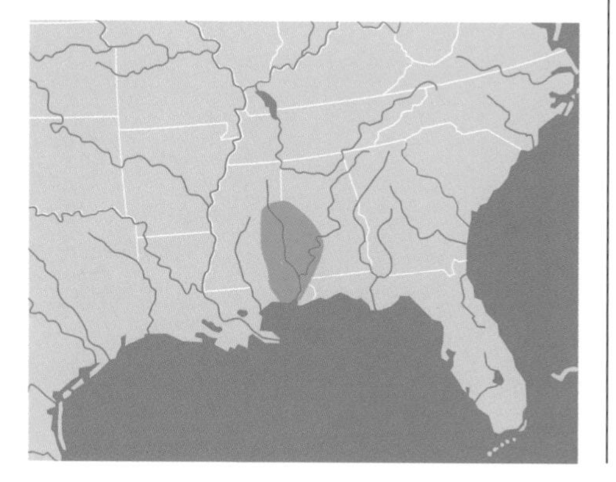

Common name

Yellow-blotched map turtle

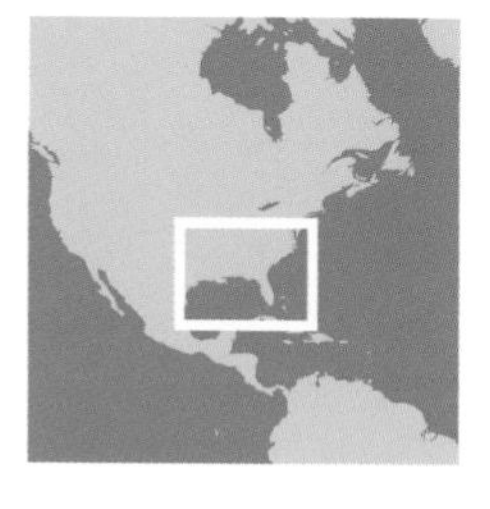

Graptemys gibbonsi Lovich and McCoy, 1992

Distribution. This species is found in Louisiana and Mississippi, in the Pascagoula, Pearl, Chickasawhay, Leaf, and Bogue Chitto Rivers.

Description. The species name *gibbonsi* is a reference to renowned herpetologist J. Whitfield Gibbons. The females reach 290 mm. The carapace is very elevated, with a well-developed keel out-

© B. Devaux

The keel is well defined and often outlined in black. But in old individuals, as with all *Graptemys,* the coloration fades and may disappear.

lined in black and with very well-developed spines on the first three vertebral scutes. The carapace is olive to brown, with discernible growth annuli on the costals. The marginals have a single yellow bar on each, edged with black and looking like an inverted Y. The lower aspect of the marginals is cream, with a dark narrow band following the seams. The skin is brown or olive, with yellow or yellow-green stripes. Between the orbits, the head shows a large olive blotch that connects two large spots of the same color behind the orbits. The females have a larger head than the males and have much more developed masticatory surfaces.

Natural history. These turtles live in open waters of the largest rivers and do not shun fast currents. They select areas with a sandy or gravel bottom where basking sites are numerous. At night, numerous individuals may sleep on branches in a small area, just a few centimeters below the surface. Male maturity is reached at the early age of four years. The males mainly eat insects, and the females mainly snails.

Common name

Pascagoula map turtle

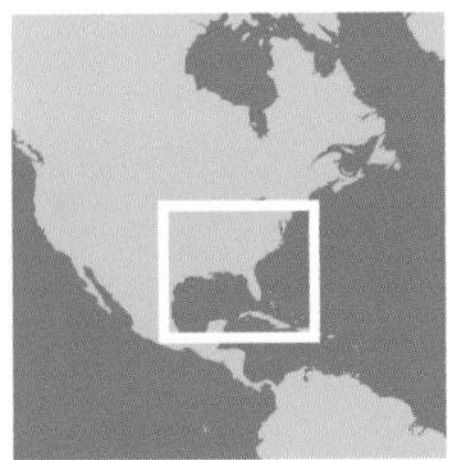

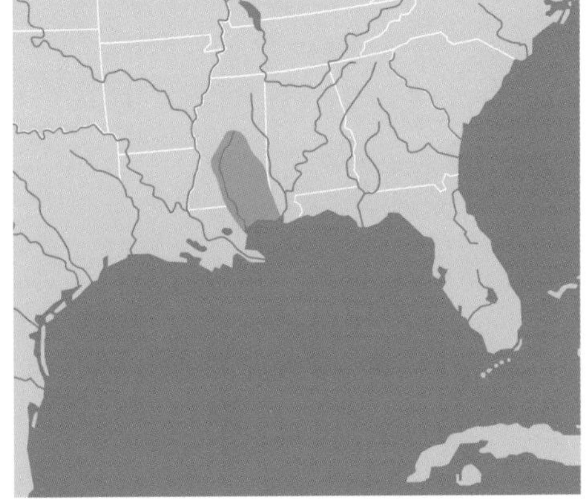

Graptemys nigrinoda Cagle, 1954

Distribution. This species is found in Alabama and Mississippi, in the Alabama, Tombigbee, Black Warrior, Coosa, Tallapoosa, and Cahaba Rivers.

Description. This is a small turtle (220 mm maximum for females, 120 mm for males). It has highly developed dorsal spines, forming blunt tubercles. The hind edge of the carapace is very strongly denticulated, and the vertebral scutes are wider than long. The carapace is olive to dark brown. A circular or semicircular narrow-edged figure, yellow or orange in color, is present on

each pleural scute and each marginal. The plastron is yellow, often tinted with red. It has a black central figure that crosses the seams transversely. The head is narrow and delicate and has a flat upper jaw, and the snout is not projecting. In general, the color is black with yellow bands. Behind each eye, yellow stripes form a Y-shaped figure. Two to four yellow bands on the neck connect with the orbits. Even in the hatchlings, the spines are very well developed.

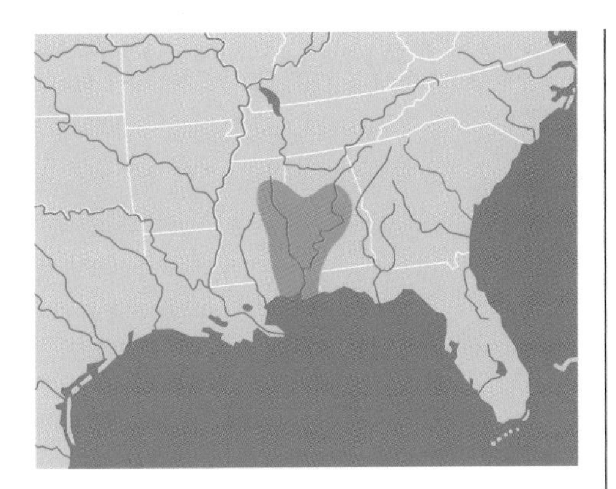

Black-knobbed map turtle

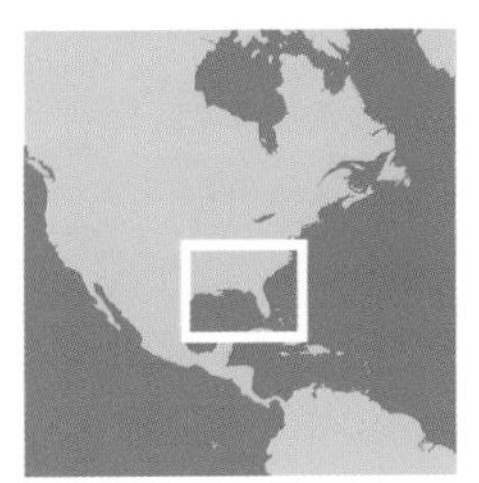

Subspecies. Two subspecies are recognized. *G. n. nigrinoda* (described above): The northern black-knobbed sawback occurs in the Alabama and Tombigbee Rivers and may be recognized by the modestly developed plastral figure, which takes up less than a third of the plastron. The mark behind the eye is in the form of a very incurved cross. The light yellow stripes that connect the eyes are rarely interrupted. *G. n. delticola* Folkerts and Mount, 1969: The southern black-knobbed sawback is confined to Alabama, in the Mobile Bay drainages on the edge of the Gulf of Mexico, notably the Tensaw River. It has a plastral figure that takes up more than two-thirds of the area of the plastron. The soft parts of the head are black, and the postorbital mark is slightly incurved laterally. Within a large series of individuals, one may see perhaps a single turtle with an interruption of the light lines that connect the eyes.

Natural history. This turtle occupies watercourses with a clay or sandy bottom, a slow current, and lots of good basking sites. It basks for much of the year and remains active even when the water cools to 1° C. The males reach maturity at the age of three years. Nesting occurs between 9 p.m. and midnight, sometimes a moderate distance from the water (up to 50 m). The females nest up to three times per season, laying about five eggs each time. These turtles mainly eat insects (dragonflies or cockroaches that fall into the water) but also consume sponges, mollusks, and various aquatic plants.

Protection. This turtle is placed on the Mississippi threatened species list. It is still abundant in the Tombigbee and Alabama Rivers. Crows and raccoons destroy many nests, and alligators consume the juveniles (probably adults too).

This youngster is a veritable jewel. The fine ocelli on the carapace go well with the black dorsal spines and the white lines on the neck and head.

Graptemys oculifera (Baur, 1890)

Distribution. This species inhabits a restricted range in southern and central Mississippi and eastern Louisiana, along the Pearl River and neighboring waterways.

Description. This turtle does not exceed a length of 220 mm and has very developed vertebral spines, especially those on the second and third vertebrals. The posterior border of the shell is sawtoothed. The carapace is green to olive, each costal having a large circular ocellus of yellow or orange. The marginals are decorated with yellow semicircular figures. The plastron is yellow or orange, with a brown motif along the seams that disappears with age. The lower edge of the marginals often has a series of yellow L-shaped figures, edged with black, and the bridges are marked with two yellow oval or circular blotches. The head is small, and the nose flat. The head, neck, and limbs are black, with numerous yellow stripes. The light markings behind the eyes are oval, rectangular, or circular and are not connected, as in *G. flavimaculata,* to the horizontal bar that extends from the neck. Two distinct yellow bands connect with the orbits. Between the orbits is a strong yellow bar. The irises are yellow, and the eye is crossed by a black stripe that connects with the top of the head. Males do not exceed 110 mm.

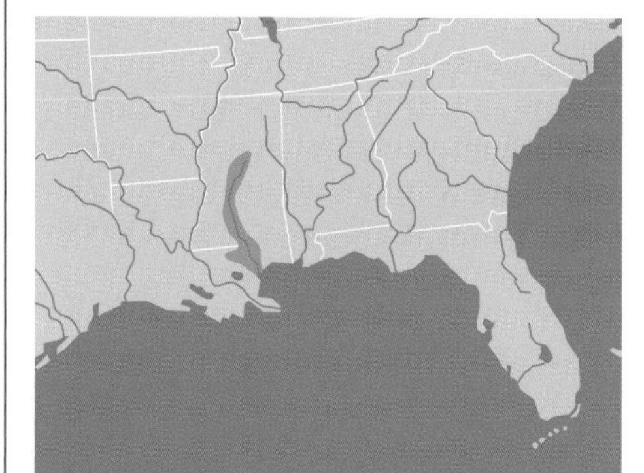

Common name

Ringed sawback turtle

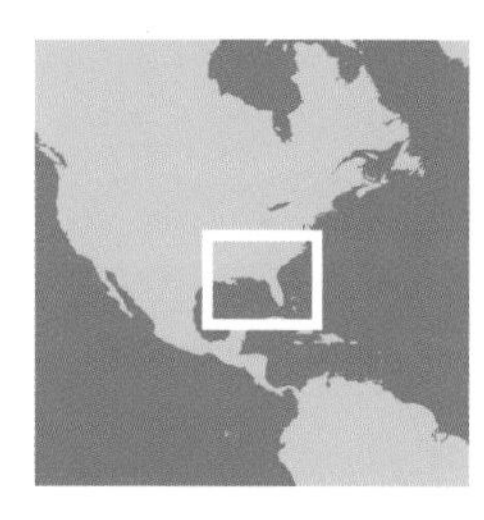

Natural history. These turtles live in fast rivers, with a sandy or clay bottom. They sun themselves on tree trunks or brush piles at the edge of the water. They are very timid and are difficult to approach in the wild, but they sometimes remain on emergent branches to spend the night. The females lay two or three eggs, each weighing about 10 g and 39 × 24 mm in size. Incubation takes 63 days at 30°C. The diet is almost entirely carnivorous: insects, mollusks, and some vegetation.

Protection. Human activities (fishing, motor boating) are the main causes of mortality in this species, which has been classified by the U.S. government as endangered since 1986. Habitat degradation and poor water quality are making things worse.

The dorsal spines of this young specimen are especially well developed and may protect it from predation. The large ocelli appear almost luminous.

© F. Bonin

Graptemys ouachitensis Cagle, 1953

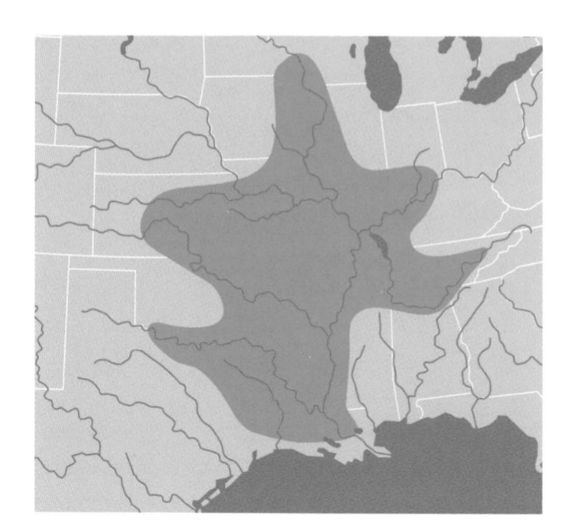

Distribution. Fred R. Cagle named a great many *Graptemys* species during the 1950s, and this one, as its name implies, was found in the Ouachita River. The range is in the central United States, from southern Louisiana to Illinois in the north and includes northern Alabama, Tennessee, Kentucky, eastern Kansas, Iowa, Minnesota, and Wisconsin.

Description. As with most *Graptemys* species, there is major sexual dimorphism: males do not exceed 140 mm, while females reach 240 mm. The dorsal keel is well developed in the males, and the spines at the rear of the vertebral scutes are black. The general coloration is brown-olive, with black markings on the costals (often one per scute), and additional orange figures may overlap from one scute to another. The plastron is yellow-white and is decorated in adults with a black figure that follows the seam lines. This pattern is very marked in the young, but it fades with age. The head is of medium size. The skin is brown to olive, with orange-yellow streaks on the limbs, tail, and neck. Behind each eye there is a small, square-shaped marking in the nominal subspecies. The claws of the males are very elongated, especially the one on the third digit, in the middle of the "hand."

Subspecies. *G. o. ouachitensis* (described above) occurs in the Ouachita River, in northern Louisiana, west as far as Oklahoma, and north to Kansas, Minnesota, Wisconsin, Indiana, Ohio, and western Virginia. It has a large rectangular mark behind each orbit, as well as two or three bands on the neck that reach to the eye. There are also two light circles on each side of the head, one under the eyes and the other on the lower jaw. *G. o. sabinensis* Cagle, 1953, is found farther south, in Texas and Louisiana, in the Sabine River and its effluents. The yellow mark behind the eye is more oval, and the neck bears five (or more) light stripes that connect to the orbit. The round light markings on each side of the head are very small.

Natural history. These turtles live in rivers with a rapid current, but also in lakes and swamps, where they prefer a sand or clay bottom and dense aquatic vegetation. They spend a great deal of the day sunning. This is a very nervous species, difficult to observe in the wild because it goes into hiding underwater at the least disturbance. Sexual maturity is reached in just three years in males. Mating occurs in April to October or even November, according to latitude. The courtship is similar to that of other *Graptemys* species, with much vibration of the elongate foreclaws of the male in front of the female's face; small bites or nibbles at the neck, forelimbs, or tail are frequent. In Wisconsin, nesting occurs from mid-May to the end of July. There may be two or three nests per year, with an average of 12 eggs, measuring about 32 ×25 mm. Incubation is brief (60–75 days).

Common name

Ouachita map turtle

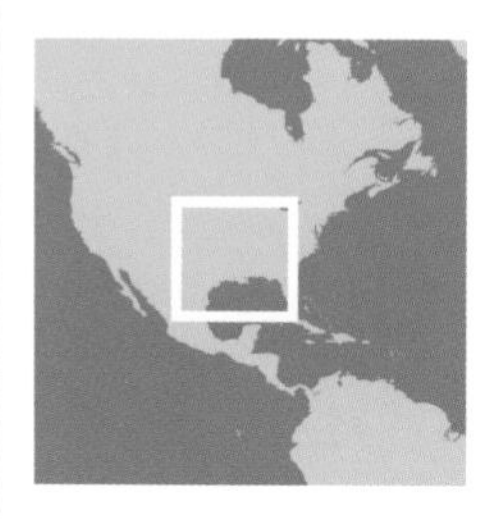

The old specimen shown here has lost the carapace colors and even the vertebral spines. Only the coloration of the head and the neck remains bright.

© F. Bonin

The diet is omnivorous, and the turtles are able to eat out of the water, enjoying insects and vegetation—the diet includes mollusks, aquatic plants, insect larvae, dead fish, and crayfish.

Graptemys pseudogeographica

(Gray, 1831)

© F. Bonin

The iris is very white, the snout is pointed and the lines on top of the head are particularly aesthetic.

Common name

False map turtle

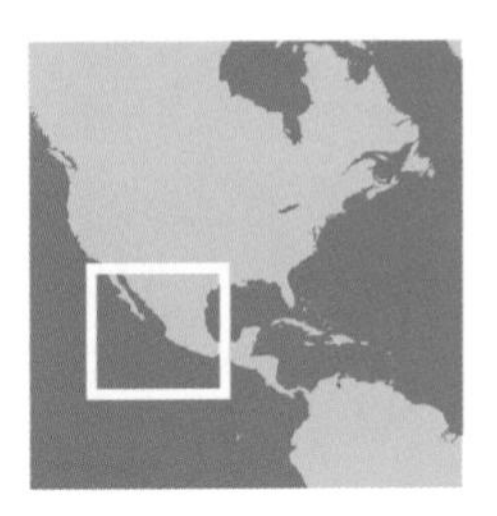

Distribution. These turtles live in the great rivers and their tributaries in the central and southern USA (Texas, Mississippi, Ohio, Wisconsin, Indiana, Illinois, Minnesota, Louisiana, and Missouri).

Description. This is a turtle of medium size (150 mm for the male, 270 mm for the female). The keel is weakly developed, and the posterior margin of the shell is serrated. The vertebral spikes are reduced but are higher and more conspicuous in the juveniles. The general color is olive brown, with an orange oval marking on each costal. These figures fade with age. The plastron is uniformly cream or yellow. The head is of medium size, and the diameter is identical to that of the neck. The skin is brown to black, with numerous rather fine yellow lines on the limbs and tail and most evident on the temples. Behind the eye on each side there is a pale boomerang-shaped marking, allowing one to distinguish the two subspecies. The head of the female is wider than that of the male.

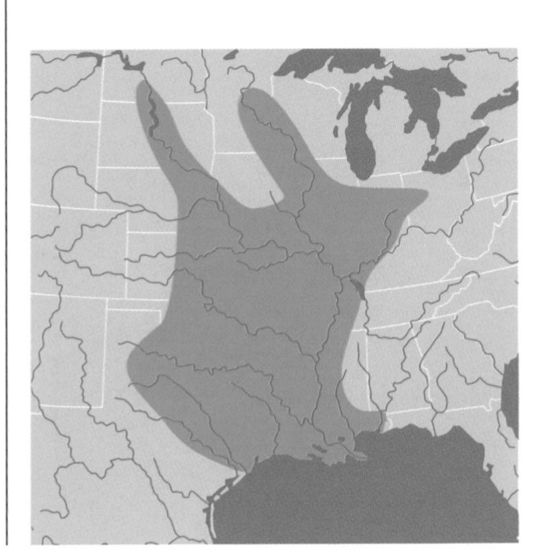

Subspecies. *G. p. pseudogeographica* (described above): The northern map turtle has two symmetrical yellow bands on the top of the head and behind the eyes. These bold yellow markings have the form of a boomerang that is extended and tapered toward the neck. The third claw on each forelimb of the male is especially elongated. *G. p. kohnii* (Baur, 1890): The southern representative of the species is called the Mississippi map turtle. It has a wide curved bar behind the eyes that extends as a band along the top of the head to the base of the neck. The irides are completely white, with a perfectly round pupil and no dark crossbar. The length does not exceed 250 mm. The keel has the most pronounced spines on the second and third vertebral scutes.

Natural history. This turtle may be found in marshes and swamps, but most often it is seen in major rivers and lakes, as well as ponds and flooded forest. It is at its best in waters with abundant vegetation and a modest current. Very cautious, it always maintains a state of alert when it is basking. It may hibernate, or it may simply spend several weeks in a lethargic condition, according to latitude. It has been seen in the north (Wisconsin) moving around at the end of winter when the water was as cold as 7°C. It does not try to bite when grasped in the hand. In May, the females look for nesting sites and gather on small islands in the middle of the big rivers. During the summer, males move to feed in calmer waters, and the females join them when they have finished laying. The turtles may congregate on sandbanks far from water, in organized, peaceful groups for very long periods (up to an hour and a half), with some individuals actually resting on top of others. Mating occurs in April and October, according to latitude. The courtship ritual involves lively move-

ments of the forelimbs of the male, claws extended, before the face of the female. Sometimes the male bobs his head rapidly up and down. For *G. p. kohnii*, nesting occurs in June. Two or three nestings per season are normal, with 14 eggs the average for *kohnii* but fewer (5 or 6) for the nominate race. The eggs measure about 35 mm × 22 mm and are larger in *kohnii*. Incubation takes 70 days. The young measure about 35 mm and weigh 7 to 8 g, according to subspecies. Mostly omnivorous, these turtles eat mollusks, insect larvae, and insects that have fallen into the water, but they may also eat some aquatic vegetation.

Protection. Ninety percent of the eggs and hatchlings are usually lost just after nesting, devoured by foxes, otters, raccoons, and also birds and catfish. The turtles are also caught by humans. In the Missouri River, the populations have diminished sharply as a result of river "management" and pollution. Also, vast numbers have been collected for sale as pets. This is one of the three species (along with *Trachemys scripta elegans* and *Pseudemys concinna*) that were exported in huge quantities from 1980 to 2000.

© F. Bonin

This adult specimen has kept the serration of the posterior marginal scutes and the black-bordered keel. The curved marking on the top of the head is characteristic.

Graptemys pulchra Baur, 1893

Distribution. In spite of its name, this species is not confined to Alabama; it also occurs in northwestern Georgia and possibly in the Pearl River in eastern Mississippi. In Alabama, it occurs in several Gulf drainages, including the Pascagoula, Black Warrior, Coosa, Cahaba, Alabama, and Tombigbee.

Description. Female may reach 270 mm, and the dorsal spines are well developed, especially on the second and third vertebral scutes. The first vertebral is small, but the other four are all wider than long. The posterior border of the shell is slightly serrated. The color is light olive, and each costal has a pattern composed of yellow to orange circles and squares, outlined in black. A black band also runs along the vertebral keel. The marginal scutes have a white or yellow semicircular streak, bordered with black. The plastron is yellow, with a narrow black border on the posterior part of each scute, along the seams. The head is rather large, with a protruding snout and a flattened upper jaw. There is a very diagnostic feature in the interorbital region—namely, a huge greenish yellow figure that narrows and extends behind the eyes and rejoins the jaw anteriorly and touches the skin of the neck posteriorly, where it takes on an

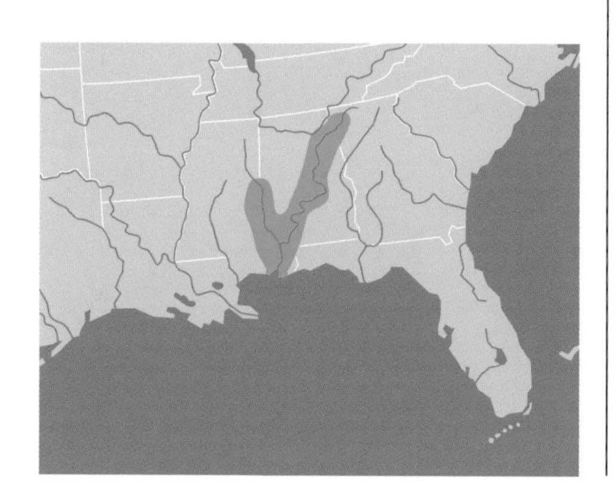

Common name

Alabama map turtle

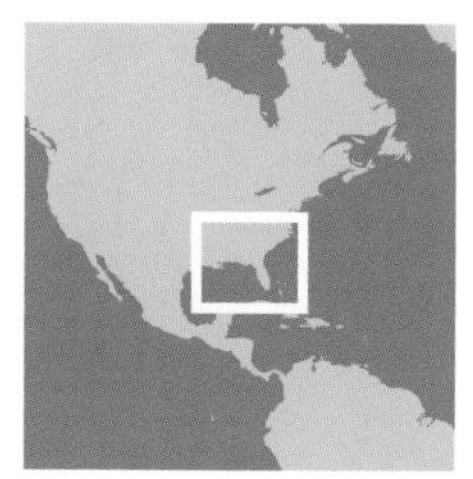

© F. Bonin

The marginals have light borders and are marked with semicircular designs that open toward the rear. The triangular marking behind the eye is wide and bright white, whereas the eye itself is dark.

orange tint. A wide yellow band extends backward from the tip of the chin. The remainder of the skin is dark brown with some yellow stripes. Males do not exceed 140 mm. Females usually have a very wide head and a well-developed jaw.

Natural history. This turtle lives in large rivers with deep waters but moderate current. The bottom may be sand or gravel. It is most active by day, from late March to late November. Very cautious, it dives into the water at the least disturbance. The female, which primarily feeds on snails and freshwater clams and mussels, has broad tomial surfaces that allow it to smash hard-shelled prey. The nesting season begins in late April and continues to August. There may be as many as five nests per season, each containing four to six soft-shelled eggs that measure about 42 × 32 mm. Incubation takes 70 to 80 days.

Graptemys versa Stejneger, 1925

Distribution. This species is found only in on the Edwards Plateau of central Texas, along the Colorado River and certain neighboring watercourses.

Description. The shell of this species does not exceed 210 mm, and it is wider behind than in front. The posterior border is sinuous rather than sharply serrated. The spines along the median keel are poorly developed and are generally black, each being preceded by a pale yellow mark. All of the carapacial scutes have a slightly "inflated," convex surface; they are olive in color, with yellow stripes. The marginals have a fine reticulated pattern of yellow lines on the upper parts and irregular yellow markings surrounded by very fine dark lines on the lower parts. The plastron is large and yellow, with a dark line marking the seams. Various fine, dark longitudinal lines are present on the bridges. The head is rather narrow, with an orange or yellow mark in the shape of a J behind each eye. Fine yellow and black stripes extend from the neck to the side of the head and rejoin at the orbits. The chin often has orange or yellow spots, bordered with black. Adult males do not exceed 110 mm.

Natural history. These turtles live in calm, shallow creeks and watercourses and feed on insects and mollusks.

Common name

Texas map turtle

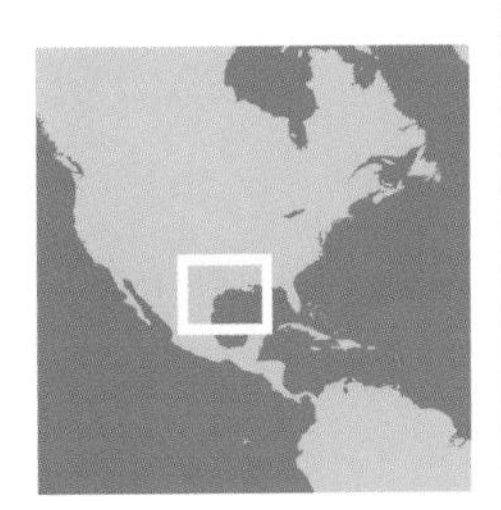

© B. Devaux

The discrete coloration of this adult *G. versa* shows the remains of ocelli on the scutes. The light figures on the head are rather small but still bright.

Malaclemys terrapin (Schoepff, 1793)

Distribution. This species is found only in the eastern United States and Bermuda. In the United States, it inhabits a narrow band along the Atlantic coast and the Gulf of Mexico, from Cape Cod, Massachusetts, to Corpus Christi, Texas, and including the Florida Keys.

Description. This is a truly beautiful little turtle, with a unique pattern, and it does not exceed 230 mm in length (males up to only 150 mm). It is identified by the strong sculpturing of the carapacial scutes, which gives rise to the vernacular name "diamondback." The carapace is widest behind the bridges and is somewhat sinuous or serrated posteriorly. There is a strong median keel, and the vertebral scutes are always wider than long. The general color varies according to subspecies and ranges from gray to light brown to black. The plastron lacks any hinge and is yellow or greenish, with dark irregular markings of varying intensity. The bridges are wide and strong. The head is rather short and narrow in the males and wider in the females. The eyes are large, black, and always protruding. The skin of the head, neck, and limbs is unique: grayish or whitish, sometimes spangled with dark spots but never with stripes or streaks. The jaws are light in color, often yellowish, with perhaps a touch of black on the chin. But there are major variations in color in this species, even within a collection from a single locality. The juveniles tend to be lighter than the adults.

Subspecies. Several subspecies are recognized.

M. t. terrapin (described above): The northern

Common name

Diamondback terrapin

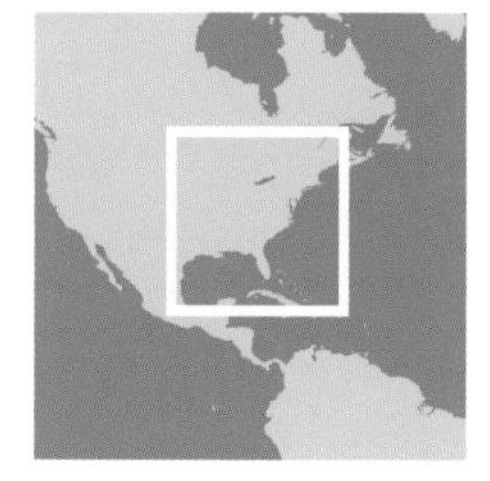

diamondback occurs from Cape Cod to Cape Hatteras. The shell ranges from light brown to uniform black. The plastron is sometimes greenish gray but often almost orange. The vertebral keel has no tubercles.

M. t. centrata (Latreille, 1801): The Carolina diamondback occurs from Cape Hatteras to northern Florida. The sides of the shell are parallel, and there are no tubercles on the median keel. The posterior marginals are curved upward.

M. t. littoralis Hay, 1904: The Texas diamondback occurs from western Louisiana to eastern Texas. The shell is rather elevated, with prominent tubercles. The scales do not have a light center. The plastron is generally whitish. The top of the head is white, while the neck and the limbs are greenish gray, always with a multitude of small black spots.

M. t. macrospilota Hay, 1904: The ornate diamondback occurs from Florida Bay to the Panhandle in northwest Florida. The median keel includes enlarged tubercles on the third and fourth vertebral scutes. The center of the scutes is always very light and usually yellow or orange. Two to four black spots scattered about on each light center.

M. t. pileata (Wied, 1865): The Mississippi diamondback occurs from the Florida Panhandle to western Louisiana. The keel forms tubercles only on the last three vertebral scutes. These excrescences are narrower and more elongate than in *M. t. macrospilota*. The scutes are of uniform color, without light centers. The marginals curve upward on both sides, and the coloration of the undersides, yellow or orange, is thus exposed. The upper surface of the limbs, head, and neck is dark brown to black. The plastron is always yellow.

M. t. rhizophorarum Fowler, 1906: The mangrove terrapin is found only in the Florida Keys. The vertebral tubercles are large. The scutes are black to brown, and their center is never light. A blackish tint marks the seam lines of the plastron. The posterior limbs are gray, with perhaps some black stripes, whereas the forelimbs and the neck are always uniformly gray.

M. t. tequesta (Schwartz, 1955): The east Florida diamondback occurs on the mid- and southern Atlantic coast of Florida. The keel has projecting tubercles that lean toward the rear. The shell is brown or bronze. The costals are large, and the center is often lighter than the edge.

Natural history. These turtles are found in brackish waters, coastal swamps, estuaries, and lagoons. The activity cycle varies with latitude. In the north, the turtles may hibernate from November to early May. They bury themselves in the mud, and one may find several individuals buried together in the same place. In the south, they are active all year round. Diurnal in habit, they spend the day feeding and basking at the edge of the water. They seem to adapt well to differences in

This extraordinary species has a very strong white head, with a degree of black spotting, like that of a leopard. Much sought after by collectors, it was formerly consumed by native tribes and the first American colonists.

salinity in the various environments that they inhabit, but when the salinity exceeds 27% of seawater, they need to drink freshwater. Even though they remain a long time in seawater, their rehydration time is very short. When it rains, they position themselves at the surface of the water and drink from the thin layer of nearly fresh water on top, even though it is only about 2 mm thick. If it rains really hard, they hold their head and neck above the water and open their mouth wide to drink the rain directly. Their lachrymal glands allow them to excrete at least a part of the salt accumulated during the sojourns in very salty water. All the tissues and body fluids of these turtles have an osmotic pressure higher than that of other turtles, with high concentrations of urea and various amino acids. Other studies have revealed that this turtle is quite sedentary and moves very little from its small home range. The males mature at a length of about 90 mm (at three to five years of age), while the females need to reach a length of 150 mm (at six to eight years of age). In Florida, courtship occurs from March to April, and in certain canals and quiet backwaters there may be a courting or mating couple every square meter. The male shakes his head against that of the female, and copulation ensues and may last for just two minutes. The females lay from April to July, most often just after the high tide. The clutch size ranges from 4 to 18, the clutches in the south containing fewer—but larger—eggs than those in the north. The eggs are pinkish white, and their texture is comparable to leather. They measure 35 ×22 mm. Incubation lasts at least 75 to 80 days, but in the north the eggs hatch later and the hatchlings may remain underground until spring.

© F. Bonin

The carapace of this young individual still has light circular markings, but the head and limbs have the luminous effect that is characteristic of the species.

These turtles have powerful jaws, adapted for the consumption of crustaceans and mollusks. They also eat crabs, bivalves, marine worms, and various aquatic plants.

Protection. Raccoons, foxes, crabs, muskrats, and birds all raid diamondback terrapin nests. This species has a long history of being eaten by humankind, and its meat is considered to be a delicacy. Today the species continues to be sold in Chinese communities in large American cities. Some regional populations have collapsed, which certainly is one way of stopping the trade. Today coastal development and the alteration of aquatic habitats, as well as incidental capture in crab traps, is causing great inroads into surviving populations, and in some areas, including southern New Jersey, the highway toll of females seeking nesting sites is disastrous.

Pseudemys concinna (Le Conte, 1830)

Distribution. This species is found throughout the southeastern United States and into extreme northeastern Mexico.

Description of the genus *Pseudemys*. These turtles form part of the group often simply called Florida turtles, and they are also called hieroglyphic turtles because of the colorful graphics displayed on their carapaces—less regular than those of the genus *Graptemys*. The size is also greater than in the latter and may reach 400 mm. They all have wide yellow stripes on the neck, the limbs, and the plastron, which is decorated with figures that evoke ancient forms of writing. The head is rather small, and the snout flat. The juveniles are circular, with a strong keel and serrations at the rear, which both disappear with growth. *Pseudemys* are less aquatic than *Graptemys* and

BELOW: Notice the pointed snout of *P. concinna,* the very conspicuous light stripes, and the Y-shaped figure behind the mouth.

© F. Bonin

spend a lot of time thermoregulating on stumps or on small islands in the middle of a watercourse. The adults are highly herbivorous.

Description of *P. concinna*. This species may attain a length of 430 mm. There is a characteristic reverse-C-shaped marking on the second costal scute. The carapace is oval, highest in the center, and slightly serrated posteriorly. The general color is brown-black, with yellow stripes. The lower face of each marginal scute has a black circular marking, each centered on an intermarginal seam and with a light spot in the center. Old animals may lose their bright colors, but they do not become melanistic as do many *Trachemys*. The plastron is orange-yellow, with a dark design that follows the seam lines and is sometimes absent or faded in the front half only. The skin, brown to olive, has light yellowish stripes. Two yellow, very fine rays extend from the orbit and run parallel to reach the neck. On the lower surface of the neck, wide yellow bands extend from the chin to the base of the neck. The external region of the lower jaw is finely serrated. The males always have elongated claws on the forelimbs. The females are always larger and more domed than the males.

Subspecies. Four subspecies are recognized.

P. c. concinna (described above): The eastern river turtle occurs from eastern Alabama to Virginia. It has two thin parallel lines behind the eyes. The small stripe that adorns the center of the chin extends backward and divides into two branches, forming a Y. The skin of old specimens is often entirely black, without light lines.

P. c. hieroglyphica (Holbrook, 1836): The hieroglyphic turtle occupies the entire Upper Mississippi Valley, from southern Illinois into Missouri and as far as Tennessee and southern Alabama. In the west, it occurs in Kansas, Oklahoma, and Texas. There is a very strong reverse-C figure on the second costal. The design on the plastron is not extensive.

P. c. metteri Ward, 1984: The Missouri turtle is found from eastern Texas to southwestern Missouri, including drainages in Kansas and Oklahoma. The C shape on the second costal in often absent, and the whole carapace fails to convey the reticulated theme of *P. c. hieroglyphica*. The plastron is rather dark and very large. This subspecies is somewhat controversial and is very similar to *P. c. hieroglyphica*.

P. c. suwanniensis Carr, 1937: The Suwannee turtle) occurs in northern and western Florida, between Hillsborough and Gulf Counties. There are no stripes on the hind limbs. The overall color is dark, essentially black, with five greenish-yellow light lines between the eyes. The other head stripes are light yellow. The plastron is orange-yellow and has a bold design. The cervical scute is small; its length is less than 2% of the carapace length. This is the largest of the subspecies.

Natural history. These turtles are found in rivers with a moderate current, dense aquatic vegetation, and sometimes with a rocky bottom. They may also be found in certain lakes, swamps, or water holes for cattle. They can tolerate saline water. They bask a great deal and may be seen in quantity on emergent tree limbs and trunks or rocky riverbanks. Very active and alert, they dive at the slightest sign of danger. They are most active from April to October, but in the south they may be active year-round. Although they are diurnal, certain females may nest at nightfall. These turtles mainly feed early in the morning or at the end of the day, spending the rest of their time basking. They are quite aquatic and rarely come on land. They can breathe without extending the head out of the water, because the edge of the nostrils is water resistant, and when they are at the surface, their nostrils form a small air pocket, which allows for respiration. On land they are somewhat clumsy, slow, and hesitant. Mating occurs in the spring. Courtship is somewhat protracted, the two sexes behaving more or less in the same manner as other American emydids. Nesting occurs from May to June and, in the north, up to the end of July. The eggs are pinkish white and about 35 ×26 mm in size. The average clutch

River cooter

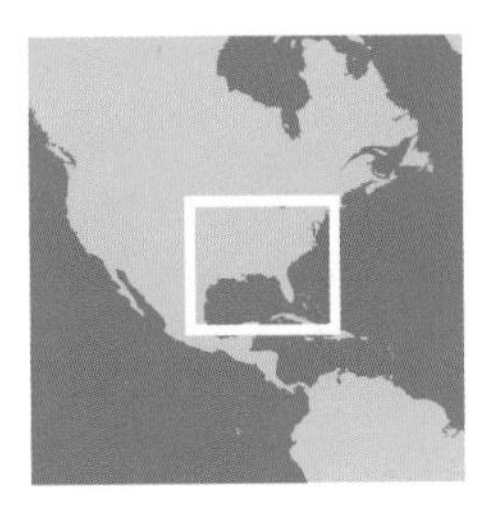

contains 20 eggs, and there may be several nestings in the season. Hatching occurs from the end of August to the end of September. The hatchlings are very colorful and have particularly attractive patterns. These are thoroughly herbivorous species, but they do not disdain insect larvae and certain aquatic mollusks. They consume much vegetation, including *Sagittaria,* water lily leaves, and filamentous algae.

Protection. Formerly, turtles of the genus *Pseudemys* were the most abundant turtles throughout their entire range, and *P. c. suwanniensis* formed basking groups of hundreds on the banks and emergent snags of certain rivers. Today the populations are considerably reduced in many places, although still abundant on the Santa Fe River, and these turtles have been placed on the Florida endangered species list. Predation on the eggs by raccoons and opossums is serious, and the young fall prey to alligators, certain birds, and numerous mammals. But the growing rarity of these turtles comes from habitat destruction and continuing exploitation of the large adults as food for humans.

This individual has hieroglyphic markings, as do all *Pseudemys* species. Refer to the text information on how to tell them apart.

Pseudemys alabamensis Baur, 1893

Distribution. This turtle is found only in Alabama, in the counties of Baldwin and Mobile, near the Gulf coast.

Description. The rugose carapace does not exceed 330 mm in length and is elongate and somewhat elevated in the middle, with some serration at the rear. The vertebral scutes are wider than long. In color the carapace is olive to black, with yellow or sometimes red details on the costals and marginals. The second costal has a wide, light-colored, often Y-shaped band in the middle. The plastron is reddish or yellow and has a central dark design that extends to the bridges and the lower face of the marginal scutes. The head is small, the upper jaw is significantly tricuspid and notched in the middle, and the snout is flat. The skin of the head is black to olive, with yellow stripes. The supratemporal stripes are well marked, as are the stripes that extend from the rear of the orbits, but they do not join up. A median band connects the top of the head to the two supratemporal stripes, creating an arrow-shaped configuration. The hatchlings are almost circular, with a strongly serrated posterior margin. In color they are green,

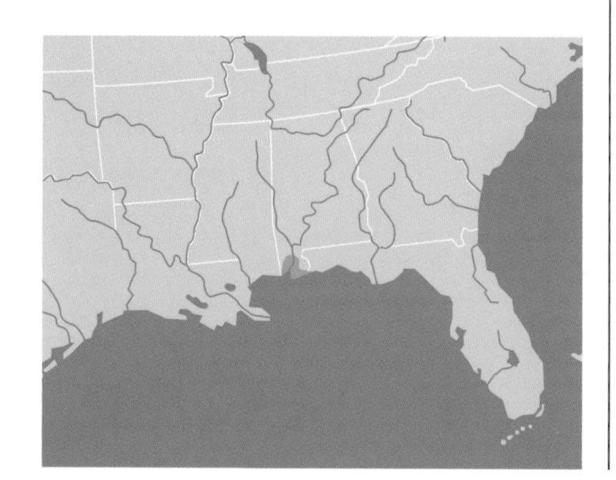

Common name

Alabama red-bellied turtle

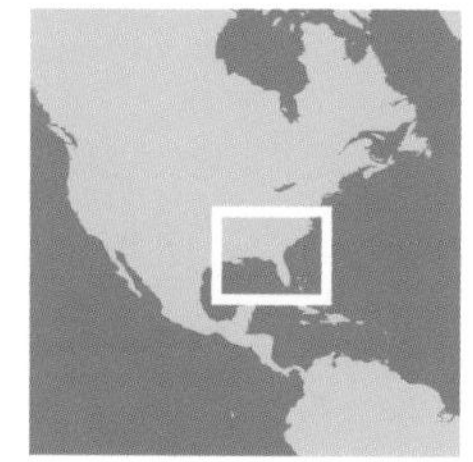

The light chevron-shaped marking on the side of the head clinches the identification of this specimen. The other lines are less obvious.

with designs composed of yellow, black-bordered stripes on each scute. Each costal scute has a yellow circle with a black center.

Natural history. These turtles like calm, shallow waters, with a muddy bottom and dense aquatic vegetation. They hibernate in the mud when the temperature gets very low, but in warmer winters they do not hibernate at all. They are diurnal, and they spend the days searching their environment for food and good basking sites. Very active and wary when basking, they jump back into the water at the slightest disturbance. Females are larger than males and have a higher carapace. The foreclaws of the males are very elongated. There may be several nestings in a season, each with about 20 eggs on average. In certain places, such as Gravine Island, some degree of aggregated nesting has been observed. The eggs hatch after about two months. This is an herbivorous species, and stomachs that have been examined contained no animal remains. However, the juveniles like earthworms and fish if they are offered.

Protection. Major predators include crows, domestic pigs (especially on Gravine Island), and raccoons. This species was formerly also eaten by local people, but this no longer occurs. This turtle is classified as endangered by the Turtle Conservation Fund (2002).

Pseudemys floridana (Le Conte, 1830)

Common cooter

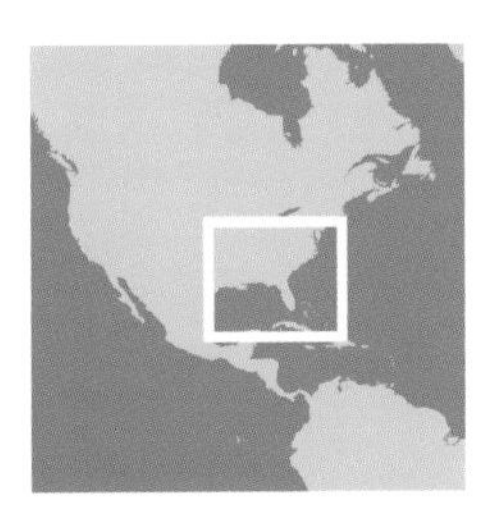

Distribution. This turtle is confined to the southeastern United States, along the Atlantic coastal plain down to south Florida and north to Virginia. The western distribution reaches as far as Mobile Bay, Alabama. The species is abundant in Georgia and South Carolina.

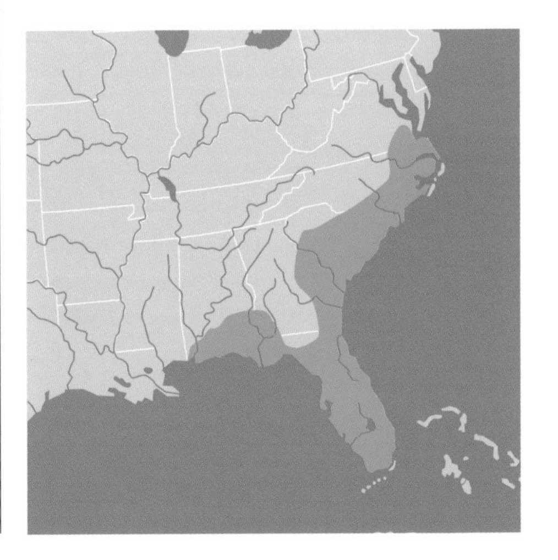

Description. The carapace may reach a length of 400 mm and is elongate and relatively narrow in the middle, but the shell widens again toward the rear. There is always a wide light-colored stripe crossing the second costal scute for its entire width. These turtles are brown, marked with various irregular yellow markings and designs. Each marginal has a central yellow stripe on its exposed side. Sometimes the lower surface of the marginals has a series of dark markings with light centers, each of which straddles a seam line. The plastron is uniformly yellow, and the inguinal scute often has a black spot. The head is rather small, and the

snout is projecting. The upper jaw is flat, without denticles. The skin of the head is black, with quite distinct yellow stripes. The paramedian and supratemporal stripes join up behind the eyes. The undersurface of the neck shows wide yellow bands and a central chin stripe, forming a Y as it bifurcates toward the rear. The young are lighter in color than the adults and have a very round shell with a well-developed median keel, especially on the second and third vertebral scutes.

Natural history. These turtles live in slow-moving waters, with a sandy or mud bottom. Their needs are for good basking sites and dense aquatic vegetation. The species is distinctly gregarious, and the turtles seem to like to be among their own kind. Sometimes as many as 20 to 30 individuals may be seen basking on the same trunk. Depending on latitude, they may hibernate at the bottom of the water. In the north, they are active from April to October, especially in the morning and the late afternoon. The rest of the time they bask. They can tolerate temperatures of up to 41°C. This species is the most terrestrial of the genus *Pseudemys*. Both males and females may be found a long way from water as they seek alternative ponds or rivers, and this unfortunately leads to an elevated level of highway mortality. Females mature in five to seven years, and males around the age of four years. Nesting starts in the afternoon and may not be finished until night. There may be three nests in a season, with an average of about 20 eggs in each. The eggs measure about 34 ×25 mm. Incubation varies according to latitude and may take between 80 and 150 days. In the north, the young remain in the nest until spring. Adults are exclusively herbivorous, eating plants and algae growing in the water where they live. The young are more omnivorous and do not disdain insect larvae or water snails.

In this brightly colored specimen, the head and neck have a lively pattern of wide light stripes. The carapace displays the complex figures on an olive to blackish background.

Protection. Although this species is widely distributed, it still suffers from predation by raccoons, opossums, certain carnivorous fish, other turtles, alligators, and birds. This is also one of the turtle species for which populations seem to be dropping as a result of pollution due to fertilizers and chemicals.

Pseudemys gorzugi Ward, 1984

Distribution. This species is found in Texas, in part of the Rio Grande and its tributaries. It also lives in watercourses pertaining to the Pecos River in western Texas and southwestern New Mexico and has been observed at 1,000 m altitude in New Mexico. In Mexico it occurs in northern Coahuila, but the populations are scattered.

Description. This rather small species, reaching 235 mm, is named after George Zug, curator of reptiles at the Smithsonian Institution. The carapace is oval, rather flattened, and wider behind than in the middle. The posterior edge of the marginals is very serrated. The color is olive or brown, with numerous black and yellow concentric rings.

This specimen from near the Rio Grande has an orange marking behind the eye. Its populations have been severely reduced.

© J. Maran

Common name

Rio Grande slider

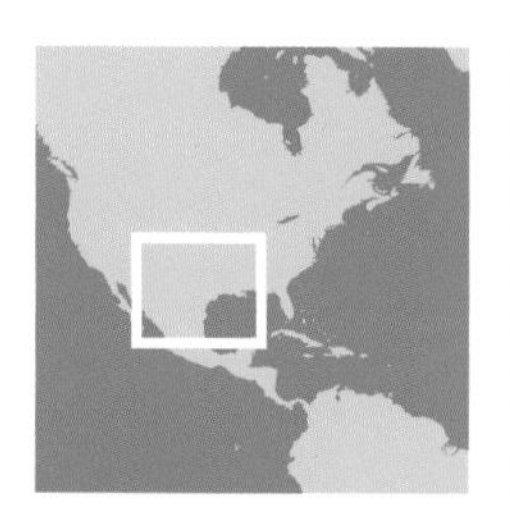

The second costal scute has an elaborate design composed of no fewer than four concentric rings, alternating between black and yellow. The bridge has two black parallel stripes on the axillary scute. The plastron is uniformly yellow, with a black design bordering the seams, and this design is especially developed in the juveniles. With age, it fades, except for the seams between the gular and humeral scutes and between the humerals and pectorals. The head is rather wide, and the mouth is full of pseudoteeth, very sharp and pointed, on the beak and the front of the palate. The general color is dark green, and there is a round yellow spot behind the eyes. A large yellow stripe is present on the sides of the head, behind the eyes. A single yellow stripe extends from the snout to the top of the head, ending at the base of the neck. Finally, the chin has a wide double yellow stripe that forms a Y shape.

Natural history. This turtle occurs in lakes and clear water bodies, as well as small rivers with a sandy or rocky bottom. Despite the dangerous-seeming spiny toothlike structures in the mouth, it seems to be mainly herbivorous, in that stomachs that have been examined appeared to contain only finely minced vegetation. It was formerly considered to be a subspecies of *P. concinna,* but little field information is available.

© F. Bonin

The young turtle shown here looks like a painted jewel. The carapace is dotted with complex volutes and ocelli.

Pseudemys nelsoni Carr, 1938

Distribution. This species is found primarily in Florida but also in the Okefenokee Swamp of southern Georgia. It does not occur in the Florida Keys, but there is an isolated colony in the Florida Panhandle, near Tallahassee.

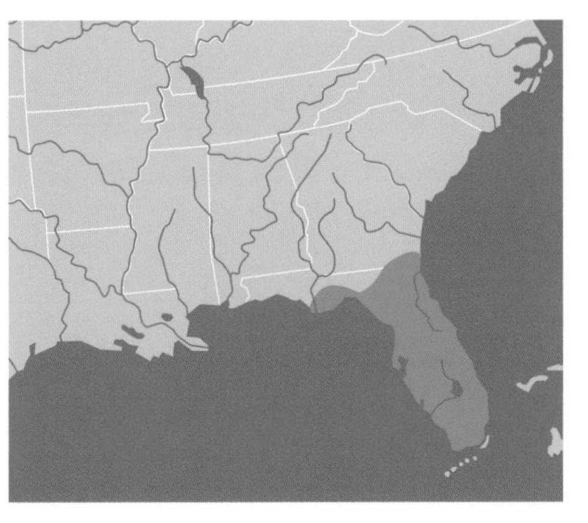

Common name

Florida red-bellied turtle

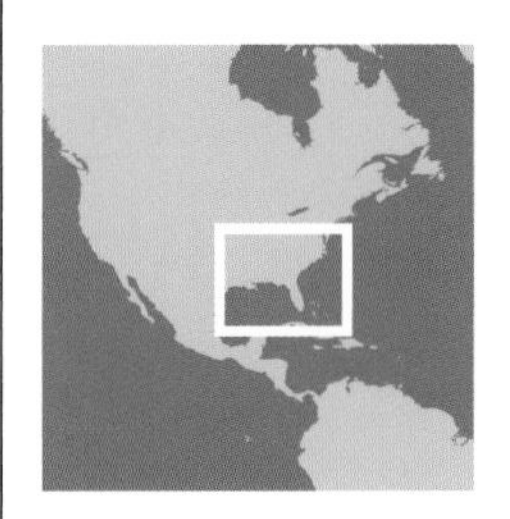

Description. The carapace does not exceed 380 mm. It is highly domed in the front, and the rear shell margin is almost smooth. The color is black, with thick, simple red markings (yellow in the young) on the costals and margins. A light central stripe on the second costal scute forms an inverted Y, which divides in the middle of the scute. All of the marginals are decorated with a reddish central stripe on the exposed face. The lower side has dark blotches, each of which straddles a seam line. Old animals become very dark. The bridges are uniformly yellow, and the plastron may be orange-red with a central design that is always lost with growth. Some individuals have a very light plastron. The head is of medium size and has a rather flat snout. The tip of the upper jaw has a significant notch, with denticulations on each side. The head is black, with numerous yellow stripes and, on the forehead, an arrow mark. The young are lighter than the adults, with a very rounded shell and an orange or red plastron.

Natural history. These turtles live in lakes, marshes, ponds, water-filled ditches, mangroves, and sluggish rivers. They are always associated with abundant aquatic vegetation. These are diurnal animals whose activity is sustained year-round, especially in the southern part of the range. They spend part of their day basking on tree trunks or masses of vegetation, often close to congeners such as *P. concinna* and *P. floridana,* but the species do not seem to hybridize. When *P. nelsoni* are looking for a basking site, they may be quite aggressive and "pushy" as they ignore the territorial rights of other turtles. They are less nervous than some of the other species, and, with care, basking animals may often be approached quite closely from a canoe. Sexual maturity in males may take three to four years, and in the females about six years. During courtship, the male swims above the female, then moves around to the rear of his companion to sniff her cloacal region. If the female continues to swim, the male follows, neck fully extended, before placing his head alongside that of his partner and touching her with his snout. There is one behavioral detail that is often observed: the neck is retracted, and the forelimbs are flapped vigorously against the sides of the head of the female. A succession of rapid caresses and vibrations of the forelimbs follows, each bout hardly lasting for more than a second. Nesting occurs from May to June, and there may be three of four nestings in a season. Interestingly, *P. nelsoni* often nests within alligator nests, which are constructed of vegetation rather than in holes in the ground as with some crocodiles. There are reports of female alligators watching over the eggs laid by *P. nelsoni* until the hatchling turtles emerge before their own offspring appear. In the alligator nests, where the temperature may attain 50°C, in-

The plastron is entirely yellow. Note the long claws on the forelimbs of this adult male.

The carapace is somewhat domed and is dark and unadorned. But the head and neck are decorated with wide, brilliant markings.

cubation of the turtle eggs lasts for just 50 days. Each nest includes 8 to 30 eggs, elliptical in shape and measuring about 35 ×24 mm. The adults are primarily herbivorous, but the young like to eat aquatic insects and other small freshwater invertebrates.

Protection. Apart from humans and the usual predators (raccoons, otters, opossums, some birds, and large common snapping turtles), the alligator may be the main predator. It has been postulated that the shape and thickness of the shell of *P. nelsoni* is an adaptation in response to this predation, and certainly many redbellies in Florida have extensive scratches on the shells that may have been made by alligators that attempted unsuccessfully to crack them. Of course, a really large alligator could break the shell of a cooter of any size.

Pseudemys peninsularis Carr, 1938

Distribution. This species is found only in the peninsular part of Florida.

Description. Formerly considered a subspecies of *P. floridana*, *P. peninsularis* is now considered to be a species in its own right. It is very similar to *P. floridana* but differs in the designs of the "hairpins" and their placement on the sides of the top of the head. These designs are incomplete or interrupted. The inguinal scute has no dark spot.

Natural history. Apparently this species is primarily a winter nester, and one may find large females on highways—often tragically already smashed—during November and December. A feature that seems to be a liability nowadays is the habit of laying one or two "decoy" eggs close to the surface and above the main clutch. Perhaps the purpose is to fool a predator into thinking that it has found all there is to find, but nature has underestimated the intelligence of raccoons, and most clutches are quickly destroyed in their entirety.

Common name

Peninsula cooter

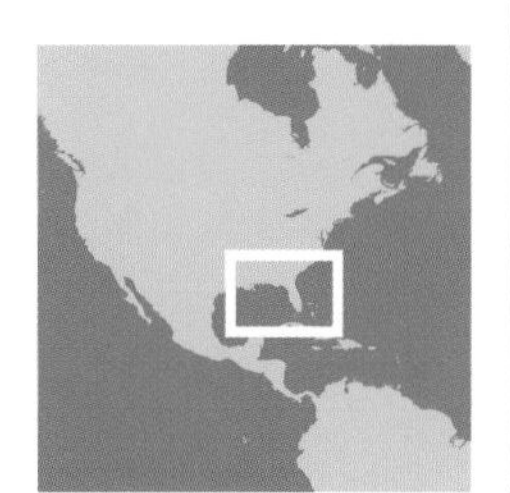

© F. Bonin

The female, on the left, is much larger than the male. The carapace is relatively uncolored, but the Y-shaped lines on the neck are clearly visible.

© B. Devaux

Pseudemys rubriventris (Le Conte, 1830)

Distribution. This turtle is found in the Atlantic coastal plain of the United States, from central New Jersey to northeastern North Carolina. Inland, it is found in the Potomac drainage, and in the east it reaches Virginia. There is an isolated colony in Plymouth County, Massachusetts.

Description. A carapace length of about 400 mm is the maximum. The shell is elongate and well domed, with the highest point near the center and almost no serration on the hind margin. Brown or black in background color, it is marked with large red or yellow blotches on the costals and marginals. The second costal has a light central band, forked at one or both ends. Each marginal has a red stripe on its exposed face, as well as a dark spot with a light dot in the center on the lower surface. Old animals are entirely black, gradually losing their red markings as they age. The plastron is red or orange, as the name indicates, with a seam-following central figure; this design fades with age. The head is of medium size, and the snout is somewhat projecting. The beak has a major notch in the middle of the upper jaw. The head is black to olive, with numerous yellow lines (six to eight extend backward from the eyes). The arrow shape on the forehead, with its point on the tip of the snout, is characteristic. Upon hatching, the juveniles are truly beautifully colored. The plastron is pink or red, with an elaborate seam-following pattern. There is a modest keel that disappears with age and growth. Sexual dimorphism is the usual—the males are narrower and flatter than the females, not to mention smaller.

Natural history. This species occurs primarily in large, deep water bodies but avoids waters with swift currents. It may also be found in flooded plains, as well as in marshes and nearby ponds. It

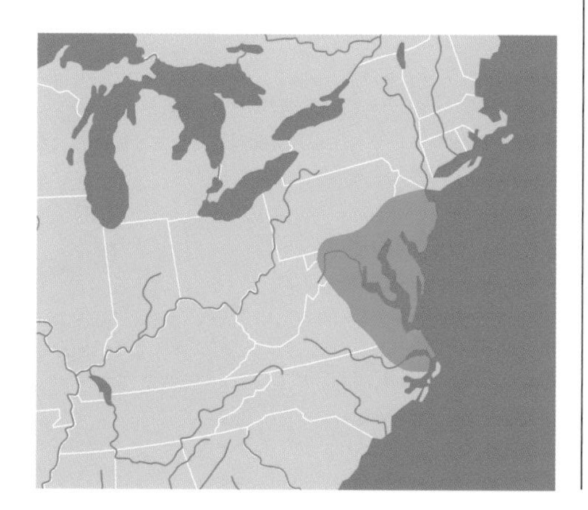

Common name

Red-bellied cooter

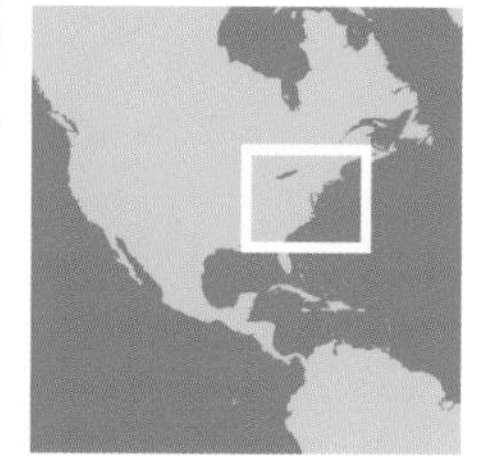

© B. Devaux

This animal is dark, with a powerful head and very few light stripes and spots. By contrast, the plastron is of a fine, rich red color, especially in young specimens.

is tolerant of brackish water at the estuaries and mouths of major rivers. Large specimens with barnacles on the shell are even seen from time to time. These turtles like a mud or clay bottom with an abundance of basking sites. The young, on the other hand, live in aquatic refuges with plenty of vegetation, and they bask very little. These diurnal turtles are active from March to November, but in the north they hibernate in the mud at the bottom of the water. They spend a great part of their day basking, and they dive at the slightest alarm. They may fight over access to the best basking places, either among themselves or with other species such as *Chrysemys picta*. Males mature at an age of about nine years. Nesting occurs from late May to July. The eggs number 6 to 25 and measure about 24 $\times$37 mm. They are white and smooth. Incubation lasts for 70 to 80 days, but often the young spend the first winter underground. These turtles are omnivorous in their youth, then become progressively more herbivorous. The young eat small fish, tadpoles, water snails, and certain aquatic plants.

Protection. The ever-present predators on this species are raccoons, but nowadays urban development and industrialization are the worst threats. In the north part of the range, industrial expansion along the Delaware River has reduced the available habitat. Pesticide pollution, swamp drainage, and suburban development are putting ever-greater pressure on this species, and it is on the brink of extinction in Pennsylvania.

Pseudemys texana Baur, 1893

Texas cooter

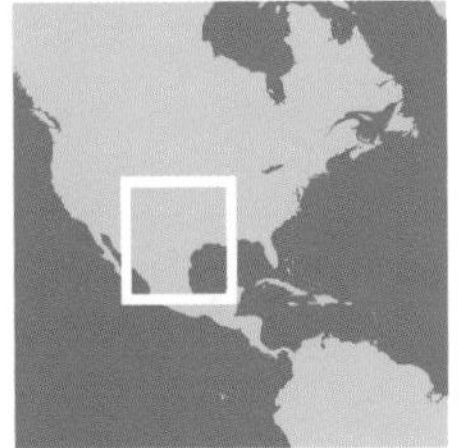

Distribution. This species is found only in central Texas, in tributaries of the Brazos, Guadalupe, San Antonio, and Colorado Rivers.

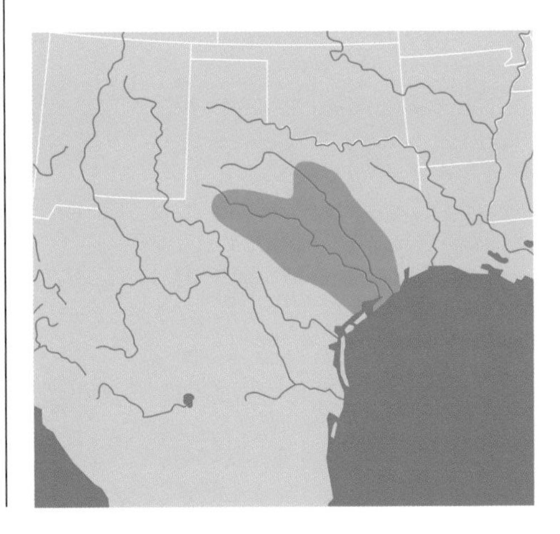

Description. This turtle has a rather flat carapace that may achieve a length of up to 330 mm. The highest point of the shell is near the middle, while the widest point is a little posterior to this. The posterior shell margin is denticulated. Minor longitudinal ridges or wrinkles are present on the costal scutes. The general color is brown to olive, with a pattern formed of small bright yellow blotches, lines, and reticulations. Each marginal scute shows a yellow vertical stripe, often at the center of the scute, while the underside of each marginal is yellow and adorned with dark ocelli. Old males may become very dark in color. The plastron is yellow, with black seam lines. The width of the plastron is always uniform. The skin is mostly black, with white and yellow stripes. The head is very black, with multiple yellow markings whose precise form varies with each individual.

© F. Bonin

The carapace of this brightly colored specimen is boldly marked with hieroglyphics, and the head is a veritable work of art, as if painted by hand.

They may be spots, small streaks, and broad bands. Two major stripes are usually present on each side of the head. The upper jaw has a median notch, and there are small, very sharp keratin denticles on the jaws and the front of the palate, as in *P. gorzugi.* Males are flatter than females and do not exceed 250 mm in length. When grasped in the hand, these turtles emit cloacal fluid and retract completely. The young—much more colorful than the adults, as in most turtles—have a minor vertebral keel, which flattens out with growth.

© B. Devaux

Natural history. *P. texana* occurs in large and small rivers and in all aquatic habitats within its range (i.e., canals, lakes, horse ponds, etc.). This species spends most of its time basking. Females mature after about six years, and males after three to four years. Nesting occurs in May and June. The eggs are finely granular and measure about 35 ×26 mm. An average clutch includes 10 to 12 eggs. Hatching starts in August and lasts until the end of September. These turtles mainly feed on aquatic mollusks but may also eat insects, crayfish, and the like.

Protection. Raccoons, opossums, and their kin remain as abundant as ever, but this turtle species does not seem to be too threatened within its present range.

Trachemys scripta (Schoepff, 1792)

Distribution. The natural range of this species is confined to the United States, from southeastern Virginia to northern Florida (see "Subspecies" below for details). It is widely introduced into other locations.

Description of the genus *Trachemys.* These turtles may be compared to species of *Pseudemys, Graptemys,* and *Chrysemys,* all generally known as Florida turtles in the European pet trade. They are of smaller size than *Pseudemys,* the males tend toward melanism when adult (except in South

Texas), and they are more aggressive and less strictly herbivorous than their cousins of the genus *Pseudemys*.

Description of *T. scripta*. Females reach 280 mm in carapace length, and the males are a little smaller (to 250 mm). The shell is oval, moderately domed and denticulated at the rear, brown to olive in color, and with vertical brushstrokes of orange-yellow on the costal scutes. The upper surface of the marginal scutes is adorned with round black markings. The bridge also has dark stripes and blotches. The plastron is yellow, sometimes with a single round black spot on each scute. The head is black, with a thick yellow, orange, or red band extending back from the rear of the orbit, descending onto the cheeks, and joining wide yellow or orange stripes on the chin. Other fine stripes start at the snout and descend toward the mouth. Several other stripes, wide but less distinct, mark the neck. The rest of the skin is brown to dark olive or almost black, with fine yellow lines from the base to the tip of each limb. The adult males are much darker than the females and in old age may be virtually black. They are so different from the females and the young that for a long time they were thought to be a different species. The males have extremely long second, third, and fourth claws on the forelimbs.

Subspecies. Many former subspecies of *T. scripta* have now been promoted to full species status. Three subspecies are currently recognized. *T. s. scripta* (described above): The yellow-bellied turtle, 270 mm in length, occurs from southern Virginia to northern Florida. It has a wide yellow band on each costal scute, and a prominent yellow crescent just behind the eye connects with a very wide yellow stripe under the chin, toward the neck. The plastron is yellow, with a black spot on each of the gulars and sometimes the humerals and even pectorals *T. s. elegans* (Wied, 1839): The red-eared slider, up to 280 mm in length, has achieved worldwide notoriety because of its ecological versatility and ability to invade natural ecosystems in many countries to which it has been exported commercially. It has a wide red band behind the eye. The stripes on the chin are narrow, and each costal has a yellow transverse band. The plastron is more extensively marked than in the other subspecies, with black blotches, spots, and ocelli, sometimes extending from one scute to its neighbor. This slider occurs in the Mississippi Valley from Illinois in the north to Louisiana in the south. *T. s. troosti* (Holbrook, 1836): The Cumberland turtle, 210 mm long, occurs in southeastern Kentucky to northeastern Alabama. It lives in the valleys of the Cumberland and Tennessee Rivers. The light stripe behind the eye is yellow to pale orange. The chin stripes are wider than those of *T. s. elegans*. Each costal has a transverse yellow band. The scutes of the plastron show a dark, almost circular design with a light center, but in *T. s. elegans* the blotches are entirely dark.

Natural history. These turtles are very opportunistic and are adapted to many aquatic habitats. They prefer calm waters with a muddy bottom, abundant vegetation, and plenty of basking sites. They can remain active even when the water drops to below 10°C. The southern populations usually remain active year-round, although in some cases they may estivate during the hottest and driest months. These are diurnal turtles, and they sleep at night, either on the surface or on the bottom of the water. Their main task in life is to bask for as long as possible, sometimes in large groups, and they tolerate very high temperatures. Their manner of sliding on their plastron to drop off the bank or stone where they have been basking gives them the common name "slider turtle." They may undertake long overland hikes, which may lead to heavy seasonal highway mortality. Males may move even farther than the females—up to 5 km from their home pond. Omnivorous in diet, they eat snails, crayfish, insect larvae, tadpoles, small fish, and various vegetative species. The young are more carnivorous, while the adults mostly eat plant life. In order to have good digestion, they need to be at a temperature of above 18°C. Sexual maturity is reached in the males at 90 to

© F. Bonin

© F. Bonin

There is great color variation in this species. The young have bright yellow blotches at the centers of the ocelli and light stripes extending diagonally from the snout and the eyes.

Common name

Common slider turtle

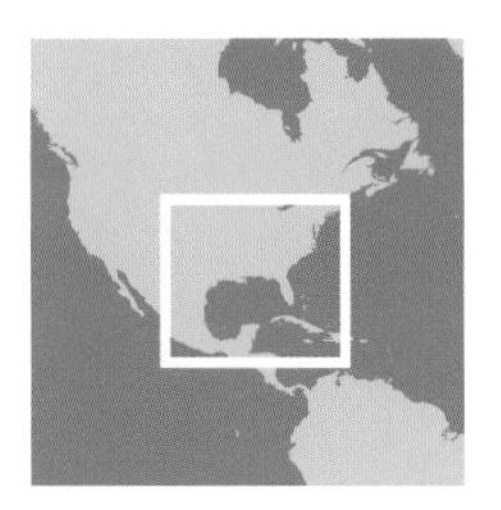

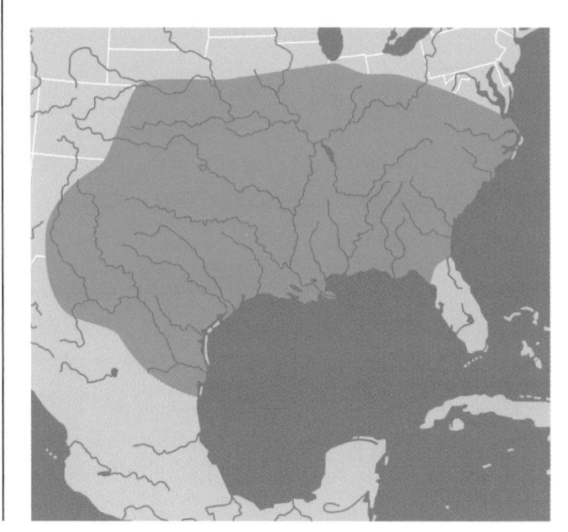

120 mm, while in the females the threshold is 200 mm. Courtship occurs in spring and summer and always in the water. The male sniffs at the cloaca of the female; then he turns around and faces her, extending his forelimbs and fluttering his elongate foreclaws over her face. When the female retracts her extremities, the male goes around to the rear of his partner and mounts her. Nesting takes place in April to July, according to latitude, and clutches consist of 2 to 23 eggs, with an average of 10. Five clutches may be laid in one season, and incubation lasts for 60 to 80 days.

Protection. This species is not really threatened, given that there are still large wild populations, but indirectly it is actually the cause of endangerment of various turtles in distant parts of the world. It is the chief "pet turtle" species raised on huge specialized farms, mainly in Louisiana, and exported to many countries, where some animals escape and others are released when they grow inconveniently large. The survivors establish themselves in ecosystems to which they do not belong. In its own environment, the main threats are urban development and highways. Around the Louisiana farms, collection of breeding adults has often eradicated the species from large areas of its range.

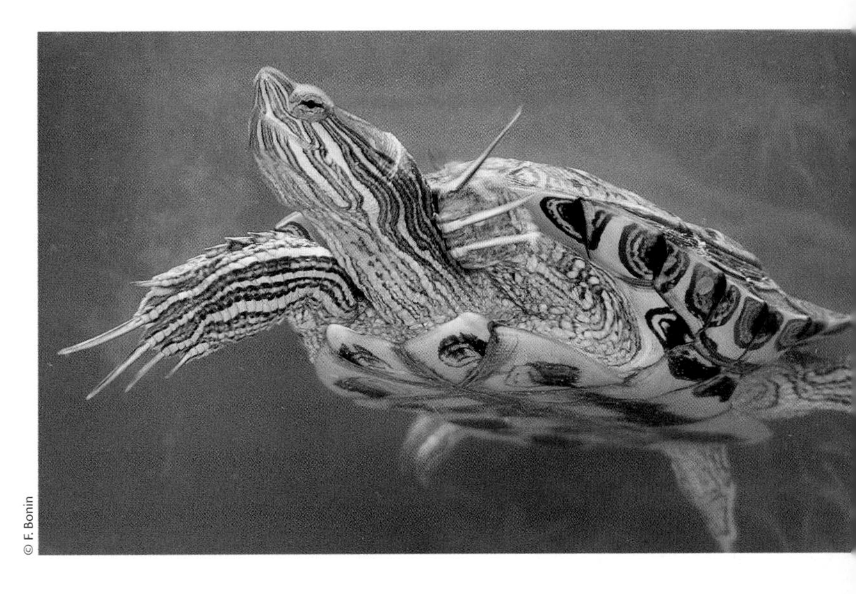

Underwater, one can see the dark markings on each scute of the yellow plastron, the very colorful undersides of the marginal scutes, and the extraordinary foreclaws of a male.

Trachemys adiutrix Vanzolini, 1995

Distribution. This species is found only in a minuscule area of northeastern Brazil, on the edge of the Lençóis Maranhenses National Park in the state of Maranhão.

Description. This is a newly described species. The rugose shell has a vertebral keel only visible toward the rear. The posterior marginals are neither denticulate nor widened. The overall color is brown, with ocelli centered around an irregular orange-red spot in the young but becoming faded and less visible in the adults. The vertebral scutes show parallel stripes of orange, brown, and black. The marginals have reddish marks above and irregular dark brown ocelli on a pale orange background below. All of these design features fade with age. The plastron is yellow, with symmetrical olive-gray stripes bordered with black. The gulars are wide and extend to each side of the plastron. The latter is twice as long as it is wide, and the two lobes have about the same width. The head lacks light lines on top. It is black above and becomes dark gray around the midpoint of the neck. On each side, a wide yellow or orange stripe extends to the temple. It starts behind the orbits and follows a sinuous path. On the upper lip, there are one or two light markings, and on the chin there

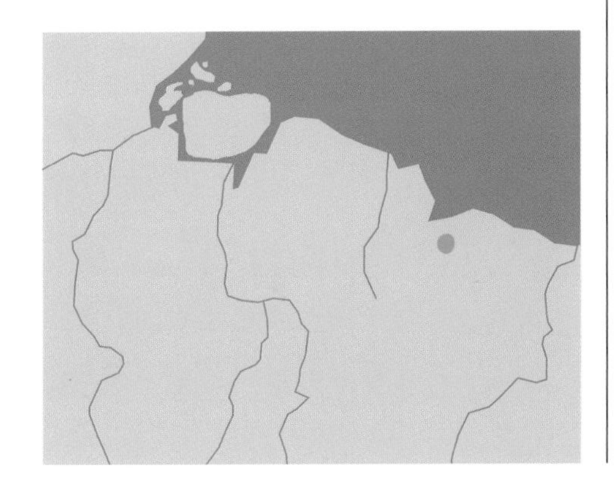

Common name

Carvalho's slider

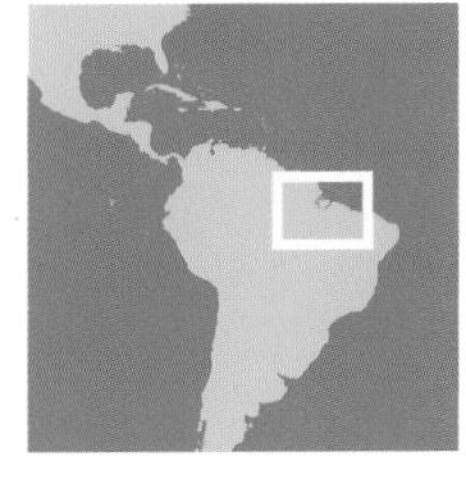

© J. Maran

Photos of this rare species are few. This one came from a Brazilian well, used by local peasants. The animal is hardly colorful, and the light lines on the neck and head are not really prominent.

is a yellow blotch encircled with black and extending toward the back and splitting so as to form an inverted Y. The throat has various ocelli and yellow or orange stripes, with a Y in the center. The limbs are well webbed, and each digit is covered with a series of scales, often quite wide, that reaches to the claw. The carapace of juveniles is decorated with large brilliant ocelli.

Natural history. This species lives in more or less permanent lakes surrounded by sandy dunes, where aquatic vegetation is sparse. The temperature is always high, averaging 26°C year-round. During the rainy season, the turtles may move from lake to lake, but during the dry season they bury themselves in the sand and await the first rains. They estivate for a long time, which seems to be a trait of the species. Nests include 10 to 12 eggs of small size, and there may be two nestings per season. This turtle lives in sympatry with a species of sideneck turtle *(Phrynops)*.

Protection. This species is listed as endangered by the Turtle Conservation Fund (2002).

Trachemys decorata (Barbour and Carr, 1940)

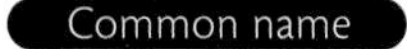

Hispaniolan slider

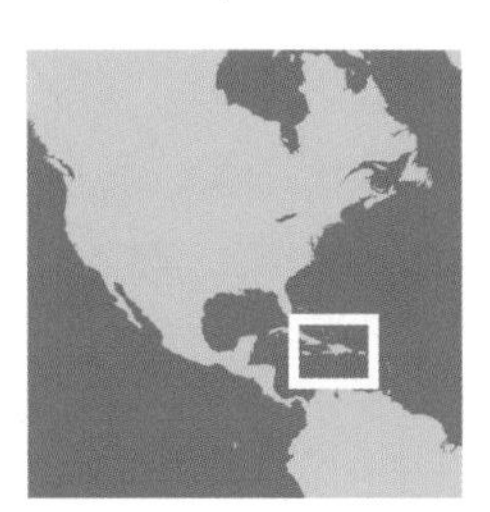

Distribution. This species is limited to the two nations of Haiti and the Dominican Republic, on the island of Hispaniola. It is especially abundant around the Tiburon Peninsula of Haiti.

Description. This turtle reaches a length of 300 mm, is moderately domed, and has a fine median keel that is denticulate toward the rear. The scales are generally smooth but may show longitudinal wrinkles that can be quite marked. The color is gray to brown, with ocelli on the costals and on the junctions between the scutes. The plastron is well developed and has a ground color of yellow or cream, with small, asymmetrical dark circles, the center of which is the same color as the background. These small circles also show up on the bridges. The head is of moderate size and has a rather projecting, conical snout. There is no notch in the upper jaw. The head is grayish to brown, marked with various yellow bands bordered with black on the sides, as well as a wide greenish yellow supratemporal stripe. The rays are more subtle on the top of the head and more contrasting on the sides. The tail and the limbs are grayish green or brown and are marked with the same black-bordered yellow stripes. The males are smaller than the females, with a more pointed snout and long, strong claws on the forelimbs. The young have very lively colors at birth, but these soon fade.

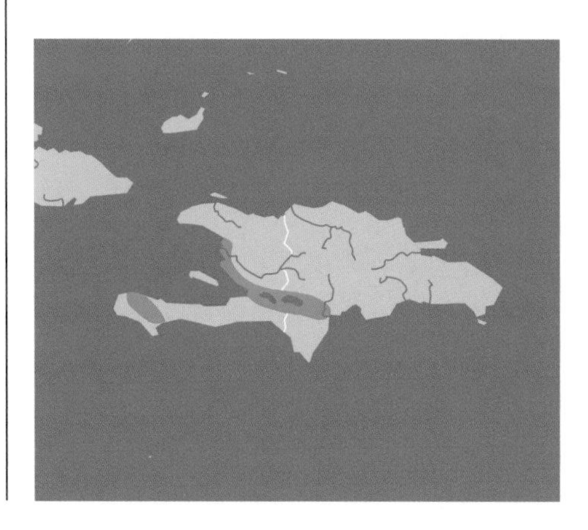

Natural history. This turtle lives in shallow lakes and ponds, freshwater or saline, with abundant vegetation and a mud bottom. The juveniles are omnivorous and do not disdain aquatic insects, as well as small fish. The adults are mostly herbivorous and consume aquatic plants and certain algal species. During courtship, the male vibrates his foreclaws against the face of the female. This procedure may be witnessed year-round, and copulation always occurs in the water. Nesting takes place from April to July, and 6 to 18 elongate eggs, usually about 40 mm long, are laid and buried in the ground. Four nestings may occur in a season, and incubation lasts for 60 to 80 days.

This is not the most decorative of the *Trachemys* species, in spite of its species name. The snout is rather pointed, and the light lines on the neck and head are reduced in number.

Trachemys decussata (Gray, 1831)

Distribution. Two subspecies occur, in Cuba (including Isla de Juventud) and on Grand Cayman.

Description. The carapace is oval and elongate and may reach a length of 390 mm. It is slightly domed and has a median keel, as well as a serrated hind margin. The first vertebral scute is somewhat convex and about as long as wide. Vertebrals 2 to 4 are each a little longer than wide, and the fifth is wider than long. The general coloration is brown or greenish olive, with no pattern in the adults. In the young, the colors are brighter, and the vertebral keel is more pronounced. There is an anal notch. The yellow background of the plastron contrasts sharply with a large black design that follows the seam lines. The bridge is yellow, with a series of longitudinal black bands or ocelli. The head is of medium size and has a projecting snout. The upper jaw has a median notch. The color of the head is olive-brown to green, with yellow stripes. Two of these stripes start at the eye and run toward the back of the head. The jaws are yellow to light brown. The limbs are dark green with yellow stripes. The males may become black or very dark, a condition known as melanism, as they age. The females have more highly domed shells than the males.

Subspecies. *T. d. decussata* (described above) is found in extreme eastern Cuba. The shell is wide and oval. The skin of the soft parts is greenish to olive. The snout is rounded, and the plastral pattern is of the seam-following type. *T. d. angusta* (Barbour and Carr, 1940) occurs in western Cuba, on Isla de Juventud, and on the Cayman Islands. The skin is gray to brownish.

Natural history. This species frequents rivers, lakes, marshes, and inundated prairies. The turtles prefer muddy soils with abundant aquatic vegetation. They spend much of the day basking on land and are often seen out of water, under tall grass or dead leaves. The adults are herbivorous and consume aquatic plants, as well as fruits that fall into the water. Courtship includes the behavioral detail of the males vibrating their elongate foreclaws against the face of the female. Nesting occurs from April to July. Incubation takes 90 days, and the eggs measure about 40.5 mm long and 25.5 mm in diameter.

Common name

North Antillean slider

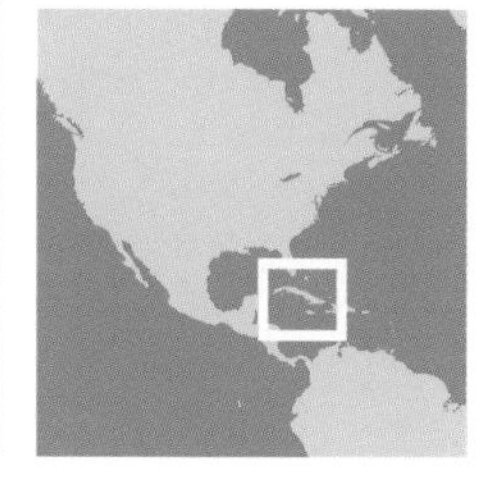

This beautiful animal has white spots and dots, from the head to the tips of the limbs and even on the carapace. But often specimens are darker or even blackish.

© P. Fidenci

Trachemys dorbigni (Duméril and Bibron, 1835)

Distribution. The two subspecies live in an area of southeastern South America near the Atlantic coast, from Brazil (Rio Guaíba, Rio Grande do Sul), in Uruguay (Paraná), and into northern Argentina.

Description. This turtle reaches a length of about 260 mm, with a strongly domed shell in the females. The median keel is well marked in juveniles. The posterior margin is serrated. The oveall color ranges from brown to olive, with markings ranging from bright red to orange to yellow. The markings are black-bordered, and each is confined to a particular scute. Their form varies greatly among individuals. Each marginal has a vertical light bar. The plastron is orange or yellow and has a large, intense central figure that follows the seams between the scutes. In the young, these designs contain multiple light areas that are eliminated as the figure fills in and become completely black with age. The head is of moderate size and has a rather pointed snout and a slightly notched upper jaw. The skin of the head is green to brown, with numerous orange and yellow stripes, each bordered with black. The supratemporal stripe is very wide and does not reach the orbit. On the sides of the head, there are three narrower bands in most specimens, which reach the neck and extend backward from the orbits. Another stripe extends from below the orbit and reaches the neck farther back. The snout has numerous narrow stripes on each side. On the chin, there are many additional longitudinal stripes. Around the mouth, there is an elongated light blotch encircled with black. The limbs and the neck are green to brown, with yellow stripes. As the males age, they become progressively darker and eventually almost black.

Common name

D'Orbigny's slider turtle

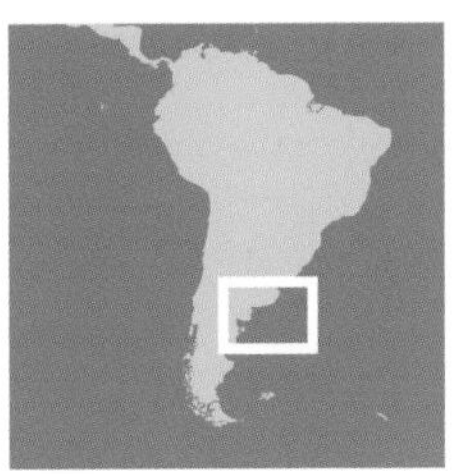

Subspecies. *T. d. dorbigni* (described above) occurs in Argentina and Uruguay. The carapace is brown, with orange stripes, and the supratemporal stripe is quite narrow. The plastron is orange. *T. d. brasiliensis* (Freiberg, 1969) occurs in Brazil, from the Rio Guaíba to Rio Grande do Sul. The shell is mostly green, with yellow or pinkish red

stripes and a wide supratemporal band. The plastron is yellow to green, and the males do not have the plastral figure.

Natural history. These turtles prefer waters with a slow current, a muddy bottom, and abundant aquatic vegetation, but also with an abundance of basking sites. They live in lakes, marshes, ponds, and sluggish rivers, often basking in small groups for hours during the day. Omnivorous and opportunistic, they spend the early morning searching the nooks and crannies of their environment for food. There may be several nestings per year, with a maximum clutch size of 17. The eggs are soft-shelled and measure about 40 ×27 mm. Nesting is mostly seen in December.

© B. Devaux

The carapace of this species is well rounded and strong, but hardly colorful. The stripes extending from the eye are of variable width and are quite striking.

Trachemys callirostris (Gray, 1855)

Distribution. This species is found both in Caribbean Colombia and in the northwestern Maracaibo Basin in Venezuela.

Description. The carapace is somewhat domed, with maximum height reached at the third vertebral scute, and the maximum length is about 250 mm. The vertebral keel is incomplete, appearing only on vertebrals 2 to 4 and becoming indistinct with age. The posterior edge of the carapace is smooth or sometimes slightly sinuous. The adults have moderate carapace rugosity, with the longitudinal wrinkles very obvious on the costal scutes. The carapace of the adult is greenish, with yellow streaks on the marginals and black spots on the vertebral and costal scutes. The plastron is large and has an anal notch. The gulars are rather thickened, and the posterior edges of the plastron are also thick and somewhat downcurved. The head is large and rounded, has a feebly developed conical snout, and is covered with smooth skin (no scales). The jaws are finely denticulate, and the upper jaw has a median notch. The head patterns are very complex, but a large reddish to ocher band is always present behind the orbit, with parallel sides and rounded ends. There are also numerous yellow stripes on the head, of which the widest extends from the orbit, past the corner of the mouth, and reaches the neck. The eyes are green to light yellow, with a horizontal dark bar. The limbs are large and well webbed, with five claws in front and four behind. The limbs are covered with small smooth scales that project slightly. They are greenish, with yellow stripes on the dorsal aspect of the forelimbs. The females are much bigger than the males, with a very large head and a more rounded snout. Also, the carapace is higher and more domed in the females. The length of the claws is the same in both sexes, in contrast to most other *Trachemys,* and the plastra are flat in both sexes. The young have a very colorful shell, with a green background and a large black spot toward the posterior lower corner. There are also numerous yellow, black, or green stripes around the light-colored figures on the scutes. The yellow plastron is covered with dark lines, which fade with adulthood.

Subspecies. *T. c. callirostris* (described above) lives in Colombia and northwestern Venezuela,

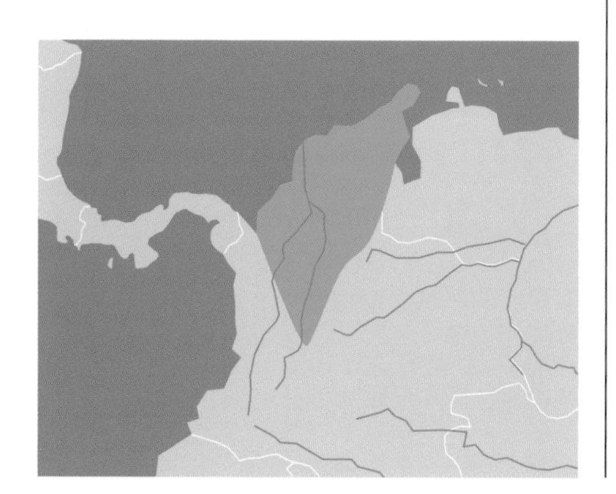

Common name

Colombian slider

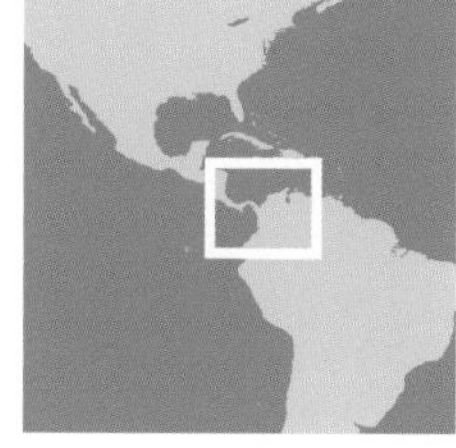

This *T. c. chichiriviche* individual was observed in the region of Chichiriviche, in Venezuela. Darker than most, it has numerous light stripes that extend from the jaws and the eye, subtle but very aesthetic.

in Caribbean drainage systems. The ocelli on the snout and on the chin are very marked. The supratemporal stripe is reddish and well separated from the orbit. The plastron is almost entirely covered by a pattern of dark lines in juveniles and young adults. *T. c. chichiriviche* (Pritchard and Trebbau, 1994) lives only in northern Venezuela, in small coastal creeks and rivers in the states of Falcón and Yaracuy, between the Río Tocuyo and Moron. It is larger than the nominate race, reaching 320 mm. There are ocelli on the chin, but on the upper jaw they are replaced by short yellow stripes edged with black. The supratemporal stripe is dark red and has very tapered ends, the anterior tip often contacting the orbit. The plastral design is narrow and dark and is centered on the median suture. Finally, the carapace is mostly olive-brown to dark brown.

Natural history. This is a freshwater turtle, found in lakes, small marshes, cattle tanks, ponds, and occasionally rivers. It prefers slow-moving waters with dense aquatic vegetation. It is often seen basking in groups. Omnivorous, like many *Trachemys,* it eats various plant species, as well as insects and their larvae, mollusks, other small invertebrates, and even small fish. Mating takes place in deep, calm water from September to December. It lasts only for two to three minutes. Females are somewhat aggressive and bite the males during courtship (Medem, 1975). The nesting period extends from late December to the end of April, during the dry season, when the nesting areas are less covered with plant growth. The nests include from 9 to 25 eggs, about 35 ×24 mm in size. Two nestings per season are usual, and incubation lasts for 70 to 90 days.

Protection. This turtle is heavily exploited for meat and eggs. Although protected by law in both Colombia and Venezuela, the harvest is especially heavy during Holy Week (Semana Santa), a tradition that has been hard to break.

Trachemys emolli (Legler, 1990)

Common name

Nicaraguan slider

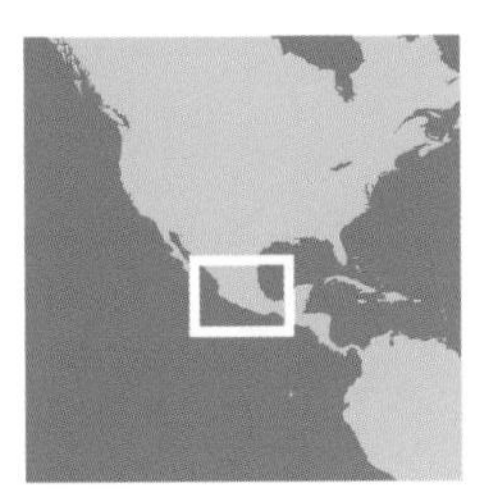

Distribution. With a rather restricted distribution in Nicaragua and Costa Rica, this species occurs in Lake Nicaragua and Lake Managua and in other lakes and streams connecting the two, including Caño Negro in Costa Rica. It occurs in the upper part of the Río San Juan but is replaced by *T. venusta* in the lower reaches.

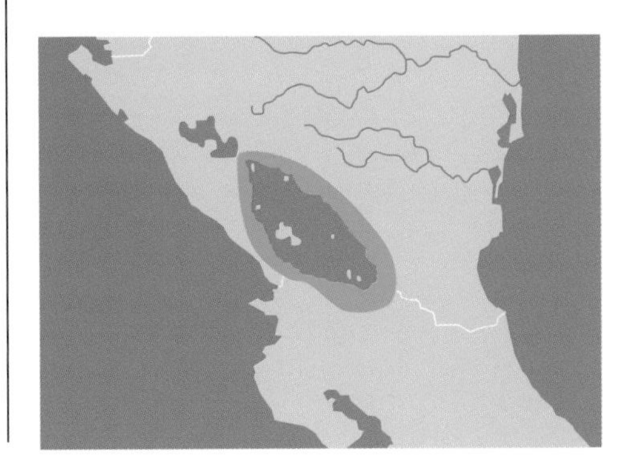

Description. Although very similar to *T. venusta,* this species has a large orange-yellow to light pink supratemporal marking that is "pinched off" or broken. The carapace bears ocelli on the costals and marginals; these are brown in the center and orange at the periphery. The carapace length of females reaches 372 mm.

Natural history. The habits of this species is generally somewhat similar to that of *T. venusta*. Estivation occurs during the dry season. The habitat includes both very large lakes and quite small rivers.

Protection. This turtle is heavily harvested by local people, and the meat is said to be delicious. The carapaces are then painted and sold to tourists. In Costa Rica, a farm or ranch for this species has been established, with the purpose not only

© F. Bonin

An orange or pink spot extends from behind the eye and allows for identification this species. The carapace is flat, dark, and faintly marked with wide hieroglyphic designs.

of boosting wild stocks but also of providing hatchlings for sale in San José, to discourage or displace the importation of *T. scripta elegans* for this purpose.

Trachemys nebulosa (Van Denburgh, 1895)

Distribution. This species is found in Baja California, from San Ignacio to San José del Cabo, and also in Texas, New Mexico, and northwestern mainland Mexico.

Description. Formerly a subspecies of *T. scripta,* this turtle differs from that species in having an orange or yellow supratemporal stripe that ends in an oval spot behind the eyes (see "Subspecies" below for details).

Subspecies. Four subspecies are currently recognized.

T. n. nebulosa (described above) is the largest of the four, reaching 370 mm in females. It occurs in the southern and central parts of Baja California. The carapace has black spots that are rather small, as well as irregular light markings. The plastron is decorated with a series of dark circles.

T. n. gaigeae (Hartweg, 1939) occurs in watercourses flowing into the Rio Grande and the Rio Conchos in Texas, New Mexico, and the Mexican states of Chihuahua and Coahuila. It does not exceed 240 mm and has a reticulated design on the carapace, very often with a small ocellus on each scute. Behind the eyes, the orange blotch is large and bordered with black, and it never touches the orbits. In the middle of the chin, there are several dark bands. Usually the buccal cavity shows sharp, pointed toothlike keratinous structures on both jaws, which would appear to have a role in capturing slippery prey.

T. n. hartwegi (Legler, 1990) occurs in the Rio Nazas, in Mexico, and does not differ significantly from the Baja race.

T. n. hiltoni (Carr, 1942) has been observed

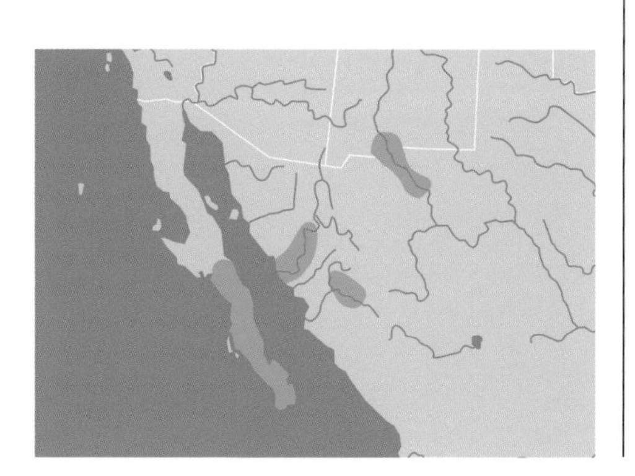

Common name

Baja California slider

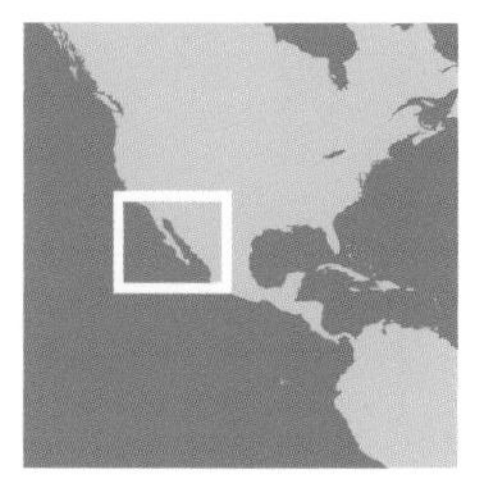

only in the region of the Rio Fuerte in the states of Sonora and Sinaloa, Mexico. The carapace may reach 280 mm and has black circles on the upper surface and on the underside of each marginal scute. The plastron has a central dark design encircling a lighter zone.

Natural history. This species is generally similar to *T. scripta* in its natural history.

Protection. Some of these subspecies, including *T. n. gaigeae*, are much sought after by collectors, and some populations have become very depleted. The main threat to these Mexican subspecies is probably human consumption, although habitat degradation and pollution are also problems.

© F. Bonin

This turtle, photographed in Mexico, has numerous light lines on the neck and a very wide orange spot at the back of the head.

Trachemys ornata (Gray, 1831)

Distribution. The range is very extensive along the western side of the American tropics, from the Mexican coastal plain in northern Sinaloa to Oaxaca, and also in Guatemala and all through central America to northern Colombia. However, this range has many major gaps (e.g., along all the Costa Rican coastal plain except for Corcovado in the extreme south), and it is unlikely that the turtles from Pacific Colombia will be truly conspecific with those from Mexico. Turtles have free communication between Caribbean and Atlantic drainages in the Panama Canal area, and it is probable that Pacific coast turtles of Central America are actually *T. venusta* (and *T. v. grayi*) rather than true *ornata*.

Description. Formerly considered to be subspecies of *T. scripta,* this turtle is now a full species. The carapace may attain a length of 380 mm and has black ocelli centered on the costal scutes. The plastron has four median stripes that do not reach the anal notch. The head is large and strong and has an orange stripe extending from the rear of the orbit onto the temples and reaching the base of the tail.

Natural history. The natural history of this species is similar to that of other tropical sliders.

Note the group of very fine stripes that extend from the jaw, as well as the wider band just behind the eye. The snout is rather pointed.

Ornate slider turtle

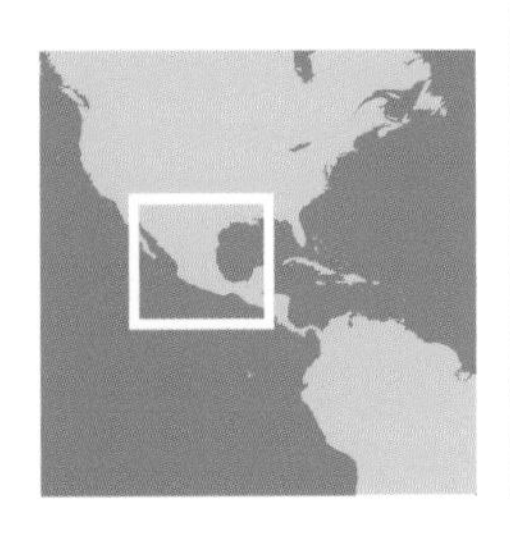

© F. Bonin

Trachemys taylori (Legler, 1960)

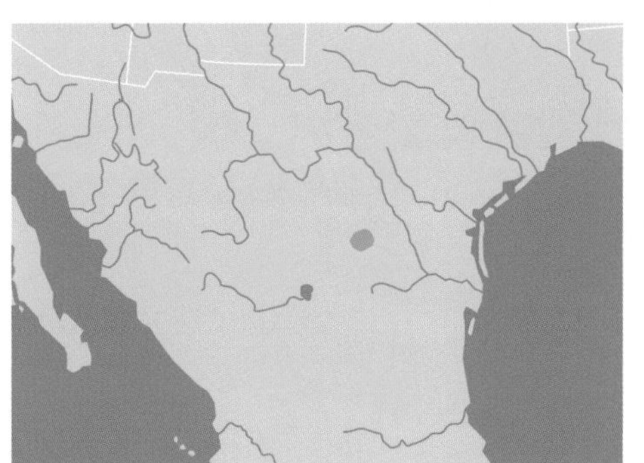

Common name

Cuatro Ciénegas slider turtle

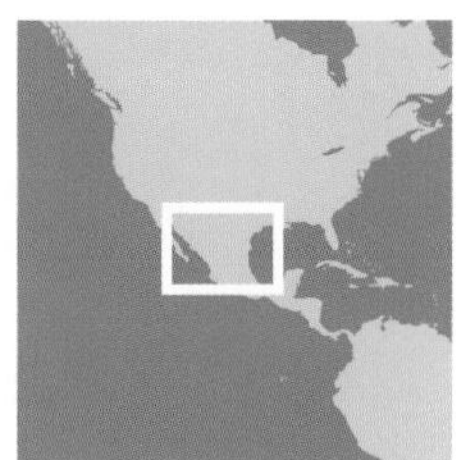

Distribution. This species is found only in the basin of Cuatro Ciénegas, in the state of Coahuila, Mexico.

Description. Formerly considered to be a subspecies of *T. scripta,* this species is a small form, not exceeding 220 mm. The carapace is green, often quite light, with numerous dark elongate spots and ovals. The plastron is cream-colored, with a black design spread over much of it. On the head, there is a red stripe that extends back from the orbit toward the base of the neck.

Natural history. This turtle lives in the basin of Cuatro Ciénegas in a very specialized environment of interconnected and separated water holes. It lives in sympatry with *Terrapene coahuila* but is less aquatic and occupies spring-fed water holes that range from freshwater to saline.

Protection. The small size of the habitat does facilitate surveillance and protection of the turtles in the basin, and these measures are undertaken by a permanent team at the biological station of Cuatro Ciénegas. *Trachemys taylori* is not really threatened, and the public bathes happily alongside the turtles in the *pozas*. On the other hand, peoples' habit of liberating turtles at Cuatro Ciénegas that are not native to the area could have serious consequences for the local fauna.

This young individual has a brightly ornamental shell, very light in color and with dark ocelli. The head is discreetly striped.

Trachemys venusta (Gray, 1855)

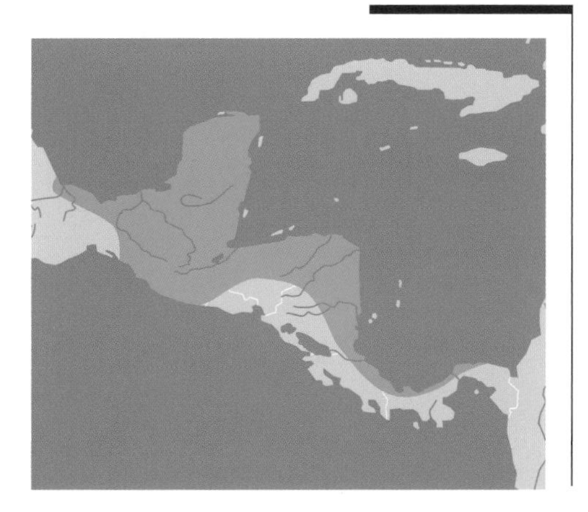

Common name

Mesoamerican slider

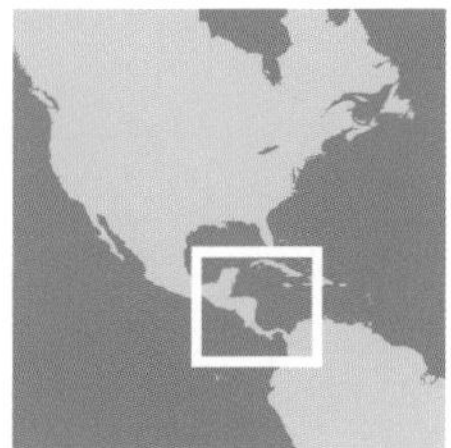

Distribution. This species is found on the Atlantic coast from Veracruz, Mexico, to Panama and on the Pacific coast from Tehuantepec to Guatemala.

Description. Formerly a subspecies of *T. scripta* but now a full species, this turtle would appear to be the largest member of the genus, reaching a maximum carapace length of nearly 50 cm (see "Subspecies" below for details).

Subspecies. Three subspecies are recognized. *T. v. venusta* (described above) may reach a

© B. Devaux

The ocelli on the carapace are regular and perfectly formed, but not very bright. On the other hand, two wide yellow stripes extend from under the eyes and from below the lower jaw.

length of 480 mm. It occurs from Veracruz in Mexico to Panama, passing through Costa Rica and Honduras. There are very large dark ocelli on the carapace, each located on the center of a costal scute. On the head, a wide orange-yellow supratemporal band always makes contact with the orbit. *T. v. cataspila* (Günther, 1885) occurs in the Mexican Gulf lowlands, from southern Tamaulipas to Punta del Morro, Veracruz. This is the smallest turtle of the genus, the females reaching about 330 mm and males 272 mm. The carapace has dark-centered ocelli on the costals and the marginals, and the supratemporal stripe on the side of the head is very wide. Old males lose their patterns and become melanistic. *T. v. grayi* (Bocourt, 1868) is found from Tehuantepec, Mexico, on the coastal plain of the Pacific to La Libertad in Guatemala. M. del Toro reports that in Mexico this form may reach a length of nearly 500 mm. The carapace also has dark ocelli on the costals and marginals, but the pattern on the plastron is diffuse, often broken, and very inconspicuous in adults. The head stripes are all narrow, and the supratemporal stripe always reaches the eye.

Natural history. In Costa Rica, these turtles are very large and often nest on sandy marine beaches, emerging to nest directly from the sea. Nevertheless, the males remain permanently in the interior and live in freshwater. The females are apparently picked up by water currents and no doubt find that the seashore makes a good place to nest. This behavior has been observed especially at Tortuguero, where the big sliders nest alongside sea turtles. The number of eggs per clutch varies from 12 to 32 (average about 20).

Trachemys stejnegeri (Schmidt, 1928)

Common name

Central Antillean slider

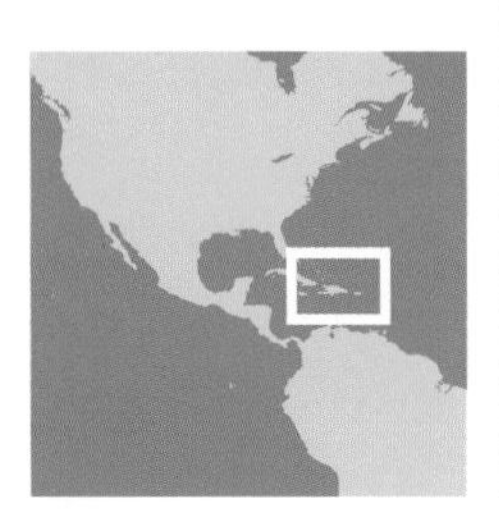

Distribution. This species is endemic to the Caribbean Islands and occurs in Hispaniola, Puerto Rico, and the Bahamas, as well as on Marie Galante (Guadeloupe), where it is presumably introduced.

Description. The shell is slightly domed and may reach 240 mm. Each scute has conspicuous longitudinal wrinkles or rugosity. The vertebral scutes are wider than long. The carapace ranges from gray to brown but is sometimes olive or black. In old males, melanism may be well developed. The plastron is large, has a small anal notch, and is usually entirely yellow, but sometimes there is system of black lines that forms a pattern of the seam-following type. Olive-colored ocelli may be present on the lower face of the marginal scutes. The head is rather short but has a pointed snout. The coloration of the head is gray to olive, with yellow or cream stripes. The supraoccipital stripe is red-brown. The upper jaw has a median notch. The neck and limbs are olive to grayish, with yellow or cream stripes. In the young, there are yellow stripes on most of the carapacial scutes.

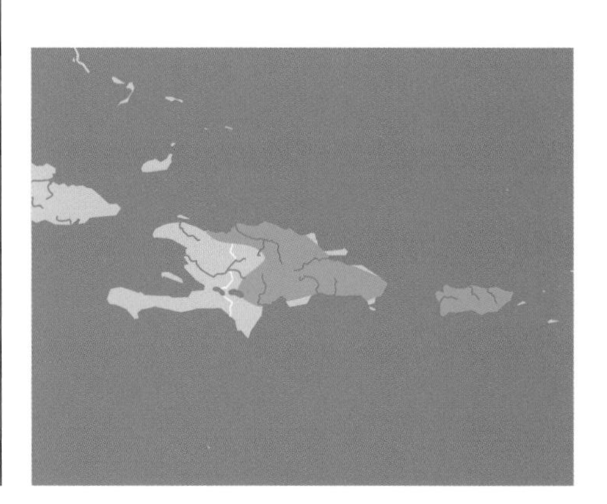

Subspecies. Three subspecies are recognized. *T. s. stejnegeri* (described above) lives on Puerto Rico and Marie Galante, where it has been introduced. The skin is brown to olive-brown. The snout

is pointed and quite elongate. A central plastral figure extends along the seams to the outer edges of the scutes. It is particularly conspicuous on the gulars and is less clear in the posterior region of the plastron. *T. s malonei* (Barbour and Carr, 1938) is originally from the Bahamas (Great Inagua Island). It has gray to olive skin and a somewhat rounded, blunt snout. The plastron is often yellow, without any pattern. If the dark pattern is present, it follows the seams and remains inconspicuous. Sometimes all that can be seen are a few dark markings on the gulars. *T. s. vicina* (Barbour and Carr, 1940) lives only on the island of Hispaniola in the Dominican Republic. The skin is olive-gray, and the nose is long and pointed. The plastral design is quite strong and well marked posteriorly, following the seams. The scutes often bear ocelli.

Natural history. These turtles live in lakes, mud-bottomed rivers, and canals with abundant aquatic vegetation. They are quite tolerant of brackish water. Herbivorous when adult, they are omnivorous during their early years. During courtship, the male vibrates his long foreclaws before the head of the female. Nesting occurs from April to July and includes 3 to 14 eggs, about 43 × 26 mm in size. There may be three nestings per season, and incubation lasts 60 to 80 days.

This massive species is dark, almost blackish, and the head has light marbling.

Trachemys terrapen (Lacepède, 1788)

Distribution. In the Caribbean, this species occurs principally in Jamaica, but it has been introduced to islands in the Bahamas, including Cat Island, southern Eleuthera, and southern Andros.

Description. This is a rather large turtle with an oval shell that is slightly domed, is wider behind than in front, and may reach a length of 320 mm. The median keel is well developed, and the posterior margin is serrated. The scales are rough and have extensive rugosity, radiating from the center of the scute to the edge. The general color ranges from dark gray to olive-beige. In the young, a yellow band marks each costal and each marginal. These bands disappear with growth. The plastron is wider at the posterior lobe, and there is a small

Common name

Jamaican slider

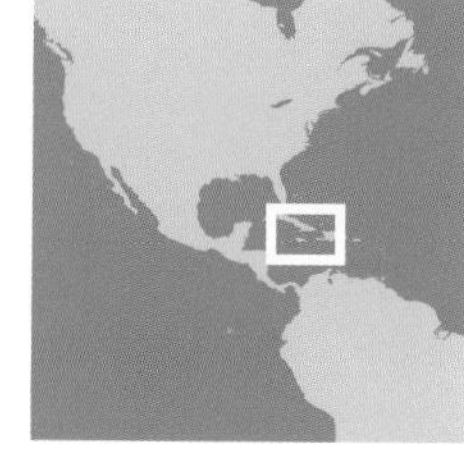

© F. Bonin

This is the least colorful of the *Trachemys* species. The carapace is strong and is striped with fine reticulations. The head is thick and has a smooth helmet, with imperceptible stripes or indistinct marbling.

anal notch. The plastron is yellow or cream and is unmarked. Sometimes traces of dark lines may be perceived along the seams, especially on the gulars and humerals. The head is of moderate size and has a rounded, blunt snout. There is a small notch on the upper jaw. The color of the head is gray to greenish, with light stripes that are feebly contrasting and may be difficult to discern. There are several white to cream stripes on the chin, as well as a wider band that extends from the nostrils to form a white "mustache" on the upper lip. The young have yellow stripes with black borders on the head, neck, limbs, and tail. In the adults, the limbs and tail are gray to green, with the forelimbs showing the stripes much more clearly.

Natural history. This turtle lives in small marshes and ponds within its habitat that have a mud bottom and abundant aquatic vegetation. This is an active and potentially aggressive turtle, especially when provoked. Omnivorous, it eats everything it can find. The courtship is similar to that of other *Trachemys* species. Nesting occurs from May to June.

Trachemys yaquia (Legler and Webb, 1970)

Common name

Yaqui slider turtle

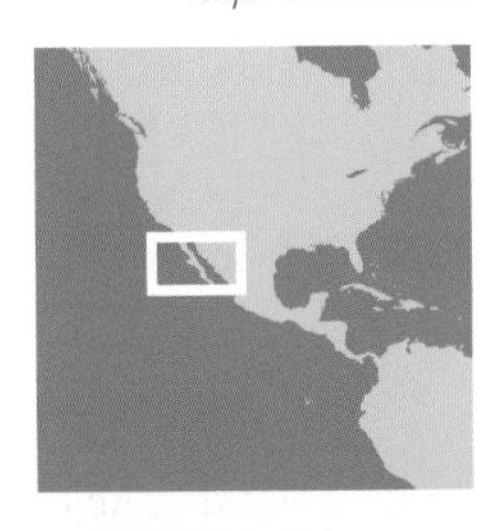

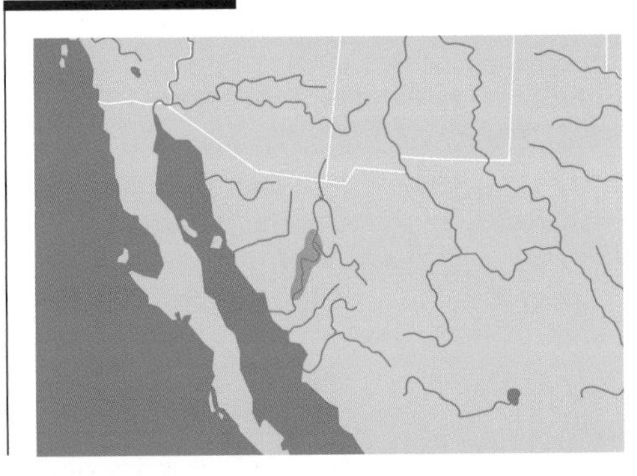

Distribution. This turtle occurs only in the state of Sonora, in northwestern Mexico, inhabiting the Yaqui, Mayo, Sonora, and Metape Rivers.

Description. This form is very close to *T. scripta,* of which it was formerly a subspecies. The females may reach 320 mm. The costals are ornamented with very faint ocelli. On the head, an orange-yellow band starts behind the eyes but does not extend along the neck. The sides of the head are marked with numerous very fine white or cream-colored lines.

Natural history. The details have not been studied specifically, but the ecology is probably similar to that of *Trachemys* in general.

Protection. This turtle is often disturbed by cattle that invade the waters where it lives, as well as by the degradation of wetlands. But the populations are still high in many small rivers and creeks in the state of Sonora.

© F. Bonin

This turtle was seen in the Yaquí River in Mexico. It has numerous fine lines on the neck, along with a large but not very contrasting orange spot at the back of the head.

References

Acuña Mesén R. A., 1993, *Las tortugas continentales de Costa Rica,* University of Costa Rica, San José.

Alderton Db 1988, *Turtles and Tortoises of the World,* Blandford Press, London.

Arvy C., Fertard B., 2001, *Pathologie des Tortues, Bull. soc. herpé. de France,* Paris.

Bonin F., Devaux B., Dupré A., 1995, *Toutes les tortues du monde,* Delachaux and Niestlé, Paris.

Bour R., 1986, L'identité des tortues terrestres européennes: spécimens-types et localités-types, *Rev. fr. aquar.* 13(4), Paris, p. 11–122.

———, 1994, *Recherches sur des animaux doublement disparus: Les tortues géantes subfossiles de Madagascar,* EPHE, Montpellier, France.

Boycott A., Bourquin O., 1986, *The Southern African Tortoise Book,* Hilton, KwaZulu-Natal, South Africa.

Branch B., 1988, *Field Guide to the Snakes and Other Reptiles of Southern Africa,* Struik Publ., Cape Town, South Africa.

Breuil M., 2002, *Histoire naturelle des amphibiens et reptiles terrestres de l'archipel guadeloupéen,* MNHN, Paris.

Burbidge A., Kirsh A., Main A., 1974, Relationship within the Chelidae of Australia and New Guinea, *Copeia* 1974, p. 392–409.

Cabrera M., 1998, *Las tortugas continentales de Sudamerica Austral,* Cordoba, Argentina.

Cann J., 1998, *Australian Freshwater Turtles,* Beaumont, Singapore.

Capula M., 1989, *Reptiles and Amphibians of the World,* Simon and Schuster, New York.

Carmichael P., Williams W., 1991, *Reptiles and Amphibians,* World Publ., Tampa, FL, USA.

Carr A., 1952, *Handbook of Turtles,* Cornell Univ. Press, Ithaca, NY, USA.

Cheylan M., 1981, *Biologie et Ecologie de la tortue d'Hermann,* EPHE, Montpellier, France.

Cobb J., 1987, *A Complete Introduction to Turtles and Terrapins,* TFH, Neptune, FL, USA.

Das I., 2002, *Amphibians and Reptiles of Tropical Asia,* Nat. Hist. Publ., Borneo, Malaysia.

David P., 1994, *Dumerilia, liste des reptiles actuels du monde,* AALRAM, MNHN, Paris.

Devaux B., 1988, *La tortue sauvage des Maures, ou tortue d'Hermann,* Sang de la Terre, Paris.

———, 2000, *The Crying Tortoise* Centrochelys sulcata, Éditions SOPTOM, France.

Dumeril A., Bibron G., 1835, *Herpétologie générale dans histoire naturelle complète des Reptiles,* Librairie Roret, Paris.

Ernst C., 1963–2006, *Catalogue of American Amphibians and Reptiles,* Soc. for Study of Amph. and Rept., USA.

Ernst C., Barbour R., 1989, *Turtles of the World,* Smithsonian Inst. Press, Washington, DC.

Ernst C., Lovich J., Barbour R., 1994, *Turtles of the United States and Canada,* Smithsonian Inst. Press, Washington, DC.

Ferry V., 2000, *Guide des tortues,* Delachaux et Niestlé, Paris.

Fitzinger L., 1826, *Neue Classification der Reptilien,* Im Verlag Von J. G. Heubner, Vienna, Austria.

Fornelino M., Martinez Silvestre A., 1999, *Tortugas de España,* Antiqvaria, C. Lagasca, Spain.

Franke J., Telecky T., 2001, *Reptiles as Pets: An Examination of the Trade in Live Reptiles in the United States,* the Humane Society, Washington, DC.

Fretey J., 2001, *Biogéographie et conservation des tortues marines de la côte atlantique de l'Afrique,* UNEP/CMS Bonn, Germany.

———, 2005, *Les tortues marines de Guyane,* Plume Verte, Cayenne, French Guiana.

Fritz U., 1998, *Mertensiella, Proceedings of the Emys Symposium Dresden 96,* DGHT, Germany.

Frye F., 1991, *Biomedical and Surgical Aspects of Captive Reptile Husbandry,* Krieger Publ., Malabar, FL.

Gerlach J., 2004, *Giant Tortoises of the Indian Ocean,* Chimaira, Frankfurt, Germany.

Goris R., Maeda N., 2004, *Guide to the Amphibians and Reptiles of Japan,* Krieger Publ., Malabar, FL.

Groombridge B., 1982, *The IUCN Amphia-Reptilia Red Data Book,* Édition IUCN, Gland, Switzerland.

Hödl W., 2000, *Die Europäische Sumpfschildkröte,* Landesmuseums, Neue Folde, Austria.

Holbrook J., 1976, *North American Herpetology,* Soc. for the Study of Amph. and Rept., Athens, OH.

Iverson J., 1992, *A Revised Checklist with Distribution Maps of the Turtles of the World,* Richmond, IN.

Kalb H., 1992, *Turtle and Tortoise Newsletter,* Evansville, IN.

Kenneth Dodd C., 2001, *North American Box Turtles,* University of Oklahoma Press, Norman.

Kuchling G., 1999, *The Reproductive Biology of the Chelonia,* Springer, Germany.

Kuzmin S., 2002, *The Turtles of Russia and Other Ex-Soviet Republics,* Chimaira, Frankfurt, Germany.

Lee J., 1996, *The Amphibians and Reptiles of the Yucatan Peninsula,* Cornell Univ. Press, Ithaca, NY.

Lehrer J., 1990, *Turtles and Tortoises: A Photographic Survey,* Friedman Group Book, London.

Loveridge A., Williams E., 1957, "Revision of the African Tortoises and Turtles of the Suborder Cryptodira," *Bull. Mus. Comp. Zool.,* Paris.

Lutz P., 1997, *The Biology of Sea Turtles,* CRC Press, New York.

Mao S., 1971, *Turtles of Taiwan,* Com. Press, Taipei, China.

Métrailler S., 1993, *Manouria,* Éd. A Cupulatta-Vignola, Corsica.

Obst F., 1988, *Turtles, Tortoises, and Terrapins,* St. Martin's Press, New York.

Orenstien R., 2001, *Turtles, Tortoises, and Terrapins: Survivors in Armor,* Firefly Books, Buffalo, NY.

Pritchard P., 1979, *Encyclopedia of Turtles,* TFH Publ., Neptune, FL.

———, 1996, *The Galapagos Tortoises: Nomenclatural and Survival Status,* Chel. Res. Found., Lunenburg, MA.

Rhodin A., 1997, *Chelonian Conservation and Biology,* Chel. Research Found., Lunenburg, MA.

Rogner M., 1995, *Schildkröten 1, 2, 3,* Heiro, Hurtgenwald, Germany.

Schaefer I., 2005, *Zacken-Erdschildkröten, Die Gattung* Geoemyda, Natur unf Tier, Munster, Germany.

Schleich H., 2001, *Das Andere Nepal,* Fuhlrott Museum, Wuppertal, Germany.

Schleich H., Kästle W., 2002, *Amphibians and Reptiles of Nepal,* Koeltz, Koenigstein, Germany.

Sobolik K., 1996, *A Turtle Atlas to Falilitate Archaeological Identifications,* Fenske Com., Rapid City, IA.

Stafford P., Meyer J., 2000, *A Guide to the Reptiles of Belize,* Nat. Hist. Museum, London.

Tyning T., 1997, *Status and Conservation of Turtles of the Northeastern United States,* Natural History Book Distributors, Lanesboro, MN.

Van Denburgh J., 1998, *The Gigantic Land Tortoises of the Galapagos Archipelago,* Soc. for Study Amphib. and Rept., USA.

Van Dijk P., 2000, *Asian Turtle Trade Phnom Penh,* Chel. Res. Found., Lunenburg, MA.

Vetter H., 2002, *Turtles of the World,* Frankfurt, Germany.

Villiers A., 1958, *Tortues et crocodiles de l'Afrique noire française,* IFAN, Dakar, Senegal.

Wegehaupt W., 2005, *Sardinien, die Insel der europäischen Schildkröten,* Kressbronn Verlag, Germany.

Xiangkui Y., 1994, *Fossil and Recent Turtles of China,* Science Press, Beijing, China.

Index of Scientific Names

Bold text indicates the page number of the species description.

BOCA RATON PUBLIC LIBRARY, FLORIDA
3 3656 0385323 5